BOLYAI SOCIETY
MATHEMATICAL STUDIES
27

More information about this series at http://www.springer.com/series/4706

BOLYAI SOCIETY MATHEMATICAL STUDIES

1. **Combinatorics, Paul Erdős is Eighty, Vol. 1**
D. Miklós, V. T. Sós, T. Szőnyi (Eds.)
2. **Combinatorics, Paul Erdős is Eighty, Vol. 2**
D. Miklós, V. T. Sós, T. Szőnyi (Eds.)
3. **Extremal Problems for Finite Sets**
P. Frankl, Z. Füredi, G. Katona, D. Miklós (Eds.)
4. **Topology with Applications**
A. Császár (Ed.)
5. **Approximation Theory and Function Series**
P. Vértesi, L. Leindler, Sz. Révész, J. Szabados, V. Totik (Eds.)
6. **Intuitive Geometry**
I. Bárány, K. Böröczky (Eds.)
7. **Graph Theory and Combinatorial Biology**
L. Lovász, A. Gyárfás, G. Katona, A. Recski (Eds.)
8. **Low Dimensional Topology**
K. Böröczky, Jr., W. Neumann, A. Stipsicz (Eds.)
9. **Random Walks**
P. Révész, B. Tóth (Eds.)
10. **Contemporary Combinatorics**
B. Bollobás (Ed.)
11. **Paul Erdős and His Mathematics I+II**
G. Halász, L. Lovász, M. Simonovits, V. T. Sós (Eds.)
12. **Higher Dimensional Varieties and Rational Points**
K. Böröczky, Jr., J. Kollár, T. Szamuely (Eds.)
13. **Surgery on Contact 3-Manifolds and Stein Surfaces**
B. Ozbagci, A. I. Stipsicz
14. **A Panorama of Hungarian Mathematics in the Twentieth Century, Vol. 1**
J. Horváth (Ed.)
15. **More Sets, Graphs and Numbers**
E. Győri, G. Katona, L. Lovász (Eds.)
16. **Entropy, Search, Complexity**
I. Csiszár, G. Katona, G. Tardos (Eds.)
17. **Horizons of Combinatorics**
E. Győri, G. Katona, L. Lovász (Eds.)
18. **Handbook of Large-Scale Random Networks**
B. Bollobás, R. Kozma, D. Miklós (Eds.)
19. **Building Bridges**
M. Grötschel, G. Katona (Eds.)
20. **Fete of Combinatorics and Computer Science**
G. Katona, A. Schrijver, T. Szőnyi (Eds.)
21. **An Irregular Mind**
I. Bárány, J. Solymosi (Eds.)
22. **Cylindric-like Algebras and Algebraic Logic**
H. Andréka, M. Ferenczi, I. Németi (Eds.)
23. **Deformations of Surface Singularities**
A. Némethi, Á. Szilárd (Eds.)
24. **Geometry – Intuitive, Discrete, and Convex**
I. Bárány, K. Böröczky, G. Fejes Tóth, J. Pach (Eds.)
25. **Erdős Centennial**
L. Lovász, I. Ruzsa, V. T. Sós (Eds.)
26. **Contact and Symplectic Topology**
F. Bourgeois, V. Colin, A. Stipsicz (Eds.)

Gergely Ambrus · Imre Bárány
Károly J. Böröczky · Gábor Fejes Tóth
János Pach
Editors

New Trends in Intuitive Geometry

BOLYAI SOCIETY

Springer

Editors
Gergely Ambrus
MTA Alfréd Rényi Institute
of Mathematics
Budapest
Hungary

Imre Bárány
MTA Alfréd Rényi Institute
of Mathematics
Budapest
Hungary

Károly J. Böröczky
MTA Alfréd Rényi Institute
of Mathematics
Budapest
Hungary

Gábor Fejes Tóth
MTA Alfréd Rényi Institute
of Mathematics
Budapest
Hungary

János Pach
MTA Alfréd Rényi Institute
of Mathematics
Budapest
Hungary

ISSN 1217-4696
Bolyai Society Mathematical Studies
ISBN 978-3-662-57412-6 ISBN 978-3-662-57413-3 (eBook)
https://doi.org/10.1007/978-3-662-57413-3

Library of Congress Control Number: 2018941233

Mathematics Subject Classification (2010): 05-06, 52-06

Cover illustration: "The 6-move M_6" by Wöden Kusner (see Fig. 10 on page 260)

This Springer imprint is published by the registered company Springer-Verlag GmbH, DE part of Springer Nature
The registered company address is: Heidelberger Platz 3, 14197 Berlin, Germany

Preface

The 7th Conference on Intuitive Geometry was held at the Alfréd Rényi Institute of Mathematics of the Hungarian Academy of Sciences, Budapest, from June 22 to 28, 2015. It commemorated the 100th anniversary of the birth of the founder of the Hungarian school of discrete geometry, the great mathematician, László Fejes Tóth. The English translation of his classic monograph, "Lagerungen in der Ebene, auf der Kugel und im Raum," written in 1953, will soon be published by Springer. It includes about 100 pages of new material, summarizing the most important developments in this area during the past 6 decades. The translation and the appendix are joint work of Gábor Fejes Tóth and Wlodzimierz Kuperberg.

The conference was a great success. There were 20 excellent plenary lectures and more than a hundred participants. The editors of this volume, members of the organizing committee of the conference, asked some of the speakers to contribute a survey on an important topic in discrete and convex geometry. Their response was overwhelmingly positive. The outcome is the present volume which contains 17 surveys that cover many recent developments in this very active research field.

Perhaps the most popular subject in this area is packing and covering, which is represented by five surveys.

K. Bezdek and M. A. Khan concentrate on the problem of determining the maximal number of contact points in a packing of congruent copies of a convex body. This is a generalization of the famous kissing number problem of Newton and Gregory.

R. Kusner, W. Kusner, J. C. Lagarias, and S. Shlosman study a related problem relevant to various questions in physics and materials science: Describe the configuration space of 12 non-overlapping equal spheres touching a central unit sphere.

O. Musin describes five classical problems that L. Fejes Tóth worked on. This includes Tammes' problem and estimating one-sided kissing numbers.

Providing non-trivial upper bounds for the packing density of congruent copies of a convex body C has been a major and notoriously difficult question in discrete geometry. F. M. Oliveira Filho and F. Vallentin utilize a modification of the linear programming method in order to obtain a general upper bound on packing densities. They illustrate their method by the special case when C is the regular pentagon.

Covering problems are in some sense dual to packing ones. Starting with some classical methods and results, M. Naszódi reviews several different approaches to translative covering problems, including the illumination conjecture.

Besides packing and covering problems, the present volume has plenty of other interesting topics to offer.

A. Barvinok describes how the tensorization trick can be applied to the following three problems: packing points in the sphere, approximating convex bodies by algebraic hypersurfaces of bounded degree, and approximation of convex bodies by polytopes.

In their article, P. M. Blagojević, A. S. D. Blagojević, and G. M. Ziegler apply the constraint method in combination with a generalized Borsuk–Ulam-type theorem and a cohomological intersection lemma to show how one can obtain many new Tverberg-type topological transversal theorems.

B. Csikós gives a detailed historical account of the famous Kneser–Poulsen conjecture about the measure of unions of balls, which is still open in dimensions at least 3.

Applications of algebraic methods to classical problems in discrete geometry have recently seen a great surge. In the present volume, three articles follow this trend.

F. de Zeeuw discusses the new developments concerning the Elekes–Rónyai problem and related questions lying at the crossroads of combinatorics, algebra, and geometry.

M. Sharir and N. Solomon give a fairly elementary and simple proof for the best upper bound on the maximal number of incidences between points and lines in $\mathbb{R}^3$. This result was the cornerstone of the argument of Guth and Katz in their breakthrough solution of the distinct distances problem of Erdős.

Finally, J. Solymosi and F. de Zeeuw prove upper bounds on another Szemerédi–Trotter-type quantity: the number of incidences between a set of algebraic curves in $\mathbb{C}^2$ and a Cartesian product $A \times B$ with finite sets $A, B \subset \mathbb{C}$.

G. Domokos and G. W. Gibson tell us the story of pebbles: In their survey article, they describe a number of attempts to lay the precise mathematical background for the abrasion of particles.

P. Hajnal and E. Szemerédi illustrate the power of semi-random methods by improving the best-known bounds on Heilbronn's quadrangle problem and the question of independent points.

A. Holmsen surveys new developments about Erdős-Szekeres type problems for convex sets.

E. León and G. M. Ziegler study the space of partitions of $\mathbb{R}^d$ into n non-empty open convex regions (n-partitions).

In his paper, P. McMullen points out a mistake in Coxeter's analysis of possible sections of the 120-cell $\{5, 3, 3\}$ by a systematic faceting procedure. As a result, six apparently new regular compounds of polytopes in $\mathbb{R}^4$ are described.

Studying unit and distinct distance problems is a classical topic in combinatorial geometry, introduced by P. Erdős. K. Swanepoel surveys related problems for normed spaces. He describes various properties of unit distance graphs, minimum distance graphs, diameter graphs as well as minimum spanning trees and Steiner minimum trees.

This book is dedicated to the memory of László Fejes Tóth. We hope that our readers will find many fascinating open problems in it and perhaps useful tools and ideas for their solution.

Budapest, Hungary
August 2018

Gergely Ambrus
Imre Bárány
Károly J. Böröczky
Gábor Fejes Tóth
János Pach

The original version of the book was revised: Copyright Holder name has been corrected throughout the book. The correction to the book is available at https://doi.org/10.1007/978-3-662-57413-3_18

Contents

The Tensorization Trick in Convex Geometry

Alexander Barvinok

Abstract The "tensorization trick" consists in proving some geometric result for a set of vectors $\{v_i\}$ in some vector space V and then applying the same result to the tensor powers $\left\{v_i^{\otimes k}\right\}$ in $V^{\otimes k}$, which in turn produces a considerably stronger version of the original result for vectors $\{v_i\}$. Our main examples concern packing vectors in the sphere, approximation of convex bodies by algebraic hypersurfaces and approximation of convex bodies by polytopes. We also discuss applications of a closely related polynomial method to constructing neighborly polytopes, bounding the Grothendieck constant, proving the polynomial ham sandwich theorem, bounding the number of equiangular lines in $\mathbb{R}^d$ and to constructing a counterexample to Borsuk's conjecture.

1991 Mathematics Subject Classification 15A69 · 52A20 · 52A45 · 52C17 · 14P05

1 Introduction

The *tensorization trick* that we are talking about in this paper (there are several other tricks with that name) can be loosely described as follows. We prove some general (usually, not very complicated) result applicable to a set $\{v_i : i \in I\}$ of vectors in a real vector space V. We then consider tensor powers $\left\{v_i^{\otimes k} : i \in I\right\}$ of those vectors in $V^{\otimes k}$ and apply the result to them. This gives us a substantially stronger version of the original result for vectors $\{v_i : i \in I\}$.

In this paper, we consider three main examples: packing points in the sphere (Sect. 2), approximating convex bodies by algebraic hypersurfaces of a small degree (Sect. 3) and approximation of convex bodies by polytopes (Sect. 4).

This research was partially supported by NSF Grant DMS 1361541.

A. Barvinok (✉)
Department of Mathematics, University of Michigan, Ann Arbor, MI 48109-1043, USA
e-mail: barvinok@umich.edu

G. Ambrus et al. (eds.), *New Trends in Intuitive Geometry*, Bolyai Society Mathematical Studies 27, https://doi.org/10.1007/978-3-662-57413-3_1

The reasons that allow us to strengthen the original result by tensorization appear to come from two sources. First, we save on dimension: tensor powers $v_i^{\otimes k}$ lie in the *symmetric part* of $V^{\otimes k}$ and, as a corollary, lie in a subspace of dimension

$$\binom{\dim V + k - 1}{k} \ll (\dim V)^k.$$

Second, a linear function in $v_i^{\otimes k}$ is a *polynomial* in v_i. Thus by proving results for linear functions on $v_i^{\otimes k}$, we essentially prove results for polynomials on v_i. Moreover, we can choose any polynomial of degree k we like by lifting v_i to $1 \oplus v_i \oplus \cdots \oplus v_i^{\otimes k}$ in the space $\mathbb{R} \oplus V \oplus \cdots \oplus V^{\otimes k}$.

In some situations saving on the dimension is not crucial and only the fact that tensor products linearize polynomials is used. This approach is often known as the polynomial method and in Sect. 5 we mention its applications to constructing neighborly polytopes, bounding the Grothendieck constant, proving the polynomial ham sandwich theorem, bounding the number of equiangular lines in $\mathbb{R}^d$ and to constructing a counterexample to Borsuk's conjecture.

In the rest of this section, we introduce our main tools.

1.1 Tensor Power

Let V be a d-dimensional real vector space and let k be a positive integer. We consider the tensor power

$$V^{\otimes k} = \underbrace{V \otimes \cdots \otimes V}_{k \text{ times}},$$

so $\dim V^{\otimes k} = d^k$. If we choose an orthonormal basis in V thus identifying $V = \mathbb{R}^d$, we can identify $V^{\otimes k}$ with the space of all k-dimensional $d \times \cdots \times d$ arrays (tensors) $A = \left(\alpha_{i_1 \ldots i_k}\right)$ of real numbers, where $1 \leq i_1, \ldots, i_k \leq d$. We call $\alpha_{i_1 \ldots i_k}$ the *coordinates* of A. We make $V^{\otimes k}$ Euclidean space by introducing the inner product

$$\langle A, B \rangle = \sum_{1 \leq i_1, \ldots, i_k \leq d} \alpha_{i_1 \ldots i_k} \beta_{i_1 \ldots i_k} \quad \text{where} \quad A = \left(\alpha_{i_1 \ldots i_k}\right) \quad \text{and} \quad B = \left(\beta_{i_1 \ldots i_k}\right).$$

For a vector $x \in V$, $x = (\xi_1, \ldots, \xi_d)$, the coordinates of

$$A = x^{\otimes k} = \underbrace{x \otimes \cdots \otimes x}_{k \text{times}}, \quad A = \left(\alpha_{i_1 \ldots i_k}\right)$$

are

$$\alpha_{i_1 \ldots i_k} = \xi_{i_1} \cdots \xi_{i_k}.$$

One can see that

$$\left\langle x^{\otimes k}, y^{\otimes k} \right\rangle = \langle x, y \rangle^k, \tag{1.1.1}$$

where $\langle x, y \rangle$ is the standard inner product in V.

Another important observation is that for every $x \in V$ the tensor $A = x^{\otimes k}$, $A = \left(\alpha_{i_1 \ldots i_k}\right)$ ends up in the *symmetric part* $\operatorname{Sym}\left(V^{\otimes k}\right)$ of $V^{\otimes k}$, that is, $\alpha_{i_1 \ldots i_k}$ depends only on the multiset $\{i_1, \ldots, i_k\}$ and not on the order of $i_1, \ldots, i_k$. Moreover, the subspace $\operatorname{Sym}\left(V^{\otimes k}\right)$ is spanned by all tensors $x^{\otimes k}$ for $x \in V$. It then follows that the dimension of $\operatorname{Sym}\left(V^{\otimes k}\right)$ is the number non-negative integer solutions of the equation $n_1 + \cdots + n_d = k$, and hence

$$\dim \operatorname{span}\left\{x^{\otimes k} : \quad x \in V\right\} = \dim \operatorname{Sym}\left(V^{\otimes k}\right) = \binom{\dim V + k - 1}{k}. \tag{1.1.2}$$

1.2 Tensor Powers and Polynomials

For a positive integer k and a space V as above, we consider the direct sum

$$W_k = \mathbb{R} \oplus V \oplus V^{\otimes 2} \oplus \cdots \oplus V^{\otimes k}.$$

We make W_k Euclidean space in the obvious way, letting

$$\langle a, b \rangle = \sum_{m=0}^{k} \langle a_m, b_m \rangle \quad \text{where} \quad a = a_0 \oplus a_1 \oplus \cdots \oplus a_k \quad \text{and} \quad b = b_0 \oplus b_1 \oplus \cdots \oplus b_k$$

and $a_m, b_m \in V^{\otimes m}$ for $m = 0, \ldots, k$ (we agree that $V^{\otimes 0} = \mathbb{R}$).

If $p(t) = \alpha_0 + \alpha_1 t + \cdots + \alpha_k t^k$ is a polynomial, for $x \in V$ we formally define $p^{\otimes}(x) \in W_k$ by

$$p^{\otimes}(x) = \alpha_0 \oplus \alpha_1 x \oplus + \cdots + \oplus \alpha_k x^{\otimes k}.$$

Then

$$\left\langle p^{\otimes}(x), q^{\otimes}(y) \right\rangle = r\left(\langle x, y \rangle\right) \quad \text{where} \quad r = p * q \tag{1.2.1}$$

is the Hadamard product defined by

$$\begin{gathered} r(t) = \sum_{m=0}^{k} \left(\alpha_m \beta_m\right) t^m \quad \text{provided} \\ p(t) = \sum_{m=0}^{k} \alpha_m t^m \quad \text{and} \quad q(t) = \sum_{m=0}^{k} \beta_m t^m. \end{gathered} \tag{1.2.2}$$

We consider the embedding

$$\phi : V \longrightarrow W_k, \quad \phi(x) = 1 \oplus x \oplus \cdots \oplus x^{\otimes k},$$

so formally we can write

$$\phi(x) = f^{\otimes}(x) \quad \text{where} \quad f(t) = 1 + t + \cdots + t^k.$$

Hence by (1.2.2) we have

$$\big\langle p^{\otimes}(x),\ \phi(y)\big\rangle = p\big(\langle x, y\rangle\big) \tag{1.2.3}$$

for any polynomial p of $\deg p \leq k$.

We note that

$$\begin{aligned}\dim \operatorname{span}\{\phi(x) :\ x \in V\} =& \sum_{m=0}^{k} \binom{\dim V + m - 1}{m} \\ =& \binom{\dim V + k}{k}.\end{aligned} \tag{1.2.4}$$

1.3 Chebyshev Polynomials

We will apply (1.2.3) to some special polynomials p.

The Chebyshev polynomial of the first kind $T_k(t)$ of degree k is defined by

$$\begin{aligned}T_k(t) =& \cos(k \arccos t) \quad \text{provided} \quad -1 \leq t \leq 1 \quad \text{and} \\ T_k(t) =& \frac{1}{2}\left(t + \sqrt{t^2-1}\right)^k + \frac{1}{2}\left(t - \sqrt{t^2-1}\right)^k \quad \text{provided} \quad |t| \geq 1.\end{aligned}$$

In particular,

$$|T_k(t)| \ \leq\ 1 \quad \text{provided} \quad |t| \leq 1. \tag{1.3.1}$$

It turns out that the polynomial $T_k(t)$ has the following extremal property that is relevant to us: for any given $\tau \in \mathbb{R}$ such that $|\tau| > 1$, the maximum value of $|p(\tau)|$ for a polynomial $p(t)$ such that $\deg p \leq k$ and $|p(t)| \leq 1$ for all $-1 \leq t \leq 1$ is attained for $p = T_k$, see, for example, Sect. 2.1 of [13].

2 Packing Points in the Sphere

2.1 Packing Points in the Sphere

Let us fix a real $0 < \epsilon < 1$. We want to obtain an upper bound for the number n of vectors $u_1, \dots, u_n \in \mathbb{R}^d$ such that

$$\begin{aligned} \langle u_i, u_i \rangle &= 1 \quad \text{for} \quad i = 1, \dots, n \quad \text{and} \\ \left| \langle u_i, u_j \rangle \right| &\leq \epsilon \quad \text{for all} \quad 1 \leq i \neq j \leq n. \end{aligned}$$

Geometrically, we want to find an upper bound for the number n of vectors $u_1, \dots, u_n$ in the unit sphere $\mathbb{S}^{d-1} \subset \mathbb{R}^d$ such that the angle between any two is at least $\arccos \epsilon$.

This, of course, is an old question, and the classical Kabatyanskii–Levenshtein bound [26] states that for a fixed angle

$$\theta = \arccos \epsilon,$$

we have

$$\begin{aligned} \frac{\ln n}{d} &\leq \frac{1 + \sin\theta}{2 \sin\theta} \ln \frac{1 + \sin\theta}{2 \sin\theta} - \frac{1 - \sin\theta}{2 \sin\theta} \ln \frac{1 - \sin\theta}{2 \sin\theta} + o(1) \\ \text{as} \quad & d \longrightarrow +\infty. \end{aligned} \tag{2.1.1}$$

For the rest of this section, we follow [1, 2], see also [36].

Lemma 2.1 *Suppose that*

$$\langle u_i, u_i \rangle = 1 \quad \textit{for} \quad i = 1, \dots, n$$

and

$$\left| \langle u_i, u_j \rangle \right| \leq \epsilon \quad \textit{for all} \quad 1 \leq i \neq j \leq n$$

and some $0 < \epsilon \leq 1$. *Then*

$$\frac{n}{1 + (n-1)\epsilon^2} \leq d.$$

Proof Let $A = (a_{ij})$ be the Gram matrix of $u_1, \dots, u_n$, that is,

$$a_{ij} = \langle u_i, u_j \rangle \quad \text{for} \quad 1 \leq i, j \leq n.$$

Let $\lambda_1, \dots, \lambda_n \geq 0$ be the eigenvalues of A, so that

$$\sum_{i=1}^{n} \lambda_i = \operatorname{tr} A = n.$$

Since rank $A \leq d$, at most d of the eigenvalues of A are non-zero and, therefore,

$$\sum_{i=1}^{n} \lambda_i^2 \ \geq \ \frac{n^2}{d}. \tag{2.1.2}$$

On the other hand,

$$\sum_{i=1}^{n} \lambda_i^2 = \sum_{i,j=1}^{n} a_{ij}^2 \ \leq \ n + n(n-1)\epsilon^2. \tag{2.1.3}$$

Comparing (2.1.2) and (2.1.3), we get the desired result. □

If we choose

$$\epsilon = \frac{1}{\sqrt{n-1}}$$

in Lemma 2.1, we obtain a non-trivial bound $n \leq 2d$. However, if $\epsilon > 0$ is fixed (independent of n and d), the inequality of Lemma 2.1 holds for all sufficiently large n and d and, by and large, useless. We obtain meaningful bounds for a wide range of ϵ if we use tensorization.

Theorem 2.2 *Suppose that*

$$\langle u_i, u_i \rangle = 1 \ \ \textit{for} \ \ i = 1, \ldots, n$$

and

$$\left| \langle u_i, u_j \rangle \right| \ \leq \ \epsilon \ \ \textit{for all} \ \ 1 \leq i \neq j \leq n$$

and some $0 < \epsilon \leq 1$. *Then, for any positive integer* k *we have*

$$\frac{n}{1 + (n-1)\epsilon^{2k}} \ \leq \ \binom{d+k-1}{k}.$$

Proof We apply Lemma 2.1 to vectors $u_1^{\otimes k}, \ldots, u_n^{\otimes k}$ and use (1.1.1) and (1.1.2). □

Choosing

$$k = \left\lceil \frac{\ln(n-1)}{2\ln(1/\epsilon)} \right\rceil$$

in Theorem 2.2, we obtain the following corollary [1, 2].

Corollary 2.3 *If*

$$\frac{1}{\sqrt{n}} \ \leq \ \epsilon \ < \ \frac{1}{2},$$

we have

$$\frac{\ln n}{d} \leq \gamma\epsilon^2 \ln \frac{1}{\epsilon}.$$

for some absolute constant $\gamma > 0$ *(one can choose any* $\gamma > 2e \approx 5.44$ *provided n is large enough).*

Asymptotically, for $\epsilon \approx 0$, the bound of Corollary 2.3 has the same order as the Kabatyanskii–Levenshtein bound (2.1.1), as the right hand side of (2.1.1) can be written as

$$\frac{1}{2}\epsilon^2 \ln \frac{1}{\epsilon} + \left(\frac{1}{4} + \frac{1}{2}\ln 2\right)\epsilon^2 + \quad \text{lower order terms.}$$

Although the constant γ in Corollary 2.3 is worse than what we get in (2.1.1), the advantage of the bound in Theorem 2.2 is that it is non-asymptotic, simple and easy to check.

As is noticed in [2], one can improve the estimate in Theorem 2.2 by using a more general construction of $p^{\otimes}(x)$ instead of just $x^{\otimes k}$, see Sect. 1.2.

Theorem 2.4 *Suppose that*

$$\langle u_i, u_i \rangle = 1 \quad \textit{for} \quad i = 1, \ldots, n$$

and

$$\left|\langle u_i, u_j \rangle\right| \leq \epsilon \quad \textit{for all} \quad 1 \leq i \neq j \leq n.$$

Then, for any positive integer k, we have

$$\frac{n}{1 + (n-1)T_k^{-2}(1/\epsilon)} \leq \binom{d+k}{k}$$

where

$$T_k(t) = \frac{1}{2}\left(t - \sqrt{t^2 - 1}\right)^k + \frac{1}{2}\left(t + \sqrt{t^2 - 1}\right)^k.$$

Proof Let $A = \left(u_{ij}\right)$ be the Gram matrix of $u_1, \ldots, u_n$, so that

$$a_{ij} = \langle u_i, u_j \rangle \quad \text{for} \quad 1 \leq i, j \leq n.$$

Let us define a polynomial $p_k(t)$ by

$$p_k(t) = \frac{T_k(t/\epsilon)}{T_k(1/\epsilon)},$$

where T_k is the Chebyshev polynomial, see Sect. 1.3. Hence

$$p_k(1) = 1 \quad \text{and} \quad |p_k(t)| \leq \frac{1}{T_k(1/\epsilon)} \quad \text{provided} \quad |t| \leq \epsilon.$$

Let us define an $n \times n$ matrix $B = (b_{ij})$ by

$$b_{ij} = p_k(a_{ij}) \quad \text{for all} \quad i, j.$$

By (1.2.3), we can write

$$b_{ij} = \langle p_k^{\otimes}(u_j), \ \phi(u_j) \rangle \quad \text{where} \quad \phi(x) = 1 \oplus x + x^{\otimes 2} + \cdots + x^{\otimes k},$$

and hence by (1.2.4), we have

$$\operatorname{rank} B \leq \binom{d+k}{k}.$$

Let $\lambda_1, \ldots, \lambda_n$ be the eigenvalues of B. Arguing as in the proof of Lemma 2.1, we conclude that

$$\sum_{i=1}^n \lambda_i = \operatorname{tr} B = n \quad \text{and hence} \quad \sum_{i=1}^n \lambda_i^2 \geq \frac{n^2}{\binom{d+k}{k}}.$$

Since

$$\sum_{i=1}^n \lambda_i^2 = \sum_{i,j} b_{ij}^2 \leq n + \frac{n(n-1)}{T_k^2(1/\epsilon)},$$

the proof follows. □

We note that for small $\epsilon > 0$,

$$T_k^{-1}(1/\epsilon) \approx 2^{-k+1}\epsilon^k,$$

so the advantage of using $T_k(t)$ instead of t^k comes from the fact that $T_k(t)$ grows roughly as as $2^{-k+1}t^k$ for large t while staying in the interval $[-1, 1]$ for $|t| \leq 1$ as t^k does. This leads to an improvement of the constant γ in Corollary 2.3 roughly by a factor of 4: one can can choose any $\gamma > e/2 \approx 1.36$ provided both n and $\frac{\ln n}{\ln(1/\epsilon)}$ are large enough.

As follows from the proof of Theorem 2.4, Theorem 2.2 and Corollary 2.3 can be extended to bound from below the rank of a small entry-wise perturbation A of the identity matrix, whether or not A is the Gram matrix of unit vectors. In [1], Alon applies this approach to bound from below the distortion of low-dimensional embedding of a regular simplex, to obtain upper bounds for the cardinality of codes with guaranteed Hamming distance between words and lower bounds for the size of finite probability spaces supporting "nearly independent" Bernoulli random variables.

3 Approximating a Norm by a Polynomial and a Convex Body by an Algebraic Hypersurface

3.1 Norms and Convex Bodies

Let V be a real vector space, $\dim V = d$, and let $\| \cdot \| : V \longrightarrow \mathbb{R}$ be an arbitrary norm. Thus we have

$\|x\| \geq 0$ for all $x \in V$ and $\|x\| = 0$ only when $x = 0$,

$\|x + y\| \leq \|x\| + \|y\|$ for all $x, y \in V$ and

$\|\lambda x\| = |\lambda| \|x\|$ for all $x \in V$ and all $\lambda \in \mathbb{R}$.

As is known, the unit ball

$$B = \{x \in V : \ \|x\| \leq 1\} \tag{3.1.1}$$

is a convex body, symmetric about the origin and containing the origin in its interior. A classical result of John [25] states that any norm $\| \cdot \| : V \longrightarrow \mathbb{R}$ can be approximated within a factor of $\sqrt{d}$ by Euclidean norm. In other words, there exists a positive definite quadratic form $q : V \longrightarrow \mathbb{R}$ such that

$$\sqrt{q(x)} \ \leq \ \|x\| \ \leq \ \sqrt{d}\sqrt{q(x)} \quad \text{for all} \quad x \in V.$$

Equivalently, for any convex body B, symmetric about the origin and containing the origin in its interior, there is an ellipsoid E, centered at the origin such that

$$\frac{1}{\sqrt{d}} E \ \subset \ B \ \subset \ E.$$

One can choose E to be the (necessarily unique) ellipsoid of the minimum volume containing B, see, for example, [3].

One can ask if one can obtain a better approximation of a norm by a root of a higher degree polynomial (equivalently, if one can achieve a better approximation of a convex body symmetric about the origin by a higher degree algebraic hypersurface). The following result is proven in [7], see also Sect. 5.3 of [5].

Theorem 3.1 *Let V be a d-dimensional vector space and let $\| \cdot \| : V \longrightarrow \mathbb{R}$ be any norm. Then, for any positive integer k there is a homogeneous polynomial $p : V \longrightarrow \mathbb{R}$ of degree $2k$ which is a sum of squares of polynomials of degree k and such that*

$$p^{\frac{1}{2k}}(x) \ \leq \ \|x\| \ \leq \ \binom{d+k-1}{k}^{\frac{1}{2k}} p^{\frac{1}{2k}}(x) \quad \textit{for all} \quad x \in V.$$

Proof Let us introduce in V an inner product $\langle \cdot, \cdot \rangle$, thus making V Euclidean space. Let B be the unit ball of the norm defined by (3.1.1) and let

$$B^\circ = \{x \in V : \langle x, y \rangle \leq 1 \text{ for all } \quad y \in B\}$$

be its polar. The standard duality argument, see, for example, Sect. 4.1 of [5], implies that

$$\|x\| = \max_{y \in B^\circ} \langle x, y \rangle \quad \text{for all} \quad x \in V,$$

and since B° is symmetric about the origin, we can further write

$$\|x\| = \max_{y \in B^\circ} |\langle x, y \rangle|.$$

Let us define a compact set $C \subset V^{\otimes k}$ by

$$C = \left\{ y^{\otimes k} : \ y \in B^\circ \right\}.$$

Using (1.1.1), we can write:

$$\|x\|^k = \max_{y \in B^\circ} \left| \left\langle x^{\otimes k}, y^{\otimes k} \right\rangle \right| = \max_{z \in C} \left| \left\langle x^{\otimes k}, z \right\rangle \right|. \tag{3.1.2}$$

Let $W = \operatorname{span} C$, so by (1.1.2),

$$\dim W \ \leq \ \binom{d+k-1}{k}. \tag{3.1.3}$$

Then the function

$$w \longmapsto \max_{z \in C} |\langle w, z \rangle|$$

is a norm in W and hence there exists a positive definite quadratic form $q : W \longrightarrow \mathbb{R}$ such that

$$q(w) \ \leq \ \max_{z \in C} |\langle w, z \rangle| \ \leq \ \sqrt{\dim W} \sqrt{q(w)} \quad \text{for all} \quad w \in W. \tag{3.1.4}$$

Let us define $p : V \longrightarrow \mathbb{R}$ by

$$p(x) = q\left(x^{\otimes k}\right).$$

It is not hard to see that p is a homogeneous polynomial of degree $2k$ in x, and, moreover, since q is a sum of squares of linear forms, p is in fact is a sum of squares of homogeneous polynomials of degree k. Combining (3.1.2)–(3.1.4), we get the desired result. □

3.2 Some Geometric Corollaries

If $d \gg k \gg 1$ then

$$\binom{d+k-1}{k}^{\frac{1}{2k}} \approx \sqrt{\frac{de}{k}},$$

and hence for any fixed $\epsilon > 1$ there is a $k = k(\epsilon)$ such that any norm $\| \cdot \| : V \longrightarrow \mathbb{R}$ can be approximated by a root of a polynomial of degree $2k$ within a factor of $\epsilon\sqrt{\dim V}$. Equivalently, any convex body $B \subset \mathbb{R}^d$, symmetric about the origin, can be approximated within a factor $\epsilon\sqrt{d}$ by a semi-algebraic set S of the type

$$S = \left\{x \in \mathbb{R}^d : \quad p(x) \leq 1\right\}, \tag{3.2.1}$$

where $p : \mathbb{R}^d \longrightarrow \mathbb{R}$ is a polynomial of degree $2k$ for some k depending on ϵ and independent on the dimension d or the convex body B.

3.3 Approximation by Convex Semi-algebraic Sets

A natural question is whether one can choose S in (3.2.1) to be convex, or, equivalently, whether one can choose a polynomial p in Theorem 3.1 such that $x \longmapsto p^{\frac{1}{2k}}(x)$ is a norm. Although in general this is not known, the answer is affirmative if the unit ball B has a sufficiently large symmetry group.

More precisely, the following result is obtained in [6], see also [9]. Let V be a finite-dimensional real vector space with inner product $\langle \cdot, \cdot \rangle$ and let G be a compact group acting in V by orthogonal linear transformations. Let $v \in V$ be a point and suppose that the orbit of v spans V:

$$\operatorname{span}\big(g(v) : \ g \in G\big) = V.$$

Suppose that the norm $\| \cdot \| : V \longrightarrow \mathbb{R}$ is defined by

$$\|x\| = \max_{g \in G} |\langle x, g(v)\rangle| .$$

In other words, the polar to the unit ball of the norm $\| \cdot \|$ in V is the convex hull of the orbit of a compact group. Then for any positive integer k one can define the polynomial p in Theorem 3.1 by

$$p(x) = \int_G \langle x, g(v)\rangle^{2k} \, dg, \tag{3.3.1}$$

where dg is the Haar probability measure on G. It is then clear that $x \longmapsto p^{\frac{1}{2k}}(x)$ is itself a norm in V.

Examples of such symmetric norms $\| \cdot \|$ include the L^1, L^∞ and L^2 norms on $V = \mathbb{R}^d$. We obtain a more interesting example when V is a space of all real homogeneous polynomials of a given degree in n variables and

$$\|f\| = \max_{x \in \mathbb{S}^{n-1}} |f(x)| \quad \text{for all} \quad f \in V,$$

where $\mathbb{S}^{n-1} \subset \mathbb{R}^n$ is the unit sphere (in the Euclidean norm) in the space of variables. In this case, (3.3.1) can be written as

$$p(f) = \int_{\mathbb{S}^{n-1}} f^{2k}(x)\, dx,$$

where dx is the rotationally invariant probability measure on $\mathbb{S}^{n-1}$. Hence $p^{\frac{1}{2k}}(f)$ is just the L^{2k} norm of a polynomial f on the unit sphere. For homogeneous polynomials f of degree m in n variables, the bound of Theorem 3.1 can be sharpened to

$$\|f\|_{2k} \leq \|f\|_\infty \leq \binom{km+n-1}{km}^{1/2} \|f\|_{2k}, \tag{3.3.2}$$

where $\|f\|_{2k}$ and $\|f\|_\infty$ are respectively the L^{2k} and L^∞ norms of a polynomial f on the unit sphere in $\mathbb{R}^n$ [6].

The inequality (3.3.2) was used by Blekherman in his proof that as the number n of variables grows and degree $m \geq 4$ remains fixed, an overwhelming majority of polynomials $f : \mathbb{R}^n \longrightarrow \mathbb{R}$ of an even degree m that are non-negative on $\mathbb{R}^n$ cannot be represented as sums of squares of polynomials [11]. Another application of Theorem 3.1 concerns a computationally efficient approximation of the joint spectral radius of matrices [33].

4 Approximation of Convex Bodies by Polytopes

4.1 Convex Bodies and Polytopes

Let V be a d-dimensional real vector space, let $B \subset V$ be a convex body symmetric about the origin. A *polytope* $P \subset V$ is the convex hull of finitely many points. Our goal is to construct a polytope $P \subset V$ with as few vertices as possible, so that

$$P \subset B \subset \tau P$$

for some given $\tau > 1$.

This is a very old question extensively studied in the literature, see, for example [15, 22] and [23], although mainly in the case of fine approximations, where

$\tau = 1 + \epsilon$ for some small $\epsilon > 0$. A classical "volumetric" or "entropy" bound going back to Kolmogorov and Tikhomirov [29] states that for $\epsilon \leq 1/2$ one can choose a polytope P with

$$N \leq \left(\frac{\gamma}{\epsilon}\right)^d \tag{4.1.1}$$

vertices for some absolute constant $\gamma > 0$, see, for example, Lemma 4.10 of [34]. One can use fewer vertices if the boundary ∂B of the body is C^2-smooth. Namely, for all sufficiently small $\epsilon < \epsilon_0(B)$, one can choose the approximating polytope with

$$N \leq \left(\frac{\gamma}{\epsilon}\right)^{(d-1)/2} \tag{4.1.2}$$

vertices for some absolute constant $\gamma > 0$ and the dependence on ϵ cannot be made better as the example of the Euclidean ball demonstrates [12, 23]. Note that the upper bound for admissible ϵ depends on the body B, more specifically, on the curvature of the boundary of B.

Much less is known when the approximation factor $\tau \gg 1$. If B is the Euclidean ball then by choosing a polytope with $N \leq c^d$ vertices for some absolute constant $c > 1$ one can achieve

$$\tau = \gamma\sqrt{\frac{d}{\ln(N/d)}} \tag{4.1.3}$$

for an absolute constant $\gamma > 0$ [28]. In particular, by allowing the number N of vertices to grow polynomially in the dimension, one can achieve the approximation factor of $\tau = \epsilon\sqrt{d/\ln d}$ for any $\epsilon > 0$, fixed in advance, and the bound is optimal up to an absolute constant [4].

Our construction is based on the following lemma, proved independently in [8, 19].

Lemma 4.1 *Let V be a d-dimensional real vector space and let $C \subset V$ be a compact set. Then there is a subset $X \subset C$ of not more than*

$$|X| \leq 4d$$

points such that for any linear function $\ell : V \longrightarrow \mathbb{R}$ we have

$$\max_{x\in X} |\ell(x)| \leq \max_{x\in C} |\ell(x)| \leq 3\sqrt{d} \max_{x\in X} |\ell(x)|.$$

Sketch of proof. Without loss of generality, we assume that span $C = V$. We choose an inner product $\langle\cdot, \cdot\rangle$ in V such that the ellipsoid E of the minimum volume among all ellipsoids centered at the origin and containing C is the Euclidean unit ball. Then

John's optimality criterion (see, for example, [3]) implies that there exist contact points $x_1, \ldots, x_n \in C \cap \partial E$ and real numbers $\alpha_1, \ldots, \alpha_n > 0$ such that

$$\sum_{i=1}^{n} \alpha_i (x_i \otimes x_i) = I, \tag{4.1.4}$$

where I is the $d \times d$ identity matrix and we identify $x_i \otimes x_i$ with a symmetric $d \times d$ matrix. Equating the traces of the left and right hand sides of (4.1.4), we also conclude that

$$\sum_{i=1}^{n} \alpha_i = d. \tag{4.1.5}$$

A linear function $\ell : V \longrightarrow \mathbb{R}$ can be written as

$$\ell(x) = \langle y, x \rangle \quad \text{for some} \quad y \in V \quad \text{and all} \quad x \in V.$$

Then (4.1.4) implies that

$$\sum_{i=1}^{n} \alpha_i \langle y, x_i \rangle^2 = \|y\|^2. \tag{4.1.6}$$

Since C is contained in the unit ball, we obtain

$$\max_{x \in C} |\langle y, x \rangle| \ \leq \ \|y\|, \tag{4.1.7}$$

and, combining (4.1.5)–(4.1.7) we conclude that

$$\max_{x \in X} |\ell(x)| \ \leq \ \max_{x \in C} |\ell(x)| \ \leq \ \sqrt{d} \max_{x \in X} |\ell(x)|,$$

which is an even stronger inequality than we attempt to prove.

Unfortunately, the number of contact points $x_1, \ldots, x_n$ can be quadratic in $\dim V$. Indeed, Carathéodory's Theorem implies that we can choose

$$n \leq \binom{d+1}{2} + 1$$

and, generally, speaking, the quadratic dependence on $\dim V$ is unavoidable [21].

To get a subset $X \subset \{x_1, \ldots, x_n\}$ of a linear in $\dim V$ cardinality, we use the sparsification result of Batson, Spielman and Srivastava [10]. The sparsification theorem of [10] implies, in particular, that one can choose a subset $J \subset \{1, \ldots, n\}$ such that $|J| \leq 4d$ and reals $\beta_j > 0$ for $j \in J$ such that instead of (4.1.4) we have

$$I \preceq \sum_{j \in J} \beta_j \left(x_j \otimes x_j\right) \preceq 9I,$$

where we write $A \preceq B$ if $B - A$ is a positive semidefinite matrix. We let $X = \{x_j : j \in J\}$ and the proof proceeds as above. □

As a corollary, we conclude that a d-dimensional origin-symmetric convex body B can be approximated within a factor of $\tau = 3\sqrt{d}$ by a polytope P with at most $8d$ vertices: we pick the vertices of P from the set $X \cup -X$, where X is the set of points constructed in Lemma 4.1 with $C = B$.

Then tensorization produces the following result [8].

Theorem 4.2 *Let k and d be positive integers and let $\tau > 1$ be a real number such that*

$$\left(\tau - \sqrt{\tau^2 - 1}\right)^k + \left(\tau + \sqrt{\tau^2 - 1}\right)^k \geq 6\binom{d+k}{k}^{1/2}.$$

Then for any symmetric convex body $B \subset \mathbb{R}^d$ there is a symmetric polytope $P \subset \mathbb{R}^d$ with at most

$$8\binom{d+k}{k}$$

vertices such that

$$P \subset B \subset \tau P.$$

Proof Let us denote $V = \mathbb{R}^d$ and let

$$W_k = \mathbb{R} \oplus V \oplus V^{\otimes 2} \oplus \cdots \oplus V^{\otimes k}.$$

We consider a polynomial map $\phi : V \longrightarrow W_k$ defined by

$$\phi(x) = 1 \oplus x \oplus x^{\otimes 2} \oplus \cdots \oplus x^{\otimes k}$$

and let $C = \phi(B)$ be the image of B. Then C is a compact set and since C ends up in the symmetric part of W_k, by (1.2.4) we have

$$\dim \operatorname{span} C \leq \binom{d+k}{k}.$$

Applying Lemma 4.1, we conclude that there is a set $X \subset B$ of $|X| \leq 4\binom{d+k}{k}$ points such that for any linear function $\mathcal{L} : W_k \longrightarrow \mathbb{R}$ we have

$$\max_{x \in X} |\mathcal{L}(\phi(x))| \leq \max_{x \in B} |\mathcal{L}(\phi(x))| \leq 3\binom{d+k}{k}^{1/2} \max_{x \in X} |\mathcal{L}(\phi(x))|. \qquad (4.1.8)$$

We let

$$P = \operatorname{conv}(X, -X).$$

Clearly, $P \subset B$ is a symmetric polytope with at most $2|X| \leq 8\binom{d+k}{k}$ vertices.

It remains to show that $B \subset \tau P$. We use a separation argument. Let $\ell : V \longrightarrow \mathbb{R}$ be a linear function such that $\ell(x) \leq 1$ for all $x \in P$ and hence $|\ell(x)| \leq 1$ for all $x \in X$. Our goal is to show that $\ell(x) \leq \tau$ for all $x \in B$. We can write $\ell(x) = \langle y, x \rangle$ for some $y \in V$ and all $x \in V$. Let T_k be the Chebyshev polynomial, see Sect. 1.3. We define a function $\mathcal{L} : W_k \longrightarrow \mathbb{R}$ by

$$\mathcal{L}(w) = \left\langle T_k^{\otimes}(y),\ w \right\rangle \quad \text{for all} \quad w \in W_k.$$

Since $|\ell(x)| \leq 1$ for all $x \in X$, we have

$$|\mathcal{L}(\phi(x))| = \left|T_k\bigl(\ell(x)\bigr)\right| \leq 1 \quad \text{for all} \quad x \in X.$$

Moreover, if $\ell(x) > \tau$ for some $x \in B$ then

$$\begin{aligned} \mathcal{L}(\phi(x)) = T_k\bigl(\ell(x)\bigr) \ > \ T_k(\tau) &= \frac{\left(\tau - \sqrt{\tau^2 - 1}\right)^k + \left(\tau + \sqrt{\tau^2 - 1}\right)^k}{2} \\ &\geq 3\binom{d+k}{k}^{1/2}, \end{aligned}$$

which contradicts (4.1.8). □

Tuning up the value of k in Theorem 4.2, we obtain different asymptotic regimes.

4.2 Fine Approximations

It follows that for some absolute constant $\epsilon_0, \gamma > 0$ and for any $\tau = 1 + \epsilon$ with $0 < \epsilon \leq \epsilon_0$, any origin-symmetric d-dimensional convex body can be approximated within a factor of τ by a symmetric polytope with

$$N \ \leq \ \left(\frac{\gamma}{\sqrt{\epsilon}} \ln \frac{1}{\epsilon}\right)^d \tag{4.2.1}$$

vertices. To obtain (4.2.1), we choose k of the order of

$$k \sim \frac{d}{\sqrt{\epsilon}} \ln \frac{1}{\epsilon}$$

in Theorem 4.2. Hence we use roughly the square root of the number of points guaranteed by the volumetric/entropy bound (4.1.1). The savings come, in particular, from the fact that roughly

$$T_k(1+\epsilon) \approx \frac{1}{2}\left(1+\sqrt{2\epsilon}\right)^k$$

for small $\epsilon > 0$. Curiously, whereas in Theorem 2.4 we gain by replacing t^k by $T_k(t)$ because $T_k(t)$ outperforms t^k for $t \gg 1$, in Theorem 4.2 we gain because $T_k(t)$ outperforms t^k for $t > 1$ that are close to 1.

It is not known whether the bound (4.2.1) is optimal. It is slightly weaker than the bound (4.1.2) for convex bodies with a C^2-smooth boundary, but the advantage of (4.2.1) is that it is uniform over $\epsilon > 0$ and convex bodies B.

4.3 Coarse Approximations

Suppose now that we want to keep the number N of points of the approximating polytope bounded by a polynomial in the dimension d. For any $\epsilon > 0$, we can find such a polytope with $N = d^{O(1)}$ vertices and

$$\tau \leq \epsilon\sqrt{d} \tag{4.3.1}$$

by choosing some constant $k = k(\epsilon)$ in Theorem 4.2. It is not known whether the bound (4.3.1) is optimal, although in the case of the Euclidean ball the best possible bound (4.1.3) is slightly better than (4.3.1) and it does sound plausible that the Euclidean ball should exhibit the worst possible approximability by polytopes (this is provably so for C^2-smooth bodies and $\tau \approx 1$ [12, 23]).

4.4 Intermediate Approximations

It was noticed in [31] that Theorem 4.2 implies the following general inequality between the number N of vertices of the approximating polytope and the approximating factor τ that the polytope can attain:

$$\tau \leq \gamma \max\left\{1, \ \sqrt{\frac{d}{\ln N} \ln \frac{d}{\ln N}}\right\} \tag{4.4.1}$$

for some absolute constant $\gamma > 0$. Again, it is not known whether (4.4.1) is optimal and in the case of the Euclidean ball the best possible bound (4.1.3) is slightly better. We also note that (4.4.1) does not quite capture the asymptotic behavior (4.2.1) for fine approximations; in fact, getting (4.2.1) is the only place where we need to use the Chebyshev polynomials T_k instead of the monomial t^k in the proof of Theorem 4.2 and is the only place where we have to resort to sparsification in Lemma 4.1.

5 The Polynomial Method

We described three examples where the strengthening of the original result is obtained through the combination of the two factors: first, that

$$\dim \operatorname{span}\left(v^{\otimes k} \, : \, v \in V\right) \ll (\dim V)^k$$

and second, that a linear function in $v^{\otimes k}$ is a polynomial in V. In many more applications, only the second factor is put to work while the first is either not used at all or is used in some limited way and the approach is often called "the polynomial method".

We give a selection of such examples below.

5.1 *Constructing Neighborly Polytopes*

Perhaps the earliest application of the polynomial method in convex geometry is the construction of polytopes $P \subset \mathbb{R}^d$ such that every $\lfloor d/2 \rfloor$ vertices of P span a face of P (neighborly polytopes) [16, 18]. We consider the embedding

$$\psi : \mathbb{R} \longrightarrow \mathbb{R}^d, \quad t \longmapsto \left(t, t^2, \ldots, t^d\right)$$

and define P as the convex hull

$$P = \operatorname{conv}\left(\psi(t_1), \ldots, \psi(t_n)\right) \quad \text{some} \quad t_1 < t_2 < \cdots < t_n.$$

If $\ell : \mathbb{R}^d \longrightarrow \mathbb{R}$ is a linear function and $\alpha \in \mathbb{R}$ is a constant then $p(t) = \ell(\psi(t)) + \alpha$ is a polynomial of $\deg p \leq d$ and every univariate polynomial p with $\deg p \leq d$ can be obtained this way. Hence to prove that P is neighborly, one has to present a polynomial p of degree at most d which is non-negative for all t and has zeros at $k = \lfloor d/2 \rfloor$ prescribed points $t_{i_1}, \ldots, t_{i_k}$. The polynomial

$$p(t) = \left(t - t_{i_1}\right)^2 \cdots \left(t - t_{i_k}\right)^2$$

obviously satisfies the required property.

5.2 Bounding the Constant in the Grothendieck Inequality

The Grothendieck inequality [20] asserts that there is an absolute constant $K_G > 0$ such that for any $m \times n$ real matrix $A = \left(a_{ij}\right)$, for any vectors $x_1, \ldots, x_m; y_1, \ldots, y_n \in \mathbb{R}^d$ of Euclidean unit length, one can find signs $\epsilon_1, \ldots, \epsilon_m; \delta_1, \ldots, \delta_n \in \{-1, 1\}$ such that

$$\sum_{\substack{i=1,\ldots,m \\ j=1,\ldots,n}} a_{ij}\langle x_i, y_j\rangle \leq K_G \sum_{\substack{1\leq i\leq m \\ 1\leq j\leq n}} a_{ij}\epsilon_i\delta_j. \tag{5.2.1}$$

Up until very recently, the best upper bound

$$K_G \leq \frac{\pi}{2\ln\left(1+\sqrt{2}\right)} \approx 1.782 \tag{5.2.2}$$

was due to Krivine [30]. It was shown in [14] that the bound *can* be improved, though it is not immediately clear by exactly how much.

Krivine's proof combines the idea of the polynomial method and that of the *randomized rounding*.

Let $\mathbb{S}^{d-1} \subset \mathbb{R}^d$ be the unit sphere. Given vectors $x_1, \ldots, x_m; y_1, \ldots, y_n \in \mathbb{S}^{d-1}$, one can try to construct signs $\epsilon_1, \ldots, \epsilon_m; \delta_1, \ldots, \delta_n$ by choosing a random vector $u \in \mathbb{S}^{d-1}$ from the uniform distribution on the sphere and defining

$$\epsilon_i = \operatorname{sgn}\langle x_i, u\rangle \quad \text{and} \quad \delta_j = \operatorname{sgn}\langle y_j, u\rangle \quad \text{for all} \quad i, j.$$

It is not hard to show that

$$\mathbf{E}\, \epsilon_i\delta_j = \frac{2\arcsin\langle x_i, y_j\rangle}{\pi}$$

and hence

$$\mathbf{E}\sum_{i,j} a_{ij}\epsilon_i\delta_j = \frac{2}{\pi}\sum_{i,j} a_{ij}\arcsin\langle x_i, y_j\rangle,$$

which is hard to compare to the target value of $\sum_{ij} a_{ij}\langle x_i, y_j\rangle$. We first need to modify the vectors $x_i \longmapsto X_i$ and $y_j \longmapsto Y_j$ so that

$$\langle X_i, Y_j\rangle = \sin c\langle x_i, y_j\rangle \quad \text{for all} \quad i, j \tag{5.2.3}$$

for some constant $c > 0$ and then construct random signs ϵ_i and δ_j as above, so that we get

$$\mathbf{E}\sum_{i,j} a_{ij}\epsilon_i\delta_j = \frac{2c}{\pi}\sum_{i,j} a_{ij}\langle x_i, y_j\rangle$$

which would prove (5.2.1) with

$$K_G = \frac{\pi}{2c}.$$

To that end, we let

$$X_i = \sum_{k=0}^{\infty} \sqrt{\frac{c^{2k+1}}{(2k+1)!}} x_i^{\otimes(2k+1)} \quad \text{and} \quad Y_j = \sum_{k=1}^{\infty} (-1)^k \sqrt{\frac{c^{2k+1}}{(2k+1)!}} y_j^{\otimes(2k+1)} \tag{5.2.4}$$

and (5.2.3) holds. We find the value of c from the constraints

$$1 = \langle X_i, X_j \rangle = \langle Y_j, Y_j \rangle = \sum_{k=0}^{\infty} \frac{c^{2k+1}}{(2k+1)!} = \frac{e^c - e^{-c}}{2},$$

from which $c = \ln(1 + \sqrt{2})$ and (5.2.2) follows.

Although vectors X_i and Y_j defined by (5.2.4) lie in some infinite-dimensional space

$$\bigoplus_{k=0}^{\infty} V^{\otimes k} \quad \text{for} \quad V = \mathbb{R}^d,$$

they span a finite-dimensional subspace there and we can restrict our computations to that subspace.

5.3 *Polynomial Ham Sandwich Theorem*

Let $S \subset \mathbb{R}^d$ be a finite set of points and let $p : \mathbb{R}^d \longrightarrow \mathbb{R}$ be a polynomial. We say that p *bisects* S if at most half of the points $x \in S$ satisfy $p(x) > 0$ and at most half of the points $x \in S$ satisfy $p(x) < 0$. In their recent breakthrough on the Erdős distinct distances problem [24], Guth and Katz make use of the following *polynomial ham sandwich theorem*, which goes back to Stone and Tukey [35]:

If $S_1, \ldots, S_m$ are finite subsets of $\mathbb{R}^d$ and $m \leq \binom{d+k}{k} - 1$ then there is a polynomial p of degree at most k which bisects each set S_i for $i = 1, \ldots, m$.

In the original ham sandwich theorem $\deg p = 1$, so the zero set of p is an affine hyperplane and the result follows from the Borsuk - Ulam Theorem, cf. [35]. If $\deg p = k > 1$ the result follows by tensorization

$$\psi(x) = x \oplus x^{\otimes 2} \oplus \cdots \oplus x^{\otimes k}$$

and applying the original ham sandwich theorem to $\psi(S)$.

5.4 Equiangular Lines

In this problem, one is interested how many lines one can find in $\mathbb{R}^d$ so that the angle between any two is the same and not equal to 0. Let us orient each line arbitrarily and choose a unit vector v_i in the direction of the ith line. Then the Gram matrix of vectors $v_i^{\otimes 2}$ has 1s on the diagonal and some number $0 \leq a < 1$ away from the diagonal and hence is positive definite and thus invertible. It follows that the vectors $v_i^{\otimes 2}$ are linearly independent and hence the number n of equiangular lines in $\mathbb{R}^d$ cannot exceed

$$\dim \operatorname{span}\left(v_i^{\otimes} : i = 1, \ldots, n\right) \leq \binom{d+1}{2}.$$

In particular, for $d = 3$ we obtain $n \leq 6$, which is an exact bound, as one can ascertain by drawing a line through each of 6 pairs of opposite vertices of a regular icosahedron, see Miniature 9 of [32].

5.5 A Counterexample to Borsuk's Conjecture

As in Sect. 5.4, the quadratic map $v \longrightarrow v \otimes v$ played a crucial role in the Kahn and Kalai [27] counterexample to Borsuk's conjecture that stated that any convex body in $\mathbb{R}^d$ can be subdivided into $d + 1$ of parts of strictly smaller diameters. We follow the exposition of Miniature 18 of [32]. Let $n = 4p$, where p is prime and for every $(2p - 1)$-subset A of the set $\{1, \ldots, 4p\}$ let us define $v(A) \in \mathbb{R}^n$ by

$$v(A) = \begin{cases} 1 & \text{if } \ i \in A \\ -1 & \text{if } \ i \notin A. \end{cases}$$

Then the polytope

$$P = \operatorname{conv}\left(v(A)^{\otimes 2} : \quad A \subset \{1, \ldots, 4p\}, \ |A| = 2p - 1\right), \quad P \subset \mathbb{R}^{n^2}.$$

is a counterexample for all sufficiently large p. It is not hard to see that

$$\begin{aligned} \left\langle v(A), \ v(B) \right\rangle &\geq 0 \quad \text{for all} \quad A, B \qquad \text{and} \\ \left\langle v(A), \ v(B) \right\rangle &= 0 \quad \text{if and only if} \quad |A \cap B| = p - 1. \end{aligned}$$

Consequently, the diameter of P is n and vectors $v(A)^{\otimes 2}$ and $v(B)^{\otimes 2}$ are the endpoints of a diameter if and only if $|A \cap B| = p - 1$. By a result of Frankl and Wilson [17] the cardinality of a family $\mathcal{F}$ of $(2p - 1)$-subsets of the set $\{1, \ldots, 4p\}$ such that $|A \cap B| \neq p - 1$ for any two $A, B \in \mathcal{F}$ is at most an exponential in n fraction of the cardinality of the family of all $(2p - 1)$ subsets of $\{1, \ldots, 4p\}$,

$$\frac{|\mathcal{F}|}{\binom{4p}{2p-1}} \leq (1.1)^{-n}.$$

Consequently, if the polytope P is split into fewer than $(1.1)^n$ parts, at least one of the parts will contain vertices $v(A)$ and $v(B)$ with $|A \cap B| = p - 1$ and hence will have diameter n.

Acknowledgements I am grateful to Terence Tao for pointing to [1, 2, 36].

References

1. N. Alon, Problems and results in extremal combinatorics. I, EuroComb'01 (Barcelona). Discret. Math. **273**(1–3), 31–53 (2003)
2. N. Alon, Perturbed identity matrices have high rank: proof and applications. Comb. Probab. Comput. **18**(1–2), 3–15 (2009)
3. K. Ball, An elementary introduction to modern convex geometry, *Flavors of Geometry*, vol. 31, Mathematical Sciences Research Institute Publications (Cambridge University Press, Cambridge, 1997), pp. 1–58
4. I. Bárány, Z. Füredi, Approximation of the sphere by polytopes having few vertices. Proc. Am. Math. Soc. **102**(3), 651–659 (1988)
5. A. Barvinok, *A Course in Convexity*, vol. 54, Graduate Studies in Mathematics (American Mathematical Society, Providence, 2002)
6. A. Barvinok, Estimating L^∞ norms by L^{2k} norms for functions on orbits. Found. Comput. Math. **2**(4), 393–412 (2002)
7. A. Barvinok, Approximating a norm by a polynomial, *Geometric Aspects of Functional Analysis*, vol. 1807, Lecture Notes in Mathematics (Springer, Berlin, 2003), pp. 20–26
8. A. Barvinok, Thrifty approximations of convex bodies by polytopes. Int. Math. Res. Not. IMRN **2014**(16), 4341–4356 (2014)
9. A. Barvinok, G. Blekherman, Convex geometry of orbits, *Combinatorial and Computational Geometry*, vol. 52, Mathematical Sciences Research Institute Publications (Cambridge University Press, Cambridge, 2005), pp. 51–77
10. J. Batson, D.A. Spielman, N. Srivastava, Twice-Ramanujan sparsifiers. SIAM J. Comput. **41**(6), 1704–1721 (2012)
11. G. Blekherman, There are significantly more nonnegative polynomials than sums of squares. Isr. J. Math. **153**, 355–380 (2006)
12. K. Böröczky Jr., Approximation of general smooth convex bodies. Adv. Math. **153**(2), 325–341 (2000)
13. P. Borwein, T. Erdélyi, *Polynomials and Polynomial Inequalities*, vol. 161, Graduate Texts in Mathematics (Springer, New York, 1995)
14. M. Braverman, K. Makarychev, Y. Makarychev, A. Naor, The Grothendieck constant is strictly smaller than Krivine's bound. Forum of Mathematics Pi **1**, e4–42 (2013)
15. E.M. Bronshtein, Approximation of convex sets by polyhedra. Sovremennaya Matematika. Fundamental'nye Napravleniya **22**, 5–37 (2007); translated in J. Math. Sci. (New York) **153**(6), 727–762 (2008)
16. C. Carathéodory, Über den Variabilitatsbereich det Fourierschen Konstanten von Positiven harmonischen Furktionen. Rendiconti del Circolo Matematico di Palermo **32**, 193–217 (1911)
17. P. Frankl, R. Wilson, Intersection theorems with geometric consequences. Combinatorica **1**, 357–368 (1981)
18. D. Gale, Neighborly and cyclic polytopes, *Proceedings of Symposia in Pure Mathematics*, vol. VII (American Mathematical Society, Providence, 1963), pp. 225–232

19. E. Gluskin, A. Litvak, A remark on vertex index of the convex bodies, *Geometric Aspects of Functional Analysis*, vol. 2050, Lecture Notes in Mathematics (Springer, Heidelberg, 2012)
20. A. Grothendieck, Résumé de la théorie métrique des produits tensoriels topologiques. Boletim da Sociedade Matemática São Paulo **8**, 1–79 (1953)
21. P.M. Gruber, Application of an idea of Voronoi to John type problems. Adv. Math. **218**(2), 309–351 (2008)
22. P.M. Gruber, Aspects of approximation of convex bodies, *Handbook of Convex Geometry*, vol. A, B (North-Holland, Amsterdam, 1993), pp. 319–345
23. P.M. Gruber, Asymptotic estimates for best and stepwise approximation of convex bodies. I. Forum Math. **5**(3), 281–297 (1993)
24. L. Guth, N.H. Katz, On the Erdős distinct distances problem in the plane. Ann. Math. Second series **181**(1), 155–190 (2015)
25. F. John, Extremum problems with inequalities as subsidiary conditions, *Studies and Essays*, vol. 1948 (Interscience Publishers Inc., New York, 1948), pp. 187–204. Presented to R. Courant on his 60th Birthday, January 8
26. G.A. Kabatyanskii, V.I. Levenshtein, Bounds for packings on the sphere and in space (Russian). Problemy Peredachi Informacii **14**(1), 3–25 (1978)
27. J. Kahn, G. Kalai, A counterexample to Borsuk's conjecture. Bull. Am. Math. Soc. New Series **29**(1), 60–62 (1993)
28. M. Kochol, Constructive approximation of a ball by polytopes. Mathematica Slovaca **44**(1), 99–105 (1994)
29. A.N. Kolmogorov, V.M. Tihomirov, ϵ-entropy and ϵ-capacity of sets in function spaces. Uspehi Matematicheskih Nauk **14**(2(86)), 3–86 (1959). translated in Am. Math. Soc. Transl. **17**(2), 277–364 (1961)
30. J.-L. Krivine, Constantes de Grothendieck et fonctions de type positif sur les sphères. Adv. Math. **31**(1), 16–30 (1979)
31. A. Litvak, M. Rudelson, N. Tomczak-Jaegermann, On approximation by projections of polytopes with few facets. Isr. J. Math. **203**(1), 141–160 (2014)
32. J. Matoušek, *Thirty-Three Miniatures. Mathematical and Algorithmic Applications of Linear Algebra*, vol. 53, Student Mathematical Library (American Mathematical Society, Providence, 2010)
33. P. Parrilo, A. Jadbabaie, Approximation of the joint spectral radius using sum of squares. Linear Algebra Appl. **428**(10), 2385–2402 (2008)
34. G. Pisier, *The Volume of Convex Bodies and Banach Space Geometry*, vol. 94, Cambridge Tracts in Mathematics (Cambridge University Press, Cambridge, 1989)
35. A.H. Stone, J.W. Tukey, Generalized "sandwich" theorems. Duke Math. J. **9**, 356–359 (1942)
36. T. Tao, A cheap version of the Kabatjanskii-Levenstein bound for almost orthogonal vectors (2013), blog post at https://terrytao.wordpress.com/2013/07/18/

Contact Numbers for Sphere Packings

Károly Bezdek and Muhammad A. Khan

Abstract In discrete geometry, the contact number of a given finite number of non-overlapping spheres was introduced as a generalization of Newton's kissing number. This notion has not only led to interesting mathematics, but has also found applications in the science of self-assembling materials, such as colloidal matter. With geometers, chemists, physicists and materials scientists researching the topic, there is a need to inform on the state of the art of the contact number problem. In this paper, we investigate the problem in general and emphasize important special cases including contact numbers of minimally rigid and totally separable sphere packings. We also discuss the complexity of recognizing contact graphs in a fixed dimension. Moreover, we list some conjectures and open problems.

MSC (2010) (Primary) 52C17 · 52C15 · (Secondary) 52C10

1 Introduction

The well-known *"kissing number problem"* asks for the maximum number $k(d)$ of non-overlapping unit balls that can touch a unit ball in the d-dimensional Euclidean space $\mathbb{E}^d$. The problem originated in the 17th century from a disagreement between Newton and Gregory about how many 3-dimensional unit spheres without overlap could touch a given unit sphere. The former maintained that the answer was 12, while the latter thought it was 13. The question was finally settled many years later [41] when Newton was proved correct. The known values of $k(d)$ are $k(2) = 6$ (trivial), $k(3) = 12$ [41], $k(4) = 24$ [39], $k(8) = 240$ [40], and $k(24) = 196560$ [40]. The problem of finding kissing numbers is closely connected to the more general problems

K. Bezdek (✉) · M. A. Khan
Department of Mathematics and Statistics, University of Calgary, Calgary, Canada
e-mail: bezdek@math.ucalgary.ca

M. A. Khan
e-mail: muhammkh@ucalgary.ca

K. Bezdek
Department of Mathematics, University of Pannonia, Veszprém, Hungary

G. Ambrus et al. (eds.), *New Trends in Intuitive Geometry*, Bolyai Society Mathematical Studies 27, https://doi.org/10.1007/978-3-662-57413-3_2

of finding bounds for spherical codes and sphere packings. For old and new results on kissing numbers we refer the interested reader to the recent survey article [16]. In this paper, we focus on a more general relative of kissing number called contact number.

Let $\mathbf{B}^d$ be the d-dimensional unit ball centered at the origin $\mathbf{o}$ in $\mathbb{E}^d$. As is well known, a *finite packing* of unit balls in $\mathbb{E}^d$ is a finite family of non-overlapping translates of $\mathbf{B}^d$ in $\mathbb{E}^d$. Furthermore, the *contact graph* of a finite unit ball packing in $\mathbb{E}^d$ is the (simple) graph whose vertices correspond to the packing elements and whose two vertices are connected by an edge if and only if the corresponding two packing elements touch each other. The number of edges of a contact graph is called the *contact number* of the underlying unit ball packing. The *"contact number problem"* asks for the largest contact number, that is, for the maximum number $c(n, d)$ of edges that a contact graph of n non-overlapping translates of $\mathbf{B}^d$ can have in $\mathbb{E}^d$.

The problem of determining $c(n, d)$ is equivalent to Erdős's repeated shortest distance problem, which asks for the largest number of repeated shortest distances among n points in $\mathbb{E}^d$. The planar case of this question was originally raised by Erdős in 1946 [19], with an answer conjectured by Reutter in 1972 and established by Harborth [25] in 1974, whereas the problem in its more general forms was popularized by Erdős and Ulam. Another way to look at the contact number problem is to think of it as the combinatorial analogue of the densest sphere packing problem, which dates back to the 17th century.

Let $\mathbf{K}$ be a convex body, i.e., a compact convex set with non-empty interior in $\mathbb{E}^d$. (If $d = 2$, then $\mathbf{K}$ is called a convex domain.) If $\mathbf{K}$ is symmetric about the origin $\mathbf{o}$ in $\mathbb{E}^d$, then one can regard $\mathbf{K}$ as the unit ball of a given norm in $\mathbb{R}^d$. In the same way as above one can talk about the largest contact number of packings by n translates of $\mathbf{K}$ in $\mathbb{E}^d$ and label it by $c(\mathbf{K}, n, d)$. Here we survey the results on $c(n, d)$ as well as $c(\mathbf{K}, n, d)$.

The notion of total separability was introduced in [21] as follows: a packing of unit balls in $\mathbb{E}^d$ is called *totally separable* if any two unit balls can be separated by a hyperplane of $\mathbb{E}^d$ such that it is disjoint from the interior of each unit ball in the packing. Finding the densest totally separable unit ball packing is a difficult problem, which is solved only in dimensions two [6, 21] and three [34]. As a close combinatorial relative it is natural to investigate the maximum contact number $c_{\text{sep}}(n, d)$ of totally separable packings of n unit balls in $\mathbb{E}^d$. In what follows, we survey the results on $c_{\text{sep}}(n, d)$ as well.

The paper is organized as follows. In Sect. 2, we briefly discuss the importance of the contact number problem in materials science. The next two sections are devoted to the known bounds on the contact number for $d = 2, 3$. Section 5, explores three computer-assisted empirical approaches that have been developed by applied scientists to estimate the contact numbers of packings of small number of unit spheres in $\mathbb{E}^3$. We analyze these approaches at length and show that despite being of interest, they fall short of providing exact values of largest contact numbers. In Sect. 6, we study contact numbers of unit sphere packings in $\mathbb{E}^2$ and $\mathbb{E}^3$ that live on the integer lattice or are totally separable. Section 7 covers recent general results on packings of congruent balls and translates of an arbitrary convex body in d-space. It also includes results on the integer lattice and totally separable packings of d-dimensional unit balls. Finally, the last section deals with the state of the contact number problem for non-congruent sphere packings.

2 Motivation from Materials Science

In addition to finding its origins in the works of pioneers like Newton, Erdős, Ulam and Fejes Tóth (see Sect. 3 for more on the role of latter two), the contact number problem is also important from an applications point of view. Packings of hard sticky spheres - impenetrable spheres with short-range attractive forces - provide excellent models for the formation of several real-life materials such as colloids, powders, gels and glasses [27]. The particles in these materials can be thought of as hard spheres that self-assemble into small and large clusters due to their attractive forces. This process, called *self-assembly*, is of tremendous interest to materials scientists, chemists, statistical physicists and biologists alike.

Of particular interest are *colloids*, which consist of particles at micron scale, dispersed in a fluid and kept suspended by thermal interactions [37]. Colloidal matter occurs abundantly around us - for example in glue, milk and paint. Moreover, controlled colloid formation is a fundamental tool used in scientific research to understand the phenomena of self-assembly and phase transition.

From thermodynamical considerations it is clear that colloidal particles assemble so as to minimize the potential energy of the cluster. Since the range of attraction between these particles is extremely small compared to their sizes, two colloidal particles do not exert any force on each other until they are infinitesimally close, at which point their is strong attraction between them. As a result, they stick together, are resistant to drift apart, but strongly resistant to move any closer [3, 27]. Thus two colloidal particles experiencing an attractive force from one another in a cluster can literally be thought of as being in contact.

It can be shown that under the force law described above, the potential energy of a colloidal cluster at reasonably low temperatures is inversely proportional to the number of contacts between its particles [3, 31, 32]. Thus the particles are highly likely to assemble in packings that maximize the contact number. This has generated significant interest among materials scientists towards the contact number problem [3, 32] and has led to efforts in developing computer-assisted approaches to attack the problem. More details will appear in Sect. 5.

3 Largest Contact Numbers in the Plane

3.1 The Euclidean Plane

Harborth [25] proved the following well-known result on the contact graphs of congruent circular disk packings in $\mathbb{E}^2$.

Theorem 3.1 $c(n, 2) = \lfloor 3n - \sqrt{12n - 3} \rfloor$, *for all* $n \geq 2$.

This result shows that an optimal way to pack n congruent disks to maximize their contacts is to pack them in a 'hexagonal arrangement'. The arrangement starts by packing 6 unit disks around a central disk in such a way that the centers of the surrounding disks form a regular hexagon. The pattern is then continued by packing hexagonal layers of disks around the first hexagon. Thus the hexagonal packing arrangement, which is known to be the densest congruent disk packing arrangement, also achieves the maximum contact number $c(n, 2)$, for all n.

Interestingly, this also means that $c(n, 2)$ equals the maximum number of sides that can be shared between n cells of a regular hexagon tiling of the plane. This connection was explored in [24], where isoperimetric hexagonal lattice animals of a given area n were explored. The connection between contact numbers and isoperimetric lattice animals is studied in detail in Sect. 6. So we skip the details here.

Despite the existence of a simple formula for $c(n, 2)$, recognizing contact graphs of congruent disk packings is a challenging problem. The difficulty of this problem is made apparent by the following complexity result from [18].

Theorem 3.2 *The problem of recognizing contact graphs of unit disk packings is NP-hard.*

Quite surprisingly, the following rather natural stability version of Theorem 3.1 is still an open problem. (See also the final remarks in [17].)

Conjecture 3.3 *There exists an* $\epsilon > 0$ *such that for any packing of n circular disks of radii chosen from the interval* $[1 - \epsilon, 1]$ *the number of touching pairs in the packing is at most* $\lfloor 3n - \sqrt{12n - 3} \rfloor$, *for all* $n \geq 2$.

In 1984, Ulam [20] proposed to investigate Erdős-type distance problems in normed spaces. Pursuing this idea, Brass [17] proved the following extension of Theorem 3.1 to normed planes.

Theorem 3.4 *Let* $\mathbf{K}$ *be a convex domain different from a parallelogram in* $\mathbb{E}^2$. *Then for all* $n \geq 2$, *one has* $c(\mathbf{K}, n, 2) = \lfloor 3n - \sqrt{12n - 3} \rfloor$. *If* $\mathbf{K}$ *is a parallelogram, then* $c(\mathbf{K}, n, 2) = \lfloor 4n - \sqrt{28n - 12} \rfloor$ *holds for all* $n \geq 2$.

The same idea inspired the first named author to investigate this question in d-space, details of which appear in Sect. 7.

Returning to normed planes, the following is a natural question.

Problem 1 Find an analogue of Theorem 3.4 for totally separable translative packings of convex domains in $\mathbb{E}^2$.

3.2 Spherical and Hyperbolic Planes

An analogue of Harborth's theorem in the hyperbolic plane $\mathbb{H}^2$ was found by Bowen in [15]. In fact, his method extends to the 2-dimensional spherical plane $\mathbb{S}^2$. We prefer to quote these results as follows.

Theorem 3.5 *Consider disk packings in $\mathbb{H}^2$ (resp., $\mathbb{S}^2$) by finitely many congruent disks, which maximize the number of touching pairs for the given number of congruent disks and of given diameter D. Then such a packing must have all of its centers located on the vertices of a triangulation of $\mathbb{H}^2$ (resp., $\mathbb{S}^2$) by congruent equilateral triangles of side length D provided that the equilateral triangle in $\mathbb{H}^2$ (resp., $\mathbb{S}^2$) of side length D has each of its angles equal to $\frac{2\pi}{N}$ for some positive integer $N \geq 3$.*

In 1984, L. Fejes Tóth [12] raised the following attractive and related problem in $\mathbb{S}^2$: Consider an arbitrary packing $\mathcal{P}_r$ of disks of radius $r > 0$ in $\mathbb{S}^2$. Let $\deg_{\mathrm{avr}}(\mathcal{P}_r)$ denote the average degree of the vertices of the contact graph of $\mathcal{P}_r$. Then prove or disprove that $\limsup_{r\to 0}\left(\sup_{\mathcal{P}_r} \deg_{\mathrm{avr}}(\mathcal{P}_r)\right) < 5$. This problem was settled in [12].

Theorem 3.6 *Let $\mathcal{P}_r$ be an arbitrary packing of disks of radius $r > 0$ in $\mathbb{S}^2$. Then*

$$\limsup_{r\to 0}\left(\sup_{\mathcal{P}_r} \deg_{\mathrm{avr}}(\mathcal{P}_r)\right) < 5.$$

We conclude this section with the still open hyperbolic analogue of Theorem 3.6 which was raised in [12].

Conjecture 3.7 *Let $\mathcal{P}_r$ be an arbitrary packing $\mathcal{P}_r$ of disks of radius $r > 0$ in $\mathbb{H}^2$. Then*

$$\limsup_{r\to 0}\left(\sup_{\mathcal{P}_r} \deg_{\mathrm{avr}}(\mathcal{P}_r)\right) < 5.$$

4 Largest Contact Numbers in 3-Space

Theorem 3.1 implies in a straightforward way that

$$\lim_{n\to+\infty} \frac{3n - c(n,2)}{\sqrt{n}} = \sqrt{12} = 3.464\ldots . \qquad (1)$$

Although one cannot hope for an explicit formula for $c(n,3)$ in terms of n, there might be a way to prove a proper analogue of (1) in $\mathbb{E}^3$.

To this end we know only what is stated in Theorem 4.1. In order to state these results we need an additional concept. Let us imagine that we generate packings of n unit balls in $\mathbb{E}^3$ in such a special way that each and every center of the n unit

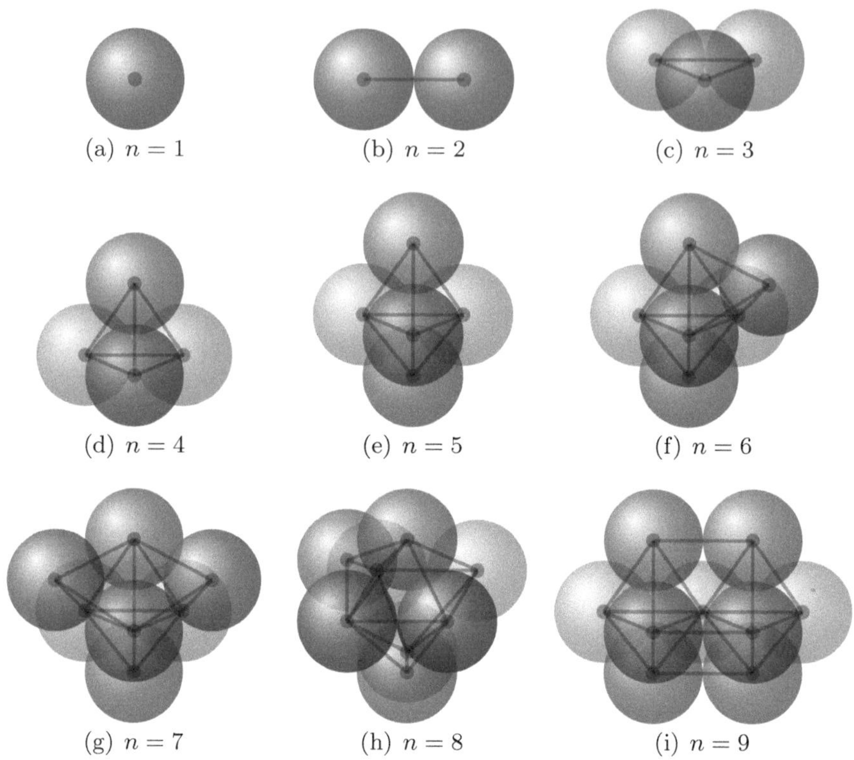

Fig. 1 Contact graphs with $c(n, 3)$ contacts, for $n = 1, 2, 3, 4, 5$ (trivial cases) and largest known number of contacts, for $n = 6, 7, 8, 9$. For $n = 1, 2, 3, 4, 5$ the maximal contact graphs are unique up to isometry. All the packings listed are minimally rigid and only for $n = 9$, the packing is not rigid as the two bipyramids can be twisted slightly about the common pivot (see Sect. 5)

balls chosen is a lattice point of the face-centered cubic lattice with shortest non-zero lattice vector of length 2. Then let $c_{\text{fcc}}(n)$ denote the largest possible contact number of all packings of n unit balls obtained in this way.

The motivation for considering $c_{\text{fcc}}(n)$ is obvious. Since in the planar case, the densest disk packing arrangement also maximizes contacts between disks and the face-centered cubic lattice is the densest for sphere packings in $\mathbb{E}^3$ [23], it makes sense to consider $c_{\text{fcc}}(n)$ as a candidate for $c(n, 3)$. Moreover, it is easy to see that $c_{\text{fcc}}(2) = c(2, 3) = 1$, $c_{\text{fcc}}(3) = c(3, 3) = 3$ and $c_{\text{fcc}}(4) = c(4, 3) = 6$ (Fig. 1a).

Theorem 4.1 *(i)* $c(n, 3) < 6n - 0.926n^{\frac{2}{3}}$, *for all* $n \geq 2$.
(ii) $c_{\text{fcc}}(n) < 6n - \frac{3\sqrt[3]{18\pi}}{\pi}n^{\frac{2}{3}} = 6n - 3.665\ldots n^{\frac{2}{3}}$, *for all* $n \geq 2$.
(iii) $6n - \sqrt[3]{486}n^{\frac{2}{3}} < 2k(2k^2 - 3k + 1) \leq c_{\text{fcc}}(n) \leq c(n, 3)$, *for all* $n = \frac{k(2k^2+1)}{3}$ *with* $k \geq 2$.

Recall that (i) was proved in [8] (using the method of [10]), while (ii) and (iii) were proved in [10]. Clearly, Theorem 4.1 implies that

$$0.926 < \frac{6n - c(n,3)}{n^{\frac{2}{3}}} < \sqrt[3]{486} = 7.862\ldots, \tag{2}$$

for all $n = \frac{k(2k^2+1)}{3}$ with $k \geq 2$.

Now consider the complexity of recognizing contact graphs of congruent sphere packings in $\mathbb{E}^3$. Just like its 2-dimensional analogue, Hliněný [29] showed the 3-dimensional problem to be NP-hard by reduction from 3-SAT. In fact, the same is true in four dimensions [29].

Theorem 4.2 *The problem of recognizing contact graphs of unit sphere packings in* $\mathbb{E}^3$ *(resp.,* $\mathbb{E}^4$*) is NP-hard.*

5 Empirical Approaches

Throughout this section, we deal with finite unit sphere packings in three dimensional Euclidean space, that is, with finite packings of unit balls in $\mathbb{E}^3$. Therefore, in this section a 'sphere' always means a unit sphere in $\mathbb{E}^3$. Taking inspiration from materials science and statistical physics, we will often refer to a finite sphere packing as a *cluster*. Our aim is to describe three computational approaches that have recently been employed in constructing putatively maximal contact graphs for packings of n spheres under certain rigidity assumptions (Table 1).

Definition 1 (*Minimal rigidity* [3]) A cluster of $n \geq 4$ unit spheres is said to be *minimally rigid* if

- each sphere is in contact with at least 3 others, and
- the cluster has at least $3n - 6$ contacts (that is, the corresponding contact graph has at least $3n - 6$ edges).

Definition 2 (*Rigidity* [31]) A cluster of n unit spheres is (nonlinearly) rigid if it cannot be deformed continuously by any finite amount and still maintain all contacts [31].

The first two approaches - which we discuss together - deal with minimally rigid clusters, while the third investigates rigid clusters. We observe that one can find minimally rigid clusters that are not rigid. The paper [3] contains such an example for $n = 9$.

5.1 *Contact Number Estimates for up to 11 Spheres*

Arkus, Manoharan and Brenner [3] made an attempt to exhaustively generate all minimally rigid packings of n spheres that are either local or global maxima of

Table 1 Bounds on the contact numbers of sphere packings in 3-space. The second column lists the lower bound when n equals an octahedral number, i.e., $n = \frac{k(2k^2+1)}{3}$, for some $k = 2, 3, \ldots$. The third column lists the upper bound for packings on the face-centered cubic (fcc) lattice for all n, while the fourth column contains the general upper bound for all n. The final column contains the trivially known exact values for $n = 2, 3, 4, 5$ and the largest contact numbers found by the empirical approaches (for $n = 6, 7, 8, 9, 10$ from [3], for $n = 11$ from [32] and for $n = 12, \ldots, 19$ from [31]). An asterisk * in the last column indicates the largest known contact number for minimally rigid clusters, while a double asterisk ** indicates the largest known contact number for rigid clusters

n	Lower bound [10] $2k(2k^2 - 3k + 1)$	fcc upper bound [10] $\left\lfloor 6n - \frac{3\sqrt[3]{18\pi}}{\pi} n^{2/3} \right\rfloor$	General upper bound [8] $\left\lfloor 6n - 0.926 n^{2/3} \right\rfloor$	(Putatively) Largest [3, 31, 32]
2		6	10	1 $(= 3n - 5)$(trivial)
3		10	16	3 $(= 3n - 6)$(trivial)
4		14	21	6 $(= 3n - 6)$(trivial)
5		19	27	9 $(= 3n - 6)$(trivial)
6	12	23	32	12* $(= 3n - 6)$
7		28	38	15* $(= 3n - 6)$
8		33	44	18* $(= 3n - 6)$
9		38	49	21* $(= 3n - 6)$
10		42	55	25* $(= 3n - 5)$
11		47	61	29* $(= 3n - 4)$
12		52	67	33** $(= 3n - 3)$
13		57	72	36** $(= 3n - 3)$
14		62	78	40** $(= 3n - 2)$
15		67	84	44** $(= 3n - 1)$
16		72	90	48** $(= 3n)$
17		77	95	52** $(= 3n + 1)$
18		82	101	56** $(= 3n + 2)$
19	60	87	107	60** $(= 3n + 3)$

the number of contacts. Here a packing is considered a global maximum if the spheres in the cluster cannot form any additional contacts or a local maximum if new contacts can only be created after breaking an existing contact. They produce a list of maximal contact minimally rigid sphere packings for $n = 2, \ldots, 9$, which is putatively complete up to possible omissions due to round off errors, and a partial list for 10 spheres. Since the number of such packings grows exponentially with n, their approach can only be implemented on a computer.

Before we delve into the details of their methodology, it would be pertinent to understand why it focuses on finding minimally rigid clusters. It seems the minimal rigidity was considered due to two reasons: First, is Maxwell's criterion [38], which is popular in physics literature and states that a rigid cluster of n spheres has at least $3n - 6$ contacts. This is false as in [31] examples of rigid clusters with $n \geq 10$ have been reported that are not minimally rigid. Second, is the intuition that any maximum contact cluster of $n \geq 4$ spheres should be minimally rigid. Up to our knowledge, there exists no proof of or counterexample to this intuition. We can, however, prove the following. The proof depends on the assumption that Arkus et al. [3] have found all minimally rigid packings of $n \leq 9$ spheres that maximize the number of contacts.

Proposition 5.1 *Assume that all maximal contact minimally rigid packings of $n \leq 9$ spheres are listed in [3], then for $n = 4, \ldots, 9$,*

$$c(n, 3) = 3n - 6,$$

and there exists a minimally rigid cluster with $c(n, 3)$ contacts.

Proof We introduce some terminology for finite sphere packings and their contact graphs. We say that any three pairwise touching spheres form a *triangle*. A triangle is called an *exposed triangle* if an additional sphere, not part of the original packing, can be brought in contact with all the three spheres in the triangle without overlapping with any sphere already in the packing. Triangles and exposed triangles can be equivalently defined in terms of contact graphs.
Claim: Any maximal contact graph on n vertices with $4 \leq n \leq 9$ has an exposed triangle and each vertex of such a graph has degree at least 3.

By checking the list of all minimally rigid packings of $4 \leq n \leq 9$ sphere given in [3][1] exhaustively, we see that the claim holds for all such sphere packings. We now proceed by induction on n.

For $n = 4$, there is only one maximal contact graph and for that Claim holds. Now suppose Claim holds for some $n \geq 4$. Consider any contact graph G that has the largest number of contacts among all contact graphs having $n + 1$ vertices. Let v be a vertex of G.

Suppose that v has degree 2. Then $G - \{v\}$, the graph obtained by deleting v and all edges incident to v from G, must be a maximal contact graph on n vertices, since if this is not the case, then replacing $G - \{v\}$ by a maximal contact graph H on n vertices and joining v to any exposed triangle of H produces a contact graph on $n + 1$ vertices with strictly more contacts than G. But then $G - \{v\}$ has an exposed triangle and joining v to that triangle produces a contact graph on $n + 1$ vertices with strictly more contacts than G. This is a contradiction and so v has degree at least 3. Thus G is minimally rigid and has an exposed triangle. This completes the proof of Claim.

Thus for $n = 4, \ldots, 9$, the list of all maximal contact graphs coincides with the list of minimally rigid maximal contact graphs that have $3n - 6$ contacts according to [3]. □

We now describe the approach of Arkus et al. [3]. Note that since we are dealing with unit spheres (as stated in the opening of this section), the distance between the centers of two touching spheres is 2. Let $n \geq 4$ be a positive integer.

[1]The complete list (up to possible omissions due to round off errors) of minimally rigid packings of $n \leq 9$ spheres and a preliminary list of $n = 10$ spheres appears on the arXiv [3]. The paper [3] only contains a partial list, so for the more complete list we refer to the arXiv version.

Procedure 1 ([3]):

Step1: List the adjacency matrices of all nonisomorphic simple graphs with n vertices and exactly $3n - 6$ edges such that each vertex has degree at least 3. Let $\mathcal{A}$ be the set of all such adjacency matrices. In [3], this step is performed using the graph isomorphism testing program *nauty* and Sage package *nice*.

Step 2: For each $A \in \mathcal{A}$, there is a corresponding simple graph G_A with vertex set (say) $V = \{v_1, \dots, v_n\}$. Denote the (i, j)-entry of A by A_{ij} and consider each vertex v_i of G_A as a point $v_i = (x_i, y_i, z_i)$ (the coordinates are yet unknown) in $\mathbb{E}^3$. Then G_A is a contact graph if and only if we can place congruent spheres centered at the vertices of G_A such that none of the spheres overlap and $A_{ij} = 1$ implies that the spheres centered at v_i and v_j touch. Use the simple geometric elimination rules derived in [3] to remove a substantial number of adjacency matrices from $\mathcal{A}$ that cannot be realized into contact graphs. These geometric rules basically detect certain patterns that cannot occur in the adjacency matrices of contact graphs. Let the resulting set of adjacency matrices be denoted by $\mathcal{B}$.

Step 3: For any $A \in \mathcal{B}$, G_A is a contact graph of a unit sphere packing if and only if for $i > j$ the system of nonlinear equations

$$\begin{aligned} D_{ij}^2 &= (x_i - x_j)^2 + (y_i - y_j)^2 + (z_i - z_j)^2 = 2, \quad A_{ij} = 1, \\ D_{ij}^2 &= (x_i - x_j)^2 + (y_i - y_j)^2 + (z_i - z_j)^2 \geq 2, \quad A_{ij} = 0. \end{aligned} \tag{3}$$

has a real solution. Note that we are only considering $i > j$ as the matrix A is symmetric. Without loss of generality, we can assume that $x_1 = y_1 = z_1 = 0$ (the first sphere is centered at the origin); $y_2 = z_2 = 0$ (the second sphere lies on the x-axis) and $z_3 = 0$ (the third sphere lies in the xy-plane). Therefore, we obtain a system with $\frac{n(n-1)}{2}$ constraints (of which $3n - 6$ are equality constraints) in $3n - 6$ unknowns. Here D_{ij} is the distance between vertices v_i and v_j. In [3], for each $A \in \mathcal{B}$, the system (3) is solved analytically for $n \leq 9$ and numerically for $n = 10$.

Step 4: Form the distance matrix $D_A = [D_{ij}]$ and let $\mathcal{D}$ be the set of all distance matrices corresponding to valid contact graphs of packings of n unit spheres. The contact number corresponding to any $D \in \mathcal{D}$ equals the number of entries of D that equal 2 and lie above (equivalently below) the main diagonal. Note that, although we started with the adjacency matrices corresponding to exactly $3n - 6$ contacts, solving system (3) yields all distance matrices with $3n - 6$ or more contacts.

Since the geometric rules used in Step 2, are susceptible to round off errors, there is a possibility that some adjacency matrices are incorrectly eliminated from $\mathcal{A}$. Also for $n = 10$, Newton's method was used to solve (3) as the computational limit of analytical methods was reached for packings of 10 spheres. Thus the list of minimally rigid sphere packings provided in [3] could potentially be incomplete. As a result, the contact number $c(n, 3)$ is still unknown for $n \geq 6$.[2] Nevertheless, it is quite reasonable to conjecture the following.

Conjecture 5.2 *For $n \geq 6$, every contact graph of a packing of n spheres with $c(n, 3)$ contacts is minimally rigid. Moreover, for $n = 6, \dots, 9$,*

$$c(n, 3) = 3n - 6.$$

Hoy et al. [32] extended Procedure 1–11 spheres. However, they employ Newton's method, which cannot guarantee to obtain a solution, whenever one exists. Also they make the erroneous assumption that the contact graph of any minimally rigid sphere packing contains a Hamiltonian path. This assumption greatly reduces the number of adjacency matrices to be considered. However, Connelly, E. Demaine and M. Demaine showed this to be false [28], providing a counterexample with 16 vertices.

5.2 *Maximal Contact Rigid Clusters*

Despite its intuitive significance, we have seen that minimal rigidity is neither sufficient nor necessary for rigidity. Holmes–Cerfon [31] developed a computational technique to potentially construct all rigid sphere packings of a small number of spheres. Her idea to consider rigid clusters comes from the intuition that in a physical system (like self-assembling colloids), a rigid cluster is more likely to form and survive than a non-rigid cluster.

Some aspects of Holmes-Cerfon's method are similar to the empirical approaches described earlier. For instance, the mathematical formulation in terms of adjacency and distance matrices, and use of system (3) to arrive at potential solutions remains unchanged. However, there are two fundamental differences.

Obviously, one is the consideration of rigidity instead of minimal rigidity. This results in the removal of the restriction that the contact graph should have $3n - 6$ edges. Instead, any solution obtained is tested for rigidity.

The second major difference lies in the way all potential solutions are reached. In the previous approaches, the method involved an exhaustive adjacency matrix search followed by filtering through some geometrical rules. Here the procedure starts with a single rigid packing $\mathcal{P}$ of n spheres and attempts to generate all other rigid packings of n spheres as follows: Break an existing contact in $\mathcal{P}$ by deleting an equation from (3). This usually leads to a single internal degree of freedom that results in a one-dimensional solution set. When this happens, one can follow the one-dimensional path numerically until another contact is formed, typically resulting in another rigid packing [31].

---

[2] According to [3], for $n \leq 7$, it is possible to solve the system (3) using standard algebraic geometry methods for all $A \in \mathcal{A}$ without filtering by geometric rules. Arkus et al. [3] attempted this using the package *SINGULAR*. Therefore, most likely for $n = 6, 7$, the maximal contact graphs as obtained in [3] are optimal for minimally rigid sphere packings.

The paper [31] provides a preliminary list for all rigid sphere packings of up to 14 spheres and maximal contact packings of up to 19 spheres. However, the use of numerical methods and approximations throughout means that the list is potentially incomplete.

6 Digital and Totally Separable Sphere Packings for $d = 2, 3$

In this section, we use the terms 'cube', 'sphere' and 'ball' to refer to two and three dimensional objects of these types. Consider the 3-dimensional (resp. 2-dimensional) integer lattice $\mathbb{Z}^3$ (resp. $\mathbb{Z}^2$), which can be thought of as an infinite space tiling array of unit cubes called *lattice cells*. For convenience, we imagine these cubes to be centered at the integer points, rather than having their vertices at these points. Two lattice cells are *connected* if they share a facet.

We refer to a packing of congruent unit diameter spheres centered at the points of $\mathbb{Z}^3$ (resp. $\mathbb{Z}^2$) as a *digital sphere packing*. These packings provide a natural means for generating totally separable sphere packings. We denote the maximal contact number of such a digital packing of n spheres by $c_{\mathbb{Z}}(n, 3)$ (resp. $c_{\mathbb{Z}}(n, 2)$). Clearly, $c_{\mathbb{Z}}(n, 2) \leq c_{\mathrm{sep}}(n, 2)$ and $c_{\mathbb{Z}}(n, 3) \leq c_{\mathrm{sep}}(n, 3)$. The question is how large the maximum digital contact number can be and whether it equals the corresponding maximum contact number of totally separable sphere packings.

A 3-dimensional (resp. 2-dimensional) *polyomino* is a finite collection of connected lattice cells of $\mathbb{Z}^3$ (resp. $\mathbb{Z}^2$). Considering the maximum volume ball contained in a cube, each polyomino corresponds to a digital sphere (circle) packing and vice versa. Moreover, since the ball (circle) intersects the cube (square) at 6 points (4 points), one on each facet, it follows that the number of facets shared between the cells of the polyomino equals the contact number of the corresponding digital packing. Figure 2 shows a portion of the cubic lattice centered at the points of $\mathbb{Z}^3$,

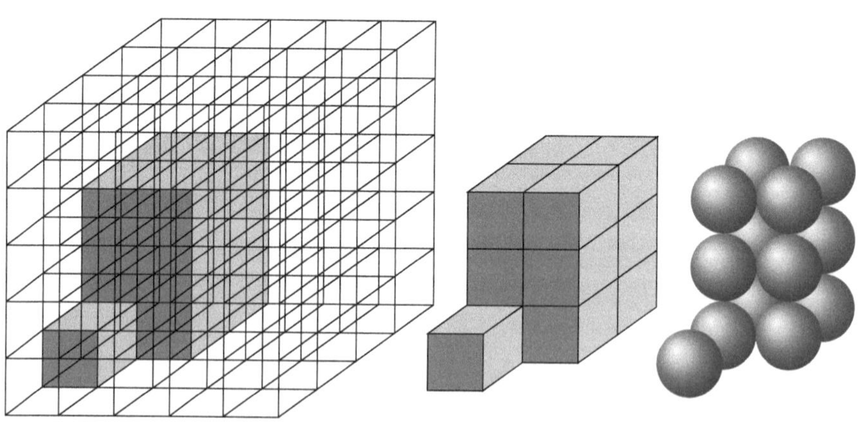

Fig. 2 A polyomino of volume 13 and the corresponding digital packing of 13 spheres

a 3-dimensional polyomino and the digital sphere packing corresponding to that polyomino.

It is easy to see that minimizing the surface area (resp., perimeter) of a 3-dimensional (resp., 2-dimensional) polyomino of volume n corresponds to finding the maximum contact number of a digital packing of n spheres. Harary and Harborth [24] studied the problem of finding isoperimetric polyominoes of area n in 2-space. Their key insight was that n squares can be arranged in a square-like arrangement so as to minimize the perimeter of the resulting polyomino. The same construction appears in [1], but without referencing [24]. The 3-dimensional case has a similar solution which first appeared in [1]. The proposed arrangement consists of forming a quasi-cube (an orthogonal box with one or two edges deficient by at the most one unit) followed by attaching as many of the remaining cells as possible in the form of a quasi-square layer. The rest of the cells are then attached to the quasi-cube in the form of a row. The main results of [1, 24] on isoperimetric polyominoes in $\mathbb{E}^2$ and $\mathbb{E}^3$ can be used to derive the following about the maximum digital contact numbers (see [14]).

Theorem 6.1 *Given $n \geq 2$, we have*

(i) $c_{\mathbb{Z}}(n,2) = \lfloor 2n - 2\sqrt{n} \rfloor$.
(ii) $c_{\mathbb{Z}}(n,3) = 3n - 3n^{\frac{2}{3}} - o(n^{\frac{2}{3}})$.

We now turn to the more general totally separable sphere packings in $\mathbb{E}^2$ and $\mathbb{E}^3$. The contact number problem for such packings was discussed in the very recent paper [14].

Theorem 6.2 *For all $n \geq 2$, we have*

(i) $c_{\text{sep}}(n,2) = \lfloor 2n - 2\sqrt{n} \rfloor$.
(ii) $3n - 3n^{\frac{2}{3}} - o(n^{\frac{2}{3}}) \leq c_{\text{sep}}(n,3) < 3n - 1.346n^{\frac{2}{3}}$.

Part (i) follows from a natural modification of Harborth's proof [25] of Theorem 3.1 (for details see [14]). The lower bound in (ii) comes from the fact that every digital sphere packing is totally separable. However, proving the upper bound in (ii) is more involved.

Theorem 6.2 can be used to generate the following analogues of relations (1) and (2).

$$\lim_{n \to +\infty} \frac{2n - c_{\text{sep}}(n,2)}{\sqrt{n}} = 2\,. \tag{4}$$

$$1.346 < \frac{3n - c_{\text{sep}}(n,3)}{n^{\frac{2}{3}}} \leq 3 + o(1)\,. \tag{5}$$

Since the bounds in (5) are tighter than (2), it is reasonable to conjecture that the limit of $\frac{3n - c_{\text{sep}}(n,3)}{n^{2/3}}$ exists as $n \to +\infty$. In fact, it can be asked if this limit equals 3. Furthermore, a comparison of Theorems 6.1 and 6.2 shows that $c_{\text{sep}}(n,2) = c_{\mathbb{Z}}(n,2)$

holds for all positive integers n. Therefore, it is natural to raise the following open problem.

Problem 2 Show that

$$\lim_{n\to+\infty} \frac{3n - c_{\mathrm{sep}}(n,3)}{n^{\frac{2}{3}}} = 3 .$$

Moreover, is it the case that $c_{\mathrm{sep}}(n,3) = c_{\mathbb{Z}}(n,3)$, for all positive integers n? If not, then characterize those values of n for which this holds.

7 On Largest Contact Numbers in Higher Dimensional Spaces

In this section, we study the contact number problem in $\mathbb{E}^d$, both for packings of $\mathbf{B}^d$ and translates of an arbitrary d-dimensional convex body $\mathbf{K}$.

7.1 Packings by Translates of a Convex Body

One of the main results of this section is an upper bound for the number of touching pairs in an arbitrary finite packing of translates of a convex body, proved in [13]. In order to state the theorem in question in a concise way we need a bit of notation. Let $\mathbf{K}$ be an arbitrary convex body in $\mathbb{E}^d$, $d \geq 3$. Then let $\delta(\mathbf{K})$ denote the density of a densest packing of translates of the convex body $\mathbf{K}$ in $\mathbb{E}^d$, $d \geq 3$. Moreover, let

$$\mathrm{iq}(\mathbf{K}) := \frac{(\mathrm{svol}_{d-1}(\mathrm{bd}\mathbf{K}))^d}{(\mathrm{vol}_d(\mathbf{K}))^{d-1}}$$

be the isoperimetric quotient of the convex body $\mathbf{K}$, where $\mathrm{svol}_{d-1}(\mathrm{bd}\mathbf{K})$ denotes the $(d-1)$-dimensional surface volume of the boundary $\mathrm{bd}\mathbf{K}$ of $\mathbf{K}$ and $\mathrm{vol}_d(\mathbf{K})$ denotes the d-dimensional volume of $\mathbf{K}$. Furthermore, let $H(\mathbf{K})$ denote the Hadwiger number of $\mathbf{K}$, which is the largest number of non-overlapping translates of $\mathbf{K}$ that can all touch $\mathbf{K}$. An elegant observation of Hadwiger [22] is that $H(\mathbf{K}) \leq 3^d - 1$, where equality holds if and only if $\mathbf{K}$ is an affine d-cube. Finally, let the one-sided Hadwiger number $h(\mathbf{K})$ of $\mathbf{K}$ be the largest number of non-overlapping translates of $\mathbf{K}$ that touch $\mathbf{K}$ and that all lie in a closed supporting halfspace of $\mathbf{K}$. In [7], using the Brunn–Minkowski inequality, it is proved that $h(\mathbf{K}) \leq 2 \cdot 3^{d-1} - 1$, where equality is attained if and only if $\mathbf{K}$ is an affine d-cube. Let $\mathbf{K_o} := \frac{1}{2}(\mathbf{K} + (-\mathbf{K}))$ be the normalized (centrally symmetric) difference body assigned to $\mathbf{K}$.

Theorem 7.1 *Let* $\mathbf{K}$ *be an arbitrary convex body in* $\mathbb{E}^d$, $d \geq 3$. *Then*

$$c(\mathbf{K}, n, d) \leq \frac{H(\mathbf{K_o})}{2} n - \frac{1}{2^d \delta(\mathbf{K_o})^{\frac{d-1}{d}}} \sqrt[d]{\frac{\mathrm{iq}(\mathbf{B}^d)}{\mathrm{iq}(\mathbf{K_o})}}\, n^{\frac{d-1}{d}} - (H(\mathbf{K_o}) - h(\mathbf{K_o}) - 1)$$
$$\leq \frac{3^d - 1}{2} n - \frac{\sqrt[d]{\omega_d}}{2^{d+1}} n^{\frac{d-1}{d}},$$

where $\omega_d = \frac{\pi^{\frac{d}{2}}}{\Gamma(\frac{d}{2}+1)} = \mathrm{vol}_d(\mathbf{B}^d)$.

Since for the most part we are interested in contact numbers of sphere packings, it would be interesting to see the form Theorem 7.1 takes when $\mathbf{K} = \mathbf{B}^d$. Recall that $k(d)$ denotes the kissing number of a unit ball in $\mathbb{E}^d$. Let δ_d stand for the largest possible density for (infinite) packings of unit balls in $\mathbb{E}^d$. The following consequence of Theorem 7.1 was reported in [10].

Corollary 7.2 *Let* $n > 1$ *and* $d \geq 3$ *be positive integers. Then*

$$c(n, d) < \frac{1}{2} k(d)\, n - \frac{1}{2^d} \delta_d^{-\frac{d-1}{d}} n^{\frac{d-1}{d}}.$$

Now, recall the well-known theorem of Kabatiansky and Levenshtein [33] that $k(d) \leq 2^{0.401d(1+o(1))}$ and $\delta_d \leq 2^{-0.599d(1+o(1))}$ as $d \to +\infty$. Together with Corollary 7.2 this gives

$$c(n, d) < \frac{1}{2} 2^{0.401d(1+o(1))} n - \frac{1}{2^d} 2^{0.599(1+o(1))(d-1)} n^{\frac{d-1}{d}},$$

for $n > 1$, as $d \to +\infty$.

In particular, for $d = 3$ we have $k(3) = 12$ [41] and $\delta_3 = \frac{\pi}{\sqrt{18}}$ [23]. Thus, by combining these with Corollary 7.2 we find that for $n > 1$,

$$c(n, 3) < 6n - \frac{1}{8} \left(\frac{\pi}{\sqrt{18}} \right)^{-\frac{2}{3}} n^{\frac{2}{3}} = 6n - 0.152 \ldots n^{\frac{2}{3}}.$$

The above upper bound for $c(n, 3)$ was substantially improved, first in [10] and then further in [8]. The current best upper bound is stated in Theorem 4.1(i).

In the proof of Theorem 7.1 published in [13], the following statement plays an important role that might be of independent interest and so we quote it as follows. For the sake of completeness we wish to point out that Theorem 7.3 and Corollary 7.4, are actual strengthenings of Theorem 3.1 and Corollary 3.1 of [5] mainly because, in our case the containers of the packings in question are highly non-convex.

Theorem 7.3 *Let* $\mathbf{K_o}$ *be a convex body in* $\mathbb{E}^d$, $d \geq 2$ *symmetric about the origin* $\mathbf{o}$ *of* $\mathbb{E}^d$ *and let* $\{\mathbf{c}_1 + \mathbf{K_o}, \mathbf{c}_2 + \mathbf{K_o}, \ldots, \mathbf{c}_n + \mathbf{K_o}\}$ *be an arbitrary packing of* $n > 1$ *translates of* $\mathbf{K_o}$ *in* $\mathbb{E}^d$. *Then*

$$\frac{n\mathrm{vol}_d(\mathbf{K_o})}{\mathrm{vol}_d(\bigcup_{i=1}^n(\mathbf{c}_i+2\mathbf{K_o}))}\leq\delta(\mathbf{K_o}).$$

The following is an immediate corollary of Theorem 7.3.

Corollary 7.4 *Let $\mathcal{P}_n(\mathbf{K_o})$ be the family of all possible packings of $n>1$ translates of the* **o***-symmetric convex body $\mathbf{K_o}$ in $\mathbb{E}^d$, $d\geq 2$. Moreover, let*

$$\delta(\mathbf{K_o},n):=\max\left\{\frac{n\mathrm{vol}_d(\mathbf{K_o})}{\mathrm{vol}_d(\bigcup_{i=1}^n(\mathbf{c}_i+2\mathbf{K_o}))}\;\middle|\;\{\mathbf{c}_1+\mathbf{K_o},\ldots,\mathbf{c}_n+\mathbf{K_o}\}\in\mathcal{P}_n(\mathbf{K_o})\right\}.$$

Then

$$\limsup_{n\to\infty}\delta(\mathbf{K_o},n)=\delta(\mathbf{K_o}).$$

Interestingly enough one can interpret the contact number problem on the exact values of $c(n,d)$ as a volume minimization question. Here we give only an outline of that idea introduced and discussed in detail in [9].

Definition 3 Let $\mathcal{P}^n:=\{\mathbf{c}_i+\mathbf{B}^d\mid 1\leq i\leq n \text{ with } \|\mathbf{c}_j-\mathbf{c}_k\|\geq 2 \text{ for all } 1\leq j<k\leq n\}$ be an arbitrary packing of $n>1$ unit balls in $\mathbb{E}^d$. The part of space covered by the unit balls of $\mathcal{P}^n$ is labelled by $\mathbf{P}^n:=\bigcup_{i=1}^n(\mathbf{c}_i+\mathbf{B}^d)$. Moreover, let $C^n:=\{\mathbf{c}_i\mid 1\leq i\leq n\}$ stand for the set of centers of the unit balls in $\mathcal{P}^n$. Furthermore, for any $\lambda>0$ let $\mathbf{P}^n_\lambda:=\bigcup\{\mathbf{x}+\lambda\mathbf{B}^d\mid \mathbf{x}\in\mathbf{P}^n\}=\bigcup_{i=1}^n(\mathbf{c}_i+(1+\lambda)\mathbf{B}^d)$ denote the outer parallel domain of $\mathbf{P}^n$ having outer radius λ. Finally, let

$$\delta_d(n,\lambda):=\max_{\mathcal{P}^n}\frac{n\omega_d}{\mathrm{vol}_d(\mathbf{P}^n_\lambda)}=\frac{n\omega_d}{\min_{\mathcal{P}^n}\mathrm{vol}_d\left(\bigcup_{i=1}^n(\mathbf{c}_i+(1+\lambda)\mathbf{B}^d)\right)}$$

and

$$\delta_d(\lambda):=\limsup_{n\to+\infty}\delta_d(n,\lambda).$$

Now, let $\mathcal{P}:=\{\mathbf{c}_i+\mathbf{B}^d\mid i=1,2,\ldots \text{ with } \|\mathbf{c}_j-\mathbf{c}_k\|\geq 2 \text{ for all } 1\leq j<k\}$ be an arbitrary infinite packing of unit balls in $\mathbb{E}^d$. Recall that the packing density δ_d of unit balls in $\mathbb{E}^d$ can be computed as follows:

$$\delta_d=\sup_{\mathcal{P}}\left(\limsup_{R\to+\infty}\frac{\sum_{\mathbf{c}_i+\mathbf{B}^d\subset R\mathbf{B}^d}\mathrm{vol}_d(\mathbf{c}_i+\mathbf{B}^d)}{\mathrm{vol}_d(R\mathbf{B}^d)}\right).$$

Hence, it is rather easy to see that $\delta_d\leq\delta_d(\lambda)$ holds for all $\lambda>0, d\geq 2$. On the other hand, it was proved in [13] (see also Corollary 7.4) that $\delta_d=\delta_d(\lambda)$ *for all* $\lambda\geq 1$ leading to the classical sphere packing problem. Now, we are ready to put forward the following question from [9].

Problem 3 Determine (resp., estimate) $\delta_d(\lambda)$ for $d\geq 2$, $0<\lambda<\sqrt{\frac{2d}{d+1}}-1$.

First, we note that $\frac{2}{\sqrt{3}} - 1 \leq \sqrt{\frac{2d}{d+1}} - 1$ holds for all $d \geq 2$. Second, observe that as $\frac{2}{\sqrt{3}}$ is the circumradius of a regular triangle of side length 2, therefore if $0 < \lambda < \frac{2}{\sqrt{3}} - 1$, then for any unit ball packing $\mathcal{P}^n$ no three of the closed balls in the family $\{\mathbf{c}_i + (1+\lambda)\mathbf{B}^d \mid 1 \leq i \leq n\}$ have a point in common. In other words, for any λ with $0 < \lambda < \frac{2}{\sqrt{3}} - 1$ and for any unit ball packing $\mathcal{P}^n$, in the arrangement $\{\mathbf{c}_i + (1+\lambda)\mathbf{B}^d \mid 1 \leq i \leq n\}$ of closed balls of radii $1+\lambda$ only pairs of balls may overlap. Thus, computing $\delta_d(n, \lambda)$, i.e., minimizing $\mathrm{vol}_d(\mathbf{P}^n_\lambda)$ means maximizing the total volume of pairwise overlaps in the ball arrangement $\{\mathbf{c}_i + (1+\lambda)\mathbf{B}^d \mid 1 \leq i \leq n\}$ with the underlying packing $\mathcal{P}^n$. Intuition would suggest to achieve this by simply maximizing the number of touching pairs in the unit ball packing $\mathcal{P}^n$. Hence, Problem 3 becomes very close to the *contact number problem* of finite unit ball packings for $0 < \lambda < \frac{2}{\sqrt{3}} - 1$. Indeed, we have the following statement proved in [9].

Theorem 7.5 *Let $n > 1$ and $d > 1$ be given. Then there exists $\lambda_{d,n} > 0$ and a packing $\widehat{\mathcal{P}}^n$ of n unit balls in $\mathbb{E}^d$ possessing the largest contact number for the given n such that for all λ satisfying $0 < \lambda < \lambda_{d,n}$, $\delta_d(n, \lambda)$ is generated by $\widehat{\mathcal{P}}^n$, i.e., $\mathrm{vol}_d(\mathbf{P}^n_\lambda) \geq \mathrm{vol}_d(\widehat{\mathbf{P}}^n_\lambda)$ holds for every packing $\mathcal{P}^n$ of n unit balls in $\mathbb{E}^d$.*

7.2 Contact Graphs of Unit Sphere Packings in $\mathbb{E}^d$

Given the NP-hardness of recognizing contact graphs of unit sphere packings for $d = 2, 3, 4$, Hliněný [29] conjectured that the problem remains NP-hard in any fixed dimension.

Conjecture 7.6 *The recognition of contact graphs of unit sphere packings is NP-hard in any fixed dimension $d \geq 2$.*

Hliněný and Kratochvil [30] made some progress towards this conjecture. They reproved Theorem 4.2 using the rather elaborate notion of a *scheme of an m-comb* and then proved Conjecture 7.6 for $d = 8, 24$. To define an m-comb we need to introduce some more terminology.

For a hyperplane h in $\mathbb{E}^d$ and $S \subseteq \mathbb{E}^d$, let S/h denote the mirror reflection of S across h. We say that a set S is a minimal-distance representation of a graph G, denoted by $G = M(S)$, if the vertices of G are the points of S, and the edges of G correspond to minimal-distance pairs of points in S. The graph G is then called the *minimal-distance graph of* S. Also, let $m(S)$ denote the minimal distance among pairs of points of S. Finally, when $m(S) = 1$, we say that the set S is *rigid* in $\mathbb{E}^d$ if for any set $S' \subseteq \mathbb{E}^d$, $m(S') = 1$, the following holds: If $\phi : M(S') \to M(S)$ is an isomorphism, then ϕ is an isometry of the underlying sets S', S. (Notice that the definition of a rigid set is slightly stronger than just saying that S has a unique representation up to isometry.) For example, the vertices of a regular tetrahedron or a regular octahedron form rigid sets in $\mathbb{E}^3$. In general, the vertices of a d-dimensional simplex or a d-dimensional cross-polytope are rigid sets in $\mathbb{E}^d$.

Definition 4 (*Scheme of an m-comb* [30]) Let T, V, W be point sets in $\mathbb{E}^d$, and let $\boldsymbol{\alpha}$, $\boldsymbol{\beta}$ be vectors in $\mathbb{E}^d$. The five-tuple $(V, W, T, \boldsymbol{\alpha}, \boldsymbol{\beta})$ is called a scheme of an m-comb in $\mathbb{E}^d$ if the following conditions are satisfied:

- The sets $V \cup W$ and T are both rigid in $\mathbb{E}^d$, and $m(V) = m(V \cup W) = m(T) = 1$.
- The set V spans a hyperplane h in $\mathbb{E}^d$. The vector $\boldsymbol{\alpha}$ is parallel to h. Let $T_0 = T \cap (T - \boldsymbol{\beta})$. Then the set T_0 spans the whole $\mathbb{E}^d$. For $i = 0, \ldots, m-1$, the set $(T_0 + i\boldsymbol{\alpha}) \cap V$ spans the hyperplane h.
- Let c be the maximal distance of W from h. Then the distance between h and $h + \boldsymbol{\beta}$ is greater than $2c + 1$. The distance between h, and $T + \boldsymbol{\beta}$ or $T - 2\boldsymbol{\beta}$, is greater than $c + 1$.
- Let p be the straight line parallel to $\boldsymbol{\beta}$ such that the maximal distance c' between p and the points of T is minimized. Then the distance of p and $p + \boldsymbol{\alpha}$ is greater than $2c' + 1$. For $j \in \mathbb{Z} - \{0, \ldots, m-1\}$, the distance between the sets $V \cup W$ and $p + j\boldsymbol{\alpha}$ is greater than $c' + 1$.
- The sets T and $(T - \boldsymbol{\beta}) \setminus T$ are non-overlapping, and the sets T and $T + 2\boldsymbol{\beta}$ are strictly non-overlapping; while the sets T and $(T/h) + \boldsymbol{\beta}$ are overlapping each other. Let $T' = T \cup (T - \boldsymbol{\beta})$. Then, for $i = 0, \ldots, m-1$, the sets V and $(T' + i\boldsymbol{\alpha}) \setminus V$ are non-overlapping.

The term 'm-comb' comes from the actual geometry of such a scheme, which is comb-like (see the illustration of an m-comb in [30]). It turns out that if $d \geq 3$ is such that for every $m > 0$, there exists a scheme of an m-comb in $\mathbb{E}^d$, then the recognition of contact graphs of unit sphere packings in $\mathbb{E}^d$ is an NP-hard problem [30].

Theorem 7.7 *The problem of recognizing contact graphs of unit sphere packings is NP-hard in $\mathbb{E}^3$, $\mathbb{E}^4$, $\mathbb{E}^8$ and $\mathbb{E}^{24}$.*

The proof relies on constructing such schemes for $d = 3, 4, 8, 24$. For $d \neq 2, 3, 4, 8, 24$, the complexity of recognizing unit sphere contact graphs is unknown, while for $d = 2$ it is NP-hard from Theorem 3.2.

7.3 Digital and Totally Separable Sphere Packings in $\mathbb{E}^d$

Let us imagine that we generate totally separable packings of unit diameter balls in $\mathbb{E}^d$ such that every center of the balls chosen, is a lattice point of the integer lattice $\mathbb{Z}^d$ in $\mathbb{E}^d$. Then, as in Sect. 6, let $c_{\mathbb{Z}}(n, d)$ denote the largest possible contact number of all packings of n unit diameter balls obtained in this way.

Theorem 7.8 $c_{\mathbb{Z}}(n, d) \leq \lfloor dn - dn^{\frac{d-1}{d}} \rfloor$, *for all $n > 1$ and $d \geq 2$.*

For the convenience of the reader, we recall here the elementary short proof of Theorem 7.8 from [14]. A union of finitely many axis parallel d-dimensional orthogonal boxes having pairwise disjoint interiors in $\mathbb{E}^d$ is called a *box-polytope*. One may call the following statement the isoperimetric inequality for box-polytopes, which together with its proof presented below is an analogue of the isoperimetric inequality for convex bodies derived from the Brunn–Minkowski inequality. (For more details on the latter see for example, [4].)

Lemma 7.9 *Among box-polytopes of given volume the cubes have the least surface volume.*

Proof Without loss of generality, we may assume that the volume $\mathrm{vol}_d(\mathbf{A})$ of the given box-polytope $\mathbf{A}$ in $\mathbb{E}^d$ is equal to 2^d, i.e., $\mathrm{vol}_d(\mathbf{A}) = 2^d$. Let $\mathbf{C}^d$ be an axis parallel d-dimensional cube of $\mathbb{E}^d$ with $\mathrm{vol}_d(\mathbf{C}^d) = 2^d$. Let the surface volume of $\mathbf{C}^d$ be denoted by $\mathrm{svol}_{d-1}(\mathbf{C}^d)$. Clearly, $\mathrm{svol}_{d-1}(\mathbf{C}^d) = d \cdot \mathrm{vol}_d(\mathbf{C}^d)$. On the other hand, if $\mathrm{svol}_{d-1}(\mathbf{A})$ denotes the surface volume of the box-polytope $\mathbf{A}$, then it is rather straightforward to show that

$$\mathrm{svol}_{d-1}(\mathbf{A}) = \lim_{\epsilon \to 0^+} \frac{\mathrm{vol}_d(\mathbf{A} + \epsilon \mathbf{C}^d) - \mathrm{vol}_d(\mathbf{A})}{\epsilon},$$

where "+" in the numerator stands for the Minkowski addition of the given sets. Using the Brunn–Minkowski inequality ([4]) we get that

$$\mathrm{vol}_d(\mathbf{A} + \epsilon \mathbf{C}^d) \geq \left(\mathrm{vol}_d(\mathbf{A})^{\frac{1}{d}} + \mathrm{vol}_d(\epsilon \mathbf{C}^d)^{\frac{1}{d}}\right)^d = \left(\mathrm{vol}_d(\mathbf{A})^{\frac{1}{d}} + \epsilon \cdot \mathrm{vol}_d(\mathbf{C}^d)^{\frac{1}{d}}\right)^d.$$

Hence,

$$\begin{aligned}\mathrm{vol}_d(\mathbf{A} + \epsilon \mathbf{C}^d) &\geq \mathrm{vol}_d(\mathbf{A}) + d \cdot \mathrm{vol}_d(\mathbf{A})^{\frac{d-1}{d}} \cdot \epsilon \cdot \mathrm{vol}_d(\mathbf{C}^d)^{\frac{1}{d}} \\ &= \mathrm{vol}_d(\mathbf{A}) + \epsilon \cdot d \cdot \mathrm{vol}_d(\mathbf{C}^d) \\ &= \mathrm{vol}_d(\mathbf{A}) + \epsilon \cdot \mathrm{svol}_{d-1}(\mathbf{C}^d).\end{aligned}$$

So,

$$\frac{\mathrm{vol}_d(\mathbf{A} + \epsilon \mathbf{C}^d) - \mathrm{vol}_d(\mathbf{A})}{\epsilon} \geq \mathrm{svol}_{d-1}(\mathbf{C}^d)$$

and therefore, $\mathrm{svol}_{d-1}(\mathbf{A}) \geq \mathrm{svol}_{d-1}(\mathbf{C}^d)$, finishing the proof of Lemma 7.9. □

Corollary 7.10 *For any box-polytope* $\mathbf{P}$ *of* $\mathbb{E}^d$ *the isoperimetric quotient of* $\mathbf{P}$ *is at least as large as the isoperimetric quotient of a cube, i.e.,*

$$\frac{\mathrm{svol}_{d-1}(\mathbf{P})^d}{\mathrm{vol}_d(\mathbf{P})^{d-1}} \geq (2d)^d.$$

Now, let $\overline{\mathcal{P}} := \{\mathbf{c}_1 + \overline{\mathbf{B}}^d, \mathbf{c}_2 + \overline{\mathbf{B}}^d, \dots, \mathbf{c}_n + \overline{\mathbf{B}}^d\}$ denote the totally separable packing of n unit diameter balls with centers $\{\mathbf{c}_1, \mathbf{c}_2, \dots, \mathbf{c}_n\} \subset \mathbb{Z}^d$ having contact number $c_{\mathbb{Z}}(n, d)$ in $\mathbb{E}^d$. ($\overline{\mathcal{P}}$ might not be uniquely determined up to congruence in which case $\overline{\mathcal{P}}$ stands for any of those extremal packings.) Let $\mathbf{U}^d$ be the axis parallel d-dimensional unit cube centered at the origin $\mathbf{o}$ in $\mathbb{E}^d$. Then the unit cubes $\{\mathbf{c}_1 + \mathbf{U}^d, \mathbf{c}_2 + \mathbf{U}^d, \dots, \mathbf{c}_n + \mathbf{U}^d\}$ have pairwise disjoint interiors and $\mathbf{P} = \cup_{i=1}^n (\mathbf{c}_i + \mathbf{U}^d)$ is a box-polytope. Clearly, $\mathrm{svol}_{d-1}(\mathbf{P}) = 2dn - 2c_{\mathbb{Z}}(n, d)$. Hence, Corollary 7.10 implies that

$$2dn - 2c_{\mathbb{Z}}(n, d) = \mathrm{svol}_{d-1}(\mathbf{P}) \geq 2d\mathrm{vol}_d(\mathbf{P})^{\frac{d-1}{d}} = 2dn^{\frac{d-1}{d}} .$$

So, $dn - dn^{\frac{d-1}{d}} \geq c_{\mathbb{Z}}(n, d)$, finishing the proof of Theorem 7.8.

Here we recall Theorem 6.1 and refer to [14] to note that the upper bound of Theorem 7.8 is sharp for $d = 2$ and all $n > 1$ and for $d \geq 3$ and all $n = k^d$ with $k > 1$. On the other hand, it is not a sharp estimate for example, for $d = 3$ and $n = 5$.

We close this section by stating the recent upper bounds of [14] for the contact numbers of totally separable unit ball packings in $\mathbb{E}^d$.

Theorem 7.11 $c_{\mathrm{sep}}(n, d) \leq dn - \frac{1}{2d^{\frac{d-1}{2}}} n^{\frac{d-1}{d}}$, *for all* $n > 1$ *and* $d \geq 4$.

8 Contact Graphs of Non-congruent Sphere Packings

So far, we have exclusively focused on contact graphs of packings of congruent spheres. In this section, we discuss what is known for general non-congruent sphere packings. Let us denote by $c^*(n, d)$ the maximal number of edges in a contact graph of n not necessarily congruent d-dimensional balls. Clearly, $c^*(n, d) \geq c(n, d)$, for any positive integers n and $d \geq 2$.

The planar case was first resolved by Koebe [35] in 1936. Koebe's result was later rediscovered by Andreev [2] in 1970 and by Thurston in [42] 1978.[3] The result is referred to as Koebe–Andreev–Thurston theorem or the circle packing theorem.

In terms of contact graphs, the result can be stated as under.

Theorem 8.1 (Koebe–Andreev–Thurston) *A graph G is a contact graph of a (not necessarily congruent) circle packing in $\mathbb{E}^2$ if and only if G is planar.*

In other words, for any planar graph G of any order n, there exist n circular disks with possibly different radii such that when these disks are placed with their centers at the vertices of the graph, the disks centered at the end vertices of each edge of G

[3] It is worth-noting that Koebe's paper was written in German and titled 'Kontaktprobleme der konformen Abbildung' (Contact problems of conformal mapping). Andreev's paper appeared in Russian. Probably, the first instance of this result appearing in English was in Thurston's lecture notes that were distributed by the Princeton University in 1980. However, the lectures were delivered in 1978–1979 [42].

touch. In addition, this cannot be achieved for any nonplanar graph. This is a rather unique result that is, as we will see shortly, highly unlikely to have an analogue in higher dimensions. It shows that $c^*(n, 2) = 3n - 6$, for $n \geq 2$, which is the number of edges in a maximal planar graph.

A similar simple characterization of contact graphs of general not necessarily congruent sphere packings cannot be found for all dimensions $d \geq 3$, unless P = NP. We briefly discuss this here. In [29, 30], the authors report that Kirkpatrick and Rote informed them of the following result in a personal communication in 1997. The proof appears in [30].

Theorem 8.2 *A graph G has a d-unit-ball contact representation if and only if the graph $G \oplus K_2$ has a $(d+1)$-ball contact representation.*

Here K_2 denotes the complete graph on two vertices, while $G \oplus H$ represents the graph formed by taking the disjoint union of G and H and then adding all edges across [30]. Theorem 8.2 provides an interesting connection between contact graphs of unit sphere packings in $\mathbb{E}^d$ and contact graphs of not necessarily congruent sphere packings in $\mathbb{E}^{d+1}$. Combining this with Theorems 3.2, 4.2 and 7.7 gives the following [30].

Corollary 8.3 *The problem of recognizing general contact graphs of (not necessarily congruent) sphere packings is NP-hard in dimensions $d = 3, 4, 5, 9, 25$.*

Not much is known about $c^*(n, d)$, for $d \geq 4$. However, for $d = 3$, an upper bound was found by Kuperberg and Schramm [36]. (Also see [26] for some elementary results on forbidden subgraphs of contact graphs of non-congruent sphere packings in $\mathbb{E}^3$.) Define the average kissing number $k^*_{\text{av}}(d)$ in dimension d as the supremum of average vertex degrees among all contact graphs of finite sphere packings in $\mathbb{E}^d$. In a packing of three dimensional congruent spheres, a sphere can touch at the most 12 others [41]. Thus a three dimensional ball B cannot touch more than 12 other balls at least as large as B. It follows that $k^*_{\text{av}}(3) \leq 2k(3) = 24$. In [36], this was improved to $12.566 \approx 666/53 \leq k^*_{\text{av}}(3) < 8 + 4\sqrt{3} \approx 14.928$. In the language of contact numbers, the Kuperberg–Schramm bound translates into the following.

Theorem 8.4 $c^*(n, 3) < (4 + 2\sqrt{3})n \approx 7.464n$.

The method of Kuperberg and Schramm relies heavily on the geometry of 3-dimensional space. As a result it seems difficult to generalize it to higher dimensions. We close this section with the following open question.

Problem 4 Find upper and lower bounds on $c^*(n, d)$ in the spirit of Kuperberg–Schramm bounds on $c^*(n, 3)$.

Acknowledgements The first author is partially supported by a Natural Sciences and Engineering Research Council of Canada Discovery Grant. The second author is supported by a Vanier Canada Graduate Scholarship (NSERC), an Izaak Walton Killam Memorial Scholarship and Alberta Innovates Technology Futures (AITF). The authors would like to thank the anonymous referee for careful reading and an interesting reference.

References

1. L. Alonso, R. Cerf, The three dimensional polyominoes of minimal area. Electr. J. Combin. **3** (1996). #R27
2. E.M. Andreev, Convex polyhedra of finite volume in Lobačevskiĭ space. Mat. Sb. (N.S.) **83**(125), 256–260 (1970). (Russian)
3. N. Arkus, V.N. Manoharan, M.P. Brenner, Deriving finite sphere packings. SIAM J. Discret. Math. **25**(4), 1860–1901 (2011), arXiv:1011.5412v2 [cond-mat.soft]
4. K. Ball, An elementary introduction to modern convex geometry, in *Flavors of Geometry*, vol. 31, Mathematical Sciences Research Institute Publications, ed. by S. Levy (Cambridge University Press, Cambridge, 1997), pp. 1–58
5. U. Betke, M. Henk, J.M. Wills, Finite and infinite packings. J. reine angew. Math. **53**, 165–191 (1994)
6. A. Bezdek, Locally separable circle packings. Studia Sci. Math. Hungar. **18**(2–4), 371–375 (1983)
7. K. Bezdek, On the maximum number of touching pairs in a finite packing of translates of a convex body. J. Combin. Theory Ser. A **98**, 192–200 (2002)
8. K. Bezdek, Contact numbers for congruent sphere packings in Euclidean 3-space. Discret. Comput. Geom. **48**(2), 298–309 (2012)
9. K. Bezdek, *Lectures on Sphere Arrangements - the Discrete Geometric Side*, vol. 32, Fields Institute Monographs (Springer, New York, 2013)
10. K. Bezdek, P. Brass, On k^+-neighbour packings and one-sided Hadwiger configurations. Beitr. Algebr. Geom. **44**, 493–498 (2003)
11. K. Bezdek, S. Reid, Contact graphs of unit sphere packings revisited. J. Geom. **104**(1), 57–83 (2013)
12. K. Bezdek, Zs. Lángi, Density bounds for outer parallel domains of unit ball packings. Proc. Steklov Inst. Math. **288/1**, 209–225 (2015)
13. K. Bezdek, R. Connelly, G. Kertész, On the average number of neighbours in spherical packing of congruent circles, *Intuitive Geometry*, vol. 48, Colloquia Mathematica Societatis János Bolyai (North Holland, Amsterdam, 1987), pp. 37–52
14. K. Bezdek, B. Szalkai, I. Szalkai, On contact numbers of totally separable unit sphere packings. Discret. Math. **339**(2), 668–676 (2015)
15. L. Bowen, Circle packing in the hyperbolic plane. Math. Phys. Electr. J. **6**, 1–10 (2000)
16. P. Boyvalenkov, S. Dodunekov, O. Musin, A survey on the kissing numbers. Serdica Math. J. **38**(4), 507–522 (2012)
17. P. Brass, Erdős distance problems in normed spaces. Comput. Geom. **6**, 195–214 (1996)
18. H. Breu, D.G. Kirkpatrick, On the complexity of recognizing intersection and touching graphs of discs, in *Graph Drawing* ed. by F.J. Brandenburg, Proceedings of Graph Drawing 95, Passau, September 1995. Lecture Notes in Computer Science, vol. 1027, Springer, Berlin, (1996), pp. 88–98
19. P. Erdős, On sets of distances of n points. Am. Math. Mon. **53**, 248–250 (1946)
20. P. Erdős, Problems and results in combinatorial geometry, in *Discrete Geometry and Convexity*, vol. 440, Annals of the New York Academy of Sciences, ed. by J.E. Goodman, et al. (vvv, bbb, 1985), pp. 1–11
21. G. Fejes Tóth, L. Fejes Tóth, On totally separable domains. Acta Math. Acad. Sci. Hungar. **24**, 229–232 (1973)
22. H. Hadwiger, Über Treffenzahlen bei translations gleichen Eikörpern. Arch. Math. **8**, 212–213 (1957)
23. T.C. Hales, A proof of the Kepler conjecture. Ann. Math. **162**(2–3), 1065–1185 (2005)
24. F. Harary, H. Harborth, Extremal animals. J. Comb. Inf. Syst. Sci. **1**(1), 1–8 (1976)
25. H. Harborth, Lösung zu problem 664A. Elem. Math. **29**, 14–15 (1974)
26. H. Harborth, L. Szabó, Z. Ujvári-Menyhárt, Regular sphere packings. Arch. Math. (Basel) **78/1**, 81–89 (2002)

27. B. Hayes, The science of sticky spheres. Am. Sci. **100**, 442–449 (2012)
28. B. Hayes, Sphere packings and hamiltonian paths (blog post posted on 13 March 2013), http://bit-player.org/2013/sphere-packings-and-hamiltonian-paths
29. P. Hliněný, Touching graphs of unit balls, in *Graph Drawing* ed. by G. DiBattista, Proceedings of Graph Drawing 97, Rome, September. Lecture Notes in Computer Science, vol. 1353 (Springer, Berlin, 1997), pp. 350–358
30. P. Hliněný, J. Kratochvíl, Representing graphs by disks and balls (a survey of recognition-complexity results). Discret. Math. **229**, 101–124 (2001)
31. M. Holmes-Cerfon, Enumerating nonlinearly rigid sphere packings. SIAM Rev. **58**(2), 229–244 (2016), arXiv:1407.3285v2 [cond-mat.soft]
32. R.S. Hoy, J. Harwayne-Gidansky, C.S. O'Hern, Structure of finite sphere packings via exact enumeration: implications for colloidal crystal nucleation. Phys. Rev. E **85**, 051403 (2012)
33. G.A. Kabatiansky, V.I. Levenshtein, Bounds for packings on a sphere and in space. Probl. Pereda. Inf. **14**, 3–25 (1978)
34. G. Kertész, On totally separable packings of equal balls. Acta Math. Hungar. **51**(3-4), 363–364 (1988)
35. P. Koebe, Kontaktprobleme der konformen Abbildung. Ber. Verh. Sächs. Akad. Leipzig **88**, 141–164 (1936). (German)
36. G. Kuberberg, O. Schramm, Average kissing numbers for non-congruent sphere packings. Math. Res. Lett. **1**, 339–344 (1994)
37. V.N. Manoharan, Colloidal matter: packing, geometry, and entropy. Science **349**, 1253751 (2015)
38. J.C. Maxwell, On the calculation of the equilibrium and stiffness of frames. Philos. Mag. **27**, 294–299 (1864)
39. O.R. Musin, The kissing number in four dimensions. Ann. Math. (2) **168/1**, 1–32 (2008)
40. A.M. Odlyzko, N.J.A. Sloane, New bounds on the number of unit spheres that can touch a unit sphere in n-dimensions. J. Comb. Theory, Ser. A **26**, 210–214 (1979)
41. K. Schütte, B.L. Van Der Waerden, Das Problem der dreizehn Kugeln. Math. Ann. **125**, 325–334 (1953)
42. W. Thurston, The geometry and topology of 3-manifolds, Princeton Lecture Notes (1980)

The Topological Transversal Tverberg Theorem Plus Constraints

Pavle V. M. Blagojević, Aleksandra S. Dimitrijević Blagojević and Günter M. Ziegler

Abstract In this paper we use the strength of the constraint method in combination with a generalized Borsuk–Ulam type theorem and a cohomological intersection lemma to show how one can obtain many new topological transversal theorems of Tverberg type. In particular, we derive a topological generalized transversal Van Kampen–Flores theorem and a topological transversal weak colored Tverberg theorem.

1 Introduction

At the Symposium on Combinatorics and Geometry in Stockholm 1989, Helge Tverberg formulated the following conjecture that in a special case coincides with his famous 1966 result [14, Theorem 1].

The research by Pavle V. M. Blagojević leading to these results has received funding from DFG via Berlin Mathematical School. Also supported by the grant ON 174024 of the Serbian Ministry of Education and Science. The research by Aleksandra Dimitrijević Blagojević leading to these results has received funding from the grant ON 174024 of the Serbian Ministry of Education and Science. The research by Günter M. Ziegler received funding from DFG via the Research Training Group "Methods for Discrete Structures" and the Collaborative Research Center TRR 109 "Discretization in Geometry and Dynamics."

The original version of this chapter was revised: The grant number above has been corrected. The correction to this chapter is available at https://doi.org/10.1007/978-3-662-57413-3_18

P. V. M. Blagojević (✉) · G. M. Ziegler
Inst. Math., FU Berlin, Arnimallee 2, 14195 Berlin, Germany
e-mail: blagojevic@math.fu-berlin.de

G. M. Ziegler
e-mail: ziegler@math.fu-berlin.de

P. V. M. Blagojević · A. S. D. Blagojević
Mat. Institut SANU, Knez Mihailova 36, 11001 Beograd, Serbia
e-mail: aleksandra1973@gmail.com

G. Ambrus et al. (eds.), *New Trends in Intuitive Geometry*, Bolyai Society Mathematical Studies 27, https://doi.org/10.1007/978-3-662-57413-3_3

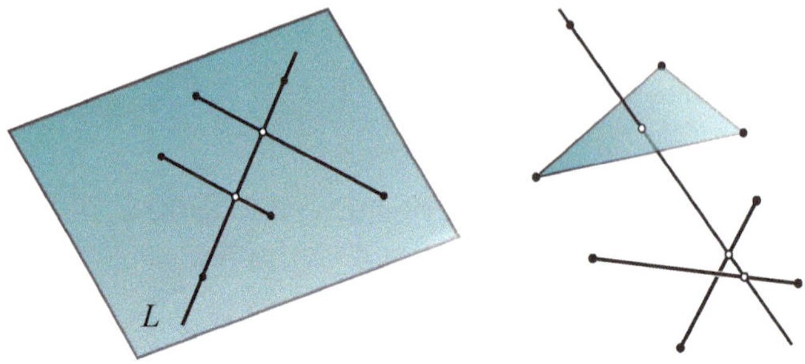

Fig. 1 The transversal Tverberg conjecture for $m = 1$, $r_0 = r_1 = 2$ and $d = 2$ or $d = 3$

Conjecture 1.1 (The transversal Tverberg conjecture) *Let*

- *m and d be integers with $0 \leq m \leq d - 1$,*
- *$r_0, \ldots, r_m \geq 1$ be integers, and*
- *$N_0 = (r_0 - 1)(d + 1 - m), \ldots, N_m = (r_m - 1)(d + 1 - m)$.*

Then for every collection of sets $X_0, \ldots, X_m \subset \mathbb{R}^d$ with $|X_0| = N_0 + 1, \ldots, |X_m| = N_m + 1$, there exist an m-dimensional affine subspace L of $\mathbb{R}^d$ and r_ℓ pairwise disjoint subsets $X_\ell^1, \ldots, X_\ell^{r_\ell}$ of X_ℓ, for $0 \leq \ell \leq m$, such that

$$\mathrm{conv}(X_0^1) \cap L \neq \emptyset, \ldots, \mathrm{conv}(X_0^{r_0}) \cap L \neq \emptyset, \ \ldots \ , \mathrm{conv}(X_m^1) \cap L \neq \emptyset, \ldots, \mathrm{conv}(X_m^{r_m}) \cap L \neq \emptyset.$$

For $m = 0$ this conjecture is Tverberg's well-known theorem. Tverberg and Vrećica published the full conjecture in 1993 [15]. They proved that it also holds for $m = d - 1$ [15, Prop. 3]. For $m = 1$ and arbitrary d they verified the conjecture only in the following three cases: $r_0 = 1$, $r_1 = 1$, and $r_0 = r_1 = 2$ [15, Prop. 1] (Fig. 1).

The classical Tverberg theorem from 1966 was extended to a topological setting by Bárány, Shlosman, and Szűcs [2] in 1981. Similarly, it is natural to consider the following extension of the transversal Tverberg conjecture.

Conjecture 1.2 (The topological transversal Tverberg conjecture) *Let*

- *m and d be integers with $0 \leq m \leq d - 1$,*
- *$r_0, \ldots, r_m \geq 1$ be integers, and*
- *$N_0 = (r_0 - 1)(d + 1 - m), \ldots, N_m = (r_m - 1)(d + 1 - m)$.*

Then for every collection of continuous maps $f_0 \colon \Delta_{N_0} \longrightarrow \mathbb{R}^d, \ldots, f_m \colon \Delta_{N_m} \longrightarrow \mathbb{R}^d$ there exist an m-dimensional affine subspace L of $\mathbb{R}^d$ and r_ℓ pairwise disjoint faces $\sigma_1^\ell, \ldots, \sigma_{r_\ell}^\ell$ such that

$$f_0(\sigma_1^0) \cap L \neq \emptyset, \ldots, f_0(\sigma_{r_0}^0) \cap L \neq \emptyset, \ \ldots \ , f_m(\sigma_1^m) \cap L \neq \emptyset, \ldots, f_m(\sigma_{r_m}^m) \cap L \neq \emptyset.$$

In 1999 using advanced methods of algebraic topology Živaljević [19, Theorem 4.8] proved this conjecture for d and m odd integers and $r_0 = \cdots = r_m$ being an odd prime.

The topological transversal Tverberg conjecture was settled for $r_0 = \cdots = r_m = 2$ by Vrećica [17, Theorem 2.2] in 2003. In 2007 Karasev [10, Theorem 1] established the topological transversal Tverberg conjecture in the cases when integers $r_0, \ldots, r_m$ are, not necessarily equal, powers of the same prime p and the product $p(d - m)$ is even.

In the same paper Karasev [10] proved a colored topological transversal Tverberg's theorem [10, Theorem 5], which for $m = 0$ coincides with the colored Tverberg theorem of Živaljević and Vrećica [20, Theorem p.1] and colored Tverberg theorem of type B of Živaljević and Vrećica [18, Theorem 4]. In 2011 Blagojević, Matschke and Ziegler gave yet another colored topological transversal Tverberg theorem [5, Theorem 1.3] that in the case $m = 0$ coincides with their optimal colored Tverberg theorem [6, Theorem 2.1].

The existence of counterexamples to the topological Tverberg conjecture for non-primepowers, obtained by Frick [3, 9] based on the remarkable work of Mabillard and Wagner [11, 12], in particular invalidates Conjecture 1.2 in the case when $m = 0$ and r_0 is not a prime power.

2 Statement of the Main Results

In 2014 Blagojević, Frick and Ziegler [4] introduced the "constraint method," by which the topological Tverberg theorem implies almost all its extensions, which had previously been obtained as substantial independent results, such as the "Colored Tverberg Theorem" of Živaljević and Vrećica [20] and the "Generalized Van Kampen–Flores Theorem" of Sarkaria [13] and Volovikov [16]. Thus the constraint method reproduced basically all Tverberg type theorems obtained during more than three decades with a single elementary idea. Moreover, the constraint method in combination with the work of Mabillard and Wagner on the "r-fold Whitney trick" [11, 12] yields counterexamples to the topological Tverberg theorem for non-prime powers, as demonstrated by Frick [3, 9].

In this paper we use the constraint method in combination with a generalized Borsuk–Ulam type theorem and a cohomological intersection lemma to show how one can obtain many new topological transversal theorems of Tverberg type. We prove in detail a new generalized transversal van Kampen–Flores theorem and a new topological transversal weak colored Tverberg theorem.

Theorem 2.1 (The topological generalized transversal Van Kampen–Flores theorem) *Let*

- *m and d be integers with $0 \leq m \leq d - 1$,*
- *$r_0 = p^{e_0}, \ldots, r_m = p^{e_m}$ be powers of the prime p, where $e_0, \ldots, e_m \geq 0$ are integers,*
- *$N_0 = (r_0 - 1)(d + 2 - m), \ldots, N_m = (r_m - 1)(d + 2 - m)$,*
- *$k_0 = \lceil \frac{r_0-1}{r_0} d \rceil, \ldots, k_m = \lceil \frac{r_m-1}{r_m} d \rceil$, and*
- *$p(d - m)$ be even, or $m = 0$.*

Then for every collection of continuous maps $f_0\colon \Delta_{N_0} \longrightarrow \mathbb{R}^d, \ldots, f_m\colon \Delta_{N_m} \longrightarrow \mathbb{R}^d$ *there exist an* m*-dimensional affine subspace* L *in* $\mathbb{R}^d$ *and* r_ℓ *pairwise disjoint faces* $\sigma_1^\ell, \ldots, \sigma_{r_\ell}^\ell$ *in the* k_ℓ*-skeleton of* Δ_{N_ℓ}*, for* $0 \le \ell \le m$*, such that*

$$f_0(\sigma_1^0) \cap L \neq \emptyset, \ldots, f_0(\sigma_{r_0}^0) \cap L \neq \emptyset, \quad \ldots \quad , f_m(\sigma_1^m) \cap L \neq \emptyset, \ldots, f_m(\sigma_{r_m}^m) \cap L \neq \emptyset.$$

The special case $r_0 = \cdots = r_m = 2$ of the previous theorem is due to Karasev [10, Cor. 4].

In order to state the next result we recall the notion of a rainbow face. Suppose that the vertices of the simplex Δ_N are partitioned into color classes $\mathrm{vert}(\Delta_N) = C_0 \sqcup \cdots \sqcup C_k$. The subcomplex $R_N := C_0 * \cdots * C_k \subseteq \Delta_N$ is called the *rainbow complex*, that is, the subcomplex of all faces that have at most one vertex of each color class $C_0, \ldots, C_k$. Faces of R_N are called *rainbow faces*.

Theorem 2.2 (The topological transversal weak colored Tverberg theorem) *Let*

- m *and* d *be integers with* $0 \le m \le d-1$,
- $r_0 = p^{e_0}, \ldots, r_m = p^{e_m}$ *be powers of the prime* p*, where* $e_0, \ldots, e_m \ge 0$ *are integers,*
- $N_0 = (r_0 - 1)(2d + 2 - m), \ldots, N_m = (r_m - 1)(2d + 2 - m)$,
- *the vertices of the simplex* Δ_{N_ℓ}*, for every* $0 \le m \le d$*, be colored by* $d+1$ *colors, where each color class has cardinality at most* $2r_\ell - 1$,
- $p(d-m)$ *be even, or* $m = 0$.

Then for every collection of continuous maps $f_0\colon \Delta_{N_0} \longrightarrow \mathbb{R}^d, \ldots, f_m\colon \Delta_{N_m} \longrightarrow \mathbb{R}^d$ *there exist an* m*-dimensional affine subspace* L *of* $\mathbb{R}^d$ *and* r_ℓ *pairwise disjoint rainbow faces* $\sigma_1^\ell, \ldots, \sigma_{r_\ell}^\ell$ *in* Δ_{N_ℓ}*, for* $0 \le \ell \le m$*, such that*

$$f_0(\sigma_1^0) \cap L \neq \emptyset, \ldots, f_0(\sigma_{r_0}^0) \cap L \neq \emptyset, \quad \ldots \quad , f_m(\sigma_1^m) \cap L \neq \emptyset, \ldots, f_m(\sigma_{r_m}^m) \cap L \neq \emptyset.$$

The proofs of Theorems 2.1 and 2.2 are almost identical; see Sects. 4.1 and 4.2. The only difference occurs in the definition of the bundles ξ_ℓ and τ_ℓ, in (1) and (2), and the bundle maps Φ_ℓ for $0 \le \ell \le m$; see Sects. 4.1 and 4.2. Using the same proof technique as for these theorems and modifying the bundles ξ_ℓ and bundle maps Φ_ℓ using recipes from [4, Lemma 4.2], one can also derive, for example, a topological transversal colored Tverberg theorem of type B, a topological transversal Tverberg theorem with equal barycentric coordinates, or mixtures of those. The most general transversal Tverberg theorem that is produced by the constraint method can be formulated using the concept of "Tverberg unavoidable subcomplexes" [4, Definition 4.1], as follows.

Let $r \ge 2$, $d \ge 1$ and $N \ge r-1$ be integers, and let $f\colon \Delta_N \longrightarrow \mathbb{R}^d$ be a continuous map with at least one Tverberg r-partition, that is, a collection of r pairwise disjoint faces $\sigma_1, \ldots, \sigma_r$ such that $f(\sigma_1) \cap \cdots \cap f(\sigma_r) \neq \emptyset$. A subcomplex Σ of the simplex Δ_N is *Tverberg unavoidable with respect to* f if for every Tverberg partition $\{\sigma_1, \ldots, \sigma_r\}$ of f there exists at least one face σ_i that lies in the subcomplex Σ.

Theorem 2.3 (A constraint topological transversal Tverberg theorem) *Let*

- *m and d be integers with $0 \leq m \leq d-1$,*
- *$c_1, \ldots, c_m \geq 0$ be integer,*
- *$r_0 = p^{e_0}, \ldots, r_m = p^{e_m}$ be powers of the prime p, where $e_0, \ldots, e_m \geq 0$ are integers,*
- *$N_0 = (r_0 - 1)(d + 1 + c_1 - m), \ldots, N_t = (r_t - 1)(d + 1 + c_m - m)$,*
- *$\Sigma_{i,j}$ be a Tverberg unavoidable subcomplex of the simplex Δ_{N_i} with respect to any continuous map $\Delta_{N_i} \longrightarrow \mathbb{R}^d$ for $1 \leq i \leq m$ and $0 \leq j \leq c_i$, assuming that $\Sigma_{i,0} = \Delta_{N_i}$, and*
- *$p(d-m)$ be even, or $m = 0$.*

Then for every collection of continuous maps $f_0 \colon \Delta_{N_0} \longrightarrow \mathbb{R}^d, \ldots, f_m \colon \Delta_{N_m} \longrightarrow \mathbb{R}^d$ there exist an m-dimensional affine subspace L of $\mathbb{R}^d$ and r_ℓ pairwise disjoint faces $\sigma_1^\ell, \ldots, \sigma_{r_\ell}^\ell$ that belong to the subcomplex $\Sigma_{\ell,0} \cap \cdots \cap \Sigma_{\ell,c_i}$, for $0 \leq \ell \leq m$, such that

$$f_0(\sigma_1^0) \cap L \neq \emptyset, \ldots, f_0(\sigma_{r0}^0) \cap L \neq \emptyset, \quad \ldots \quad , f_m(\sigma_1^m) \cap L \neq \emptyset, \ldots, f_m(\sigma_{rm}^m) \cap L \neq \emptyset.$$

3 A Generalized Borsuk–Ulam Type Theorem and Two Lemmas

In this section, we present the topological methods, developed in [5, 10], that we will use in the proofs of Theorems 2.1 and 2.2. In particular, we will review and slightly modify a generalized Borsuk–Ulam type theorem [5, Theorem 4.1], give an intersection lemma [5, Lemma 4.3] and recall the Euler class computation of Dol'nikov [7, Lemma p. 2], Živaljević [19, Proposition 4.9], and Karasev [10, Lemma 8].

3.1 Fadell–Husseini Index

In 1988 Fadell and Husseini [8] introduced an ideal-valued index theory for the category of G-space, or more general for the category of G-equivariant maps to a fixed space with a trivial G-action. We give an overview of the index theory adjusted to the needs of this paper.

Let G be a finite group, let R be a commutative ring with unit, and let B be a space with the trivial G action. For a G-equivariant map $p \colon X \longrightarrow B$ and a ring R, the *Fadell–Husseini index* of p is defined to be the kernel ideal of the map in the equivariant Čech cohomology with coefficients in the the ring R induced by p:

$$\begin{aligned}\mathrm{Index}_G^B(p; R) &= \ker\big(p^* \colon H^*(\mathrm{E}G \times_G B; R) \longrightarrow H^*(\mathrm{E}G \times_G X; R)\big)\\ &= \ker\big(p^* \colon H_G^*(B; R) \longrightarrow H_G^*(X; R)\big).\end{aligned}$$

The equivariant cohomology of a G-space X is assumed to be the Čech cohomology of the Borel construction $\mathrm{E}G \times_G X$ associated to the space X.

The basic properties of the index are:

- *Monotonicity*: If $p\colon X \longrightarrow B$ and $q\colon Y \longrightarrow B$ are G-equivariant maps, and $f\colon X \longrightarrow Y$ is a G-equivariant map such that $p = q \circ f$, then
$$\mathrm{Index}_G^B(p;\, R) \supseteq \mathrm{Index}_G^B(q;\, R).$$
- *Additivity*: If $(X_1 \cup X_2, X_1, X_2)$ is an excisive triple of G-spaces and $p\colon X_1 \cup X_2 \longrightarrow B$ is a G-equivariant map, then
$$\mathrm{Index}_G^B(p|_{X_1};\, R) \cdot \mathrm{Index}_G^B(p|_{X_2};\, R) \subseteq \mathrm{Index}_G^B(p;\, R).$$
- *General Borsuk–Ulam–Bourgin–Yang theorem:* Let $p\colon X \longrightarrow B$ and $q\colon Y \longrightarrow B$ be G-equivariant maps, and let $f\colon X \longrightarrow Y$ be a G-equivariant map such that $p = q \circ f$. If $Z \subseteq Y$ then
$$\mathrm{Index}_G^B(p|_{f^{-1}(Z)};\, R) \cdot \mathrm{Index}_G^B(q|_{Y\setminus Z};\, R) \subseteq \mathrm{Index}_G^B(p;\, R).$$

In the case when B is a point and $p\colon X \longrightarrow B$ is a G-equivariant map we simplify notation and write $\mathrm{Index}_G^B(p;\, R) = \mathrm{Index}_G^{\mathrm{pt}}(X;\, R)$. With this, the next property of the index can be formulated as follows.

If X is a G-space and $p\colon B \times X \longrightarrow B$ is the projection on the first factor, then
$$\mathrm{Index}_G^B(p;\, R) = \mathrm{Index}_G^{\mathrm{pt}}(X;\, R) \otimes H^*(B;\, R).$$

3.2 A Generalized Borsuk–Ulam Type Theorem

The cohomology of the elementary abelian groups $(\mathbb{Z}/p)^e$, where p is a prime and $e \geq 1$ is an integer, is given by

$$\begin{aligned} H^*((\mathbb{Z}/2)^e;\mathbb{F}_2) &= \mathbb{F}_2[t_1,\ldots,t_e], && \deg t_j = 1\\ H^*((\mathbb{Z}/p)^e;\mathbb{F}_p) &= \mathbb{F}_p[t_1,\ldots,t_e]\otimes\Lambda[u_1,\ldots,u_e], && \deg t_j = 2, \deg u_i = 1 \text{ for } p \text{ odd}. \end{aligned}$$

The following theorem and its proof is just a slight modification of [5, Theorem 4.1].

Theorem 3.1 (Borsuk–Ulam type theorem) *Let*

- $G = (\mathbb{Z}/p)^e$ *be an elementary abelian group where* p *is a prime and* $e \geq 1$,
- B *be a connected space with the trivial* G*-action,*
- $q\colon E \longrightarrow B$ *be a* G*-equivariant vector bundle where all fibers carry the same* G*-representation,*
- $q|_{E^G}\colon E^G \longrightarrow B$ *be the fixed-point subbundle of the vector bundle* $q\colon E \longrightarrow B$,

- *$q|_C: C \longrightarrow B$ be its G-invariant orthogonal complement subbundle $(E = C \oplus E^G)$,*
- *F be the fiber of the vector bundle $q_C: C \longrightarrow B$ over the point $b \in B$,*
- *$0 \neq \alpha \in H^*(G; \mathbb{F}_p)$ be the Euler class of the vector bundle $F \longrightarrow \mathrm{E}G \times_G F \longrightarrow \mathrm{B}G$, and*
- *K be a G-CW-complex such that $\alpha \notin \mathrm{Index}_G^{\mathrm{pt}}(K; \mathbb{F}_p)$.*

Assume that

- *$\pi_1(B)$ acts trivially on $H^*(F; \mathbb{F}_p)$, and*
- *we are given a G-equivariant map $\Phi: B \times K \longrightarrow E$ such that the following diagram commutes*

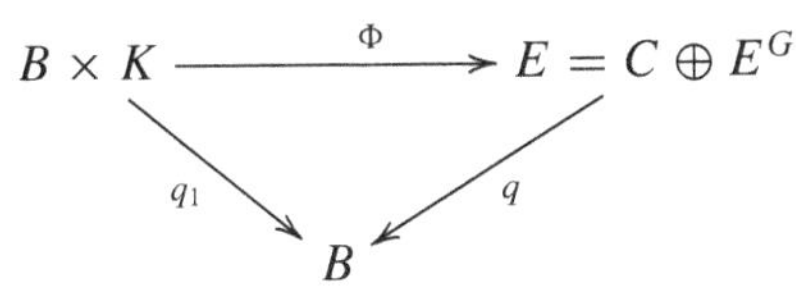

where $q_1: B \times K \longrightarrow B$ is the projection on the first coordinate.

Then for

$$S := \Phi^{-1}(E^G) \quad \text{and} \quad T := \Phi(S) = \mathrm{im}(\Phi) \cap E^G$$

the following maps, induced by the projections q_1 and q, are injective:

$$(q_1|_S)^*: H^*(B; \mathbb{F}_p) \longrightarrow H^*_G(S; \mathbb{F}_p) \quad \text{and} \quad (q|_T)^*: H^*(B; \mathbb{F}_p) \longrightarrow H^*(T; \mathbb{F}_p).$$

3.3 Two Lemmas

In this section we recall two facts: an intersection lemma from [5, Lemma 4.3] and the computation of a particular Euler class from [10, Lemma 8].

Lemma 3.2 (The intersection lemma) *Let*

- *$k \geq 1$ be an integer, and p a prime,*
- *B be an $\mathbb{F}_p$-orientable compact m-manifold,*
- *$q: E \longrightarrow B$ be an n-dimensional real vector bundle whose mod-p Euler class $e \in H^n(B; \mathbb{F}_p)$ satisfies $e^k \neq 0$, and*
- *$T_0, \ldots, T_k \subseteq E$ be compact subsets with the property that the induced maps*

$$(q|_{T_i})^*: H^m(B; \mathbb{F}_p) \longrightarrow H^m(T_i; \mathbb{F}_p),$$

for all $0 \leq i \leq k$, are injective.

Then

$$T_0 \cap \cdots \cap T_k \neq \emptyset.$$

Let $G_n(\mathbb{R}^d)$ denote the Grassmann manifold of all n-dimensional subspaces in $\mathbb{R}^d$, and let $\gamma^n(\mathbb{R}^d)$ be the corresponding canonical vector bundle over $G_n(\mathbb{R}^d)$. Furthermore, let $\widetilde{G}_n(\mathbb{R}^d)$ denote the oriented Grassmann manifold of all n-dimensional oriented subspaces in $\mathbb{R}^d$, and let $\widetilde{\gamma}^n(\mathbb{R}^d)$ be the corresponding canonical vector bundle over $\widetilde{G}_n(\mathbb{R}^d)$. Then the "forgetting orientation" map $\widetilde{G}_n(\mathbb{R}^d) \longrightarrow G_n(\mathbb{R}^d)$ is a double cover, and it induces a vector bundle map $\widetilde{\gamma}^n(\mathbb{R}^d) \longrightarrow \gamma^n(\mathbb{R}^d)$ that is an isomorphism on fibers.

Lemma 3.3 (Euler classes of the canonical bundles of real Grassmannians) *Let d and m be positive integers with $0 \le m \le d-1$, and let p be a prime.*

(1) If $p = 2$ and $\gamma := \gamma^{d-m}(\mathbb{R}^d)$, then the m-th power of the Euler class of γ does not vanish, that is

$$0 \neq e(\gamma)^m = w_{d-m}(\gamma)^m \in H^{(d-m)m}(G_{d-m}(\mathbb{R}^d); \mathbb{F}_2).$$

(2) If p is an odd prime, $d-m$ is even, and $\widetilde{\gamma} := \widetilde{\gamma}^{d-m}(\mathbb{R}^d)$, then the m-th power of the mod-p Euler class of $\widetilde{\gamma}$ does not vanish, that is

$$0 \neq e(\widetilde{\gamma})^m \in H^{(d-m)m}(\widetilde{G}_{d-m}(\mathbb{R}^d); \mathbb{F}_p).$$

(3) If p is an odd prime, $d-m$ is even, and $\gamma := \gamma^{d-m}(\mathbb{R}^d)$, then the m-th power of the mod-p Euler class of γ does not vanish, that is

$$0 \neq e(\gamma)^m \in H^{(d-m)m}(G_{d-m}(\mathbb{R}^d); \mathbb{F}_p).$$

The third part of the lemma is a consequence of the second part and the naturality property of the Euler class. The case $p = 2$ of this lemma was proved by Dol'nikov in [7, Lemma, p. 112]. For p an odd prime, $d \ge 3$ an odd integer, and $d - m$ even the lemma was first proved by Živaljević [19, Proposition 4.9].

4 Proofs

Now, combining the methods presented in Sect. 1, Theorem 3.1 and Lemmas 3.2 and 3.3, we prove our main results, Theorems 2.1 and 2.2.

4.1 Proof of the Topological Generalized Transversal Van Kampen–Flores Theorem

Let $B := G_{d-m}(\mathbb{R}^d)$ be the Grassmann manifold, and let $\gamma := \gamma^{d-m}(\mathbb{R}^d)$ be the canonical bundle. Without loss of generality we can assume that $m \ge 1$. The proof of Theorem 2.1 is done in several steps.

4.1.1

Fix an integer $0 \leq \ell \leq m$, and define $K_\ell := (\Delta_{N_\ell})^{\times r_\ell}_{\Delta(2)}$ to be the r_ℓ-fold 2-wise deleted product of the simplex Δ_{N_ℓ}. According to [2, Lemma 1] the complex K_ℓ is an $(N_\ell - r_\ell + 1)$-dimensional and $(N_\ell - r_\ell)$-connected CW complex. The symmetric group $\mathfrak{S}_{r_\ell}$ acts freely on K_ℓ by permuting factors in the product, that is $\pi \cdot (x_1, \dots, x_{r_\ell}) := (x_{\pi(1)}, \dots, x_{\pi(r_\ell)})$, for $\pi \in \mathfrak{S}_{r_\ell}$ and $(x_1, \dots, x_{r_\ell}) \in K_\ell$.

Consider the regular embedding reg : $(\mathbb{Z}/p)^{e_\ell} \longrightarrow \mathfrak{S}_{r_\ell}$ of the elementary abelian group $(\mathbb{Z}/p)^{e_\ell}$, as explained in [1, Exercise 2.7, p. 100]. It is given by the left translation action of $(\mathbb{Z}/p)^{e_\ell}$ on itself. To every element $g \in (\mathbb{Z}/p)^{e_\ell}$ we associate the permutation $L_g\colon (\mathbb{Z}/p)^{e_\ell} \longrightarrow (\mathbb{Z}/p)^{e_\ell}$ from $\mathrm{Sym}((\mathbb{Z}/p)^{e_\ell}) \cong \mathfrak{S}_{r_\ell}$ given by $L_g(x) = g + x$. Thus, the elementary abelian group $G_\ell := (\mathbb{Z}/p)^{e_\ell}$ is identified with subgroup im(reg) of the symmetric group $\mathfrak{S}_{r_\ell}$. Consequently, K_ℓ is a free G_ℓ-space.

Furthermore, let $\mathbb{R}^{r_\ell}$ be a vector space with the (left) action of the symmetric group $\mathfrak{S}_{r_\ell}$ given by permutation of coordinates. Then the subspace $W_{r_\ell} := \{(t_1, \dots, t_{r_\ell}) \in \mathbb{R}^{r_\ell} : \sum t_i = 0\}$ is a $\mathfrak{S}_{r_\ell}$-invariant subspace. The group G_ℓ acts on both $\mathbb{R}^{r_\ell}$ and W_{r_ℓ} via the regular embedding.

Let τ_ℓ be the the trivial vector bundle $B \times W_{r_\ell} \longrightarrow B$. The action of G_ℓ on W_{r_ℓ} makes τ_ℓ into a G_ℓ-equivariant vector bundle. Next, $\gamma^{\oplus r_\ell}$ is also a G_ℓ-equivariant vector bundle where the action is given by permutation of summands in the Whitney sum. Then the vector bundle

$$\xi_\ell := \tau_\ell \oplus \gamma^{\oplus r_\ell} \tag{1}$$

inherits the structure of a G_ℓ-equivariant vector bundle via the diagonal action. Let $E(\cdot)$ denote the total space of a vector bundle. Since the G_ℓ fixed point set of W_{r_ℓ} is just zero, that is $W_{r_\ell}^{G_\ell} = \{0\}$, the fixed point set of the total space of ξ_ℓ is

$$E(\xi_\ell)^{G_\ell} = E(\tau_\ell \oplus \gamma^{\oplus r_\ell})^{G_\ell} \cong E(\gamma^{\oplus r_\ell})^{G_\ell} \cong E(\gamma).$$

4.1.2

We define a continuous G_ℓ-equivariant bundle map $\Phi_\ell\colon B \times K_\ell \longrightarrow E(\xi_\ell)$ as follows: For the point $(b, (x_1, \dots, x_{r_\ell})) \in B \times K_\ell$ let

$$\begin{aligned}\Phi_\ell(b, (x_1, \dots, x_{r_\ell})) &:= \\ &\Big(b, \big(\mathrm{dist}(x_1, \mathrm{sk}_{k_\ell}(\Delta_{N_\ell})) - a(x_1, \dots, x_{r_\ell}), \dots, \mathrm{dist}(x_{r_\ell}, \mathrm{sk}_{k_\ell}(\Delta_{N_\ell})) - a(x_1, \dots, x_{r_\ell})\big)\Big) \oplus \\ &\big((q_b \circ f_\ell)(x_1) \oplus \cdots \oplus (q_b \circ f_\ell)(x_{r_\ell})\big),\end{aligned}$$

where

- $q\colon \mathbb{R}^d \longrightarrow b$ is the orthogonal projection onto the $(d-m)$-dimensional subspace $b \in B$ of $\mathbb{R}^d$,

- $\operatorname{dist}(\,\cdot\,,\operatorname{sk}_{k_\ell}(\Delta_{N_\ell}))$ denotes the distance function to the k_ℓ-skeleton of the simplex Δ_{N_ℓ}, and
- $a(x_1,\dots,x_{r_\ell})=\frac{1}{r_\ell}\big(\operatorname{dist}(x_1,\operatorname{sk}_{k_\ell}(\Delta_{N_\ell}))+\cdots+\operatorname{dist}(x_{r_\ell},\operatorname{sk}_{k_\ell}(\Delta_{N_\ell}))\big)$.

Next we consider the compact subsets

$$S_\ell := \Phi_\ell^{-1}(E(\xi_\ell)^{G_\ell}) \qquad \text{and} \qquad T_\ell := \Phi_\ell(S_\ell)=\operatorname{im}(\Phi_\ell)\cap E(\xi_\ell)^{G_\ell}.$$

where $T_\ell \subseteq E(\xi_\ell)^{G_\ell}\cong E(\gamma)$. The set S_ℓ contains of all points $(b,(x_1,\dots,x_{r_\ell}))\in B\times K_\ell$ such that

$$\operatorname{dist}(x_1,\operatorname{sk}_{k_\ell}(\Delta_{N_\ell}))=\cdots=\operatorname{dist}(x_{r_\ell},\operatorname{sk}_{k_\ell}(\Delta_{N_\ell})) \quad \text{and} \quad (q_b\circ f_\ell)(x_1)=\cdots=(q_b\circ f_\ell)(x_{r_\ell}).$$

Since $(x_1,\dots,x_{r_\ell})\in K_\ell$, then there exist r_ℓ pairwise disjoint faces $\sigma_1^\ell,\dots,\sigma_{r_\ell}^\ell$ of $\Delta_{N-\ell}$ such that

$$(x_1,\dots,x_{r_\ell})\in \operatorname{relint}\sigma_1^\ell\times\cdots\times\operatorname{relint}\sigma_{r_\ell}^\ell,$$

and at least one of the faces $\sigma_1^\ell,\dots,\sigma_{r_\ell}^\ell$ belongs to the k_ℓ-skeleton of the simplex Δ_{N_ℓ} [4, Lemma 4.2 (iii)]. Indeed, if this would no be true all the faces $\sigma_1^\ell,\dots,\sigma_{r_\ell}^\ell$ would be at least $(k_\ell+1)$-dimensional, implying the following contradiction

$$N_\ell+1=|\Delta_{N_\ell}|\ge|\sigma_1^\ell|+\cdots+|\sigma_{r_\ell}^\ell|\ge r_\ell(k_\ell+2)\ge r_\ell\big(\lceil\tfrac{r_\ell-1}{r_\ell}d\rceil+2\big)\ge(r_\ell-1)(d+2)+2=N_\ell+2.$$

Therefore, at least one of the faces $\sigma_1^\ell,\dots,\sigma_{r_\ell}^\ell$ lies in $\operatorname{sk}_{k_\ell}(\Delta_{N_\ell})$ and consequently

$$\operatorname{dist}(x_1,\operatorname{sk}_{k_\ell}(\Delta_{N_\ell}))=\cdots=\operatorname{dist}(x_{r_\ell},\operatorname{sk}_{k_\ell}(\Delta_{N_\ell}))=0,$$

implying that all the faces $\sigma_1^\ell,\dots,\sigma_{r_\ell}^\ell$ lies in $\operatorname{sk}_{k_\ell}(\Delta_{N_\ell})$.

Thus, in order to conclude the proof of Theorem 2.1 we need to show that

$$\emptyset\neq T_0\cap\cdots\cap T_m\subseteq E(\gamma),$$

and for that we would like to use Lemma 3.2.

4.1.3

First, let $0\le\ell\le m$ and let $e_\ell=0$. Then $r_\ell=1$, $N_\ell=0$, $K_\ell=\Delta_{N_\ell}$ is a point, G_ℓ is the trivial group, and $S_\ell=B\times K_\ell$. Consider the commutative diagram induced by the bundle map $\Phi_\ell\colon B\times K_\ell\longrightarrow E(\xi_\ell)$ and the corresponding diagram in cohomology:

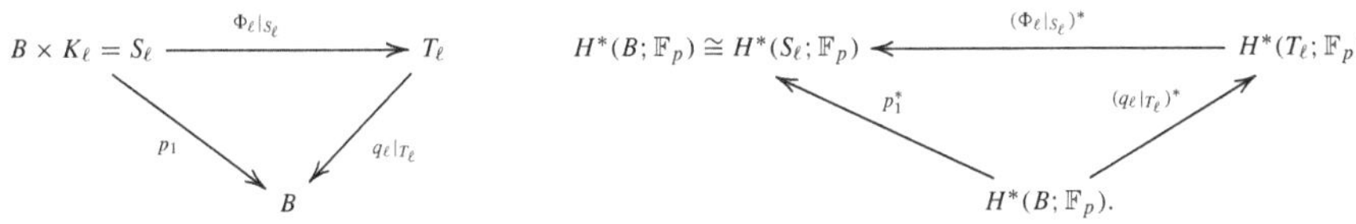

Since K_ℓ is a point the map p_1^* induced by the projection p_1 is the identity map. Consequently, the map

$$(q_\ell|_{T_\ell})^*\colon H^*(B;\mathbb{F}_p) \longrightarrow H^*(T_\ell;\mathbb{F}_p).$$

is an injection.

4.1.4

Next, let $0 \le \ell \le m$, and let $e_\ell > 0$. Now we apply Theorem 3.1 to the G_ℓ-equivariant bundle map $\Phi_\ell\colon B \times K_\ell \longrightarrow E(\xi_\ell)$. In order to do so we check the necessary assumptions. Since

- $G_\ell = (\mathbb{Z}_p)^{e_\ell}$ is an elementary abelian group,
- $B = G_{d-m}(\mathbb{R}^d)$ is a connected space with the trivial G_ℓ-action,
- $q_\ell\colon E(\xi_\ell) \longrightarrow B$ is a G_ℓ-equivariant vector bundle where all fibers carry the same G_ℓ-representation,
- $q_\ell|_{E(\xi_\ell)^{G_\ell}}\colon E(\xi_\ell)^{G_\ell} \longrightarrow B$ is the fixed-point subbundle with the G_ℓ-invariant orthogonal complement subbundle $q_\ell|_{C_\ell}\colon C_\ell \longrightarrow B$, $(E(\xi_\ell) = C_\ell \oplus E(\xi_\ell)^{G_\ell})$,
- F_ℓ is the fiber of the vector bundle $q_\ell|_{C_\ell}\colon C_\ell \longrightarrow B$ over the point $b \in B$,
- $\pi_1(B)$ acts trivially on the cohomology of the sphere $H^*(S(F_\ell);\mathbb{F}_p)$,
- the Euler class $0 \ne \alpha_\ell \in H^{(r_\ell-1)(d-m+1)}(G_\ell;\mathbb{F}_p)$ of the vector bundle $F_\ell \longrightarrow \mathrm{E}G_\ell \times_{G_\ell} F_\ell \longrightarrow \mathrm{B}G_\ell$ does not vanish, more precisely

$$\alpha_\ell = \Big(\prod_{(a_1,\dots,a_{e_\ell})\in\mathbb{F}_p^{e_\ell}\setminus\{0\}} (a_1t_1+\cdots+a_{e_\ell}t_{e_\ell})\Big)^{\frac{d-m+1}{2}},$$

- $\mathrm{Index}^{\mathrm{pt}}_{G_\ell}(K_\ell;\mathbb{F}_p) \subseteq H^{\ge (r_\ell-1)(d-m+1)+1}(G_\ell;\mathbb{F}_p)$ because the cell complex K_ℓ is $((r_\ell-1)(d-m+1)-1)$-connected,

we have that $\alpha_\ell \notin \mathrm{Index}^{\mathrm{pt}}_{G_\ell}(K_\ell;\mathbb{F}_p)$ and Theorem 3.1 can be applied on the G_ℓ-equivariant bundle map $\Phi_\ell\colon B \times K_\ell \longrightarrow E(\xi_\ell)$. Thus, the following map in Čech cohomology induced by q_ℓ is injective:

$$(q_\ell|_{T_\ell})^*\colon H^*(B;\mathbb{F}_p) \longrightarrow H^*(T_\ell;\mathbb{F}_p).$$

4.1.5

Finally, Lemma 3.2 comes into play. Since,

- T_ℓ is a compact subset of $E(\gamma)$ for every $0 \le \ell \le m$,
- $(q_\ell|_{T_\ell})^*\colon H^*(B;\mathbb{F}_p) \longrightarrow H^*(T_\ell;\mathbb{F}_p)$ is injective for every $0 \le \ell \le m$, and
- $0 \ne e(\gamma)^m \in H^{(d-m)m}(B;\mathbb{F}_p)$ according to $p(d-m)$ being even and Lemma 3.3,

we can apply Lemma 3.2 and get that

$$T_0 \cap \cdots \cap T_m \ne \emptyset.$$

This concludes the proof of Theorem 2.1. □

4.2 *Proof of the Topological Transversal Weak Colored Tverberg Theorem*

Let $B := G_{d-m}(\mathbb{R}^d)$ be the Grassmann manifold, and let $\gamma := \gamma^{d-m}(\mathbb{R}^d)$ be the canonical bundle. Without lost of generality we can assume that $m \ge 1$. The proof of Theorem 2.2 is done in the footsteps of the proof of Theorem 2.1. The only difference will occur in the definition of the bundles τ_ℓ and consequently bundle maps Φ_ℓ.

4.2.1

Again, fix an integer $0 \le \ell \le m$, and define $K_\ell := (\Delta_{N_\ell})^{\times r_\ell}_{\Delta(2)}$. As we have seen the complex K_ℓ is an $(N_\ell - r_\ell + 1)$-dimensional and $(N_\ell - r_\ell)$-connected CW complex. The symmetric group $\mathfrak{S}_{r_\ell}$ acts freely on K_ℓ by permuting factors in the product.

The regular embedding $\mathrm{reg}\colon (\mathbb{Z}/p)^{e_\ell} \longrightarrow \mathfrak{S}_{r_\ell}$ of the elementary abelian group $(\mathbb{Z}/p)^{e_\ell}$ identifies elementary abelian group $G_\ell := (\mathbb{Z}/p)^{e_\ell}$ with a subgroup $\mathrm{im(reg)}$ of the symmetric group $\mathfrak{S}_{r_\ell}$.

Once more, $\mathbb{R}^{r_\ell}$ is a vectors space with the (left) action of the symmetric group $\mathfrak{S}_{r_\ell}$ given by permutation of coordinates. The subspace $W_{r_\ell} := \{(t_1, \dots, t_{r_\ell}) \in \mathbb{R}^{r_\ell} : \sum t_i = 0\}$ is a $\mathfrak{S}_{r_\ell}$-invariant subspace, and G_ℓ acts on both $\mathbb{R}^{r_\ell}$ and W_{r_ℓ} via the regular embedding.

Let τ_ℓ be the the trivial vector bundle $B \times W_{r_\ell}^{\oplus d+1} \longrightarrow B$. The action of G_ℓ on $W_{r_\ell}^{\oplus d+1}$ is diagonal and makes τ_ℓ into a G_ℓ-equivariant vector bundle. As we have seen, $\gamma^{\oplus r_\ell}$ is also a G_ℓ-equivariant vector bundle. Thus the vector bundle

$$\xi_\ell := \tau_\ell \oplus \gamma^{\oplus r_\ell} \tag{2}$$

inherits the structure of a G_ℓ-equivariant vector bundle via the diagonal action. Since $(W_{r_\ell}^{\oplus d+1})^{G_\ell} = \{0\}$, the fixed point set of the total space of ξ_ℓ is

$$E(\xi_\ell)^{G_\ell} = E(\tau_\ell \oplus \gamma^{\oplus r_\ell})^{G_\ell} \cong E(\gamma^{\oplus r_\ell})^{G_\ell} \cong E(\gamma).$$

4.2.2

The vertices of the simplex Δ_{N_ℓ} are colored by $d+1$ colors, where each color class has cardinality at most $2r_\ell - 1$. Set $\mathrm{vert}(\Delta_N) = C_0 \sqcup \cdots \sqcup C_d$ where $|C_i| \leq 2r_\ell - 1$ for all $0 \leq i \leq d$. Following the idea from [4, Lemma 4.2 (ii)] we define Σ_i^ℓ, $0 \leq i \leq d$, to be the subcomplex of Δ_{N_ℓ} consisting of all faces with at most one vertex in C_i. Then the rainbow subcomplex R_{N_ℓ} coincides with the intersection $\Sigma_0^\ell \cap \cdots \cap \Sigma_d^\ell$.

Now we define a continuous G_ℓ-equivariant bundle map $\Phi_\ell \colon B \times K_\ell \longrightarrow E(\xi_\ell)$ as follows: For the point $(b, (x_1, \ldots, x_{r_\ell})) \in B \times K_\ell$ let

$$\begin{aligned}\Phi_\ell(b, (x_1, \ldots, x_{r_\ell})) := &\big(b, \mathrm{dist}(x_1, \Sigma_0^\ell) - a_0(x_1, \ldots, x_{r_\ell}), \ldots, \mathrm{dist}(x_{r_\ell}, \Sigma_0^\ell) - a_0(x_1, \ldots, x_{r_\ell})\big) \oplus \\ &\cdots \\ &\big(b, \mathrm{dist}(x_1, \Sigma_d^\ell) - a_d(x_1, \ldots, x_{r_\ell}), \ldots, \mathrm{dist}(x_{r_\ell}, \Sigma_d^\ell) - a_d(x_1, \ldots, x_{r_\ell})\big) \oplus \\ &\big((q_b \circ f_\ell)(x_1) \oplus \cdots \oplus (q_b \circ f_\ell)(x_{r_\ell})\big)\end{aligned}$$

where

- $q \colon \mathbb{R}^d \longrightarrow b$ is the orthogonal projection onto the $(d-m)$-dimensional subspace $b \in B$ of $\mathbb{R}^d$,
- $\mathrm{dist}(\,\cdot\,, \Sigma_i^\ell)$ denotes the distance function to the subcomplex Σ_i^ℓ where $0 \leq i \leq d$, and
- $a_i(x_1, \ldots, x_{r_\ell}) = \frac{1}{r_\ell}\big(\mathrm{dist}(x_1, \Sigma_i^\ell) + \cdots + \mathrm{dist}(x_{r_\ell}, \Sigma_i^\ell)\big)$ for $0 \leq i \leq d$.

Again, we consider the compact subsets

$$S_\ell := \Phi_\ell^{-1}(E(\xi_\ell)^{G_\ell}) \qquad \text{and} \qquad T_\ell := \Phi_\ell(S_\ell) = \mathrm{im}(\Phi_\ell) \cap E(\xi_\ell)^{G_\ell}.$$

where $T_\ell \subseteq E(\xi_\ell)^{G_\ell} \cong E(\gamma)$. The set S_ℓ contains of all points $(b, (x_1, \ldots, x_{r_\ell})) \in B \times K_\ell$ such that

$$\mathrm{dist}(x_1, \Sigma_0^\ell) = \cdots = \mathrm{dist}(x_{r_\ell}, \Sigma_0^\ell), \quad \ldots \quad, \mathrm{dist}(x_1, \Sigma_d^\ell) = \cdots = \mathrm{dist}(x_{r_\ell}, \Sigma_d^\ell),$$

and

$$(q_b \circ f_\ell)(x_1) = \cdots = (q_b \circ f_\ell)(x_{r_\ell}).$$

Since the point $(x_1, \ldots, x_{r_\ell}) \in K_\ell$, then we can find r_ℓ unique pairwise disjoint faces $\sigma_1^\ell, \ldots, \sigma_{r_\ell}^\ell$ with the property that

$$(x_1, \ldots, x_{r_\ell}) \in \mathrm{relint}\, \sigma_1^\ell \times \cdots \times \mathrm{relint}\, \sigma_{r_\ell}^\ell.$$

Moreover, for every i in the range $0 \leq i \leq d$ there exists at least one of the faces $\sigma_1^\ell, \ldots, \sigma_{r_\ell}^\ell$ that belongs to the subcomplex Σ_i of the simplex Δ_{N_ℓ}, [4, Lemma 4.2 (ii)]. Indeed, if this would no be true for an index i then all the faces $\sigma_1^\ell, \ldots, \sigma_{r_\ell}^\ell$ would have at least two vertices from the set C_i and we have the contradiction

$$2r_\ell - 1 \geq |C_i| \geq |\sigma_1^\ell \cap C_i| + \cdots + |\sigma_{r_\ell}^\ell \cap C_i| \geq 2r_\ell.$$

Thus for every index i at least one of the faces $\sigma_1^\ell, \ldots, \sigma_{r_\ell}^\ell$ lies in Σ_i and consequently

$$\operatorname{dist}(x_1, \Sigma_0^\ell) = \cdots = \operatorname{dist}(x_{r_\ell}, \Sigma_0^\ell) = 0, \quad \ldots \quad , \operatorname{dist}(x_1, \Sigma_d^\ell) = \cdots = \operatorname{dist}(x_{r_\ell}, \Sigma_d^\ell) = 0,$$

implying that all the faces $\sigma_1^\ell, \ldots, \sigma_{r_\ell}^\ell$ lie in every subcomplex Σ_i, meaning they belong to the intersection $\Sigma_0^\ell \cap \cdots \cap \Sigma_d^\ell$ – the rainbow subcomplex.

Therefore, in order to finalize the proof of Theorem 2.2 we need to show, as in the proof of the previous theorem, that

$$\emptyset \neq T_0 \cap \cdots \cap T_m \subseteq E(\gamma).$$

4.2.3

Again first assume that $0 \leq \ell \leq m$ and that $e_\ell = 0$. We proceed as in Sect. 4.1. Now $r_\ell = 1$, $N_\ell = 0$, $K_\ell = \Delta_{N_\ell}$ is a point, G_ℓ is the trivial group, and $S_\ell = B \times K_\ell$. We consider the commutative diagrams induced by the bundle map $\Phi_\ell \colon B \times K_\ell \longrightarrow E(\xi_\ell)$:

$$\begin{array}{ccc} B \times K_\ell = S_\ell & \xrightarrow{\Phi_\ell|_{S_\ell}} & T_\ell \\ & {\scriptstyle p_1} \searrow \quad \swarrow {\scriptstyle q_\ell|_{T_\ell}} & \\ & B & \end{array} \qquad \begin{array}{ccc} H^*(B;\mathbb{F}_p) \cong H^*(S_\ell;\mathbb{F}_p) & \xleftarrow{(\Phi_\ell|_{S_\ell})^*} & H^*(T_\ell;\mathbb{F}_p) \\ & {\scriptstyle p_1^*} \nwarrow \quad \nearrow {\scriptstyle (q_\ell|_{T_\ell})^*} & \\ & H^*(B;\mathbb{F}_p). & \end{array}$$

The map p_1^* induced by the projection p_1 is the identity map. Hence, the map in cohomology

$$(q_\ell|_{T_\ell})^* \colon H^*(B;\mathbb{F}_p) \longrightarrow H^*(T_\ell;\mathbb{F}_p).$$

is an injection.

4.2.4

Let $0 \leq \ell \leq m$ and $e_\ell > 0$. We apply Theorem 3.1 to the G_ℓ-equivariant bundle map $\Phi_\ell \colon B \times K_\ell \longrightarrow E(\xi_\ell)$. For that we check the necessary assumptions. Since

- $G_\ell = (\mathbb{Z}_p)^{e_\ell}$ is an elementary abelian group,
- $B = G_{d-m}(\mathbb{R}^d)$ is a connected space with the trivial G_ℓ-action,
- $q_\ell \colon E(\xi_\ell) \longrightarrow B$ is a G_ℓ-equivariant vector bundle where all fibers carry the same G_ℓ-representation,
- $q_\ell|_{E(\xi_\ell)^{G_\ell}} \colon E(\xi_\ell)^{G_\ell} \longrightarrow B$ is the fixed-point subbundle with the G_ℓ-invariant orthogonal complement subbundle $q_\ell|_{C_\ell} \colon C_\ell \longrightarrow B$, $(E(\xi_\ell) = C_\ell \oplus E(\xi_\ell)^{G_\ell})$,

- F_ℓ is the fiber of the vector bundle $q_\ell|_{C_\ell}\colon C_\ell \longrightarrow B$ over the point $b \in B$,
- $\pi_1(B)$ acts trivially on the cohomology of the sphere $H^*(S(F_\ell); \mathbb{F}_p)$,
- the Euler class $0 \neq \alpha_\ell \in H^{(r_\ell-1)(2d-m+1)}(G_\ell; \mathbb{F}_p)$ of the vector bundle $F_\ell \longrightarrow \mathrm{E}G_\ell \times_{G_\ell} F_\ell \longrightarrow \mathrm{B}G_\ell$ does not vanish, more precisely

$$\alpha_\ell = \Big(\prod_{(a_1,\dots,a_{e_\ell})\in\mathbb{F}_p^{e_\ell}\setminus\{0\}} (a_1t_1 + \cdots + a_{e_\ell}t_{e_\ell})\Big)^{\frac{2d-m+1}{2}},$$

- $\mathrm{Index}^{\mathrm{pt}}_{G_\ell}(K_\ell; \mathbb{F}_p) \subseteq H^{\geq (r_\ell-1)(2d-m+1)+1}(G_\ell; \mathbb{F}_p)$ because the cell complex K_ℓ is $((r_\ell - 1)(2d - m + 1) - 1)$-connected,

we conclude that $\alpha_\ell \notin \mathrm{Index}^{\mathrm{pt}}_{G_\ell}(K_\ell; \mathbb{F}_p)$, and therefore Theorem 3.1 can be applied on the G_ℓ-equivariant bundle map $\Phi_\ell\colon B \times K_\ell \longrightarrow E(\xi_\ell)$. Consequently, the following map induced by q_ℓ is injective:

$$(q_\ell|_{T_\ell})^*\colon H^*(B; \mathbb{F}_p) \longrightarrow H^*(T_\ell; \mathbb{F}_p).$$

4.2.5

In the final step we apply Lemma 3.2. Since,

- T_ℓ is a compact subset of $E(\gamma)$ for every $0 \leq \ell \leq m$,
- $(q_\ell|_{T_\ell})^*\colon H^*(B; \mathbb{F}_p) \longrightarrow H^*(T_\ell; \mathbb{F}_p)$ is injective for every $0 \leq \ell \leq m$, and
- $0 \neq e(\gamma)^m \in H^{(d-m)m}(B; \mathbb{F}_p)$ does not vanish according to $p(d - m)$ being even and Lemma 3.3,

we can apply Lemma 3.2 and get that

$$T_0 \cap \cdots \cap T_m \neq \emptyset.$$

This concludes the proof of Theorem 2.2. □

Acknowledgements We are grateful to Florian Frick and to the referee for very good observations and useful comments.

References

1. A. Adem, J. R. Milgram, *Cohomology of Finite Groups*, vol. 309, 2nd ed., Grundlehren der Mathematischen Wissenschaften (Springer, Berlin, 2004)
2. I. Bárány, S.B. Shlosman, A. Szűcs, On a topological generalization of a theorem of Tverberg. J. Lond. Math. Soc. **23**, 158–164 (1981)
3. P.V.M. Blagojević, F. Frick, G.M. Ziegler, *Barycenters of Polytope Skeleta and Counterexamples to the Topological Tverberg Conjecture, Via Constraints* (2015), p. 6, arXiv:1510.07984 [Preprint]

4. P.V.M. Blagojević, Tverberg plus constraints. Bull. Lond. Math. Soc. **46**(5), 953–967 (2014)
5. P.V.M. Blagojević, B. Matschke, G.M. Ziegler, Optimal bounds for a colorful Tverberg–Vrećica type problem. Adv. Math. **226**(6), 5198–5215 (2011)
6. P.V.M. Blagojević, B. Matschke, G.M. Ziegler, Optimal bounds for the colored Tverberg problem. J. Eur. Math. Soc **17**(4), 739–754 (2015)
7. V.L. Dol'nikov, Transversals of families of sets in $\mathbb{R}^n$ and a relationship between Helly and Borsuk theorems. Math. Sb. **184**(5), 111–132 (1993)
8. E. Fadell, S. Husseini, An ideal-valued cohomological index theory with applications to Borsuk–Ulam and Bourgin–Yang theorems, Ergod. Theory Dynam. Syst. **8*** (1988), 73–85 [Charles Conley memorial issue]
9. F. Frick, Counterexamples to the topological Tverberg conjecture. Oberwolfach Rep. **12**(1), 318–322 (2015), arXiv:1502.00947
10. R.N. Karasev, Tverberg's transversal conjecture and analogues of nonembeddability theorems for transversals. Discrete Comput. Geom. **38**(3), 513–525 (2007)
11. I. Mabillard, U. Wagner, Eliminating Tverberg points, I. an analogue of the Whitney trick, in *Proceedings of the 30th Annual Symposium on Computational Geometry (SoCG)* (ACM, Kyoto, 2014), pp. 171–180
12. I. Mabillard, U. Wagner, Eliminating higher-multiplicity intersections, I. A Whitney trick for Tverberg-type problems (2015), p. 46, arXiv:1508.02349 [Preprint]
13. K.S. Sarkaria, A generalized van Kampen–Flores theorem. Proc. Am. Math. Soc. **11**, 559–565 (1991)
14. H. Tverberg, A generalization of Radon's theorem. J. Lond. Math. Soc. **41**, 123–128 (1966)
15. H. Tverberg, S. Vrećica, On generalizations of Radon's theorem and the ham sandwich theorem. Eur. J. Comb. **14**(3), 259–264 (1993)
16. A.Y. Volovikov, On the van Kampen–Flores theorem. Math. Notes **59**(5), 477–481 (1996)
17. S. Vrećica, Tverberg's conjecture. Discret. Comput. Geom. **29**(4), 505–510 (2003)
18. S. Vrećica, R.T. Živaljević, New cases of the colored Tverberg theorem, in *Jerusalem Combinatorics '93*, ed. by H. Barcelo, G. Kalai, Contemporary Mathematics, vol. 178. (American Mathematical Society, 1994), pp. 325–334
19. R.T. Živaljević, The Tverberg–Vrećica problem and the combinatorial geometry on vector bundles. Israel J. Math. **111**, 53–76 (1999)
20. R.T. Živaljević, S. Vrećica, The colored Tverberg's problem and complexes of injective functions. J. Combin. Theory Ser. A **61**, 309–318 (1992)

On the Volume of Boolean Expressions of Balls – A Review of the Kneser–Poulsen Conjecture

Balázs Csikós

Abstract In 1954–55, E. T. Poulsen and M. Kneser formulated the conjecture that if some congruent balls of the Euclidean space are rearranged in such a way that the distances between the centers of the balls do not increase, then the volume of the union of the balls does not increase as well. Our goal is to give a survey of attempts to prove this conjecture, to discuss possible generalizations, and to collect some relevant open questions.

2010 Mathematics Subject Classification 52A40 · 52A38 · 26B20

1 Introduction

In 1954–55 E. Thue Poulsen [1] and M. Kneser [2] posed the following conjecture.

Kneser–Poulsen Conjecture (KPC). If the points $x_1, \dots, x_k$ and $y_1, \dots, y_k$ of the n-dimensional Euclidean space $\mathbb{E}^n$ satisfy the inequalities $d(x_i, x_j) \geq d(y_i, y_j)$ for all $1 \leq i, j \leq k$, then

$$\lambda_n \left(\bigcup_{i=1}^{k} B(x_i, r)\right) \geq \lambda_n \left(\bigcup_{i=1}^{k} B(y_i, r)\right) \tag{1}$$

for any $r > 0$, where $B(x, r)$ is the closed ball of radius r about x and λ_n denotes the n-dimensional Lebesgue measure.

In Kneser's paper, the conjecture emerged naturally from the study of the Minkowski content. In the first half of the 20th century many geometric measures were constructed, (see Federer [3]). To handle different geometric measures in a unified way, A. Kolmogorov [4] proposed an axiomatic definition for an s-dimensional

The author was supported by the Hungarian Scientific Research Fund (OTKA), Grant No. K112703.

B. Csikós (✉)
Institute of Mathematics, Eötvös University, Pázmány Péter sétány 1/C, Budapest 1117, Hungary
e-mail: csikos@math.elte.hu

G. Ambrus et al. (eds.), *New Trends in Intuitive Geometry*, Bolyai Society Mathematical Studies 27, https://doi.org/10.1007/978-3-662-57413-3_4

measure function μ defined on Suslin subsets of the n-dimensional Euclidean space and proved that *for any natural number* $s \leq n$, *there is a minimal and a maximal measure among all* s *-dimensional measure functions*. Kolmogorov's third axiom, which required that *if the set* E' *is the image of the set* E *under a map which does not increase the distances between the points, then* $\mu(E') \leq \mu(E)$, emphasized that monotonicity of measures under contractions is a fundamental property for geometric measures defined on subsets of a Euclidean space. In view of this, it was natural that studying the basic properties of the Minkowski content, M. Kneser paid attention to the question whether the Minkowski content satisfies Kolmogorov's third axiom.

Recall the definition of the Minkowski content. Set $\omega_s = \pi^{s/2}/\Gamma(1+s/2)$ for $s \geq 0$. If s is integer, then ω_s is the volume of the s-dimensional Euclidean unit ball.

If $A \subset \mathbb{E}^n$ is a subset of the Euclidean space, and $\varepsilon > 0$, then denote by $B^o(A, \varepsilon)$ the open ε-neighborhood of A. The *lower* and *upper* s*-dimensional Minkowski contents of a bounded set* $A \subset \mathbb{E}^n$ are defined as the limits

$$\underline{\mu}_n^s(A) = \liminf_{\epsilon \to +0} \frac{\lambda_n(B^o(A, \epsilon))}{\omega_{n-s}\epsilon^{n-s}} \quad \text{and} \quad \overline{\mu}_n^s(A) = \limsup_{\epsilon \to +0} \frac{\lambda_n(B^o(A, \epsilon))}{\omega_{n-s}\epsilon^{n-s}},$$

respectively.

As it was observed by M. Kneser [2], if the KPC were true, then for any bounded set $A \subset \mathbb{E}^n$ and for any contraction A' of it, we would have $\lambda_n(B^o(A, \epsilon)) \geq \lambda_n(B^o(A', \epsilon))$ for any positive ϵ, which would imply the inequalities $\underline{\mu}_n^s(A) \geq \underline{\mu}_n^s(A')$ and $\overline{\mu}_n^s(A) \geq \overline{\mu}_n^s(A')$ for any $0 \leq s \leq n$. The KPC would also imply that the lower s-dimensional Minkowski content of any compact set is not less than Kolmogorov's minimal s-dimensional measure of the set.

Since the work of Kneser the KPC has been popularized by several authors ([5–9]) and has attracted the attention of many mathematicians. Although many important results have been obtained, the conjecture is still not settled in dimensions $n \geq 3$. The goal of the present paper is to show the ideas that led to the present status of the conjecture.

2 First Results

Let (M, d) be a metric space. If the k-tuples of points $\mathbf{x} = (x_1, \ldots, x_k) \in M^k$ and $\mathbf{y} = (y_1, \ldots, y_k) \in M^k$ satisfy the inequalities $d(x_i, x_j) \geq d(y_i, y_j)$ for $1 \leq i < j \leq k$, then we shall say that the configuration $\mathbf{y}$ is a *contraction* of the configuration $\mathbf{x}$, or the configuration $\mathbf{x}$ is an *expansion* of the configuration $\mathbf{y}$ and we shall denote this by $\mathbf{x} \succ \mathbf{y}$ or $\mathbf{y} \prec \mathbf{x}$.

Let $\mathbb{R}_+$ denote the set of nonnegative real numbers. Given k points $\mathbf{x} = (x_1, \ldots, x_k) \in M^k$ and k numbers $\mathbf{r} = (r_1, \ldots, r_k) \in \mathbb{R}_+^k$, we set

$$B(\mathbf{x}, \mathbf{r}) = \bigcup_{i=1}^k B(x_i, r_i).$$

M. Kneser proved a relaxed version of his conjecture.

Theorem 2.1 (M. Kneser [2]) *If* $\mathbf{x}, \mathbf{y} \in (\mathbb{E}^n)^k$, $\mathbf{x} \succ \mathbf{y}$, *and* $r > 0$, *then*

$$3^n \lambda_n\Big(\bigcup_{i=1}^k B(x_i, r)\Big) \geq \lambda_n\Big(\bigcup_{i=1}^k B(y_i, r)\Big).$$

Kneser's proof works also in finite dimensional normed spaces, but it cannot be applied to noncongruent balls directly.

In 1956 H. Hadwiger wrote a short note [5] on the conjecture and listed the special cases in which the proof of the conjecture was known at that time. In one of the cases, the notion of continuous contractions appears.

Definition 2.1 If (M, d) is a metric space, then we shall say that the configuration $\mathbf{y} \in M^k$ is a *continuous contraction* of the configuration $\mathbf{x} \in M^k$ *(in* M*)*, if there is a continuous map $\mathbf{z} : [0, 1] \to M^k$ such that $\mathbf{z}(0) = \mathbf{x}$, $\mathbf{z}(1) = \mathbf{y}$, and $\mathbf{z}(t_1) \succ \mathbf{z}(t_2)$ for any $0 \leq t_1 < t_2 \leq 1$. This relation will be denoted by $\mathbf{x} \succcurlyeq \mathbf{y}$ or $\mathbf{y} \preccurlyeq \mathbf{x}$. The map $\mathbf{z}$ will be called a *contracting homotopy* connecting the two configurations.

Theorem 2.2 (see [5]) *Inequality* (1) *holds for the configurations* $\mathbf{x} \succ \mathbf{y}$ *in the following special cases:*

(i) if $n = 1$;
(ii) if the number k *of the balls is at most* $n + 1$;
(iii) if the configurations $\mathbf{x} \succ \mathbf{y}$ *are similar;*
(iv) if $n = 2$, *and* $\mathbf{x} \succcurlyeq \mathbf{y}$.

Case (i) can be proved by a simple induction on the number of segments. The proof of case (iii) is due to G. Bouligand [10], a simple proof was published also by D. Avis, B.K. Batthacharya, and H. Imai [11]. Case (iii) was generalized for translates of convex bodies by W. Rehder [12]. Both Kneser and Hadwiger mentions, that W. Habicht has proved the special case (iv), but Habicht's proof remained unpublished.

The relation $\mathbf{x} \succ \mathbf{y}$ does not imply $\mathbf{x} \succcurlyeq \mathbf{y}$ in general, but it does imply $\mathbf{x} \succcurlyeq \mathbf{y}$ if the number of points is at most $(n + 1)$. For this reason, the condition in case (ii) is stronger than the continuos contractibility condition $\mathbf{x} \succcurlyeq \mathbf{y}$.

A proof of the continuous contraction case (iv) in the plane was published by B. Bollobás [13]. The proof is based on the observation that if we denote by $A(r)$ and $K(r)$ the area and the perimeter of the union of disks of radius r about a fixed system of centers, then $K = A'$. Thus, it is enough to show that if the system of the centers is contracted continuously, then the perimeter of the union of the disks of radius r about the points is decreasing weakly. The latter statement follows from a

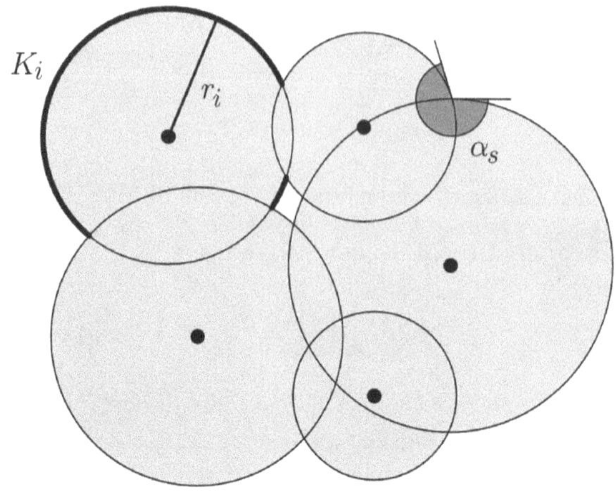

Fig. 1 The definition of K_i and the inner angles α_s

formula expressing the quotient $K(r)/r$ in terms of the inner angles and the Euler characteristic of the union of the disks (cf. Theorem 2.3 (ii) below).

The author's first contributions to the KPC were published originally in Hungarian [14] and some years later in English [15]. Finding a suitable modification of the reasoning of Bollobás, the continuous contraction case (iv) of the planar KPC was generalized to noncongruent disks.

Theorem 2.3 (B. Cs., [14, 15]) *For* $\mathbf{x} \in (\mathbb{E}^2)^k$ *and* $\mathbf{r} = (r_1, \dots, r_k) \in \mathbb{R}^k_+$ *denote by* $A(\mathbf{x}, \mathbf{r})$ *the area of the union* $B(\mathbf{x}, \mathbf{r})$, *and by* $K_i(\mathbf{x}, \mathbf{r})$ *the length of the common part of the* i*th disk and the boundary of the union* $B(\mathbf{x}, \mathbf{r})$ *if* $B(x_i, r_i)$ *does not coincide with any of the disks* $B(x_j, r_j)$ *with* $j < i$, *otherwise set* $K_i(\mathbf{x}, \mathbf{r}) = 0$. *(see Fig. 1). Then*

(i) $\dfrac{\partial A}{\partial r_i}(\mathbf{x}, \mathbf{r}) = K_i(\mathbf{x}, \mathbf{r})$, *if* $B(x_i, r_i)$ *does not coincide with any of the other disks;*

(ii) $\sum_{i=1}^k \frac{K_i(\mathbf{x},\mathbf{r})}{r_i} = 2\pi\chi + \sum_s (\alpha_s - \pi)$, *where* α_s *is running over the inner angles of the union* $B(\mathbf{x}, \mathbf{r})$, *considered to be a polygonal domain bounded by circle arcs,* χ *is the Euler characteristic of the domain* $B(\mathbf{x}, \mathbf{r})$.

(iii) If $\mathbf{z} : [0, 1] \to (\mathbb{E}^2)^k$ *is an arbitrary contracting homotopy, then the functions* $\sum_{i=1}^k \frac{K_i(\mathbf{z}(t),\mathbf{r})}{r_i}$ *and* $A(\mathbf{z}(t), \mathbf{r})$ *are weakly decreasing.*

The paper [14] presented also an example of a contraction $\mathbf{x} \succ \mathbf{y}$ such that for a given $r > 0$, $\mathbf{r} = (r, \dots, r)$, the perimeter of the union $B(\mathbf{x}, \mathbf{r})$ is less than that of the union $B(\mathbf{y}, \mathbf{r})$. In the initial configuration of the example, shown in Fig. 2, the centers are located on the two long sides of a rectangle of side lengths $2r$ and $l \gg 2r$, and on the midline of the rectangle parallel to the long sides. The points on the sides are chosen very densely, the points on the midline follow one another at equal distances $2r$. The contracted system is obtained fom the initial configuration by reflecting in the long midline of the rectangle the points lying on one of the sides of the midline. If l is much larger than $2r$, then the perimeter of the union of the disks about the

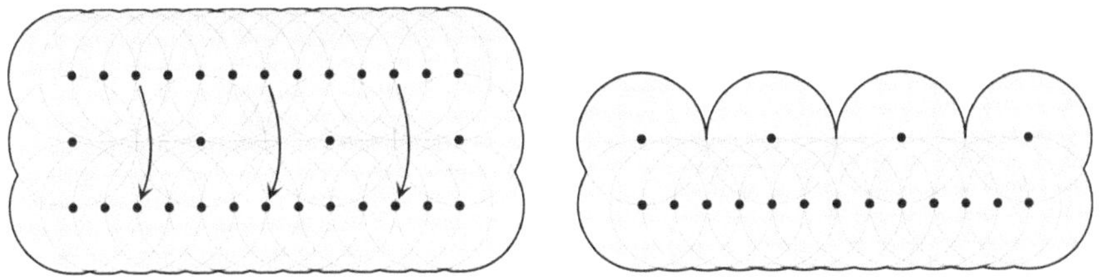

Fig. 2 An example when $\mathbf{x} \succ \mathbf{y}$, but the perimeter of $B(\mathbf{x}, \mathbf{r})$ is less than that of $B(\mathbf{y}, \mathbf{r})$. In particular, $\mathbf{x} \not\succ \mathbf{y}$

points of the initial configuration is approximately $2l$, while the perimeter after the contraction is essentially $(1 + \pi/2)l > 2l$.

In 1998 M. Bern and A. Sahai [16] published a new proof of the monotonicity of the area of the union of not necessarily congruent disks under continuous contractions of the system of the centers. The proof of Bern and Sahai involves some formulae for the derivative of the area of the union of two or three smoothly moving disks. It turned out that these formulae are special cases of a general formula, valid in any dimension, expressing the derivative of the volume of the union of a finite number of smoothly moving balls in terms of the derivatives of the distances between the centers. To present the general formula, we introduce some notations.

Let $\mathbf{x} : [0, 1] \to (\mathbb{E}^n)^k$ be a differentiable map, $\mathbf{r} \in \mathbb{R}^k_+$ be fixed. Set $V(t) = \lambda_n(B(\mathbf{x}(t), \mathbf{r}))$ and $d_{ij}(t) = d(x_i(t), x_j(t))$. If at a certain moment of time $t \in [0, 1]$ no two of the centers coincide, then denote by $W_{ij}(t)$ the wall between the Dirichlet–Voronoi (DV for short) cells of the ith and jth balls in the DV decomposition of the union $B(\mathbf{x}(t), \mathbf{r})$. $W_{ij}(t)$ is a bounded subset of the power hyperplane of the balls

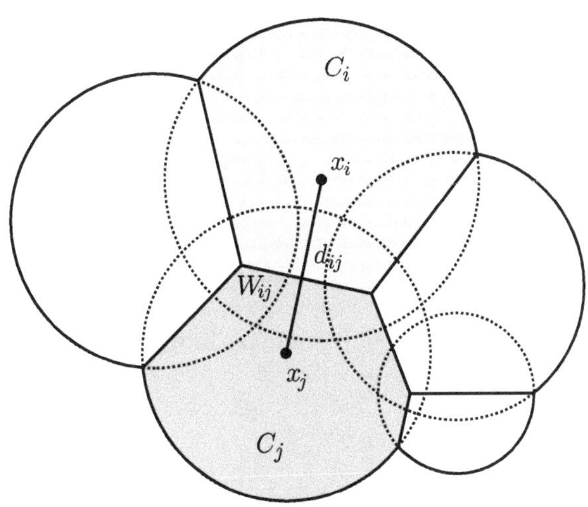

Fig. 3 DV decomposition of the union $B(\mathbf{x}, \mathbf{r})$

$B(x_i(t), r_i)$ and $B(x_j(t), r_j)$, thus it is an $(n-1)$-dimensional domain that can be empty as well (Fig. 3).

Theorem 2.4 (B. Cs. [17]) *If $n \geq 2$ and at a given moment of time $t \in (0, 1)$ no two of the moving balls have the same center, then the function V is differentiable at t, and*

$$V'(t) = \sum_{1 \leq i < j \leq k} vol_{n-1}(W_{ij}(t)) d'_{ij}(t). \tag{2}$$

The theorem implies relatively easily that in the case of a differentiable contracting homotopy, the continuous function V is differentiable everywhere except at a finite number of points, and its derivative is nonpositive, therefore V is weakly decreasing. With some extra work, one can extend this result from differentiable to continuos contractions.

Theorem 2.5 (B. Cs. [17]) *If $\mathbf{z} : [0, 1] \to (\mathbb{E}^n)^k$ is a continuous contracting homotopy, $\mathbf{r} \in \mathbb{R}^k_+$, then the function $t \mapsto \lambda_n(B(\mathbf{z}(t), \mathbf{r}))$ is weakly decreasing.*

3 Unions and Intersections of Balls in Spaces of Constant Curvature

In 1987 M. Gromov [18] proved that the KPC is true (for not necessarily congruent balls) not only in the Euclidean space $\mathbb{E}^n$ but also in the spherical space $\mathbb{S}^n$, assuming that the number of balls is at most $n + 1$. The complement of a ball in the sphere is another spherical ball and the complement of the union of the balls is the intersection of the complementary balls. Thus, the KPC in the spherical space is equivalent to the statement that if some spherical balls are rearranged in such a way that the distances between their centers do not increase, then the volume of their intersection does not decrease.

Approximating the Euclidean space with spheres of radius tending to infinity, we obtain as a corollary, that if $k \leq n + 1$ balls in $\mathbb{E}^n$ are rearranged so that their centers move closer to one another, then the volume of the intersection of the balls does not decrease.

Gromov's results were extended by Y. Gordon and M. Meyer [19] to more complicated domains built from balls using the intersection and union operations.

The elements of the lattice generated by balls with lattice operations $\cup$ and $\cap$ were called *flowers* in [19]. Flowers can be obtained as evaluations of lattice polynomials on balls. A flower can always be obtained by the following procedure. Consider a rooted tree and orient its edges so that the edges point toward the root. Call a vertex a *union vertex* if its (graph) distance from the root is even, otherwise call it an *intersection vertex*. A vertex of indegree 0 is called a *leaf* of the tree. Assign to each leaf w a ball B_w. The *evaluation of the tree with leaves labelled by balls* is a flower obtained by assigning a flower B_w to each vertex w of the tree recursively

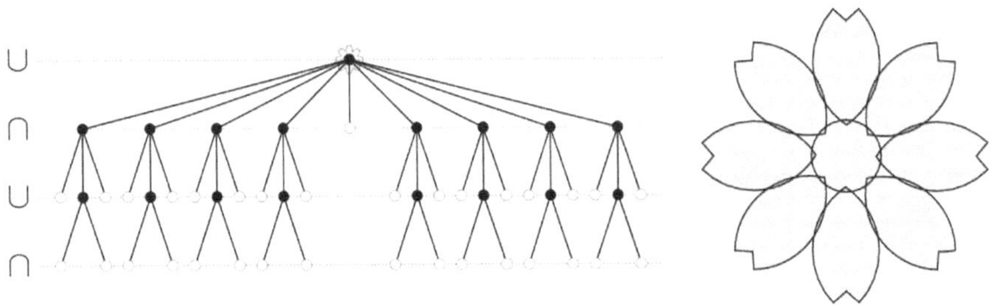

Fig. 4 A rooted tree and a flower obtained by its evaluation

as follows. The flower assigned to a leaf w is just the ball B_w. If we have already assigned a flower B_v to all vertices for which (v, w) is an edge incoming w, then we define B_w as the union or intersection of all these flowers B_v depending on whether w is a union vertex or an intersection vertex. The evaluation of the tree is the flower assigned to the root (Fig. 4).

If the leaves are labelled by distinct variable symbols $x_1, \dots, x_k$, then a similar evaluation of the tree ends up with a lattice polynomial $f(x_1, \dots, x_k)$ that is a formal lattice expression built from the variables $x_1, \dots, x_k$ and the operations $\cup$ and $\cap$. Each variable occurs in f exactly once. When the ith leaf is labelled by the ball B_i, the evaluation of the tree is the flower $f(B_1, \dots, B_k)$.

Given the lattice polynomial f, we can assign to any pair of variables a sign as follows. The paths going from the leaves labelled by x_i and x_j to the root have a first meeting point. If this first meeting point is a union vertex, then we set $\epsilon_{ij}^f = 1$, otherwise let ϵ_{ij}^f be -1.

The main result of [19] is the following.

Theorem 3.1 (Y. Gordon, M. Meyer, [19]) *Let M be $\mathbb{E}^n$ or $\mathbb{S}^n$ and $f(x_1, \dots, x_k)$ be a lattice polynomial of $k \le n+1$ variables, in which each variable occurs exactly once. Consider the volume of a flower obtained by the evaluation of f on some balls in M. If the balls are rearranged so that for all $i < j$, the distance between the centers of the ith and jth balls do not increase if $\epsilon_{ij}^f = 1$ and do not decrease if $\epsilon_{ij}^f = -1$, then the volume of the flower does not increase.*

Theorem 3.1 raises the natural question whether formula (2) and its corollaries can be extended to flowers in the spherical, Euclidean, or hyperbolic spaces. To obtain such an extension, the main difficulty is to find the right definition of the system of DV cells for a flower in one of these constant curvature spaces.

In order to handle the three different space types in a unified way, we consider their quadric models. Let $\langle\, , \rangle$ be the standard dot product on $\mathbb{R}^n$, and introduce the symmetric bilinear function

$$\{(\mathbf{x}, x_{n+1}), (\mathbf{y}, y_{n+1})\} = \langle \mathbf{x}, \mathbf{y} \rangle + \epsilon x_{n+1} y_{n+1}$$

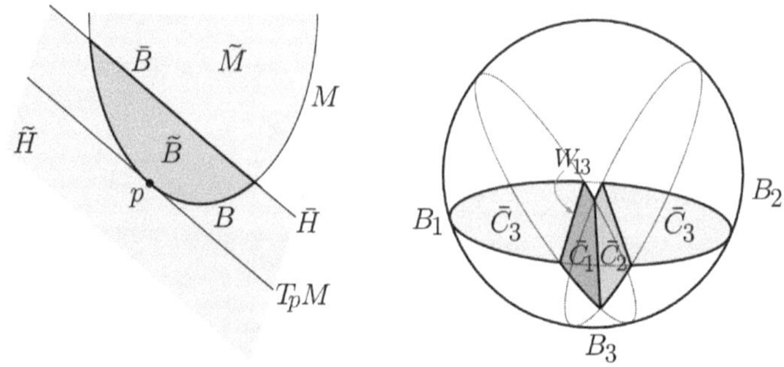

Fig. 5 Construction of the flat DVcells of a flower. (On the right: the flat DVcells of the flower $(B_1 \cup B_2) \cap B_3$)

on the linear space $\mathbb{R}^{n+1} = \mathbb{R}^n \times \mathbb{R}$, where $\epsilon \in \{-1, 0, 1\}$. Define the hypersurface M_ϵ by cases relative to ϵ. If $\epsilon = 1$, then let M_ϵ be the sphere $\{\xi \in \mathbb{R}^{n+1} \mid \{\xi, \xi\} = 1\}$. If $\epsilon = -1$, then let M_ϵ be the sheet of the two-sheeted hyperboloid $\{\xi \in \mathbb{R}^{n+1} \mid \{\xi, \xi\} = -1\}$ lying in the half-space $x_{n+1} > 0$. Finally, if $\epsilon = 0$, then let M_ϵ be the paraboloid represented by the equation $x_{n+1} = \langle \mathbf{x}, \mathbf{x} \rangle$. The restriction of the bilinear function $\{\,,\}$ to the tangent space of M_ϵ at any of its points is positive definite, therefore these restrictions provide a Riemannian metric on M_ϵ. The Riemann spaces M_1, M_0, and M_{-1} are the n-dimensional spherical, Euclidean, and hyperbolic spaces of sectional curvature $1, 0$, and -1 respectively. An isomorphism between M_0 and the standard model of the Euclidean space is given by the projection $(\mathbf{x}, x_{n+1}) \mapsto \mathbf{x} \in \mathbb{R}^n$. In the following let M denote one of the spaces M_1, M_0, M_{-1}, and let $\widetilde{M}$ be the convex hull of M.

Balls in M centered at $p \in M$ are obtained by intersecting the hypersurface M with a half-space containing p bounded by a hyperplane parallel to the tangent space T_pM of M at p. If B is the ball cut out of M by the half-space $\widetilde{H}$ bounded by the hyperplane $\bar{H}$, then $\bar{H}$ inherits a Euclidean structure from $\mathbb{R}^{n+1}$ because of $\bar{H} || T_pM$. The boundary sphere $\partial B = \bar{H} \cap M$ of B is a sphere also in the Euclidean space $\bar{H}$. We shall denote the Euclidean ball $\bar{H} \cap \widetilde{B}$ by $\bar{B}$. B and $\bar{B}$ together bound a half-lense shaped domain $\widetilde{H} \cap \widetilde{M}$ of $\mathbb{R}^{n+1}$, denote this by $\widetilde{B}$ (Fig. 5).

Let $f(x_1, \dots, x_k)$ be a lattice polynomial of k variables. In order to define DVcells for the flower $B_f = f(B_1, \dots, B_k)$ built from the balls $B_1, \dots, B_k$ of M, we have to assume some nondegeneracy conditions on the system of balls. It would be enough to assume that no two of the boundary spheres of the balls B_i coincide, but to avoid technical difficulties, we make here also the additional assumption that no three of the hyperplanes H_i corresponding to the balls B_i contain an $(n-1)$-dimensional affine subspace of $\mathbb{R}^{n+1}$.

Evaluate the polynomial f on the convex hulls $\widetilde{B}_i$ of the balls B_i. The boundary of the obtained domain $\widetilde{B}_f = f(\widetilde{B}_1, \dots, \widetilde{B}_k)$ consists of the flower B_f and some flat domains. The union of the flat pieces of the boundary is covered by the union of the flat balls $\bar{B}_i$. The contribution of the flat ball $\bar{B}_i$ to the boundary of $\widetilde{B}_f$, that is the intersection $\bar{C}_i = \bar{B}_i \cap \partial \widetilde{B}_f$ will be called the *(flat, restricted)* DV*cell of the flower corresponding to the ith ball*. We shall call the intersection $W_{ij} = \bar{C}_i \cap \bar{C}_j$ of the ith and jth cells the *wall between the ith and jth flat* DV*cells*. The wall W_{ij} lies in the subspace $\bar{H}_i \cap \bar{H}_j$, which is either an $(n-1)$-dimensional Euclidean space, or empty if $\bar{H}_i \parallel \bar{H}_j$.

We remark that in contrast to the cells of the ordinary DVdecomposition, the system of the flat DVcells itself is not a decomposition of the flower as the flat DVcells do not even live inside the space M. Nevertheless, it is not difficult to produce a decomposition of the flower in the usual sense from the flat DVcells by projecting the cells into M from the origin of $\mathbb{R}^{n+1}$ in the case of $\epsilon = \pm 1$ and parallel to the $(n+1)$st coordinate axis if $\epsilon = 0$. This projection works properly on the sphere only if the projection of the flat ball $\bar{B}_i$ is the spherical ball for all i, which happens if and only if the radius of every ball B_i is less than $\pi/2$.

Take a rooted tree with associated lattice polynomial $f(x_1, \ldots, x_k)$ and assign to each leaf a ball $B_i(t) = B(\mathbf{x}_i(t), r_i)$ of the space M, the center $\mathbf{x}_i(t)$ of which is a smooth function of the parameter $t \in (a, b)$. Denote by $V(t)$ the volume of the flower $f(B_1(t), \ldots, B_k(t))$, and by $d_{ij}(t)$ the distance between the centers $\mathbf{x}_i(t)$ and $\mathbf{x}_j(t)$. Define the signs $\epsilon^f_{ij} = \pm 1$ as in Theorem 3.1.

Theorem 3.2 (B. Cs. [20]) *Using the above notations, if $n \geq 2$ and if the balls $B_i(t_0)$ are positioned for a given $t_0 \in (a, b)$ in such a way that the intersection of any $k \leq 3$ of the hyperplanes $\bar{H}_i(t_0)$ is either empty or has dimension $(n-k)$, then the function V is differentiable at t_0 and*

$$V'(t_0) = \sum_{i<j} \epsilon^f_{ij} vol_{n-1}(W_{ij}(t_0)) d'_{ij}(t_0). \tag{3}$$

We refer to [20] for a more general version of formula (3), valid under the weaker assumption that no two of the boundary spheres of the balls B_i coincide. Theorem 3.2 implies the following theorem.

Theorem 3.3 (B. Cs. [20]) *If the balls B_i move continuously so that the signed distances $\epsilon^f_{ij} d_{ij}$ do not increase during the motion, then neither does the volume of the flower $f(B_1, \ldots, B_k)$ increase.*

It is not difficult to obtain the case of differentiable motions from Eq. (3), the continuous motion case is less trivial. Theorem 3.3 supports the following generalization of the Kneser–Poulsen conjecture.

KPC for Flowers in Constant Curvature Spaces. Let f be a lattice polynomial in k variables, in which each variable occurs exactly once, ϵ^f_{ij} be the signs defined above. Let M be the n-dimensional spherical, Euclidean, or hyperbolic space with volume function vol_n, and assume that the k-tuples of points $(x_1, \ldots, x_k), (y_1, \ldots, y_k) \in M^k$ satisfy the inequalities $\epsilon^f_{ij} d(x_i, x_j) \geq \epsilon^f_{ij} d(y_i, y_j)$ for all $1 \leq i < j \leq k$. Then for any choice of the radii $(r_1, \ldots, r_k) \in \mathbb{R}^k_+$, we have

$$\mathrm{vol}_n(f(B(x_1, r_1), \ldots, B(x_k, r_k))) \geq \mathrm{vol}_n(f(B(x_1, r_1), \ldots, B(x_k, r_k))).$$

4 Jumping into Higher Dimensions – The Leapfrog Lemma

The theorems in the previous section deal with the case of continuous or smooth motions of balls. There is a trick which, in certain cases, enables us to apply results on continuous motions of balls to discrete rearrangements. It is based on different versions of the so-called leapfrog lemma.

Leapfrog Lemma *Any two point configurations* $\mathbf{x}, \mathbf{y} \in (\mathbb{E}^n)^k$ *can be connected by an analytic homotopy* $\mathbf{z} = (z_1, \ldots, z_k) : [0, 1] \to (\mathbb{E}^{2n})^k$, $(\mathbf{z}(0) = \mathbf{x},\ \mathbf{z}(1) = \mathbf{y})$ *in* $\mathbb{E}^{2n}$ *such that all the distances* $d_{ij} = d(z_i, z_j)$ *are weakly monotonous functions on* $[0, 1]$. *(We think of* $\mathbb{E}^n$ *as a subspace of* $\mathbb{E}^m$ *for* $n \leq m$.*)*

Indeed, identify $\mathbb{E}^n$ with $\mathbb{R}^n$, $\mathbb{E}^{2n}$ with $\mathbb{R}^{2n} = \mathbb{R}^n \times \mathbb{R}^n$ fixing Cartesian coordinate systems. Then the homotopy $z_i(t) = (\cos(t\pi/2)x_i, \sin(t\pi/2)y_i)$ connects the configuration $\mathbf{x}$ embedded into $\mathbb{R}^n \times \{\mathbf{0}\}$ to the configuration $\mathbf{y}$ embedded into $\{\mathbf{0}\} \times \mathbb{R}^n$ with weakly monotonous distances between the points.

The dimension $2n$ in the lemma is sharp. H. Cheng, S. P. Tan, and Y. Zheng [21] showed that for any $n \geq 2$, there exist configurations $\mathbf{x}, \mathbf{y} \in (\mathbb{E}^n)^{(n+1)^2}$ of $(n+1)^2$ points in $\mathbb{E}^n$ such that $\mathbf{x} \succ \mathbf{y}$, but there is no continuous contraction from $\mathbf{x}$ to $\mathbf{y}$ in $\mathbb{E}^{2n-1}$. However, we have the following

Leapfrog Lemma for Few Points *If* $k \leq 2n$, *then for any two point configurations* $\mathbf{x}, \mathbf{y} \in (\mathbb{E}^n)^k$, *there is an analytic homotopy* $\check{\mathbf{z}} = (\check{z}_1, \ldots, \check{z}_k)\colon [0, 1] \to (\mathbb{E}^{k-1})^k$ *in* $\mathbb{E}^{k-1}$ *such that* $\check{\mathbf{z}}(0)$ *is congruent to* $\mathbf{x}$, $\check{\mathbf{z}}(1)$ *is congruent to* $\mathbf{y}$, *and all the distances* $d_{ij} = d(\check{z}_i, \check{z}_j)$ *are weakly monotonous functions on* $[0, 1]$.

Proof We may assume without loss of generality that $x_k = y_k$. Choose a Cartesian coordinate system with origin at x_k and identify $\mathbb{E}^n$ with $\mathbb{R}^n$ with the help of it. Then $z_k \equiv \mathbf{0}$ for the analytic homotopy constructed in the proof of the leapfrog lemma. Up to congruence, the configuration $\mathbf{z}(t)$ is uniquely determined by the Gram matrix $M(t) = (\langle z_i(t), z_j(t)\rangle)_{i,j=1}^{k-1}$. By the perturbation theory of symmetric operators, there exist real analytic functions $\lambda_i\colon \mathbb{R} \to \mathbb{R}$ and $\mathbf{e}_i\colon \mathbb{R} \to \mathbb{R}^{k-1}$ for $i = 1, \ldots, k-1$ such that the *column* vectors $\mathbf{e}_1(t), \ldots, \mathbf{e}_{k-1}(t)$ form an orthonormal basis consisting of eigenvectors of the matrix $M(t)$ and $M(t)\mathbf{e}_i(t) = \lambda_i(t)\mathbf{e}_i(t)$ for all $1 \leq i \leq k-1$ and $t \in \mathbb{R}$ (see [22, Chap. 2. Sect. 6.]). Since $M(t)$ is positive semidefinite, $\lambda_i \geq 0$ for all i. A real analytic nonnegative function on $\mathbb{R}$ always has a real analytic square root function, in particular, we can find real analytic functions $\mu_i\colon \mathbb{R} \to \mathbb{R}$ such that $\mu_i^2 = \lambda_i$. (The sign of μ_i should be chosen so that μ_i changes sign exactly at the roots of λ_i having multiplicity 2 modulo 4.) The matrix M has the spectral decomposition

$$M = \sum_{i=1}^{k-1} \lambda_i \mathbf{e}_i \mathbf{e}_i^T = \sum_{i=1}^{k-1} (\mu_i \mathbf{e}_i)(\mu_i \mathbf{e}_i)^T.$$

This means, that if we denote by $\check{z}_i(t)$ the ith row of the matrix the columns of which are $\mu_1(t)\mathbf{e}_1(t), \dots, \mu_{k-1}(t)\mathbf{e}_{k-1}(t)$, then $\langle \check{z}_i(t), \check{z}_j(t)\rangle = \langle z_i(t), z_j(t)\rangle$ for all $1 \le i, j \le k-1$. Thus, the real analytic homotopy $\check{\mathbf{z}} = (\check{z}_1, \dots, \check{z}_{k-1}, \mathbf{0})\colon [0,1] \to (\mathbb{R}^{k-1})^k$ will satisfy the requirements. □

Augmenting spherical configurations with the center of the sphere, the Euclidean leapfrog lemmata imply a leapfrog lemma for spherical configurations.

Spherical Leapfrog Lemma *Let* $\mathbf{x}, \mathbf{y} \in (\mathbb{S}^n)^k$ *be two configurations of* k *points in the sphere* $\mathbb{S}^n$, $m = \min\{2n+1, k-1\}$. *Then there is an analytic homotopy* $\mathbf{z} = (z_1, \dots, z_k) : [0,1] \to (\mathbb{S}^m)^k$, *such that* $\mathbf{z}(0)$ *is congruent to* $\mathbf{x}$, $\mathbf{z}(1)$ *is congruent to* $\mathbf{y}$, *and all the distances* $d_{ij} = d(z_i, z_j)$ *are weakly monotonous functions on* $[0,1]$.

It is not known whether the dimension $2n+1$ in the spherical leapfrog lemma can be lowered. As for the hyperbolic space, the following question seems to be open.

Question 4.1 *Is there a natural number* $m(n,k)$ *for any pair of natural numbers* n *and* k *such that any configuration* $\mathbf{x} \in (\mathbb{H}^n)^k$ *and any of its contraction* $\mathbf{y} \in (\mathbb{H}^n)^k$ *can be connected by a contracting homotopy in* $\mathbb{H}^{m(n,k)}$? *If yes, then what is the minimal value of* $m(n,k)$ *for given* n *and* k?

A typical application of the leapfrog lemma goes along the following scheme. If we want to prove that a function $\Phi\colon (\mathbb{E}^n)^k \to \mathbb{R}$ defined on point configurations in $\mathbb{E}^n$ has the monotonicity property that $\mathbf{x} \succ \mathbf{y}$ implies $\Phi(\mathbf{x}) \ge \Phi(\mathbf{y})$, it suffices to find an extension $\widetilde{\Phi}$ of Φ to point configurations lying in $\mathbb{E}^{2n}$, for which $\mathbf{x} \succcurlyeq \mathbf{y}$ implies $\widetilde{\Phi}(\mathbf{x}) \ge \widetilde{\Phi}(\mathbf{y})$.

R. Alexander [23] and V. Capoyleas and J. Pach [24] used the leapfrog lemma method to show that the perimeter of the convex hull of a set of points in the Euclidean plane, or more generally, the mean width of a point set in $\mathbb{E}^n$ cannot increase when the set is contracted. The relation of the monotonicity of the mean width to the KPC is illuminated by the paper [24], the results of which were sharpened by I. Gorbovickis [25].

Theorem 4.1 (I. Gorbovickis [25]) *Let* $X = \{x_1, \dots, x_k\} \subset \mathbb{E}^n$ *be a finite set in the Euclidean space. Then the volume functions* $V_\cup(r) = \lambda_n\big(\bigcup_{i=1}^k B(x_i, r)\big)$ *and* $V_\cap(r) = \lambda_n\big(\bigcap_{i=1}^k B(x_i, r)\big)$ *are piecewise analytic on the halfline* $[0,\infty)$. *If* r *is large, then*

$$V_\cup(r) = \omega_n\big(r^n + \frac{n}{2} w_n(X) r^{n-1}\big) + O(r^{n-2}) \text{ and } V_\cap(r) = \omega_n\big(r^n - \frac{n}{2} w_n(X) r^{n-1}\big) + O(r^{n-2}),$$

where $w_n(X)$ *is the mean value of the widths of the set* X *in all directions of* $\mathbb{E}^n$.

For a planar set X, $\pi w_2(X)$ is the perimeter of the convex hull of X. If $n < m$ and $X \subset \mathbb{E}^n \subset \mathbb{E}^m$, then $\frac{n\omega_n}{\omega_{n-1}} w_n(X) = \frac{m\omega_m}{\omega_{m-1}} w_m(X)$. This equation, the leapfrog lemma, Theorems 2.5 and 4.1 imply that if a set $Y \subset \mathbb{E}^n$ is a contraction of the finite set $X \subset \mathbb{E}^n$, then

$$w_n(X) = \frac{2\omega_{n-1}\omega_{2n}}{\omega_n\omega_{2n-1}} w_{2n}(X) \geq \frac{2\omega_{n-1}\omega_{2n}}{\omega_n\omega_{2n-1}} w_{2n}(Y) = w_n(Y).$$

Using rigidity theory, Gorbovickis proved that in fact $w_n(X) > w_n(Y)$ if X and Y are not congruent. This sharp inequality together with Theorem 4.1 yields the following corollary.

Corollary 4.1 (I. Gorbovickis [25]) *If* $\mathbf{x}, \mathbf{y} \in (\mathbb{E}^n)^k$ *are two fixed noncongruent configurations of points such that* $\mathbf{x} \succ \mathbf{y}$, *then for any sufficiently large* r, *the volume and the surface volume of* $\bigcup_{i=1}^k B(x_i, r)$ *is strictly bigger than the volume and the surface volume of* $\bigcup_{i=1}^k B(y_i, r)$ *respectively, furthermore, the volume and the surface volume of* $\bigcap_{i=1}^k B(y_i, r)$ *is strictly bigger than the volume and the surface volume of* $\bigcap_{i=1}^k B(x_i, r)$ *respectively.*

We remark that in contrast to the mean width, the width of a contraction of a set can be larger than the width of the original set. However, D. Gale [26] proved that if $f: C \to C'$ is a contracting homeomorphism between the compact convex sets C and C', then the width of C' is not greater than the width of C.

As another application of the leapfrog lemma, we prove the following monotonicity property of the Minkowski content.

Proposition 4.1 *Assume that* $A \subset \mathbb{E}^n$ *is a* μ_n^s*-measurable bounded set, where* $0 \leq s \leq n$. *If* B *is a contraction of* A *and* A *is a contraction of the bounded set* C, *then* $\underline{\mu}_n^s(B) \leq \mu_n^s(A) \leq \bar{\mu}_n^s(C)$.

Proof A bounded set $K \subset \mathbb{E}^n$ has a lower and upper Minkowski content both as a subset of $\mathbb{E}^n$ and as a subset of $\mathbb{E}^m$ for any $m > n$. These numbers are known to satisfy the inequalities

$$\underline{\mu}_n^s(K) \leq \underline{\mu}_m^s(K) \leq \bar{\mu}_m^s(K) \leq \bar{\mu}_n^s(K),$$

where all the inequalities are strict if K is not μ_n^s-measurable and all of them become equalities if K is μ_n^s-measurable (see [2]).

Then the leapfrog lemma, the definition of the Minkowski content and Theorem 2.5 give

$$\underline{\mu}_n^s(B) \leq \underline{\mu}_{2n}^s(B) \leq \underline{\mu}_{2n}^s(A) = \mu_n^s(A) = \bar{\mu}_{2n}^s(A) \leq \bar{\mu}_{2n}^s(C) \leq \bar{\mu}_n^s(C).$$

□

The following Kneser–Poulsen-type result was also proved by an application of the leapfrog lemma.

Theorem 4.2 (K. Bezdek, R. Connelly [27]) *If some hemispheres (i.e., balls of radius* $\pi/2$ *of the unit sphere* $\mathbb{S}^n$ *are rearranged in such a way that the distances between the centers of the hemispheres do not increase, then the volume of the spherical polytope obtained as the intersection of the hemispheres does not decrease.*

Proof If for a spherical configuration $\mathbf{x} = (x_1, \ldots, x_k) \in \mathbb{S}^n$, we denote by $\Phi_n(\mathbf{x})$ the volume of the intersection of the n-dimensional hemispheres centered at the points $x_1, \ldots, x_k$, then, thinking of $\mathbb{S}^n$ as a great subsphere in $\mathbb{S}^m$ for $n < m$, we have $\Phi_n(\mathbf{x}) = \frac{(n+1)\omega_{n+1}}{(m+1)\omega_{m+1}} \Phi_m(\mathbf{x})$. As for $\mathbf{x}, \mathbf{y} \in \mathbb{S}^n$, the relation $\mathbf{x} \succ \mathbf{y}$ implies $\mathbf{x} \succcurlyeq \mathbf{y}$ in $\mathbb{S}^{2n+1}$, Theorem 3.3 yields

$$\Phi_n(\mathbf{x}) = \frac{\omega_{n+1}}{2\omega_{2n+2}} \Phi_{2n+1}(\mathbf{x}) \geq \frac{\omega_{n+1}}{2\omega_{2n+2}} \Phi_{2n+1}(\mathbf{y}) = \Phi_n(\mathbf{y}).$$

□

K. Bezdek [28] proved an analogue of Theorem 4.2 for nonobtuse-angled polytopes of the hyperbolic space.

Theorem 4.3 (K. Bezdek [28]) *Let P and Q be nonobtuse-angled compact convex polyhedra of the same simple combinatorial type in hyperbolic 3-space. If each (inner) dihedral angle of Q is at least as large as the corresponding (inner) dihedral angle of P, then the volume of P is at least as large as the volume of Q. This result holds also for nonobtuse-angled hyperbolic simplices of any dimension.*

The proof does not require a leapfrog into higher dimensions since under the assumptions of the theorem, P can be deformed to Q through polytopes of the same combinatorial type with weakly decreasing dihedral angles. Then Schläfli's formula completes the proof.

5 Proof of the Kneser–Poulsen Conjecture in the Euclidean Plane

It was a breakthrough in the topic when K. Bezdek and R. Connelly [29] succeeded to prove the KPC in the Euclidean plane even for noncongruent disks. They also proved the monotonicity of the area of the intersection of the disks.

For the sake of simplicity, we outline only the case of unions. The proof rests on three main ideas. The first idea is that the monotonicity of the weighted perimeter of the union of the disks, presented in Theorem 2.3 (iii) can be extended to any dimension.

Theorem 5.1 (K. Bezdek, R. Connelly [29]) *For $\mathbf{x} \in (\mathbb{E}^n)^k$, $\mathbf{r} \in \mathbb{R}^k_+$, and $1 \leq i \leq k$, denote by $K_i(\mathbf{x}, \mathbf{r})$ the $(n-1)$-dimensional volume of the contribution $\partial B(\mathbf{x}, \mathbf{r}) \cap \partial B(x_i, r_i)$ of the ith ball to the boundary of the union of the balls. Then for any differentiable contracting homotopy $\mathbf{z} : [0, 1] \to (\mathbb{E}^n)^k$, during which no two balls coincide, the weighted surface volume $\sum_{i=1}^k K_i(\mathbf{z}, \mathbf{r})/r_i$ is weakly decreasing.*

The proof of this theorem makes use of the following lemma relating the volume of the union of the balls to the weighted surface volume of the union.

Lemma 5.1 *For a given system of radii* $\mathbf{r} = (r_1, \ldots, r_k) \in \mathbb{R}^k_+$, *define the variation* $\tilde{\mathbf{r}} = (\tilde{r}_1, \ldots, \tilde{r}_k)\colon \mathbb{R}_+ \to \mathbb{R}^k_+$ *by* $\tilde{r}_i(s) = \sqrt{r_i^2 + s}$. *Then for any* $\mathbf{x} \in (\mathbb{E}^n)^k$ *for which the balls* $B(x_i, r_i)$ *are pairwise distinct, we have*

$$\frac{\mathrm{d}}{\mathrm{d}s}\left(\lambda_n\left(\bigcup_{i=1}^{k} B(\mathbf{x}_i, \tilde{r}_i(s))\right)\right) = \frac{1}{2}\sum_{i=1}^{k}\frac{1}{\tilde{r}_i(s)}K_i(\mathbf{x}, \tilde{\mathbf{r}}(s)).$$

The second important observation is that the volume of the union of some n-dimensional balls with centers in $\mathbb{E}^n$ is proportional to the weighted surface volume of the union of the $(n+2)$-dimensional balls about the same centers.

Theorem 5.2 (K. Bezdek, R. Connelly [29]) *Let* $\mathbb{E}^n$ *be an n-dimensional subspace of the* $(n+2)$*-dimensional Euclidean space* $\mathbb{E}^{n+2}$, $\mathbf{x} \in (\mathbb{E}^n)^k$. *Denote by* $V^n(\mathbf{x}, \mathbf{r})$ *the volume of the union* $B^n(\mathbf{x}, \mathbf{r})$ *of the n-dimensional balls* $B^n(x_i, r_i) \subset \mathbb{E}^n$, *and let* $\sum_{i=1}^{k} K_i^{n+2}(\mathbf{x}, \mathbf{r})/r_i$ *be the weigthed surface volume of the union* $B^{n+2}(\mathbf{x}, \mathbf{r})$ *of the* $(n+2)$*-dimensional balls* $B^{n+2}(x_i, r_i) \subset \mathbb{E}^{n+2}$. *Then we have*

$$V^n(\mathbf{x}, \mathbf{r}) = \frac{1}{2\pi}\sum_{i=1}^{k}\frac{K_i^{n+2}(\mathbf{x}, \mathbf{r})}{r_i}$$

if the balls are pairwise distinct.

Theorems 5.1 and 5.2 show that the KPC is true for any two configurations of balls in $\mathbb{E}^n$, for which the centers can be connected by a contracting homotopy in $\mathbb{E}^{n+2}$. In fact, the proofs can also be applied to flowers and give the following.

Theorem 5.3 *Let f be a lattice polynomial in k variables, in which each variable occurs exactly once,* ϵ_{ij}^f *be the signs defined above. Assume that the k-tuples of points* $(x_1, \ldots, x_k), (y_1, \ldots, y_k) \in (\mathbb{E}^n)^k$ *satisfy the inequalities* $\epsilon_{ij}^f d(x_i, x_j) \geq \epsilon_{ij}^f d(y_i, y_j)$ *for all* $1 \leq i < j \leq k$, *and can be connected by a piecewise analytic continuos homotopy* $(z_1, \ldots, z_k)\colon [0, 1] \to (\mathbb{E}^{n+2})^k$ *so that* $z_i(0) = x_i$, $z_i(1) = y_i$ *and the distances* $d(z_i(t), z_j(t))$ *are weakly monotonous function for all* $1 \leq i, j \leq k$. *Then for any choice of the radii* $(r_1, \ldots, r_k) \in \mathbb{R}^k_+$, *we have*

$$\mathrm{vol}_n(f(B(x_1, r_1), \ldots, B(x_k, r_k))) \geq \mathrm{vol}_n(f(B(x_1, r_1), \ldots, B(x_k, r_k))).$$

The third key idea is that we can apply the leapfrog lemma to the problem. As for $n = 2$ the leapfrog lemma provides an analytic homotopy required in Theorem 5.3, the planar case of the conjecture follows (even for flowers and for noncongruent disks). We remark that Theorems 5.1–5.3 will be generalized to the spherical and hyperbolic spaces in Sect. 8. So the reason why the KPC is not fixed in $\mathbb{S}^2$ and $\mathbb{H}^2$ is the lack of the leapfrog lemma for these spaces jumping only into a four-dimensional space.

Applying the leapfrog lemma for few points, we obtain the following extension of Theorems 2.2(ii) and 3.1.

Theorem 5.4 *The generalized KPC for flowers is true in $\mathbb{E}^n$ and $\mathbb{S}^n$ if the number of the balls is at most $n+3$.*

I. Gorbovickis [30] observed, that according to Theorem 5.2 and Lemma 5.1, if $\mathbf{r} \in \mathbb{R}_+^k$ is fixed, then embedding $\mathbb{E}^n$ into $\mathbb{E}^{n+2}$, the functional $\Phi(\mathbf{x}) = V^n(\mathbf{x}, \mathbf{r})$ defined for $\mathbf{x} \in (\mathbb{E}^n)^k$, can also be written as $\Phi(\mathbf{x}) = \frac{1}{\pi}\frac{\mathrm{d}}{\mathrm{d}s}V^{n+2}(\mathbf{x}, \tilde{\mathbf{r}}(s))|_{s=0}$. Iterating this formula m times we see that the functional

$$\tilde{\Phi}(\mathbf{x}) = \frac{1}{\pi^m}\left(\frac{\mathrm{d}}{\mathrm{d}s}\right)^m V^{n+2m}(\mathbf{x}, \tilde{\mathbf{r}}(s))\bigg|_{s=0}$$

extends Φ to arbitrary configurations lying in $\mathbb{E}^{n+2m}$. Choosing $m \geq n/2$, the general philosophy of the leapfrog lemma tells us that the KPC would be true if the functional $\tilde{\Phi}$ were monotonous under continuous contractions. Unfortunately, this does not seem to be the case in general. However, one can prove monotonictiy of $\tilde{\Phi}$ under continuous contractions during which the balls do not intersect heavily.

Theorem 5.5 (I. Gorbovickis [30]) *The KPC is true for the union of not necessarily congruent balls in the Euclidean space $\mathbb{E}^n$, if in the contracted configuration of the balls, the intersection of any two balls has common point with no more than $n+1$ other balls.*

6 Monotonicity of the Volume of Weighted Flowers

If a Kneser–Poulsen type inequality holds for flowers with two given configurations of the centers and any choice of the radii of the balls, then it implies further inequalities between the integrals of certain weight functions depending on the distances of a point from the centers.

Definition 6.1 A *weight function* on a set M is a map $m\colon M \to \mathbb{R}_+$.

Weight functions are generalizations of fuzzy subsets of M, for which the weight function, called the *membership function*, takes values in $[0, 1]$. An ordinary (crisp) subset $S \subseteq M$ is identified with the fuzzy subset weighted by the indicator function χ_S of S.

Definition 6.2 A *weighted ball* $\mathcal{B}(p, r)$ centered at p in a metric space (M, d) is a weight function of the form $[\mathcal{B}(p, r)](x) = r(d(x, p))$, where $r\colon \mathbb{R}_+ \to \mathbb{R}_+$ is a decreasing function.

There are several ways to define lattice operations on weight functions. In general, a k-ary operation $\bar{Q}$ on weight functions is given by a k-ary operation $Q\colon \mathbb{R}_+^k \to \mathbb{R}_+$ on the half-line $\mathbb{R}_+$. Given Q, the result of the operation $\bar{Q}$ on the weight functions $m_1, \ldots, m_k$ is the weight function $Q \circ (m_1, \ldots, m_k)$.

We can construct weighted analogues of a lattice polynomial $f(x_1, \ldots, x_k)$ in k variables as follows. For each index $1 \le i \le k$, and each nonnegative number m, define the subset $S_i(m) \subset \mathbb{R}_+^k$ by

$$S_i(m) = \{(x_1, \ldots, x_k) \in \mathbb{R}_+^k \mid x_i \le m\}.$$

Fix a Borel measure μ on the space $\mathbb{R}_+^k$ and consider the map $Q_{f,\mu} \colon \mathbb{R}_+^k \to \mathbb{R}_+ \cup \{\infty\}$ given by

$$Q_{f,\mu}(m_1, \ldots, m_k) = \mu(f(S_1(m_1), \ldots, S_k(m_k))).$$

If μ has the property that $Q_{f,\mu}$ takes only finite values, then $Q_{f,\mu}$ induces a k-ary operation $\bar{Q}_{f,\mu}$ on weight functions.

Theorem 6.1 *Let (M, d) be a metric space with a σ-finite Borel measure ν such that spheres of M have ν-measure 0. Assume that f is a lattice polynomial in k variables, and that $(x_1, \ldots, x_k)$, $(y_1, \ldots, y_k) \in M^k$ are two configurations of centers, for which the inequality*

$$\nu(f(B(x_1, \rho_1), \ldots, B(x_k, \rho_k))) \ge \nu(f(B(y_1, \rho_1), \ldots, B(y_k, \rho_k))) \tag{4}$$

holds for any choice of the radii $(\rho_1, \ldots, \rho_k) \in \mathbb{R}_+^k$. Then for any σ-finite Borel measure μ on $\mathbb{R}_+^k$ for which the operation $\bar{Q}_{f,\mu}$ on weight functions is properly defined, we have the inequality

$$\int_M \bar{Q}_{f,\mu}(\mathcal{B}(x_1, r_1), \ldots, \mathcal{B}(x_k, r_k))\, \mathrm{d}\nu \ge \int_M \bar{Q}_{f,\mu}(\mathcal{B}(y_1, r_1), \ldots, \mathcal{B}(y_k, r_k))\, \mathrm{d}\nu \tag{5}$$

for any choice of the decreasing functions $r_i \colon \mathbb{R}_+ \to \mathbb{R}_+$.

Proof The lattice polynomial f can be evaluated on the logical values TRUE and FALSE using the lattice operations $\cup =$ OR and $\cap =$ AND. This way, identifying the logical values TRUE and FALSE with 1 and 0 respectively, f defines a logical operator $f^{\vee} \colon \{0, 1\}^k \to \{0, 1\}$. Using the definition of $\bar{Q}_{f,\mu}$, we have

$$\begin{aligned}
&\int_M \bar{Q}_{f,\mu}(\mathcal{B}(x_1, r_1), \ldots, \mathcal{B}(x_k, r_k))\, \mathrm{d}\nu \\
&\qquad = \int_M \mu(f(S_1(r_1(d(x, x_1))), \ldots, S_k(r_k(d(x, x_k)))))\, \mathrm{d}\nu(x) \\
&\qquad = \int_M \int_{\mathbb{R}_+^k} f^{\vee}([\rho_1 \le r_1(d(x, x_1))], \ldots, [\rho_k \le r_k(d(x, x_k))])\, \mathrm{d}\mu(\boldsymbol{\rho})\, \mathrm{d}\nu(x),
\end{aligned}$$

where $\boldsymbol{\rho} = (\rho_1, \ldots, \rho_k)$, and the inequality $[\rho_i \le r_i(d(x, x_i))]$ in the place of the ith variable of $f^{\vee}$ should be evaluated as 1 when the inequality is true and as 0 otherwise.

Define the function $g_i \colon \mathbb{R}_+ \to \mathbb{R}_+ \cup \{\infty\}$ by $g_i(a) = \sup\{b \in \mathbb{R}_+ \mid a \le r_i(b)\} \cup \{0\}$, (see Fig. 6). Then inequality $\rho_i \le r_i(d(x, x_i))$ is equivalent to the inequality

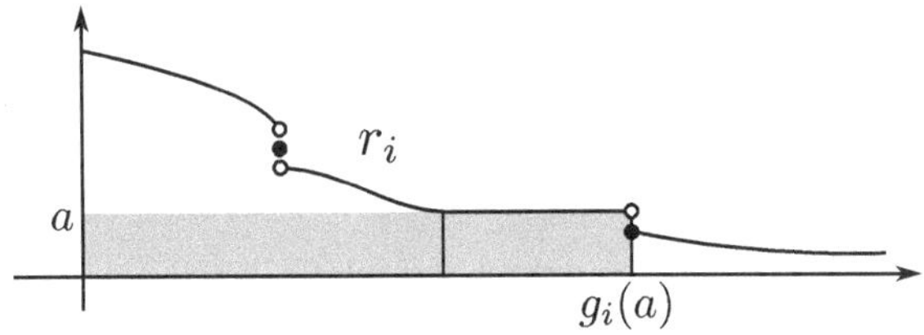

Fig. 6 The definition of the function g_i

$g_i(\rho_i) \geq d(x, x_i)$ if $r_i(g_i(\rho_i)) = \rho_i$ and to the strict inequality $g_i(\rho_i) > d(x, x_i)$ if $r_i(g_i(\rho_i)) < \rho_i$ if $r_i(g_i(\rho_i)) < \rho_i$. The integrand in the last integral is the indicator function of a Borel subset of $M \times \mathbb{R}_+^k$, and since the ν-measures of spheres in M are equal to 0, the integrand is equal to $f^\vee([d(x, x_1) \leq g_1(\rho_1)], \ldots, [d(x, x_k) \leq g_k(\rho_k])$ almost everywhere. From Fubini's theorem we obtain

$$\begin{aligned}
&\int_M \bar{Q}_{f,\mu}(\mathcal{B}(x_1, r_1), \ldots, \mathcal{B}(x_k, r_k))\, \mathrm{d}\nu \\
&\quad = \int_{\mathbb{R}_+^k} \int_M f^\vee([d(x, x_1) \leq g_1(\rho_1)], \ldots, [d(x, x_k) \leq g_k(\rho_k])\, \mathrm{d}\nu(x)\, \mathrm{d}\mu(\boldsymbol{\rho}) \\
&\quad = \int_{\mathbb{R}_+^k} \nu(f(B(x_1, g_1(\rho_1)), \ldots, B(x_k, g_k(\rho_k))))\, \mathrm{d}\mu(\boldsymbol{\rho}).
\end{aligned} \tag{6}$$

Inequality (5) follows easily from formula (6) and inequality (4).

We mention two special cases of the theorem.

- If $f(x_1, \ldots, x_k) = x_1 \cap \cdots \cap x_k$, and μ is the restriction of the Lebesgue measure onto $\mathbb{R}_+^k$, then $Q_{f,\mu}(m_1, \ldots, m_k) = \prod_{i=1}^k m_i$. In this case, Theorem 6.1 provides a Slepian type inequality, the special case $r_1 = \cdots = r_k$ of which appears in the paper [18] by M. Gromov.
- Denote by Δ the diagonal half-line $\Delta = \{(x_1, \ldots, x_k) \in \mathbb{R}_+^k \mid x_1 = \cdots = x_k\}$ in $\mathbb{R}_+^k$ and by λ_1 the 1-dimensional Lebesgue measure on it. For a Borel subset $B \subset \mathbb{R}_+^k$, set $\mu(B) = \lambda_1(B \cap \Delta)/\sqrt{k}$. Then for any lattice polynomial f in k variables, we have $Q_{f,\mu}(m_1, \ldots, m_k) = f(m_1, \ldots, m_k)$, where on the right hand side f is evaluated on nonnegative numbers by the operations $a \cup b = \max\{a, b\}$ and $a \cap b = \min\{a, b\}$. This special case of Theorem 6.1 was considered by K. Bezdek and R. Connelly [31].

7 A Schläfli-Type Formula for Polytopes with Curved Faces

The classical Schläfli formula says that the derivative of the volume V of a varying simplex in an n-dimensional Euclidean, spherical or hyperbolic space of constant sectional curvature K can be expressed by the formula

$$(n-1)KV' = \sum_{1\le i<j\le n+1} \mathrm{vol}_{n-1}(W_{ij})\alpha'_{ij}, \tag{7}$$

where W_{ij} is the intersection of the ith and jth facets, α_{ij} is the dihedral angle of the simplex at W_{ij}.

This formula resembles formulae (2) and (3). Extending results of I. Rivin and J. M. Schlenker [32] and R. Souam [33], the author [34] proved a Schläfli-type formula for polytopes with curved faces in pseudo-Riemannian Einstein manifolds, which implies formulae (2), (3), and (7) as special cases, and gave some important applications of the formula to Kneser–Poulsen-type problems. We briefly review the content of this paper, but for the sake of simplicity, we restrict our attention to the Riemannian case.

7.1 *Polytopes with Curved Faces in a Manifold and Their Variations*

Intuitively, a polytope with curved faces in a manifold is a compact domain bounded by a finite number of smooth hypersurfaces. This picture can be captured by many different mathematical models. Having in mind flowers built from balls, the approach of Constructive Solid Geometry (CSG) seems to be the most practical one. Following this approach, we define polytopes with curved faces as compact sets that can be obtained as the result of the application of regularized Boolean operations to regular domains bounded by smooth hypersurfaces of a manifold.

The *regular closure* of a subset $A \subset X$ of a topological space X is the set $\rho_X(A) = \overline{\mathrm{int}(A)}$. The regularized Boolean operations on the subsets of X are defined by

$$A \cup^* B = \rho_X(A \cup B), \quad A \cap^* B = \rho_X(A \cap B), \quad A \setminus^* B = \rho_X(A \setminus B).$$

A Boolean expression f is a formal expression (a finite sequence of symbols) such that f is either a symbol denoting a variable or has the form $(f_1 * f_2)$, where f_1 and f_2 are Boolean expressions, $*$ is one of the operation symbols $\cup$, $\cap$, $\setminus$. Every Boolean expression corresponds to a regularized Boolean expression f^* obtained from f by replacing the operation symbols with their regularized versions. Two Boolean expressions are considered to be the same if they are equal as symbol sequences. If $f(x_1, \ldots, x_k)$ is a Boolean expression of k variables, $A_1, \ldots, A_k \subseteq X$ are subsets of X, then $f(A_1, \ldots, A_k)$ and $f^*(A_1, \ldots, A_k)$ will denote the evaluations of the expression f and f^* on the sets $A_1, \ldots, A_k$. If the subsets $A_1, \ldots, A_k$ of X are contained in the subset Y of X, then we shall denote by $f^*_Y(A_1, \ldots, A_k)$ the evaluation of f^* in Y, in which the Boolean operations are regularized by the operator ρ_Y instead of ρ_X.

Let M be an n-dimensional smooth manifold. A *regular domain* in M is a set of the form $\{p \in M \mid f(p) \leq 0\}$, where $f : M \to \mathbb{R}$ is a smooth function for which 0 is a regular value.

A *CSG solid* in M is a compact subset that can be obtained as the evaluation of a regularized Boolean expression on some regular domains of M. A *CSG representation* of a CSG solid P is a regularized Boolean expression $f^*(x_1, \dots, x_k)$ together with a system of regular domains $P_1, \dots, P_k$, called the *primitive objects*, for which $P = f^*(P_1, \dots, P_k)$. The CSG representation of a CSG solid is not unique.

Definition 7.1 A *polytope with curved faces lying in a smooth manifold M* or simply a *polytope* is a CSG solid given by a fixed CSG representation $f^*(P_1, \dots, P_k)$, such that each of the variables $x_1, \dots, x_k$ occurs exactly once in $f^*(x_1, \dots, x_k)$, and the primitives P_i are regular domains in M, satisfying that any $l \leq 3$ of the hypersurfaces ∂P_i, $(1 \leq i \leq k)$ intersect transversally.

Definition 7.2 Let $P = f^*(P_1, \dots, P_k)$ be a polytope in M. The *ith facet* F_i of P is the regularized contribution $F_i = \rho_{\Sigma_i}(\Sigma_i \cap \partial P)$ of the hypersurface $\Sigma_i = \partial P_i$ to the boundary of P.

The relative boundary of the ith facet of the polytope P is covered by the intersection manifolds $\Sigma_i \cap \Sigma_j$, $j \neq i$. If $j \neq i$, then the *wall* W_{ij} *between the facets* F_i *and* F_j is the regularized contribution of the submanifold $\Sigma_i \cap \Sigma_j$ to the relative boundary of the facet F_i, that is $W_{ij} = \rho_{\Sigma_i \cap \Sigma_j}(\Sigma_i \cap \Sigma_j \cap \partial_{\Sigma_i} F_i)$.

W_{ij} is an $(n-2)$-dimensional CSG solid in $\Sigma_i \cap \Sigma_j$. It can be checked that $W_{ij} = W_{ji}$.

Assume that the manifold M is equipped with a Riemannian metric $\{\,,\}$. Then there is a properly defined exterior smooth unit normal vector field $\mathbf{N}_i$ along the ith facet F_i of the polytope $P = f^*(P_1, \dots, P_k)$, furthermore, there is a uniquely defined exterior unit normal vector field $\mathbf{n}_{ij} \in \Gamma(T\Sigma_i|_{W_{ij}})$ of F_i along the wall W_{ij}.

The dihedral angle of the polytope P along the wall W_{ij} is the smooth function $\alpha_{ij} : W_{ij} \to (0, 2\pi)$ defined by

$$\mathbf{n}_{ji} = \cos(\alpha_{ij})\mathbf{n}_{ij} + \sin(\alpha_{ij})\mathbf{N}_i.$$

Let I denote an open interval about 0. If $H : X \times I \to Y$ is an arbitrary homotopy and $t \in I$, then we shall denote by $H_t : X \to Y$ the map defined by $H_t(p) = H(p, t)$.

Definition 7.3 A *variation of a regular domain* $P = P_0$ in M is a one-parameter family P_t, $t \in I$ of regular domains in M, such that

(i) there exists an isotopy $\Phi : \partial P \times I \to M$ with the property that Φ_0 is the embedding of ∂P into M and $\Phi_t(\partial P) = \partial P_t$ for all $t \in I$;
(ii) for any $p \in M$, the sets $\{t \in I \mid p \in \operatorname{int} P_t\}$ and $\{t \in I \mid p \in \operatorname{ext} P_t\}$ are open in I.

A vector field $X \in \Gamma(TM|_{\Sigma_0})$ along Σ_0 is called an *infinitesimal variation of the hypersurface* Σ_0 *compatible with the given variation* if the isotopy Φ in the previous definition can be chosen in such a way that $X(p) = \left.\frac{\partial \Phi(p,t)}{\partial t}\right|_{t=0}$ holds for all points $p \in \Sigma_0$.

Definition 7.4 A *variation of a polytope*

$$P = f^*(P_1, \ldots, P_k) \tag{8}$$

consists of variations $P_{1,t}, \ldots, P_{k,t}$, $t \in I$ of the primitives of P satisfying the following conditions:

(i) $P_t = f^*(P_{1,t}, \ldots, P_{k,t})$ is a polytope for all $t \in I$.
(ii) For any $l \leq 3$ and $1 \leq i_1 < \cdots < i_l \leq k$, there exists an isotopy Φ_t, $t \in I$ of the intersection $\bigcap_{j=1}^{l} \partial P_{i_j}$ in M such that Φ_0 is the inclusion map and $\Phi_t(\bigcap_{j=1}^{l} \partial P_{i_j}) = \bigcap_{j=1}^{l} \partial P_{i_j,t}$ for all t.
(iii) There is a compact set $K \subseteq M$ such that $P_t \subseteq K$ for all $t \in I$.

Consider a variation of the polytope $P = f^*(P_1, \ldots, P_k)$ in M. Denote by W_{ij} the wall between the ith and jth facets of P. A vector field $X_{ij} \in \Gamma(TM|_{W_{ij}})$ along the wall W_{ij} is said to be an *infinitesimal variation of the wall W_{ij} compatible with the given variation* if the isotopy Φ_t of the intersection $\partial P_i \cap \partial P_j$ described in condition (ii) can be chosen in such a way that $X_{ij}(p) = \frac{\mathrm{d}\Phi_t(p)}{\mathrm{d}t}\Big|_{t=0}$ holds for all $p \in W_{ij}$.

7.2 Generalized Schläfli Formula in Einstein Manifolds

Before presenting the formula, we introduce some notations. Let Σ be a smooth hypersurface in the Riemannian manifold $(M, \{\,,\})$, $\mathbf{N} \in \Gamma(TM|_\Sigma)$ a fixed unit normal vector field along Σ. Let $\Phi_t : \Sigma \to M$, $t \in (-\varepsilon, \varepsilon)$ be an arbitrary isotopy for which Φ_0 is the embedding of Σ into M, and $X \in \Gamma(TM|_\Sigma)$, $X_p = \frac{\mathrm{d}}{\mathrm{d}t}\Phi_t(p)\big|_{t=0}$ be the vector field induced by it. Denote by I_t, II_t and H_t the pull-backs of the first and second fundamental forms and the Minkowski curvature of the hypersurface $\Phi_t(\Sigma)$ to the hypersurface Σ via the diffeomorphism $\Phi_t \colon \Sigma \to \Phi_t(\Sigma)$. Let $I = I_0$ and $II = II_0$ be the first and second fundamental forms of Σ, I', II' and H' be the derivatives $\frac{\mathrm{d}I_t}{\mathrm{d}t}(0)$, $\frac{\mathrm{d}II_t}{\mathrm{d}t}(0)$ and $\frac{\mathrm{d}H_t}{\mathrm{d}t}(0)$ respectively.

Consider a variation $P_t = f^*(P_{1,t}, \ldots, P_{k,t})$ of a polytope P in M. Denote by $V(t)$ the volume of P_t with respect to the volume measure induced by the Riemannian metric. Let Σ_i denote the boundary of the primitive P_i and let $X_i \in \Gamma(TM|_{\Sigma_i})$ be an infinitesimal variation of the hypersurface Σ_i which is compatible with the variation of P_i. Choose for each wall W_{ij} an infinitesimal variation X_{ij} compatible with the given variation of the polytope. By condition (ii) of Definition 7.4, we can choose an isotopy $\Phi : (\partial P_i \cap \partial P_j) \times (-\varepsilon, \varepsilon) \to M$, with initial velocity vector field X_{ij}. Extend the function α_{ij} to a smooth function $\alpha_{ij}^{\Phi} : (\partial P_i \cap \partial P_j) \times (-\varepsilon, \varepsilon) \to \mathbb{R}$ defining $\alpha_{ij}^{\Phi}(p, t)$ as the dihedral angle of the polytope P_t at the point $\Phi(p, t)$ assuming that $\Phi(p, t)$ is lying on the wall between the ith and jth facets of the polytope P_t. The value of the derivative $X_{ij}\alpha_{ij} : W_{ij} \to \mathbb{R}$ of the dihedral angle α_{ij} with respect to the vector field X_{ij} at a point $p \in W_{ij}$ is defined as the value of the partial derivative $\partial\alpha_{ij}^{\Phi}(p,t)/\partial t$ at $t=0$. The definition of $X_{ij}\alpha_{ij}$ is correct since $X_{ij}\alpha_{ij}$ does not depend on the choice of the isotopy Φ.

Now we are ready to present the generalized Schläfli formula for polytopes in Einstein manifolds. See [35] as a standard reference on Einstein manifolds.

Theorem 7.1 (B. Cs. [34]) *Let P_t be a variation of a polytope P lying in an n-dimensional Einstein manifold M of scalar curvature s, $V(t)$ be the volume of P_t. Then using the above notations, we have*

$$\begin{aligned}\frac{s}{n}V'(0)=&\sum_{i=1}^{k}\int_{F_i}(H_i'+\frac{1}{2}\{I_i',II_i\})d\sigma_i+\sum_{1\le i<j\le k}\int_{W_{ij}}(X_{ij}\alpha_{ij})d\sigma_{ij}\\&+\sum_{i=1}^{k}\sum_{\substack{j=1\\j\neq i}}^{k}\int_{W_{ij}}\{L_i(X_{ij}-X_i),\mathbf{n}_{ij}\}d\sigma_{ij},\end{aligned}\tag{9}$$

where σ_i and σ_{ij} are the volume measures induced by the Riemannian metric on the facets and the walls between the facets, L_i is the Weingarten operator of the ith facet.

8 Some Applications of the Generalized Schläfli Formula

In this section, let M be an n-dimensional hyperbolic, Euclidean or spherical space of constant sectional curvature $K=s/(n^2-n)$. Complete connected umbilical hypersurfaces in M will be called $*$*-spheres*. In the spherical space, $*$-spheres are ordinary spheres. In the Euclidean space, they are spheres or hyperplanes, while in the hyperbolic space, they can be spheres, horospheres, hyperplanes or hyperspheres. For $K\le 0$, the expression $*$*-ball* will always mean a convex regular domain in M the boundary of which is a $*$-sphere. For $K>0$, $*$*-balls* are defined as the closed (not necessarily convex) balls of the spherical space. Polytopes with curved faces built from $*$-balls as primitives are generalizations of 3-transversal flowers, call them $*$*-flowers*.

Let $P=f^*(B_1,\dots,B_k)$ be a $*$-flower in M, where the boundary of B_i is a $*$-sphere of constant normal curvature κ_i. Consider a variation P_t of P obtained by rigid motions of the primitives B_i. In this case, we can choose Killing fields for the infinitesimal variations X_i. Doing so, the integrals $\int_{F_i}(H_i'+\frac{1}{2}\{I_i',II_i\})d\sigma_i$ in Eq. (9) disappear. The intersection angle of two $*$-spheres is the same at any point of the intersection, thus the function α_{ij} depends only on t, and the value of the derivative $\alpha_{ij}'(0)=X_{ij}\alpha_{ij}$ does not depend on the actual choice of the vector field X_{ij}. As ∂B_i is an umbilical hypersurface, its Weingarten map L_i is simply the multiplication with the number κ_i. This means that Eq. (9) reduces to the form

$$\frac{s}{n}V'(0) = \sum_{1\le i<j\le k} \alpha'_{ij}(0)\sigma_{ij}(W_{ij}) + \sum_{i=1}^{k}\sum_{\substack{j=1\\ j\ne i}}^{k} \int_{W_{ij}} \kappa_i\{(X_{ij} - X_i), \mathbf{n}_{ij}\}d\sigma_{ij}.$$

Since X_i is a Killing field and X_{ij} is an infinitesimal variation compatible with the variation of the wall W_{ij}, $\sum_{\substack{j=1\\ j\ne i}}^{k} \int_{W_{ij}} \{(X_{ij} - X_i), \mathbf{n}_{ij}\}d\sigma_{ij}$ is the derivative of the $(n-1)$-dimensional volume of the ith facet at the moment $t = 0$. This way we obtain the following special case of Theorem 7.1.

Theorem 8.1 *We have the variational formula*

$$\frac{d}{dt}\left(\frac{s}{n}V - \sum_{i=1}^{k}\kappa_i\sigma_i(F_i)\right)\Bigg|_{t=0} = \sum_{1\le i<j\le k} \alpha'_{ij}(0)\sigma_{ij}(W_{ij}) \tag{10}$$

for the volume of a $*$-flower built from rigidly moving $*$-balls.

This gives the following monotonicity result.

Theorem 8.2 *If we change the shape of a $*$-flower $P = f^*(B_1, \ldots, B_k)$ by moving the $*$-balls B_i smoothly and rigidly in such a way that the dihedral angles α_{ij} of the polytope P do not increase during the motion, then the quantity $(sV/n - \sum_i \kappa_i\sigma_i(F_i))$ does not increase as well.*

The functional $(sV/n - \sum_i \kappa_i\sigma_i(F_i))$ can be extended by the formula $sV(P)/n - \int_{\partial P} \kappa d\sigma$ to any CSG solid built from arbitrary $*$-balls, not necessarily satisfying the transversality assumption, where κ assigns to a smooth point of the boundary ∂P the constant value of the normal curvature of the boundary at that point with respect to the outer unit normal.

Consider now $*$-flowers made of ordinary balls. Extending our earlier definition of flowers, we shall call any CSG solid made of balls a flower. This means that when we construct a flower we are allowed to take not only unions and intersections but also regularized differences of sets.

For two intersecting ordinary balls B_i and B_j, the dihedral angle $\alpha(B_i, B_j)$ of the union $B_i \cup B_j$ is an increasing function of the distance of the centers of the balls. If f^* is a regularized Boolean expression in which each variable occurs exactly once, we can define a sign $\epsilon_{ij}^{f^*} \in \{\pm 1\}$ depending only on f^* such that for any 3-transversal family of balls B_i, the dihedral angle α_{ij} of the flower $f^*(B_1, \ldots, B_k)$ differs from $\epsilon_{ij}^{f^*}\alpha(B_i, B_j)$ by a constant multiple of π. This means that the dihedral angle α_{ij} of a 3-transversal flower built from balls of given radii is an increasing function of the signed distance $\epsilon_{ij}^{f^*} d(P_i, P_j)$ between the centers P_i and P_j of the balls B_i and B_j.

Thus, Theorem 8.2 leads to the following generalization of Theorem 5.1.

Theorem 8.3 *If the shape of the flower $P = f^*(B_1, \ldots, B_k)$ is changed by a piecewise analytic rigid motion of the balls $B_i = B(P_i, r_i)$ in such a way that the signed distances $\epsilon_{ij}^{f^*} d_{ij}(t) = \epsilon_{ij}^{f^*} d(P_i(t), P_j(t))$ are weakly decreasing during the motion, then the quantity $sV(P_t)/n - \int_{\partial P_t} \kappa d\sigma$ is weakly decreasing as well.*

Let N be an $(n-2)$-dimensional subspace in M. The group G of orientation preserving isometries of M keeping points of N fixed consists of rotations of M about N, thus G is isomorphic to the circle group S^1.

Theorem 8.4 *Assume that P is a G-invariant CSG solid in M such that the singular points of the boundary of P form a neglectible set in the sense of [36], and that the boundaries of the primitives from which P is built intersect N transversally. If the nonsingular boundary point $p \in \partial P$ is not in N, then denote by $k_G(p)$ the normal curvature of ∂P in the direction of the tangent line of the circle Gp with respect to the outer unit normal of P. Then we have the identity*

$$\frac{s}{n} V_n(P) - \int_{\partial P} k_G d\sigma = 2\pi V_{n-2}(P \cap N), \tag{11}$$

where V_n, V_{n-2} and σ are the volume measures induced by the Riemannian metric of M on M, N and the smooth part of ∂P respectively.

We remark that Eq. (11) is a generalization of the Archimedean formula for the surface area of surfaces of revolutions. When it is applied to $*$-flowers, it gives the following result.

Theorem 8.5 *Assume that $P = f^*(B_1, \dots, B_k)$ is a CSG solid built from G-invariat $*$-balls of M. Denote by κ the principal curvature of the smooth part of the boundary ∂P, and by σ the $(n-1)$-dimensional volume measure on ∂P. Then we have*

$$\frac{s}{n} V_n(P) - \int_{\partial P} \kappa d\sigma = 2\pi V_{n-2}(f^*(B_1 \cap N, \dots, B_k \cap N)).$$

Theorems 8.3 and 8.5 give the following generalization of Theorem 5.3.

Theorem 8.6 *Let f^* be a regularized Boolean expression as above, $P_1, \dots, P_k$ and $Q_1, \dots, Q_k$ be points in an $(n-2)$-dimensional subspace N. If there exist piecewise analytic curves $\gamma_i : [0, 1] \to M$ connecting the points $\gamma_i(0) = P_i$ to the points $\gamma_i(1) = Q_i$ in such a way that the signed distances $\epsilon_{ij}^{f^*} d(\gamma_i(t), \gamma_j(t))$ weakly decrease as t increases, then for any choice of the radii r_i, we have the inequality*

$$V_{n-2}(f_N^*(\hat{B}_1^{n-2}, \dots, \hat{B}_k^{n-2})) \geq V_{n-2}(f_N^*(B_1^{n-2}, \dots, B_k^{n-2}))$$

between the volumes of the flowers made of the balls $\hat{B}_i^{n-2} = B^n(P_i, r_i) \cap N$ and $B_i^{n-2} = B^n(Q_i, r_i) \cap N$.

9 Application to a Problem of M. Kneser

The definition of polytopes with curved faces involved the condition that in the Boolean expression $f^*(x_1, \dots, x_k)$ which is evaluated on the primitives, each variable occurs exactly once. This was a convenient assumption to eliminate some technical difficulties. Nevertheless, the generalized Schläfli formula can be applied to

CSG solids obtained as the evaluation $P = f^*(P_1, \dots, P_k)$ of an *arbitrary* Boolean expression f^*, provided that any $l \leq 3$ of the boundaries of $P_1, \dots, P_k$ intersect transversally. To this end, we can decompose the solid $f^*(P_1, \dots, P_k)$ into a non-overlapping union of *atomic* solids which have the form $\left(\bigcap^*_{i\in A} P_i\right) \cap^* \left(\bigcup^*_{i\notin A} P_i\right)$, where A is a subset of the index set $\{1, \dots, k\}$. Writing the generalized Schläfli type formula for each atomic component and adding these formulae, we obtain a Schläfli-type formula for P.

In 1997, after the appearance of the paper [15], M. Kneser sent the author a copy of the private letter he wrote to B. Bollobás in 1968. In this letter, he raised the following generalization of the original KPC.

Question 9.1 *(M. Kneser) Assume that $B_1, \dots, B_k$ are balls in the n-dimensional Euclidean space, $t_1 \geq \dots \geq t_k$ are real numbers. For $1 \leq s \leq k$, define the set E_s as the set of those points that belong to at least s of the balls. (E_1 is the union of the balls, E_k is their intersection.) Is it true that the linear combination $\sum_{s=1}^k t_s \mathrm{vol}_n(E_s)$ of the volumes of the sets E_s cannot increase when the balls are rearranged in such a way that the distances between their centers do not increase?*

The question makes sense also in the spherical and hyperbolic spaces. Let M be the n-dimensional hyperbolic, Euclidean, or spherical space of scalar curvature s, and assume that $B_1, \dots, B_k$ are balls in M. The set E_s is a flower since

$$E_s = f_s(B_1, \dots, B_k) := \bigcup_{1\leq i_1<\dots<i_s\leq k} \bigcap_{j=1}^{s} B_{i_j}.$$

However, it is not possible to obtain this flower as the evaluation of a Boolean expression in which each variable occurs exactly once. For this reason, we cannot define the signs $\epsilon_{ij}^{f^*}$ properly. This is related to the geometrical symptom that there are some points of the wall $W_{s,ij}$ between the ith and jth facets of E_s in a small neighborhood of which E_s coincides with the union $B_i \cup B_j$ and there are also points around which E_s coincides locally with the intersection $B_i \cap B_j$. According to the local shape of E_s, we can introduce the disjoint subsets

$$W^+_{s,ij} := \{p \in W_{s,ij} \mid \exists \text{ a neighborhood } U \text{ of } p \text{ such that } U \cap E_s = U \cap (B_i \cup B_j)\},$$
$$W^-_{s,ij} := \{p \in W_{s,ij} \mid \exists \text{ a neighborhood } U \text{ of } p \text{ such that } U \cap E_s = U \cap (B_i \cap B_j)\}$$

of the wall $W_{s,ij}$. Let α_{ij} be the dihedral angle of the union $B_i \cup B_j$ along the intersection of the boundary spheres. Applying the generalized Schläfli formula to E_s we obtain the following results.

Theorem 9.1 *If the balls $B_1, \dots, B_k$ are moved smoothly, then at any moment when the boundary spheres of any $m \leq 3$ of the balls B_i intersect transversally, we have*

(i)

$$\left(\frac{s}{n}V_s - \int_{\partial E_s} \kappa d\sigma\right)' = \sum_{1\le i<j\le k} (\mathrm{vol}_{n-2}(W^+_{s,ij}) - \mathrm{vol}_{n-2}(W^-_{s,ij}))\alpha'_{ij},$$

where $V_s = \mathrm{vol}_n(E_s)$.

(ii) $W^-_{1,ij} = W^+_{k,ij} = \emptyset$ *and* $W^+_{s,ij} = W^-_{s+1,ij}$ *for all* $1 \le s < k$ *and for all* $1 \le i < j \le k$. *Furthermore,*

(iii)

$$\sum_{s=1}^{k} t_s \left(\frac{s}{n}V_s - \int_{\partial E_s} \kappa d\sigma\right)' = \sum_{s=2}^{k} \sum_{1\le i<j\le k} (t_{s-1} - t_s) vol_{n-2}(W^-_{s,ij}))\alpha'_{ij}.$$

The theorem together with Theorem 8.5 implies the following.

Theorem 9.2 *If the balls $B_1, \dots, B_k$ of an n-dimensional simply connected space M of constant curvature are rearranged so that the new system of centers can be obtained from the original system of centers by a piecewise analytic contracting homotopy in an ambient $(n+2)$-dimensional simply connected space of the same constant curvature, then the value of the functional $\sum_{s=1}^{k} t_s vol_n(E_s)$ is not increasing.*

Theorem 9.2 and the leapfrog lemma yield that the answer to Question 9.1 is affirmative in the Euclidean plane.

10 Alexander's Conjecture

The counterexample depicted in Fig. 2 shows that the surface volume of the boundary of the union of some balls can increase when the balls are contracted as the boundary of the union of the contracted system of balls can become bumpier. As opposed to the union, the intersection of balls is always convex, so one cannot expect a similar simple counterexample for the monotonicity of the surface volume of the intersection of balls. This might have led R. Alexander to the following conjecture.

Conjecture (R. Alexander *[23]*) *If the planar configurations* $\mathbf{x} = (x_1, \dots, x_k)$, $\mathbf{y} = (y_1, \dots, y_k) \in (\mathbb{E}^2)^k$ *satisfy* $\mathbf{x} \succ \mathbf{y}$ *and* $r > 0$, *then the perimeter of the intersection* $\bigcap_{i=1}^{k} B(x_i, r)$ *is not greater than that of* $\bigcap_{i=1}^{k} B(y_i, r)$.

Alexander's conjecture was studied by K. Bezdek, the author, and R. Connelly [37]. The conjecture was proved in some special cases that implied that the conjecture is true for four disks. The paper also explains that Alexander's conjecture is a direct generalization of the theorem that the perimeter of the convex hull of a bounded planar set cannot increase when the set is contracted. To describe the connection, for a set $X \subset \mathbb{E}^2$ that can be covered by a disk of radius r, denote by $I_r(X)$ the intersection of the disks of radius r centered at a point of X, and by $C_r(X)$ the intersection of all the disks of radius r that contain X. We call $C_r(X)$ the r-convex hull of X. When r tends to infinity, $C_r(X)$ tends to the closure of the convex hull of X. It can be seen that the Minkowski sum of $I_r(X)$ and $C_r(X)$ has constant width $2r$, so the sum of their perimeters is $2r\pi$. Alexander's conjecture is equivalent to the statement that the perimeter of the r-convex hull of a set $X \subset \mathbb{E}^2$ contained in a disk of radius r is not greater than the perimeter of the r-convex hull of its contraction.

Alexander's conjecture fails to be true for noncongruent circles. There is an example in [37], in which 3 disks in the plane are contracted *continuously* so that the perimeter of their intersection decreases in the meantime. In addition to that, two of the circles are congruent, and the radius of the third one can be chosen arbitrarily close to the radius of the congruent ones. Nevertheless, the following generalization of Alexander's conjecture might be true.

Conjecture *If M is the n-dimensional hyperbolic, Euclidean, or spherical space of scalar curvature s, then the value of the functional $\frac{s}{n}V(P) - \int_{\partial P} \kappa d\sigma$ (cf. Theorem 8.3) on the intersection P of some (not necessarily congruent) balls does not decrease when the balls are contracted.*

11 The Conjecture in More General Spaces

The original KPC is about balls of the Euclidean space but it makes sense in any metric measure space, i.e., in any metric space with a fixed Borel measure, in particular, in any normed space, or in any connected Riemannian manifold.

If the KPC is true for any k closed balls of a metric measure space, then one can show using the sieve formula and induction on k that the measure of the intersection of $l \leq k$ closed balls can depend only on the distances between the centers of the balls and the radii of the balls. This observation motivates the following definition.

Definition 11.1 We say that a metric measure space has the *KP_k property* if the measure of the intersection of k closed balls depends only on the distances between the centers and the radii of the balls. The space is said to have the *$KP_k^=$ property* if the KP_k property holds for balls of equal radius.

It is clear that property KP_k implies $KP_k^=$, and if $k \geq k'$, then $KP_k^{(=)}$ implies $KP_{k'}^{(=)}$.

There are some special cases of the KPC that are true in normed spaces as well. For example, Kneser's proof of Theorem 2.1 works also in arbitrary finite dimensional normed spaces. The special case, considered by Bouligand, when the systems of the centers are similar to one another was extended to normed spaces by W. Rehder [12]. However, M. Meyer, S. Reisner and M. Schmuckenschläger [38] showed that the $KP_2^=$ property characterizes Euclidean spaces in the family of finite dimensional normed spaces, so the original conjecture can hold in a finite dimensional normed space only if the norm is Euclidean. Thus, the following problem seems to be a reasonable modification of the KPC for normed spaces.

Problem For a given finite dimensional normed space, find the smallest number $1 \leq c \leq 3^n$ such that the volume of the neighborhood of radius r of a contraction of a finite set is at most c times the volume of the neighborhood of radius r of the original set.

Consider now the question which Riemannian manifolds the KPC can hold in. Since many special cases of the KPC are known to be true in the hyperbolic and spherical spaces as well, it seems to be reasonable to extend the conjecture to these spaces. This extension is supported also by recent results of I. Gorbovickis [39], who proved the KPC in $\mathbb{S}^2$ and $\mathbb{H}^2$ under the assumption that the union of the disks in the expanded system is simply connected.

The study of the geometrical consequences of the KP_k and $KP_k^=$ properties revealed that within the family of complete connected Riemannian manifolds, these are the only spaces in which the KPC can be true.

Property $KP_1 = KP_1^=$ means simply that the volume of a geodesic ball depends only on the radius of the ball. Property KP_1 is closely related to the notion of ball-homogeneity, introduced O. Kowalski and L. Vanhecke [40]. Recall that a Riemannian manifold is called *ball-homogeneous* if the volume of "small" geodesic balls depends only on the radius of the ball. Ball-homogeneous spaces have been studied extensively, see e.g. the papers [41, 42] and the references therein. Using the asymptotical formula

$$\mathrm{vol}_n(B(p,r)) = \omega_n r^n \Big(1 - \frac{s(p)}{6(n+2)} r^2 + O(r^4)\Big)$$

for the volume of geodesic balls, where $s(p)$ is the scalar curvature at p (see [43]), we obtain that a Riemannian manifold with the KP_1-property must be of constant scalar curvature. In particular, for 2-dimensional manifolds, property KP_1 implies that the surface is of constant sectional curvature.

The author and M. Horváth [44, 45] proved, that within the family of simply connected, connected, and complete Riemannian manifolds, the KP_2 and $KP_2^=$ properties are equivalent and hold if and only if the manifold is harmonic. Harmonic manifolds were introduced by E. T. Copson and H. S. Ruse [46]. A Riemannian manifold is harmonic if small geodesic spheres have constant mean curvature. The two known classes of harmonic manifolds, 2-point homogeneous symmetric spaces and Damek–Ricci spaces provide several examples of Riemannian manifolds with the KP_2 property that are not of constant sectional curvature.

The KP_3 and $KP_3^=$ properties are more restrictive. The author with D. Kunszenti-Kovács [47], and M. Horváth [48] proved that if the KP_3 or $KP_3^=$ property holds for small balls in a Riemannian manifold, then the manifold must have constant sectional curvature. Counterexamples in [47–49] show also that for connected and complete manifolds of constant curvature, property KP_3 implies that the manifold is simply connected.

The author and G. Moussong [49] studied the KPC in the elliptic space $\mathbb{S}^n/\mathbb{Z}_2$. They gave an example of some smoothly moving balls in the elliptic space such that the distances between the centers of the balls are weakly decreasing while the volume of the union of the balls is increasing. This shows that even the continuos contraction case of the KPC fails to be true in the elliptic space. However, there is a special case of the KPC which happens to be true. If the metric of the elliptic space is scaled so that the sectional curvature of the space is 1, then the diameter of the space is $\pi/2$, and one can find $n+1$ points, say $P_1, \dots P_{n+1}$, the pairwise distances of which are equal to $\pi/2$. For these points, we have the following.

Theorem 11.1 (B. Cs., G. Moussong [49]) *If $Q_1, \dots, Q_{n+1}$ are points of the elliptic space $\mathbb{S}^n/\mathbb{Z}_2$, and $r_1, \dots, r_{n+1} \in (0, \pi/2)$ are arbitrary real numbers, then the volume of the union of the balls $B(P_i, r_i)$ is not less than the volume of the union of the balls $B(Q_i, r_i)$.*

References

1. E.T. Poulsen, Problem 10. Math. Scand. **2**, 346 (1954)
2. M. Kneser, Einige Bemerkungen über das Minkowskische Flächenmass. Arch. Math. (Basel) **6**, 382–390 (1955)
3. H. Federer, in *Geometric Measure Theory*, Die Grundlehren der mathematischen Wissenschaften, Band 153 (Springer, New York Inc., 1969)
4. A. Kolmogoroff, Beiträge zur Maßtheorie. Math. Ann. **107**, 351–366 (1932)
5. H. Hadwiger, Ungelöste probleme Nr. 11. Elem. Math. **11**, 51–60 (1956)
6. V. Klee, Some unsolved problems in plane geometry. Math. Mag. **52**(3), 131–145 (1979)
7. W. Moser, Problem 32. Pushing disks around, in *Research Problems in Discrete Geometry*, 6th edn. (1981), pp. 1–32[mimeographed notes]
8. V. Klee, S. Wagon, in *Old and new unsolved problems in plane geometry and number theory*, vol. 11, The Dolciani mathematical expositions (Mathematical Association of America, Washington, 1991)
9. K. Bezdek, in *Classical Topics in Discrete Geometry*, CMS Books in Mathematics/Ouvrages de Mathématiques de la SMC (Springer, New York, 2010)
10. K. Bezdek, Ensembles impropres et nombre dimensionnel I. Bull. Sci. Math., II. Sér. **52**, 320–344 (1928)
11. D. Avis, B.K. Bhattacharya, H. Imai, Computing the volume of the union of spheres. Vis. Comput. **3**, 323–328 (1988)
12. W. Rehder, On the volume of unions of translates of a convex set. Am. Math. Mon. **87**(5), 382–384 (1980)
13. B. Bollobás, Area of the union of disks. Elem. Math. **23**, 60–61 (1968)
14. B. Csikós, "Körrendszer által lefedett tartomány területének megváltozása a körök mozgatása esetén," in *A XVI. Országos Tudományos Diákköri Konferencia kiemelkedő pályamunkái III.*, vol. 6 of *Bolyai Soc. Math. Stud.*, pp. 209–212, Művelődési Minisztérium Tudományszervezési és Informatikai Intézete (1984)
15. B. Csikós, On the Hadwiger–Kneser–Poulsen conjecture, in *Intuitive Geometry (Budapest, 1995)*. Bolyai Society Mathematical Studies, vol. 6 (János Bolyai Mathematical Society, Budapest, 1997), pp. 291–299
16. M. Bern, A. Sahai, Pushing disks together–the continuous-motion case. Discret. Comput. Geom. **20**(4), 499–514 (1998)

17. B. Csikós, On the volume of the union of balls. Discret. Comput. Geom. **20**(4), 449–461 (1998)
18. M. Gromov, Monotonicity of the volume of intersection of balls, in *Geometrical Aspects of Functional Analysis (1985–86)*. Lecture Notes in Mathematics, vol. 1267 (Springer, Berlin, 1987), pp.1–4
19. Y. Gordon, M. Meyer, On the volume of unions and intersections of balls in Euclidean space, in *Geometric Aspects of Functional Analysis (Israel, 1992–1994)*, Operator theory advance application, vol. 77 (Birkhäuser, Basel, 1995), pp. 91–101
20. B. Csikós, On the volume of flowers in space forms. Geom. Dedicata **86**(1–3), 59–79 (2001)
21. H. Cheng, S.P. Tan, Y. Zheng, On continuous expansions of configurations of points in Euclidean space, arXiv:1107.0140v1 [math.MG] (2011), pp. 1–9
22. T. Kato, *Perturbation Theory for Linear Operators*. Classics in Mathematics (Springer, Berlin, 1995) [Reprint of the 1980 edition]
23. R. Alexander, Lipschitzian mappings and total mean curvature of polyhedral surfaces I. Trans. Am. Math. Soc. **288**(2), 661–678 (1985)
24. V. Capoyleas, J. Pach, On the perimeter of a point set in the plane, in *Discrete and Computational Geometry (New Brunswick, NJ, 1989–1990)*, DIMACS series discrete mathematics and theoretical computer science, vol. 6 (American Mathematical Society, Providence, RI, 1991), pp. 67–76
25. I. Gorbovickis, Strict Kneser–Poulsen conjecture for large radii. Geom. Dedicata **162**, 95–107 (2013)
26. D. Gale, On Lipschitzian mappings of convex bodies, in *Proceedings of Symposia in Pure Mathematics*, vol. VII (American Mathematical Society, Providence, 1963), , pp. 221–223
27. K. Bezdek, R. Connelly, The Kneser–Poulsen conjecture for spherical polytopes. Discret. Comput. Geom. **32**(1), 101–106 (2004)
28. K. Bezdek, On the monotonicity of the volume of hyperbolic convex polyhedra. Beiträge Algebr. Geom. **46**(2), 609–614 (2005)
29. K. Bezdek, R. Connelly, Pushing disks apart–the Kneser–Poulsen conjecture in the plane. J. Reine Angew. Math. **553**, 221–236 (2002)
30. I. Gorbovickis, Kneser–Poulsen conjecture for a small number of intersections. Contrib. Discret. Math. **9**(1), 1–10 (2014)
31. K. Bezdek, R. Connelly, On the weighted Kneser–Poulsen conjecture. Period. Math. Hungar. **57**(2), 121–129 (2008)
32. I. Rivin, J.-M. Schlenker, The Schläfli formula in Einstein manifolds with boundary. Electron. Res. Announc. Am. Math. Soc. **5**, 18–23 (1999). electronic
33. R. Souam, The Schläfli formula for polyhedra and piecewise smooth hypersurfaces. Differ. Geom. Appl. **20**(1), 31–45 (2004)
34. B. Csikós, A Schläfli-type formula for polytopes with curved faces and its application to the Kneser–Poulsen conjecture. Monatsh. Math. **147**(4), 273–292 (2006)
35. A.L. Besse, in *Einstein Manifolds*, Ergebnisse der Mathematik und ihrer Grenzgebiete (3), vol. 10 [Results in Mathematics and Related Areas (3)] (Springer, Berlin, 1987)
36. S. Lang, in *Fundamentals of Differential Geometry*. Graduate Texts in Mathematics , vol. 191 (Springer, New York, 1999)
37. K. Bezdek, R. Connelly, B. Csikós, On the perimeter of the intersection of congruent disks. Beiträge Algebr. Geom. **47**(1), 53–62 (2006)
38. M. Meyer, S. Reisner, M. Schmuckenschläger, The volume of the intersection of a convex body with its translates. Mathematika **40**(2), 278–289 (1993)
39. I. Gorbovickis, The central set and its application to the Kneser–Poulsen conjecture. Discrete Comput. Geom. **59**(4), 784–801 (June 2018), arXiv:1511.08134v2 [math.MG]
40. O. Kowalski, L. Vanhecke, Ball-homogeneous and disk-homogeneous Riemannian manifolds. Math. Z. **180**(4), 429–444 (1982)
41. G. Calvaruso, L. Vanhecke, Special ball-homogeneous spaces. Z. Anal. Anwend. **16**(4), 789–800 (1997)
42. G. Calvaruso, P. Tondeur, L. Vanhecke, Four-dimensional ball-homogeneous and C-spaces. Beiträge Algebr. Geom. **38**(2), 325–336 (1997)

43. A. Gray, L. Vanhecke, Riemannian geometry as determined by the volumes of small geodesic balls. Acta Math. **142**(3–4), 157–198 (1979)
44. B. Csikós, M. Horváth, On the volume of the intersection of two geodesic balls. Differ. Geom. Appl. **29**(4), 567–576 (2011)
45. B. Csikós, M. Horváth, A characterization of harmonic spaces. J. Differ. Geom. **90**(3), 383–389 (2012)
46. E. Copson, H. Ruse, Harmonic Riemannian spaces. Proc. R. Soc. Edinb. **60**, 117–133 (1940)
47. B. Csikós, D. Kunszenti-Kovács, On the extendability of the Kneser–Poulsen conjecture to Riemannian manifolds. Adv. Geom. **10**(2), 197–204 (2010)
48. B. Csikós, M. Horváth, A characterization of spaces of constant curvature by minimum covering radius of triangles. Indag. Math. (N.S.) **25**(3), 608–617 (2014)
49. B. Csikós, G. Moussong, On the Kneser–Poulsen conjecture in elliptic space. Manuscr. Math. **121**(4), 481–489 (2006)

A Survey of Elekes-Rónyai-Type Problems

Frank de Zeeuw

Abstract We give an overview of recent progress around a problem introduced by Elekes and Rónyai. The prototype problem is to show that a polynomial $f \in \mathbb{R}[x, y]$ has a large image on a Cartesian product $A \times B \subset \mathbb{R}^2$, unless f has a group-related special form. We discuss this problem and a number of variants and generalizations. This includes the Elekes-Szabó problem, which generalizes the Elekes-Rónyai problem to a question about an upper bound on the intersection of an algebraic surface with a Cartesian product, and curve variants, where we ask the same questions for Cartesian products of finite subsets of algebraic curves. These problems lie at the crossroads of combinatorics, algebra, and geometry: They ask combinatorial questions about algebraic objects, whose answers turn out to have applications to geometric questions involving basic objects like distances, lines, and circles, as well as to sum-product-type questions from additive combinatorics. As part of a recent surge of algebraic techniques in combinatorial geometry, a number of quantitative and qualitative steps have been made within this framework. Nevertheless, many tantalizing open questions remain.

1 The Elekes-Rónyai Problem

1.1 Sums, Products, and Expanding Polynomials

Erdős and Szemerédi [22] introduced the following problem in 1983. Given a finite set A in some ring, is it true that the *sumset* $A + A$ or the *productset* $A \cdot A$ must be large? The rationale is that for an arithmetic progression the sumset is small, but the productset is large, while for a geometric progession, the reverse is true. Erdős and Szemerédi proved for $A \subset \mathbb{Z}$ that

F. de Zeeuw (✉)
Department of Mathematics, École Polytechnique Fédérale de Lausanne, Lausanne, Switzerland
e-mail: fdezeeuw@gmail.com

G. Ambrus et al. (eds.), *New Trends in Intuitive Geometry*, Bolyai Society Mathematical Studies 27, https://doi.org/10.1007/978-3-662-57413-3_5

$$\max\{|A+A|, |A\cdot A|\} = \Omega\left(|A|^{1+c}\right) \tag{1}$$

for a very small $c > 0$. This statement was later generalized to $\mathbb{R}$, and the constant has over the years been improved to $4/3 + c'$ for a small $c' > 0$ [29, 54].

The intuition behind this statement is that a set cannot have many "coincidences" for both addition and multiplication. A statement like (1) is not the only way to capture this intuition. Elekes [14] suggested that, since polynomials combine addition and multiplication, for most polynomials $f \in \mathbb{R}[x, y]$ we should have

$$|f(A\times A)| = \Omega\left(|A|^{1+c}\right), \tag{2}$$

for any finite $A \subset \mathbb{R}$, with a constant $c > 0$ that may depend on f. More generally, we should have

$$|f(A\times B)| = \Omega\left(n^{1+c}\right), \tag{3}$$

for $A, B \subset \mathbb{R}$ with $|A| = |B| = n$, and a similar bound when A and B have different sizes.

Of course, (3) cannot hold for all polynomials, as it fails when $f(x, y) = x + y$ and A and B are arithmetic progressions with the same difference, or when $f(x, y) = xy$ and A and B are geometric progressions with the same ratio. More generally, if f has the additive form

$$f(x, y) = g(h(x) + k(y)) \tag{4}$$

with univariate polynomials g, h, k, then one has $|f(A \times B)| = O(n)$ if one chooses A and B in such a way that $h(A)$ and $k(B)$ are arithmetic progressions. Similarly, if f has the multiplicative form

$$f(x, y) = g(h(x)\cdot k(y)), \tag{5}$$

then $|f(A \times B)| = O(n)$ if $h(A)$ and $k(B)$ are geometric progressions. We will call a polynomial *additive* if it has the form in (4), and *multiplicative* if it has the form in (5).

Elekes [14] conjectured that the additive form (4) and the multiplicative form (5) are the only exceptions to the bound (3). Elekes proved a weaker form of this statement in [14], and he collaborated with Rónyai [17] to prove this conjecture in full. We state an improved version of the result due to Raz, Sharir, and Solymosi [43].

Theorem 1.1 (Elekes-Rónyai, Raz-Sharir-Solymosi) *Let $f \in \mathbb{R}[x, y]$ be a polynomial of degree d that is not additive or multiplicative. Then for all $A, B \subset \mathbb{R}$ with $|A| = |B| = n$ we have*

$$|f(A\times B)| = \Omega_d\left(n^{4/3}\right).$$

To be precise, the bound stated in [17] was $|f(A \times B)| = \omega(n)$, but inspection of the proof leads to a bound of the form $|f(A \times B)| = \Omega(n^{1+c_d})$ with a constant $c_d > 0$ depending on the degree of f. Raz, Sharir, and Solymosi [43] showed that this dependence on the degree of f is not necessary, and moreover improved the constant significantly. Their proof used a setup inspired by Sharir, Sheffer, and Solymosi [47], which gave a similar improvement on a special case (see Theorem 1.4). The same bound $|f(A \times B)| = \Omega(n^{4/3})$ was obtained by Hegyvári and Hennecart [28, Proposition 8.3] for the polynomial $f(x, y) = xy(x + y)$.[1]

The exceptional role played by the additive and multiplicative forms suggests that groups play a special role in this type of theorem. We will touch on this in Sect. 2, but see Elekes and Szabó [19, Sect. 1.2] for further discussion. Other (older) expositions of Theorem 1.1 can be found in Elekes's magnificent survey [10], and in Matoušek's book [33, notes to Sect. 4.1].

1.2 Extensions

It was observed in [44] that a bound like (3) should also hold when A and B have different sizes, and this was proved in a weak sense. In [43] such an "unbalanced" form of Theorem 1.1 was proved: If f is not additive or multiplicative, and $A, B \subset \mathbb{R}$, then

$$|f(A \times B)| = \Omega_d \left(\min\left\{|A|^{2/3}|B|^{2/3}, |A|^2, |B|^2\right\}\right). \tag{6}$$

Another way in which Theorem 1.1 can be extended is by replacing polynomials by *rational functions*. This was indeed done in [17], but not in [43]. Here the exceptions include the same forms (4) and (5) with g, h, k rational functions, but surprisingly, a third special form shows up here, namely

$$f(x, y) = g\left(\frac{k(x) + l(y)}{1 - k(x)l(y)}\right) \tag{7}$$

with rational functions g, k, l. It was pointed out in [7] that over $\mathbb{C}$ this can be seen as a multiplicative form, because $g((k(x) + l(y))/(1 - k(x)l(y))) = G(K(x)L(y))$ if we set $G(z) = (z - 1)/(i(z + 1))$, $K(x) = (1 + ik(x))/(1 - ik(x))$, and $L(y) = (1 + il(y))/(1 - il(y))$ (and if we do some tedious computation).

It remains an open problem to improve the bound $|f(A \times B)| = \omega(n)$ of [17] for rational functions f. For one special case, the rational function $f(x, y) = (x - y)^2/(1 + y^2)$, the bound $|f(A \times A)| = \Omega(|A|^{4/3})$ was proved in [45]. This was done in the context of the distinct distance problem for distances between points and lines;

[1] The result in [28, Proposition 8.3] has the same proof setup as [47], but it appears to be somewhat isolated; it makes no reference to [17], and in turn is not referred to in [43, 47]. This may be because [28] primarily concerns expansion bounds over finite fields. The arXiv publication date of [47] is half a year before that of [28].

f gives the distance between the point $(x, 0)$ and the line spanned by the points $(y, 0)$ and $(0, 1)$.

Problem 1.2 Prove Theorem 1.1 for rational functions $f \in \mathbb{R}[x, y]$ that are not additive, multiplicative, or of the form (7).

Yet another way to extend Theorem 1.1 is from $\mathbb{R}$ to $\mathbb{C}$. Most of the proof in [43] extends easily to $\mathbb{C}$, with the exception of the incidence bound used (see Sect. 1.4), which can be replaced by the bound over $\mathbb{C}$ proved in [53] (see Theorem 1.6); the details are written down in [9]. Thus Theorem 1.1 holds also over $\mathbb{C}$.

The exponent 4/3 in Theorem 1.1 is most likely not optimal. Elekes [15] in fact conjectured that the bound in Theorem 1.1 can be improved as far as $\Omega(n^{2-\varepsilon})$, but no exponent better than 4/3 has been established for any polynomial. Elekes [15] noted that for $f(x, y) = x^2 + xy + y^2$ (and many other polynomials) and $A = B = \{1, \ldots, n\}$ we have $|f(A \times B)| = \Theta(n^2/\sqrt{\log n})$ (see [48, Chap. 6] for details), so perhaps the bound in Theorem 1.1 can even be improved to $\Omega(n^2/\sqrt{\log n})$.

Let us combine the above extensions to conjecture the ultimate Elekes-Rónyai-type theorem (although see Sect. 4 for further variants).

Conjecture 1.3 *Let $f \in \mathbb{C}(x, y)$ be a rational function of degree[2] d that is not additive or multiplicative. Then for all $A, B \subset \mathbb{C}$ with $|A| = |B| = n$ we have*

$$|f(A \times B)| = \Omega_{d,\varepsilon}\left(n^{2-\varepsilon}\right).$$

1.3 Applications

Sum-product bounds. For a first consequence of Theorem 1.1, we return to the sum-product problem mentioned at the start of Sect. 1.1. The following generalization of the bound (1) was proved by Shen [51]: If $A \subset \mathbb{R}$ and $f \in \mathbb{R}[x, y]$ is a polynomial of degree d that is not of the form $g(\ell(x, y))$, with g a univariate polynomial and ℓ a linear bivariate polynomial, then

$$\max\{|A + A|, |f(A \times A)|\} = \Omega_d\left(|A|^{5/4}\right). \tag{8}$$

For many polynomials, Theorem 1.1 improves this bound, and it also shows that in those cases one does not need to consider $|A + A|$ to conclude that $|f(A \times A)|$ is large.

On the other hand, there are many polynomials that have the special form of Theorem 1.1, but that do not have the special form of Shen. Even in those cases, it may be possible to obtain a bound on $|f(A \times A)|$ independent of $|A + A|$; for instance, Elekes, Nathanson, and Ruzsa [20] prove $|f(A \times A)| = \Omega_d(|A|^{5/4})$ for $f(x, y) = x + y^2$ (and many similar functions). Note that for this bound it is crucial

[2] The maximum of the degrees of the numerator and denominator, assuming that these do not have a common factor.

that the Cartesian product is of the form $A \times A$ rather than $A \times B$. It may be that such a bound holds for any f that is not of the form $g(h(x) + h(y))$ or $g(h(x) \cdot h(y))$; this question does not seem to have been studied.

Distances between lines. As a corollary of their result, Elekes and Rónyai [17] made progress on the following problem of Purdy (see [5, Sect. 5.5]): Given two lines with n points each, what is the minimum number of distances occurring between the two point sets? This problem is a simpler variant of the distinct distances problem of Erdős [23], which asks for the minimum number of distinct distances determined by a point set in the plane. Erdős's problem was almost completely solved by Guth and Katz [26] using new algebraic methods.

In Purdy's problem there are two exceptional situations, when the two lines are parallel or orthogonal. Indeed, if on two parallel lines one places two arithmetic progressions of size n with the same common difference, then the number of distinct distances is linear in n. On two orthogonal lines, say the x-axis and the y-axis, one can take the sets $\{(\sqrt{i}, 0) : 1 \le i \le n\}$ and $\{(0, \sqrt{j}) : 1 \le j \le n\}$ to get a linear number of distances. The following theorem states that for all other pairs of lines there are considerably more distances. Given $P_1, P_2 \subset \mathbb{R}^2$, we write $D(P_1, P_2)$ for the set of Euclidean distances between the points of P_1 and the points of P_2.

Theorem 1.4 (Elekes-Rónyai, Sharir-Sheffer-Solymosi). *Let L_1, L_2 be two lines in $\mathbb{R}^2$ that are not parallel or orthogonal, and let $P_1 \subset L_1$, $P_2 \subset L_2$ be finite sets of size n. Then the number of distinct distances between P_1 and P_2 satisfies*

$$|D(P_1, P_2)| = \Omega\left(n^{4/3}\right).$$

Proof Sketch We can assume that the lines are $y = 0$ and $y = mx$, with $m \neq 0$. The squared distance between $(s, 0)$ and (t, mt) is

$$f(s,t) = (s-t)^2 + m^2 t^2.$$

It is easy to verify that the polynomial $f(s, t)$ is not additive or multiplicative, so Theorem 1.1 implies the stated bound. □

Elekes and Rónyai first proved a superlinear bound in the case $|P_1| = |P_2|$ as a consequence of their result in [17], thus solving Purdy's problem, in the qualitative sense of distinguishing the special pairs of lines. Elekes [15] then[3] quantified the proof from [17] in this special case to obtain a short proof of the explicit bound $\Omega(n^{5/4})$, noting [10] that Brass and Matoušek asked for such a "gap theorem". An unbalanced form was proved in [44]. The bound in Theorem 1.4 was obtained by Sharir, Sheffer, and Solymosi [47], using a proof inspired by that of Guth and Katz [26].[4]

[3]The chronology is somewhat confusing here. The paper [17] was published in 2000, and [15] in 1999. However, [15] refers back to [17] and makes it clear that [15] is an improvement on a special case of [17].

[4]A preprint version of [26] became available in 2010.

An earlier version of [47] used the Elekes-Sharir transformation from [18] that was crucial in [26] to connect distances with incidences; it was then observed that in Purdy's problem a considerably easier incidence problem can be obtained, and also that the Elekes-Sharir transformation can be bypassed. The result was a proof that is even simpler than that of [15], and its simplicity allowed for many generalizations, including Theorem 1.1 and many other results in this survey.

Directions on curves. The distinct directions problem asks for the minimum number of distinct directions determined by a non-collinear point set in the plane. It is superficially similar to the distinct distances problem, in the sense that it asks for the minimum number of distinct values of a function of pairs of points in the plane. However, it was solved exactly by Ungar [59] in 1982: Any non-collinear set P in $\mathbb{R}^2$ determines at least $|P| - 1$ distinct directions.

This leaves the more difficult *structural* question: What is the structure of sets that determine few distinct directions? Let us write $S(P)$ for the set of directions (or slopes) determined by $P \subset \mathbb{R}^2$. Elekes [16] conjectured that if $|S(P)| = O(|P|)$, then P must have many points on a conic; even in the weakest form, where "many" is six, this is unknown. Elekes [16] showed that a (very) restricted version of this conjecture follows from Theorem 1.1: If P lies on the graph of a polynomial of degree at most d and has $|S(P)| = O_d(|P|)$, then the polynomial must be linear or quadratic. See [10, Sect. 3.3] for a detailed discussion. We state here the improvement of this result from [43], obtained as a consequence of Theorem 1.1. In Sect. 3.4 we will discuss the same question for points sets on arbitrary algebraic curves.

Corollary 1.5 *Let P be a finite point set that is contained in the graph $y = g(x)$ of a polynomial $g \in \mathbb{R}[x]$ of degree $d \geq 3$. Then the number of distinct directions determined by P satisfies*

$$|S(P)| = \Omega_d\left(|P|^{4/3}\right).$$

Proof Sketch The direction determined by the points $(s, g(s)), (t, g(t))$ is given by the polynomial

$$f(s,t) = \frac{g(s) - g(t)}{s - t}.$$

It is not hard to verify that this polynomial is not additive or multiplicative (except when g is linear or quadratic), so Theorem 1.1 gives the stated bound. □

When g has degree less than three, the number of directions can be linear. Take for instance the parabola $y = x^2$ and the point set $\{(i, i^2) : 1 \leq i \leq n\}$. Then the determined directions are $(i^2 - j^2)/(i - j) = i + j$, so there are $O(n)$ distinct directions. It is not hard to see that a similar example can be constructed for any other quadratic g.

1.4 About the Proof of Theorem 1.1

We now discuss the proof of Theorem 1.1. We certainly do not give a full account of the proof in [43], but we introduce the setup without too much technical detail. The proof of the weaker bound in [17] used related techniques but was inherently different. The proof setup in [43] originated in [47], and seems to have been discovered independently (and somewhat later) in [28, Proposition 8.3]; a glimpse of this setup can also be seen in the (earlier) proof of [19, Theorem 27]. We use the word "setup" to refer to the overall counting scheme of [43], which is only the surface layer of the proof. The real achievement in [43] was the treatment of high-multiplicity curves, which we discuss only briefly at the end of this subsection.

The main tool for obtaining the bound in Theorem 1.1 is incidence theory, specifically an incidence bound for points and curves of Pach and Sharir [36].[5] This incidence bound is a generalization of the classical theorem of Szemerédi and Trotter [56] that bounds incidences between po ints and lines. The following version is particularly convenient for the applications in this survey. It assumes that the point set is a Cartesian product, which allows for a significantly simpler proof over $\mathbb{R}$, and makes it easier to prove an analogue over $\mathbb{C}$ (where the corresponding bound has not yet been established without extra assumptions; see [49]). This theorem was proved by Solymosi and De Zeeuw [53], although the real case was probably folklore. We write $|I(\mathcal{P}, \mathcal{C})|$ for the set of incidences between the points $\mathcal{P}$ and the curves $\mathcal{C}$, i.e., the set of pairs $(p, C) \in \mathcal{P} \times \mathcal{C}$ such that $p \in C$. We give a quick sketch of the proof, to show that proving this tool does not require any heavy machinery (at least over $\mathbb{R}$).

Theorem 1.6 *Let $\mathcal{P} = A \times B$ be a Cartesian product in $\mathbb{R}^2$ or $\mathbb{C}^2$, and let $\mathcal{C}$ be a set of algebraic curves of degree at most d in the same plane. Assume that any two points of $\mathcal{P}$ are contained in at most M curves of $\mathcal{C}$. Then*

$$|I(\mathcal{P}, \mathcal{C})| = O_{d,M}\left(|\mathcal{P}|^{2/3}|\mathcal{C}|^{2/3} + |\mathcal{P}| + |\mathcal{C}|\right).$$

Proof sketch in $\mathbb{R}^2$ Let us assume that $|A| = |B| = n$ and $\mathcal{C} = n^2$ (which roughly holds in most of the statements in this survey). We can partition $\mathbb{R}^2$ using $O(r)$ horizontal and vertical lines, in such a way that each of the $O(r^2)$ resulting rectangles contains roughly $O(n^2/r^2)$ points of $A \times B$. Moreover, we can ensure that the partitioning lines do not contain any points of $A \times B$, and are not contained in any of the curves in $\mathcal{C}$. We observe that an algebraic curve of degree at most d intersects $O_d(r)$ rectangles, since it can intersect a partitioning line in at most d points.

We split the incidences as follows: I_1 is the set of incidences (p, C) such that p is the only incidence on C in the rectangle that p lies in, and I_2 is the remaining set of incidences. Since each curve hits $O_d(r)$ rectangles, we have $|I_1| = O_d(rn^2)$. On the other hand, if a curve has an incidence from I_2 in a certain rectangle, then it has at least one additional incidence in that same rectangle. By the assumption of the theorem, any two points are together contained in at most M curves. Thus the $O(n^2/r^2)$ points in a rectangle are involved in at most $O_M(n^4/r^4)$ incidences from I_2, which altogether gives $|I_2| = O_M(n^4/r^2)$. Choosing $r = n^{2/3}$ optimizes

[5] The proof in [43] did not directly use the bound from [36], but rather adapted the proof from [36] to the incidence situation in [43].

$$|I(\mathcal{P},\mathcal{C})| = |I_1| + |I_2| = O_{d,M}\left(rn^2 + n^4/r^2\right) = O\left(n^{8/3}\right),$$

which is the stated bound when $|\mathcal{P}| = |\mathcal{C}| = n^2$. □

The proof of Theorem 1.1 is based on an upper bound for the size of the following set of quadruples:

$$Q = \{(a, b, a', b') \in A \times B \times A \times B : f(a, b) = f(a', b')\}.$$

Given such an upper bound, the Cauchy–Schwarz inequality gives a lower bound on the size of the image set $f(A \times B)$, using the following calculation:

$$|Q| = \sum_{c \in f(A\times B)} |f^{-1}(c)|^2 \geq \frac{1}{|f(A \times B)|}\left(\sum_{c \in f(A\times B)} |f^{-1}(c)|\right)^2 = \frac{n^4}{|f(A \times B)|}. \tag{9}$$

Specifically, when f is not additive or multiplicative we obtain the upper bound $|Q| = O_d(n^{8/3})$, and then (9) implies $|f(A \times B)| = \Omega_d(n^{4/3})$. A similar application of Cauchy–Schwarz played a central role in [26].

To obtain an upper bound on $|Q|$, we define a set of curves and a set of points based on the given polynomial f and the given sets A, B, and then we apply Theorem 1.6. For each $(a, a') \in A \times A$, define

$$C_{aa'} = \{(x, y) \in \mathbb{R}^2 : f(a, x) = f(a', y)\}.$$

This is an algebraic curve of degree at most d (the degree of f). Note that for $(b, b') \in B \times B$, we have $(b, b') \in C_{aa'}$ if and only if $(a, b, a', b') \in Q$. Thus, if we set

$$\mathcal{P} = B \times B \quad \text{and} \quad \mathcal{C} = \{C_{aa'} : (a, a') \in A \times A\},$$

then $|I(\mathcal{P},\mathcal{C})| = |Q|$.

If we could apply Theorem 1.6 to $\mathcal{P}$ and $\mathcal{C}$, then we would immediately get the desired bound $|Q| = O_d(n^{8/3})$. However, $\mathcal{P}$ and $\mathcal{C}$ need not satisfy the degrees-of-freedom condition of Theorem 1.6 that two points are contained in a bounded number of curves. This is to be expected, since the bound should fail when f is additive or multiplicative. In fact, even when f is not additive or multiplicative, the degrees-of-freedom condition may be violated. However, it was shown in [43] that when f is not additive or multiplicative, the condition is only violated in a weak sense. Specifically, one can remove negligible subsets of the points and curves so that the remainder does satisfy the condition.

A key insight in the proof is that the curves $C_{aa'}$ satisfy a kind of *duality*. Indeed, we can define "dual curves" of the form $C^*_{bb'} = \{(s, t) \in \mathbb{R}^2 : f(s, b) = f(t, b')\}$, so that the point (a, a') lies on the dual curve $C^*_{bb'}$ if and only if the point (b, b') lies on the curve $C_{aa'}$. Thus, to check the degrees-of-freedom condition of Theorem 1.6 that two points $(b_1, b_1'), (b_2, b_2')$ lie on a bounded number of curves $C_{aa'}$, we can instead look at the number of points (a, a') in the intersection of the curves $C^*_{b_1b_1'}, C^*_{b_2b_2'}$. Such an intersection is easily bounded by Bézout's inequality, unless the curves $C^*_{b_1b_1'}$ and $C^*_{b_2b_2'}$ have a common component. Thus the degrees-of-freedom condition comes down to showing that when many of the curves $C^*_{bb'}$ have many common components, with high multiplicity, then f must be additive or multiplicative. By symmetry, we may as well consider this question for the original curves $C_{aa'}$.

Let us see what happens when f is additive or multiplicative. First consider the case where $f(x, y) = h(x) + k(y)$. Then $C_{aa'}$ is defined by $k(x) - k(y) = h(a') - h(a)$. Thus $C_{a_1a_1'}$ and $C_{a_2a_2'}$ are the same curve whenever $h(a_1') - h(a_1) = h(a_2') - h(a_2)$; this means that as many as $\Theta(|A|)$ pairs (a, a') may define the same curve $C_{aa'}$. Similarly, when $f(x, y) = h(x)k(y)$, then $C_{aa'}$ is defined by $k(x) = k(y) \cdot (h(a')/h(a))$, and again we can have high multiplicity. Finally, when $f(x, y) = g(h(x) + k(y))$, then $f(a, x) - f(a', y)$ has the factor $h(a) + k(x) - h(a') - h(y)$, which corresponds to a component that can have high multiplicity (and the same happens for $f(x, y) = g(h(x) \cdot k(y))$).

The key challenge in the proof of Theorem 1.1 is to obtain the converse, i.e., to show that when many curves have high multiplicity, then there must be polynomials g, h, k that explain the multiplicity in one of the ways above. In [43] this is done by algebraically prying out g, h, k from specific coefficients of the polynomial f. For instance, roughly speaking, when the curves $C_{aa'}$ have many common components and the coefficient of the leading term of $f(a, x) - f(a', y)$ is not constant as a polynomial in a, then this polynomial turns out to be the h in the multiplicative form $f(x, y) = g(h(x)k(y))$. On the other hand, if only the constant term of $f(a, x) - f(a', y)$ depends on a and a', then this leads to the polynomial h in the additive form $f(x, y) = g(h(x) + k(y))$.

2 The Elekes-Szabó Problem

2.1 Intersecting Varieties with Cartesian Products

To introduce a generalization of the Elekes-Rónyai problem due to Elekes and Szabó, we take a step back and approach from a different direction; after a while we will see what the connection between the problems is. In this section we work primarily over $\mathbb{C}$, which is the most natural setting for the Elekes-Szabó problem and the relevant proofs.

Let us consider the Schwartz–Zippel lemma (see [31] for the curious history of this lemma). The simplest non-trivial case is the following bound on the intersection of a curve with a Cartesian product.[6] If $F \in \mathbb{C}[x, y]$ is a polynomial of degree d and

[6]The word "grid" is often used in this context, but may lead to confusion with integer grids.

$A, B \subset \mathbb{C}$ are finite sets of size n, then[7]

$$|Z(F) \cap (A \times B)| = O_d(n). \tag{10}$$

This statement is "tight" in the sense that, for any fixed polynomial, there are sets A, B for which the bound is best possible. Indeed, we can arbitrarily choose n points on $Z(F)$, let A be the projection of this set to the x-axis, and let B be the projection to the y-axis; then $A \times B$ shares at least n points with $Z(F)$.

Now let us consider the tightness for the next case of the Schwartz–Zippel lemma (see Sect. 4.1 for the general statement), which says that for $F \in \mathbb{C}[x, y, z]$ and $A, B, C \subset \mathbb{C}$ of size n we have

$$|Z(F) \cap (A \times B \times C)| = O_d\left(n^2\right). \tag{11}$$

It is not so clear if this bound is tight, since the trick used above to show that (10) is tight does not work here. The best we could try is to choose A, B of size n, take n^2 points on $Z(F)$ above $A \times B$, and then project to the z-axis to get C; but the resulting C is likely to have many more than n points.

Nevertheless, for certain special polynomials the bound in (11) is tight. Take for instance $F = x + y - z$ and $A = B = C = \{1, \dots, n\}$; then $|Z(F) \cap (A \times B \times C)| = \Theta(n^2)$. Of course, one can construct similar examples for any polynomial of the form $F = f(g(x) + h(y) + k(z))$ with f, g, h, k univariate polynomials. In analogy with Theorem 1.1, one might guess that these are the only special polynomials, but this is not quite true. It turns out that the right class of special polynomials consists of those of the form $F = f(g(x) + h(y) + k(z))$ with f, g, h, k *analytic* functions that are defined in a *local* way.

Theorem 2.1 (Elekes-Szabó, Raz-Sharir-De Zeeuw) *Let $F \in \mathbb{C}[x, y, z]$ be an irreducible polynomial of degree d with with each of F_x, F_y, F_z not identically zero. Then one of the following holds.*
(i) For all $A, B, C \subset \mathbb{C}$ with $|A| = |B| = |C| = n$ we have

$$|Z(F) \cap (A \times B \times C)| = O_d(n^{11/6}).$$

(ii) There exists a one-dimensional subvariety $Z_0 \subset Z(F)$, such that every $v \in Z(F)\backslash Z_0$ has an open neighborhood $D_1 \times D_2 \times D_3$ and analytic functions $\varphi_i : D_i \to \mathbb{C}$, such that for every $(x, y, z) \in D_1 \times D_2 \times D_3$ we have

$$(x, y, z) \in Z(F) \quad \textit{if and only if} \quad \varphi_1(x) + \varphi_2(y) + \varphi_3(z) = 0.$$

Qualitatively, this result was proved by Elekes and Szabó [19],[8] who proved that $|Z(F) \cap (A \times B \times C)| = O(n^{2-\eta_d})$ for a constant η_d depending only on the degree

[7] We write $Z(F)$ for the *zero set* of a polynomial F, i.e., the set of points at which F vanishes.

[8] The publication year of [19] is 2012, but an essentially complete version of the paper existed much earlier; Elekes [10] referred to the result in 2002.

d of F,[9] unless F has the special form described in Theorem 2.1(ii) below. Using the new proof setup in [43, 47], this bound was improved to $O(n^{11/6})$ by Raz, Sharir, and De Zeeuw [40]. An exposition of the algebraic geometry underlying the proof in [19] was written by Wang [61].

The statement also holds if $\mathbb{C}$ is replaced by $\mathbb{R}$ (and the analytic $\varphi_i : D \to \mathbb{C}$ are replaced by real-analytic $\varphi_i : D \to \mathbb{R}$). We can also allow A, B, C to have different sizes. This does not affect the description in condition (ii), and the bound in condition (i) becomes

$$|Z(F) \cap (A \times B \times C)| = O_d\left(|A|^{2/3}|B|^{2/3}|C|^{1/2} + |A||C|^{1/2} + |B||C|^{1/2} + |C|\right); \tag{12}$$

of course the same bound holds for any permutation of A, B, C. We can again conjecture that the bound can be significantly improved, perhaps as far as $O_\varepsilon(n^{1+\varepsilon})$ (in the balanced case), or even $O(n\sqrt{\log n})$. No better lower bound is known than $\Omega(n\sqrt{\log n})$, for instance for $F = x^2 + xy + y^2 - z$ (which comes directly from the polynomial $f = x^2 + xy + y^2$ that provides the best known upper bound for Theorem 1.1, as mentioned in Sect. 1.2).

It is not likely that condition (ii) can be replaced by a purely polynomial condition, i.e., without mentioning analytic functions. This can be seen from the fact that the group law on an elliptic curve gives constructions for which the bound in (i) does not hold, while on the other hand, it is well-known that parametrizing elliptic curves as in (ii) requires analytic functions. We will give more details on the connection with elliptic curves at the end of Sect. 3.3.

In [19], condition (ii) of Theorem 2.1 is formulated in a somewhat stronger "global" form, although for all applications in this survey, the formulation in Theorem 2.1 seems to be more convenient. Specifically, the local functions φ_i can be replaced by *analytic multi-functions* from $\mathbb{C}$ to a *one-dimensional connected algebraic group* $\mathcal{G}$, so that $Z(F)$ is the image of the variety $\{(x, y, z) \in \mathcal{G}^3 : x \oplus y \oplus z = e\}$; we refer to [19, 61] for definitions.

2.2 A Derivative Test for Special F

In applications, it may not be easy to determine whether a given polynomial F satisfies condition (ii). For relatively simple polynomials, we have the following derivative condition. It is mentioned in [17, Sect. 1.1] and stated in [19, Lemma 33]; in [17] the sufficiency of the condition is ascribed to Jarai, although no proof is provided in [17] or [19]. We give a short sketch of the proof; a detailed proof can be found in [39].

[9] Earlier versions of [19] claimed a bound of the form $O(n^{2-\eta})$ for an absolute $\eta > 0$, and this was restated in [10]. But the published version states the theorem with η_d depending on d.

Lemma 2.2 *Let $f : \mathbb{R}^2 \to \mathbb{R}$ be a twice-differentiable function with $f_y \not\equiv 0$. There exist differentiable functions $\psi, \varphi_1, \varphi_2 : \mathbb{R} \to \mathbb{R}$ such that*

$$f(x, y) = \psi(\varphi_1(x) + \varphi_2(y)) \tag{13}$$

if and only if

$$\frac{\partial^2(\log|f_x/f_y|)}{\partial x \partial y} \equiv 0. \tag{14}$$

The same holds for analytic $f : \mathbb{C}^2 \to \mathbb{C}$ with $\psi, \varphi_1, \varphi_2 : \mathbb{C} \to \mathbb{C}$ analytic.

Proof Sketch If $f(x, y) = \psi(\varphi_1(x) + \varphi_2(y))$, then $f_x/f_y = \varphi_1'(x)/\varphi_2'(y)$, which gives $\log|f_x/f_y| = \log|\varphi_1'(x)| - \log|\varphi_2'(y)|$. Differentiating with respect to x and y gives 0.

Conversely, if $\partial^2(\log|f_x/f_y|)/\partial x \partial y \equiv 0$, integrating gives $\log|f_x/f_y| = g_1(x) - g_2(y)$. Then, setting $\varphi_1(x) = \int e^{g_1(x)}dx$ and $\varphi_2(y) = \int e^{g_2(y)}dy$, we have $f_x/f_y = \varphi_1'(x)/\varphi_2'(y)$. We express f in terms of the new variables $u = \varphi_1(x) + \varphi_2(y)$ and $v = \varphi_1(x) - \varphi_2(y)$, so that the chain rule gives $f_x = \varphi_1'(x)(f_u + f_v)$ and $f_y = \varphi_2'(y)(f_u - f_v)$. Combining these equations gives $0 = \frac{f_x}{\varphi_1'(x)} - \frac{f_y}{\varphi_2'(y)} = 2f_v$. Thus f depends only on the variable u, which means that we can write it as $f(x, y) = \psi(\varphi_1(x) + \varphi_2(y))$. □

To apply this lemma to a polynomial $F(x, y, z)$, we need to locally write the implicit surface $Z(F)$ as an explicit surface $z = f(x, y)$, for an analytic function f. Then the expression $\varphi_1(x) + \varphi_2(y) + \varphi_3(z) = 0$ in condition (ii) of Theorem 2.1 is equivalent to $f(x, y) = \varphi_3^{-1}(\varphi_1(x) + \varphi_2(y))$. In theory, such an f exists by the implicit function theorem, but in practice we can only calculate f when F has low degree (in one of the variables).

2.3 Applications

Expanding polynomials. Given $f \in \mathbb{C}[x, y]$, we can set $F(x, y, z) = f(x, y) - z$ and apply Theorem 2.1 with $|A| = |B| = n$ and $C = f(A \times B)$. If condition (i) applies, we get from the unbalanced bound (12) that

$$n^2 = |Z(f(x, y) - z) \cap (A \times B \times C)| = O_d(n^{4/3}|C|^{1/2}),$$

so $|f(A \times B)| = |C| = \Omega(n^{4/3})$. Otherwise, condition (ii) tells us that locally we have

$$f(x, y) = \psi(\varphi_1(x) + \varphi_2(y)), \tag{15}$$

with $\psi, \varphi_1, \varphi_2$ analytic functions.

Note that the multiplicative form of f also falls under (15), since we can (locally) write

$$g(h(x) \cdot k(y)) = (g \circ \log^{-1})(\log |h(x)| + \log |k(y)|)$$

with all functions analytic. We thus have almost deduced Theorem 1.1, except that the special form of f is local, and we do not know that $\varphi_1, \varphi_2, \psi$ are polynomials. It would be interesting to find a way to deduce the full Theorem 1.1 from this local analytic form. Something close to that is done by Tao in [58, Theorem 41] using arguments from complex analysis.

Distances from three points. The following result follows from Theorem 2.1.

Theorem 2.3 (Elekes-Szabó, Sharir-Solymosi). *Given three non-collinear points p_1, p_2, p_3 and a point set P in $\mathbb{R}^2$, there are $\Omega(|P|^{6/11})$ distinct distances from p_1, p_2, p_3 to P.*

Proof Sketch Let D denote the set of squared distances between p_1, p_2, p_3 and the points in P. A point $q \in P$ determines three squared distances to p_1, p_2, p_3, given by

$$a = (x_q - x_{p_1})^2 + (y_q - y_{p_1})^2, \quad b = (x_q - x_{p_2})^2 + (y_q - y_{p_2})^2, \quad c = (x_q - x_{p_3})^2 + (y_q - y_{p_3})^2.$$

The variables x_q and y_q can be eliminated from these equations to yield a quadratic equation $F(a, b, c) = 0$ with coefficients depending on p_1, p_2, p_3 (in [19] F can be seen written out). By construction, for each point $q \in P$, the corresponding squared distances a, b, c belong to D. The resulting triples (a, b, c) are all distinct, so F vanishes at $|P|$ triples of $D \times D \times D$.

The polynomial $F(x, y, z)$ turns out to be quadratic in z, so we can locally express it as $z = f(x, y)$. Then we can apply Lemma 2.2 to f to see that, if p_1, p_2, p_3 are not collinear, then f does not have the form in (13), which implies that F does not satisfy property (ii) of Theorem 2.1. Then property (i) gives $|P| = O(|D|^{11/6})$, or $|D| = \Omega(|P|^{6/11})$. When p_1, p_2, p_3 are collinear, F becomes a linear polynomial, so it does satisfy property (ii). □

This problem was introduced by Elekes [12], who showed that if p_1, p_2, p_3 are collinear (and equally spaced), then one can place P so that there are only $O(|P|^{1/2})$ distances from p_1, p_2, p_3 to P. Elekes and Szabó [19] proved Theorem 2.3 with the weaker bound $\Omega(|P|^{1/2+\eta})$ for some small absolute constant $\eta > 0$, as a consequence of their version of Theorem 2.1 (although they formulated the result in terms of triple points of circles; see the next application). Sharir and Solymosi [46] used the setup of [47] and ad hoc arguments to improve this η to $1/22$; their work preceded [40] and was the first extension of [47] that does not follow from [43].

Theorem 2.5 provides curious new information on Erdős's distinct distances problem (see [5]), which asks for the minimum number of distinct distances determined by a point set in $\mathbb{R}^2$. Erdős conjectured that this minimum is $\Theta(n/\sqrt{\log n})$, and this was almost matched by Guth and Katz [26],[10] who established $\Omega(n/\log n)$. Theorem 2.3 suggests that something stronger is true: There are many distances that occur just from three fixed non-collinear points. If, as conjectured, the bound in Theorem 2.1 can be improved from $O(n^{11/6})$ to $O(n^{1+\varepsilon})$, or even $O(n\sqrt{\log n})$, then Erdős's conjectured bound would already hold if one only considers distances from three non-collinear points in the point set (note that if the entire point set is collinear, there are $\Omega(n)$ distances from any given point).

While waiting for improvements in the bound of Theorem 2.3, we could instead consider distances from more than three points. Given k points in a suitable non-degenerate configuration and a point set P in $\mathbb{R}^2$, the number of distances from the k points to P should be $\Omega(|P|^{1/2+\alpha_k})$, where we would expect α_k to grow with k. Let us pose the first unknown step as a problem.

Problem 2.4 Let $P \subset \mathbb{R}^2$, and consider four points in $\mathbb{R}^2$ such that no three are collinear (or such that some stronger condition holds). Then the number of distinct distances from the four points to P is $\Omega(|P|^{1/2+\alpha})$, with $\alpha > 1/22$.

Triple points of circle families. Elekes and Szabó [19] formulated the problem of Theorem 2.3 in a different way. They considered three points p_1, p_2, p_3 in $\mathbb{R}^2$ and three families of n concentric circles centered at the three points, and they looked for an upper bound on the number of *triple points* of these families, i.e., points covered by one circle from each family. Theorem 2.3 states that if p_1, p_2, p_3 are not collinear, then the number of triple points is $O(n^{11/6})$; a construction in [12] shows that if p_1, p_2, p_3 are collinear, there can be as many as $\Omega(n^2)$ triple points.

One can ask the same question for any three one-dimensional families of circles, or even more general curves. Such statements were studied by Elekes, Simonovits, and Szabó [21], with a special interest in the case of concurrent unit circles, i.e., unit circles passing through a fixed point. This case was improved by Raz, Sharir, and Solymosi [42], again using the setup of [47], and ad hoc analytic arguments. By the arguments from [21], the improvement also follows from Theorem 2.1.

Theorem 2.5 (Elekes-Simonovits-Szabó, Raz-Sharir-Solymosi). *Three families of n concurrent unit circles (concurrent at three distinct points) determine $O(n^{11/6})$ triple points.*

This result shows an interesting distinction between lines and unit circles, because for three families of concurrent lines it is possible to determine $\Omega(n^2)$ triple points. We can for instance take the horizontal and vertical lines of an $n \times n$ integer grid, and n lines at a 45° angle that cover $\Omega(n^2)$ points of the grid.

It is natural to extend the question to ask for k-fold points of k families of concurrent unit circles (or other curves). This is largely unexplored when k is small. For large $k < \sqrt{n}$ and an arbitrary set of n unit circles, it follows from the incidence bound of Pach and Sharir [36] (the general case of Theorem 1.6) that there are at most $O(n^2/k^3)$ points where at least k circles meet.

---

[10] The new algebraic methods introduced in [26] indirectly led to the improvement of [47] in Theorem 1.4, and thus to many of the recent results in this survey.

2.4 About the Proof of Theorem 2.1

Let us briefly discuss the proof of Theorem 2.1, although again we will not go into too much detail. The proof is based on that of Theorem 1.1, as described in Sect. 1.4, but now we have to deal with $F(x, y, z) = 0$ instead of $f(x, y) = z$.

The first challenge is that we can no longer define the quadruples and curves using the equation $f(a, b) = f(a', b')$. Instead, we define the quadruples by

$$Q = \{(a, b, a', b') \in A \times B \times A \times B : \exists c \in C \text{ such that } F(a, b, c) = F(a', b', c) = 0\}.$$

Using the Cauchy–Schwarz inequality we get

$$\begin{aligned} |Z(F) \cap (A \times B \times C)| &= \sum_{c \in C} |\{(a, b) \in A \times B : F(a, b, c) = 0\}| \\ &\le |C|^{1/2} \left(\sum_{c \in C} |\{(a, b) \in A \times B : F(a, b, c) = 0\}|^2 \right)^{1/2} \\ &= O_d\left(|C|^{1/2}|Q|^{1/2}\right). \end{aligned}$$

In the last step we use the fact that for $(a, b) \in A \times B$, there are at most d values of $c \in \mathbb{C}$ for which $F(a, b, c) = 0$ (unless $F(x, y, z)$ contains a vertical line, but this happens at most $O_d(1)$ times). Again the goal is to obtain the upper bound $|Q| = O_d(n^{8/3})$ using an incidence bound, which will result in $|Z(F) \cap (A \times B \times C)| = O_d(n^{1/2} \cdot (n^{8/3})^{1/2}) = O_d(n^{11/6})$.

The set Q can be viewed as the *projection* of a *fiber product*. A fiber product[11] of a set with itself has the form $X \times_\varphi X = \{(x, x') \in X \times X : \varphi(x) = \varphi(x')\}$ for some function $\varphi : X \to Y$. This type of product is useful for counting, because the Cauchy–Schwarz inequality gives $|X| \le |X \times_\varphi X|^{1/2}|Y|^{1/2}$. In the calculation above, we have $X = |Z(F) \cap (A \times B \times C)|$, $Y = C$, and $\varphi(a, b, c) = c$. Then Q is the projection of $X \times_\varphi X$ to the coordinates (a, b, a', b'), and has essentially the same size as $X \times_\varphi X$. See [7, 58] for similar uses of fiber products, as well as further discussion of the technique.

The step in which we project from the fiber product to Q is necessary to make the next step work; specifically, we need Q to lie on a codimension one subvariety of $\mathbb{C}^4$, in order to be able to define the algebraic curves that we apply the incidence bound to. Unfortunately, this projection brings in the problem of *quantifier elimination*. The set Q lies on the set

$$\{(x, y, x', y') \in \mathbb{C}^4 : \exists z \in \mathbb{C} \text{ such that } F(x, y, z) = F(x', y', z) = 0\},$$

[11] This is a special case of a more general object from category theory; what we call a fiber product here is sometimes called a *set-theoretic* fiber product, or also a *relative product*.

which is not quite a variety, but only a *constructible set* (see [40] for details and references). We can eliminate the quantifier in the sense that there is a variety $Z(G) \subset \mathbb{C}^4$ that contains the constructible set, and it differs only in a lower-dimensional set. However, we have little grip on G other than that its degree is bounded in terms of that of F. Note that in the proof of Theorem 1.1 in Sect. 1.4 we had $F = f(x, y) - z$, so that we could easily eliminate z to get $f(x, y) = f(x', y')$.

To obtain $|Q| = O_d(n^{8/3})$, we define curves as in Sect. 1.4, but this becomes more complicated due to the quantifier elimination. We set

$$C_{aa'} = \{(x, y) \in \mathbb{C}^2 : \exists z \in \mathbb{C} \text{ such that } F(a, x, z) = F(a', y, z) = 0\}.$$

This set is a one-dimensional constructible set, i.e., an algebraic curve with finitely many points removed. This leads to many technical complications, but we can basically still apply an incidence bound to the points and curves

$$\mathcal{P} = B \times B, \quad \mathcal{C} = \{C_{aa'} : (a, a') \in A \times A\},$$

and we essentially have $|Q| = |I(\mathcal{P}, \mathcal{C})|$. As in Sect. 1.4, we can use Theorem 1.6 to obtain $|I(\mathcal{P}, \mathcal{C})| = O_d(n^{8/3})$, unless the curves badly violate the degrees-of-freedom condition. The hardest part of the proof is then to connect the failure of the degrees-of-freedom condition to the special form (ii) in Theorem 2.1.

3 Elekes-Rónyai Problems on Curves

In this section we discuss some variants of the Elekes-Rónyai and Elekes-Szabó problems for point sets contained in algebraic curves. We still work with Cartesian products of "one-dimensional" finite sets, but instead of finite subsets of $\mathbb{R}$ or $\mathbb{C}$, we take finite subsets of algebraic curves. We start with the first known instance of an Elekes-Rónyai problem on curves, where the function is the Euclidean distance. After that we discuss more general polynomial functions on curves, and finally we look at Elekes-Szabó problems on curves.

3.1 Distances on Curves

We have already seen one instance of the Elekes-Rónyai problem for distances on curves in Theorem 1.4, which concerned distances between two point sets on two lines. A natural generalization is to consider distances between two point sets on two algebraic curves in $\mathbb{R}^2$. More precisely, given algebraic curves $C_1, C_2 \subset \mathbb{R}^2$ and finite point sets $P_1 \subset C_1$, $P_2 \subset C_2$ with $|P_1| = |P_2| = n$, can we get a superlinear lower bound on $|D(P_1 \times P_2)|$?[12]

[12] As in Sect. 1.3, $D(p, q) = (p_x - q_x)^2 + (p_y - q_y)^2$ is the squared Euclidean distance function.

Observe that there are pairs of curves for which we cannot expect a superlinear lower bound on $|D(P_1 \times P_2)|$, as we saw in Sect. 1.3 for parallel lines and orthogonal lines. There is one more construction involving curves other than lines. If we take two concentric circles with equally spaced points, then there is also only a linear number of distances between the two point sets. It was proved by Pach and De Zeeuw [35] that these three constructions are the only exceptions to a superlinear lower bound; the proof used (once again) the setup in [47], together with ad hoc arguments.

Theorem 3.1 (Pach-De Zeeuw) *Let $C_1, C_2 \subset \mathbb{R}^2$ be irreducible algebraic curves of degree at most d. For finite subsets $P_1 \subset C_1$, $P_2 \subset C_2$ of size n we have*

$$|D(P_1 \times P_2)| = \Omega_d(n^{4/3}),$$

unless the curves are parallel lines, orthogonal lines, or concentric circles.

The proof in [35] allows C_1 and C_2 to be the same curve, which leads to a statement for distances on a single curve that is interesting in its own right. A version of this corollary was proved earlier by Charalambides [8], but with a weaker bound $\Omega_d(n^{5/4})$ (the same exponent as in [15], coming from the same counting scheme). The proof in [8] relied on an interesting connection with *graph rigidity*.

Corollary 3.2 (Charalambides, Pach-De Zeeuw) *Let $C \subset \mathbb{R}^2$ be an irreducible algebraic curve of degree d. For a finite subset $P \subset C$ we have*

$$|D(P \times P)| = \Omega_d(n^{4/3}),$$

unless C is a line or a circle.

Let us discuss some of the new issues involved in the proofs of these results. In the case of lines, a bound could be deduced from the Elekes-Rónyai theorem (or proved directly), unless the lines are parallel or orthogonal. For general algebraic curves, this does not seem possible. If the curves happen to be parametrized by polynomials, then plugging that parametrization into $D(p, q)$ would give a polynomial in two variables, and we could apply Theorem 1.1 (although it requires some work to translate the exceptional form of the polynomial to exceptional curves). This was done in [43] to prove that if C is polynomially parametrizable, then the bound in Corollary 3.2 holds, unless C is a line (a circle is not polynomially parametrizable).

If the curves are not polynomially parametrizable, then this approach will not work. The curve $y^2 = x^3 + 1$, for instance, has no parametrization with polynomials or rational functions. We could locally write $y = \sqrt{x^3 + 1}$ and plug that into $D(p, q)$, but this gives an algebraic function in two variables; moreover, most curves do not even have such an explicit solution by radicals, and the best we can do is to use the implicit function theorem to locally write y as an analytic function of x. This suggests that Theorem 2.1 may provide a larger framework for Theorem 3.1. We will see in Sect. 3.3 how one can manipulate a question about curves to fit it into the framework of Theorem 2.1. Even then, extracting the exceptional forms of the curves from property (ii) is not straightforward; for distances on curves this was done by Raz and Sharir [38], who used considerations from graph rigidity (similar to those in [8]) to deduce Theorem 3.1 from Theorem 2.1.

A natural way to extend Theorem 3.1 or Corollary 3.2 is to consider curves in higher dimensions. This was done for Corollary 3.2 by Charalambides [8] with the bound $\Omega(n^{5/4})$, using the counting scheme of [15], together with various tools from analysis. He determined that in $\mathbb{R}^3$ the exceptional curves are again lines and circles, while in $\mathbb{R}^4$ and above, the class of exceptional curves consists of so-called *algebraic helices*; see [8, 37] for definitions and exact statements. The bound in $\mathbb{R}^D$ was improved to $\Omega(n^{4/3})$ in [43] for the case of polynomially parametrizable curves using Theorem 1.1, and finally for all curves except algebraic helices by Raz [37], using both Theorem 2.1 and the analysis of Charalambides [8]. Bronner, Sharir, and Sheffer [6] considered variants involving curves in $\mathbb{R}^D$ that are not necessarily algebraic.

Another variant of Theorem 3.1 was studied by Sheffer, Zahl, and De Zeeuw [50]: Suppose that P_1 is contained in a curve C, and P_2 is an arbitrary set in $\mathbb{R}^2$. In general it is difficult to obtain bounds in this situation, but [50] showed that if C is a line or a circle, then the number of distances determined by $P_1 \cup P_2$ is reasonably large. This result was used to show that a point set that determines $o(n)$ distinct distances cannot have too many points on a line or a circle, which is a small step towards the conjecture of Erdős that a set of n points with $o(n)$ distinct distances must resemble an integer grid.

3.2 *Other Polynomials on Curves*

We can ask the same questions for any polynomial function $\mathbb{R}^2 \times \mathbb{R}^2 \to \mathbb{R}$, considered as a function $C_1 \times C_2 \to \mathbb{R}$. For clarity, we focus on the case $C_1 = C_2$. We also switch from $\mathbb{R}$ to $\mathbb{C}$, because some of the statements become more natural over $\mathbb{C}$.

Charalambides [8] also considered the function $A(p, q) = p_x q_y - p_y q_x$, which (in $\mathbb{R}^2$) gives twice the signed area of the triangle spanned by p, q, and the origin. He proved that, for P contained in an irreducible algebraic curve C in $\mathbb{R}^2$, we have $|A(P \times P)| = \Omega_d(|P|^{5/4})$, unless C is a line, an ellipse centered at the origin, or a hyperbola centered at the origin. This result was generalized by Valculescu and De Zeeuw [60], to any bilinear form over C, with a broader class of exceptional curves. The condition on the curve is tight in the sense that for any excluded curve there is a bilinear form that can take a linear number of values on that curve. To the authors, it was surprising that such curves can have large degree, whereas previous evidence (namely [8, 11]) suggested that the exceptional curves in such problems have degree at most three.

Theorem 3.3 *Let C be an irreducible algebraic curve in $\mathbb{C}^2$ of degree d, and consider the bilinear form $B_A(p, q) = p^T Aq$ for a nonsingular 2×2 matrix A. For $P \subset C$ we have*

$$|B_A(P \times P)| = \Omega_d\left(|P|^{4/3}\right),$$

unless C is a line, or linearly equivalent[13] *to a curve of the form $x^k = y^\ell$, with $k, \ell \in \mathbb{Z}\backslash\{0\}$.*

The description of the exceptional curves in Theorem 3.3 is very succinct but requires some clarification. Lines through the origin are linearly equivalent to a curve of the form $x^k = y^\ell$, but other lines are not, which is why they are listed separately. When k or ℓ is negative, one obtains a more natural polynomial equation after multiplying by an appropriate monomial. Thus hyperbola-like curves of the form $x^k y^\ell = 1$ with coprime $k, \ell \geq 1$ are included, since they can also be defined by $x^k = y^{-\ell}$. Ellipses centered at the origin are also included, since these are linearly equivalent to the unit circle $(x - iy)(x + iy) = 1$, which is linearly equivalent to $xy = 1$. Thus all the exceptional curves of Charalambides are special. Note that these curves are all *rational*, i.e., they have a parametrization by rational functions.

The reason that these curves are exceptional in the proof of Theorem 3.3 is that they have infinitely many *linear automorphisms*. Here an *automorphism* of a curve C is a map $T : \mathbb{C}^2 \to \mathbb{C}^2$ such that $T(C) = C$, and it is *linear* if it is a linear transformation (and similarly one can define affine, projective, or rational automorphisms). It was proved in [60] that the algebraic curves with infinitely many linear automorphisms are exactly those excluded in Theorem 3.3. From the proof of Theorem 3.3 in [60], it appears that if one considers more general polynomial functions instead of B_A, the exceptional curves will be those that have infinitely many rational automorphisms. By a theorem of Hurwitz (see for instance [27, Exercise IV.2.5]), a nonsingular curve with infinitely many rational automorphisms must have genus zero or one, or in other words, it must be rational or elliptic.

If we consider the statement of Theorem 3.3 for more general polynomials, then we also encounter exceptional polynomials that can take a linear number of values on *any* curve. Already for the bilinear forms B_A, this occurs when A is a singular matrix; in that case we can write $B_A(p, q) = L_1(p) \cdot L_2(q)$ with linear polynomials L_1, L_2, which is reminiscent of the multiplicative form in Theorem 1.1. More generally, for functions of the form $G(H(p) + K(q))$ or $G(H(p) \cdot K(q))$ there are exceptional constructions. We arrive at the following conjecture (where again we could conjecture a larger exponent).

Conjecture 3.4 *Let $C \subset \mathbb{C}^2$ be an algebraic curve of degree at most d and $F : C \times C \to \mathbb{C}$ a polynomial of degree at most d. Then for any $P \subset C$ we have*

$$|F(P \times P)| = \Omega_d(|P|^{4/3}),$$

unless $F(p, q) = G(H(p) + K(q))$ or $F(p, q) = G(H(p) \cdot K(q))$, or unless C is rational.

[13] We say that two curves are *linearly equivalent* if there is a linear transformation $(x, y) \mapsto (ax + by, cx + dy)$ that gives a bijection between the point sets of the curves.

3.3 Elekes-Szabó Problems on Curves

An Elekes-Szabó theorem on curves would take the following form. Let $C_1, C_2, C_3 \subset \mathbb{C}^2$ be algebraic curves of degree at most d, and let $G \in \mathbb{C}[x, y, s, t, u, v]$ be a polynomial of degree at most d. Then for point sets $P_1 \subset C_1$, $P_2 \subset C_2$, $P_3 \subset C_3$ of size n, we would want to bound $|Z(G) \cap (P_1 \times P_2 \times P_3)|$. We would expect exceptions for certain G, of a form related to that in Theorem 2.1(ii), and we would expect exceptions for certain curves, including low-degree curves and those in Conjecture 3.4. We won't state a full conjecture here, but we will discuss some of the instances that have been considered.

One possible choice of polynomial is

$$G(x, y, s, t, u, v) = \frac{1}{2}\begin{vmatrix} x & s & u \\ y & t & v \\ 1 & 1 & 1 \end{vmatrix} - 1,$$

for which $G(x, y, s, t, u, v) = 0$ if and only if the triangle determined by the points (x, y), (s, t), (u, v) has unit area. The problem of determining the maximum number of unit area triangles determined by n points in $\mathbb{R}^2$ is ascribed to Oppenheim in [24]. The best known lower bound is $\Omega(n^2 \log\log n)$, due to Erdős and Purdy [24], and the best known lower bound is $O(n^{20/9})$, due to Raz and Sharir [39]. It is easy to show that for n points on a curve, or three sets of n points on three curves, the bound $O(n^2)$ holds. The following problem was suggested by Solymosi and Sharir [46].

Problem 3.5 Given n points on an algebraic curve C in $\mathbb{R}^2$, prove a bound better than $O(n^2)$ on the number of unit area triangles, or show that there are point sets on C with $\Omega(n^2)$ unit area triangles.

The only case of this problem that has been studied is the one where C is the union of three distinct lines. Surprisingly, Raz and Sharir [39] showed that on any three distinct lines there are non-trivial constructions that determine $\Omega(n^2)$ unit area triangles. They discovered this using the derivative criterion in Lemma 2.2.

Another choice of G (which we won't try to write out) is the polynomial such that $G(x, y, s, t, u, v) = 0$ if the points (x, y), (s, t), (u, v) lie on a common unit circle, or in other words, the circle determined by the three points has radius equal to one. Since the radius of the circle determined by three points can be written as a rational function, multiplying out denominators gives a polynomial G with the property above. Raz, Sharir, and Solymosi showed that for three distinct unit circles in $\mathbb{R}^2$ with three finite point sets of size n, the number of determined unit circles is $O(n^{11/6})$ (this statement is equivalent to Theorem 2.5). One can generalize this problem as follows.

Problem 3.6 Given three algebraic curves $C_1, C_2, C_3 \subset \mathbb{R}^2$ containing three sets of n points, prove a bound better than $O(n^2)$ on the number of unit circles containing one point from each of the three sets. Are there curves for which $\Omega(n^2)$ unit circles is possible?

Another problem of this form is to bound *collinear triples* on curves. More is known on this problem, and we will discuss it in detail in the next subsection. Note that it can be seen as a degenerate case of the unit area triangle problem, where instead of unit area we ask for *zero* area.

Let us see how the Elekes-Szabó problem on curves is related to the Elekes-Szabó problem in $\mathbb{C}^3$. We can think of the three curves $C_1, C_2, C_3 \subset \mathbb{C}^2$ as a Cartesian product $C_1 \times C_2 \times C_3 \subset \mathbb{C}^6$. We can choose a generic projection $\varphi : \mathbb{C}^2 \to \mathbb{C}$, in such a way that the three-fold product $\pi = (\varphi \times \varphi \times \varphi) : \mathbb{C}^6 \to \mathbb{C}^3$ maps a product $P_1 \times P_2 \times P_3 \subset C_1 \times C_2 \times C_3$ with $|P_1| = |P_2| = |P_3| = n$ to a product $\varphi(P_1) \times \varphi(P_2) \times \varphi(P_3) \subset \mathbb{C}^3$ with $\varphi(P_1) = \varphi(P_2) = \varphi(P_3) = n$.

The variety $X = Z(F) \cap (C_1 \times C_2 \times C_3)$ is two-dimensional (unless G happens to vanish on $C_1 \times C_2 \times C_3$, in which case the problem is trivial). Again by choosing φ generically, we get that $\pi(X) \subset \mathbb{C}^3$ is also a two-dimensional variety, and thus can be written as $X = Z(F)$. Now we have

$$|Z(G) \cap (P_1 \times P_2 \times P_3)| \leq |Z(F) \cap (\varphi(P_1) \times \varphi(P_2) \times \varphi(P_3))|,$$

so if the upper bound of Theorem 2.1(i) applies to F, then we also have that upper bound for G on $C_1 \times C_2 \times C_3$. Otherwise, F satisfies property (ii) of Theorem 2.1.

Unfortunately, it is not clear how to transfer back property (ii) for F to the original setting. Note that F not only encodes information about G, but also about the curves C_1, C_2, C_3. Thus property (ii) for F should imply that either G has an exceptional form, or the curves C_1, C_2, C_3 have an exceptional form. This was made to work for the problem of collinear triples (see Sect. 3.4), but it remains difficult to do this in general.

The projection from $\mathbb{C}^6$ to $\mathbb{C}^3$ above lets us connect Theorem 2.1 to the group law on elliptic curves (irreducible nonsingular algebraic curves of degree three). We refer to [52] for an introduction to the group structure of an elliptic curve, but we summarize it here as follows: On any elliptic curve E there are an operation $\oplus$ and an identity element $\mathcal{O}$ that turn the point set E into a group, with the special property that $P, Q, R \in E$ are collinear if and only if $P \oplus Q \oplus R = \mathcal{O}$.

This group structure on elliptic curves allows us to construct finite point sets with many collinear triples. If we take a finite subgroup H of size n of an elliptic curve E (which exists for any n), then $|Z(G) \cap (H \times H \times H)| = \Omega(n^2)$, where G is the polynomial that represents collinearity. This follows from the fact that for any two distinct elements $P, Q \in H$, the line spanned by P and Q intersects E in $R = \ominus(P \oplus Q)$ (the group element R such that $(P \oplus Q) \oplus R = \mathcal{O}$), which must also be in the subgroup H. We could have $R = P$ or $R = Q$, but there are only $O(n)$ pairs that satisfy $P \oplus P \oplus Q = \mathcal{O}$, so we get $\Omega(n^2)$ collinear triples of distinct points.

Applying the generic projection $\pi : \mathbb{C}^6 \to \mathbb{C}^3$, we get a polynomial $F \in \mathbb{C}[x, y, z]$ such that $|Z(F) \cap (\varphi(H) \times \varphi(H) \times \varphi(H))| = \Omega(n^2)$, which implies that G satisfies property (ii) of Theorem 2.1. It follows that, locally, there are $\varphi_i : \mathbb{C} \to \mathbb{C}$ such that $P, Q, R \in E$ are collinear if and only if $\varphi_1(\varphi(P)) + \varphi_2(\varphi(Q)) + \varphi_3(\varphi(R)) = 0$. This is why it appears that the local analytic nature of property (ii) is necessary;

it would be a big surprise if such maps existed globally, or could be described by polynomial or rational functions.

3.4 Collinear Triples on Curves

The one instance of an Elekes-Szabó problem on curves that has been studied in some detail (and is not an Elekes-Rónyai problem) is the problem where F represents collinearity. This question was first considered by Elekes and Szabó [11]; they proved a weaker bound in $\mathbb{R}^2$ using the main bound from [19], which was then improved in [40] to the following statement.

Theorem 3.7 (Elekes-Szabó, Raz-Sharir-De Zeeuw) *Let C be an irreducible algebraic curve in $\mathbb{C}^2$ of degree d and let $P \subset C$ be a finite set. Then P determines $O_d(n^{11/6})$ proper collinear triples, unless C is a line or a cubic curve.*

As we saw at the end of Sect. 3.3, elliptic curves must be exceptions to this statement, because their group structure gives constructions with a quadratic number of collinear triples. In fact, on any cubic curve (including the union of a conic and a line, or a union of three lines) there is a "quasi-group law"; see Green and Tao [25, Proposition 7.3]. This law does not quite give the whole point set a group structure (and there may not be an identity element), but comes close enough to allow for a construction with $\Omega(n^2)$ collinear lines. These constructions are worked out in [11].

Green and Tao [25] proved the Dirac-Motzkin conjecture for large point sets, which states that any non-collinear point set $P \subset \mathbb{R}^2$ of size n determines at least $n/2$ *ordinary lines* (lines containing exactly two points of P). As a by-product, they also solved Sylvester's "orchard problem" for large n, which asks for the maximum number of lines with at least three points (*triple lines*) from a set of n points. It is easy to see that this number is at most $\frac{1}{3}\binom{n}{2}$, and Sylvester noted (in 1868; see [25]) that there are constructions on elliptic curves with almost exactly this number of triple lines. Green and Tao showed that any set with at least $\frac{1}{6}n^2 - O(n)$ triple lines must have most of its points on a cubic curve. Elekes made the bolder conjecture that any set with $\Omega(n^2)$ collinear triples has many points on a cubic; to underline our ignorance he suggested to show that ten of the points are on a cubic (which is one more than the trivial number). The conjecture was stated in [11] but already present in disguise in [10].

Conjecture 3.8 (Elekes) *If $P \subset \mathbb{R}^2$ determines $\Omega(n^2)$ collinear triples, then at least ten points of P lie on a cubic.*

An interesting application of Theorem 3.7, pointed out in [11], is to the problem of *distinct directions* mentioned in Sect. 1.3. Theorem 2.5 allows us to generalize Corollary 1.5 to any algebraic curve. The connection with triple lines is that if two pairs of points determine lines in the same direction, then these lines intersect the line at infinity in the same point. Thus a point set on a curve C that has few directions determines few such points at infinity, or conversely, adding these points at infinity gives a point set with many collinear triples. An unbalanced form of Theorem 3.7 then gives the following lower bound on the number of directions. There is an exception when C together with the line at infinity is a cubic curve, which means that C is a conic.

Corollary 3.9 *Let C be an irreducible algebraic curve in $\mathbb{C}^2$ of degree d and let $P \subset C$ be a finite set. Then P determines $\Omega_d(n^{4/3})$ distinct directions, unless C is a conic.*

4 Other Variants

4.1 Longer One-Dimensional Products

One way to extend Theorems 1.1 and 2.1 is to consider "longer" Cartesian products, i.e., products with more factors. The reference point for such bounds is the Schwartz–Zippel lemma, which we now state in full generality. This version was proved by Lang and Weil [30, Lemma 1][14]; see also Tao [57] for a proof sketch (both focus on finite fields, but their proofs work also over $\mathbb{C}$). The bound in fact holds over any field, and can be modified to allow factors A_i of different sizes.

Theorem 4.1 *Let $X \subset \mathbb{C}^D$ be a variety. Then for finite sets $A_1, \ldots, A_D \subset \mathbb{C}$ of size n we have*

$$|X \cap (A_1 \times \cdots \times A_D)| = O_{D,\deg(X)}(n^{\dim(X)}).$$

Theorem 2.1 told us that in the case with $D = 3$ and $\dim(X) = 2$, this bound can be improved on, unless the defining polynomial of X is special. Analogously, one would expect that the bound in Theorem 4.1 can be improved on, unless the variety is of some special type. The only other case that has so far been studied is $D = 4$ and $\dim(X) = 3$, for which Raz, Sharir, and De Zeeuw [41] proved the following.

Theorem 4.2 (Raz-Sharir-De Zeeuw) *Let $F \in \mathbb{C}[x, y, s, t]$ be an irreducible polynomial of degree d with with each of F_x, F_y, F_s, F_t not identically zero. Then one of the following holds.*
(i) For all $A, B, C, D \subset \mathbb{C}$ of size n we have

$$|Z(F) \cap (A \times B \times C \times D)| = O_d(n^{8/3}).$$

(ii) There exists a one-dimensional subvariety $Z_0 \subset Z(F)$, such that every $v \in Z(F)\backslash Z_0$ has an open neighborhood $D_1 \times D_2 \times D_3 \times D_4$ and analytic functions $\varphi_i : D_i \to \mathbb{C}$, such that for every $(x, y, s, t) \in D_1 \times D_2 \times D_3 \times D_4$ we have

[14]This brings into question if "Schwartz–Zippel" is the right name, but it has become standard in combinatorics.

$$(x, y, s, t) \in Z(F) \quad \textit{if and only if} \quad \varphi_1(x) + \varphi_2(y) + \varphi_3(s) + \varphi_4(t) = 0.$$

The proof of Theorem 4.2 mostly uses the same techniques as that of Theorem 2.1, and the setup turns out to be simpler. The reason is that (because four is an even number) we can define curves by

$$C_{cd} = \{(x, y) : F(x, y, c, d) = 0\}, \tag{16}$$

which avoids quantifier elimination and thus many technical complications. The theorem is in fact closely related to the incidence bound in Theorem 1.6: In most applications of that bound, the curves are defined as in (16), for some given polynomial F (for instance, the proof of Theorem 1.1 in Sect. 1.4 uses $F = f(s, x) - f(t, y)$). Theorem 4.2 thus gives the same bound as Theorem 1.6, but it replaces the combinatorial condition (two points being contained in a bounded number of curves) by an algebraic condition (that F is not of the form in (ii)). Much like condition (ii) in Theorem 2.1, condition (ii) in Theorem 4.2 can often be checked using derivatives.

A consequence of Theorem 4.2 is an expansion bound for three-variable polynomials $f \in \mathbb{C}[x, y, z]$ on sets $A, B, C \subset \mathbb{C}$ with $|A| = |B| = |C| = n$, namely

$$|f(A \times B \times C)| = \Omega(n^{3/2}),$$

unless f is in a local sense of the form $\psi(\varphi_1(x) + \varphi_2(y) + \varphi_3(s))$ with analytic $\psi, \varphi_1, \varphi_2, \varphi_3$. A weaker bound for this question was proved in [44], with the more precise condition that f is not of the form $g(h(x) + k(y) + l(z))$ or $g(h(x) \cdot k(y) \cdot l(z))$ with g, h, k, l polynomials. It should be possible to use techniques similar to those in [43] to prove the stronger bound with the precise condition.

We make the following conjecture for the general case.

Conjecture 4.3 *There is a constant $c > 0$ such that one of the following holds for any irreducible $F \in \mathbb{C}[x_1, \ldots, x_D]$ of degree d with each of $F_{x_1}, \ldots, F_{x_D}$ not identically zero.*

(i) For all $A_1, \ldots, A_D \subset \mathbb{C}$ of size n we have

$$|Z(F) \cap (A_1 \times \cdots \times A_D)| = O_d(n^{D-1-c}).$$

(ii) In a local sense we have

$$(x_1, \ldots, x_D) \in Z(F) \quad \textit{if and only if} \quad \sum_{i=1}^{D} \varphi_i(x_i) = 0.$$

A largely unexplored question is how the bound of Theorem 1.6 can be improved for a variety $X \subset \mathbb{C}^D$ with $\dim(X) < D - 1$. When $\dim(X) = 1$, the bound $O(n)$ is tight: We can place n points on X and project in the coordinate directions to get A_i such that $|X \cap (A_1 \times \cdots \times A_D)| = n$. Thus the first open case is $D = 4$, $\dim(X) = 2$.

Problem 4.4 Determine for which two-dimensional varieties $X \subset \mathbb{C}^4$ the Schwartz–Zippel-type bound $|X \cap (A_1 \times A_2 \times A_3 \times A_4)| = O_{\deg(X)}(n^2)$ is tight.

4.2 Two-Dimensional Products

A different way of extending Theorems 1.1 and 2.1 would be to consider Cartesian products of "two-dimensional" sets instead of "one-dimensional" sets, i.e., finite subsets of $\mathbb{C}^2$ instead of finite subsets of $\mathbb{C}$. For instance, the two-dimensional analogue of the Elekes-Rónyai problem would be: Given a map $\mathcal{F} : \mathbb{C}^2 \times \mathbb{C}^2 \to \mathbb{C}^2$, defined by $\mathcal{F}(x, y, s, t) = (F_1(x, y, s, t), F_2(x, y, s, t))$ for polynomials $F_1, F_2 \in \mathbb{C}[x, y, s, t]$, and given finite sets $P, Q \subset \mathbb{C}^2$, can we obtain a non-trivial lower bound on $|\mathcal{F}(P \times Q)|$?

The study of such questions was initiated by Nassajian Mojarrad et al. [34]. Even analogues of the Schwartz–Zippel lemma become more complicated: The easiest case would be a bound of the form $|X \cap (P \times Q)|$ for a variety $X \subset \mathbb{C}^4$ and finite sets $P, Q \subset \mathbb{C}^2$, but there are varieties for which no non-trivial bound holds. Let us call a polynomial $F \in \mathbb{C}[x, y, s, t]$ *Cartesian* if it can be written as

$$F(x, y, s, t) = G(x, y)H(x, y, s, t) + K(s, t)L(x, y, s, t), \qquad (17)$$

with $G \in \mathbb{C}[x, y]\backslash\mathbb{C}$, $K \in \mathbb{C}[s, t]\backslash\mathbb{C}$, and $H, L \in \mathbb{C}[x, y, s, t]$. We say that a variety $X \subset \mathbb{C}^4$ is *Cartesian* if every polynomial vanishing on X is Cartesian with the same G, K. For a Cartesian variety X, we can have $|X \cap (P \times Q)| = |P||Q|$, since we can take $P \subset Z(G)$ and $Q \subset Z(K)$ to get $P \times Q \subset X$.

It is proved in [34] that if X is not Cartesian, then one can obtain non-trivial upper bounds on $|X \cap (P \times Q)|$. As a consequence, we obtain the following lower bound on the number of distinct values of a polynomial map $\mathcal{F} = (F_1, F_2) : \mathbb{C}^2 \times \mathbb{C}^2 \to \mathbb{C}^2$.

Theorem 4.5 (Nassajian Mojarrad et al.) *Let $F_1, F_2 \in \mathbb{C}[x, y, s, t]$ be polynomials of degree d and $\mathcal{F} = (F_1, F_2)$. Then for $P, Q \subset \mathbb{C}^2$ with $|P| = |Q| = n$ we have*

$$|\mathcal{F}(P \times Q)| = \Omega_d(n),$$

unless $F_1 = \varphi_1 \circ \psi$ and $F_2 = \varphi_2 \circ \psi$ for a nonlinear polynomial ψ, or F_1 and F_2 are both Cartesian with the same G, K.

This theorem provides a starting point for the study of Elekes-Rónyai problems for two-dimensional polynomial maps. To compare with the one-dimensional case, for any $f \in \mathbb{C}[x, y]\backslash\mathbb{C}$ and $A, B \subset \mathbb{C}$ of size n, we can deduce $|f(A \times B)| = \Omega(n)$ from the Schwartz–Zippel lemma (Theorem 4.1), and Theorem 1.1 improves on that bound for polynomials that are not additive or multiplicative. The question is thus for which polynomial maps (other than those already excluded in Theorem 4.5) the bound $\Omega(n)$ can be improved on.

The bound of Theorem 4.5 is tight for certain polynomial maps. For instance, if we take the vector addition map $\mathcal{F} = (x + s, y + t)$ and we take P to be any arithmetic progression on a line, then we have $|\mathcal{F}(P \times P)| = O(n)$; note that $x + s$ and $y + t$ are both Cartesian, but not with the same G, K. We can do something similar for any map of the form $\mathcal{F} = (f_1(x, s), f_2(y, t))$, where f_1, f_2 are additive or multiplicative in the sense of Theorem 1.1. Just like for the Elekes-Rónyai problem, composing these maps with other polynomials gives more exceptions: If we write $(x, y) \oplus (s, t) = (x + s, y + t)$, and we set $\mathcal{F} = \psi(\varphi_1(x, y) \oplus \varphi_2(s, t))$ with reasonable maps $\psi, \varphi_1, \varphi_2 : \mathbb{C}^2 \to \mathbb{C}^2$, then again the image of $\mathcal{F}$ can have linear size.

Problem 4.6 Determine for which polynomial (or even rational) maps $\mathcal{F} : \mathbb{C}^2 \times \mathbb{C}^2 \to \mathbb{C}^2$ there exists an $\alpha > 0$ such that

$$|\mathcal{F}(P \times Q)| = \Omega_d(n^{1+\alpha})$$

for all $P, Q \subset \mathbb{C}^2$ of size n, perhaps with further restrictions on P and Q.

Superlinear bounds like in Problem 4.6 are known for only a few functions (which happen to be rational maps), and only over $\mathbb{R}$. Beck's "two extremities" theorem [2, Theorem 3.1] can be phrased in terms of the rational map $L : \mathbb{R}^2 \times \mathbb{R}^2 \to \mathbb{R}^2$ that maps a pair of points to the line spanned by the pair (for instance represented by the point in $\mathbb{R}^2$ whose coordinates are the slope and intercept of the line). Beck's theorem says that $|L(P \times P)| = \Omega_c(|P|^2)$, unless P has at least $c|P|$ points on a line. Raz and Sharir [39] work with the map that sends a pair of points to the line that consists of all the points that span a unit area triangle (of a fixed orientation) with the pair. They prove a superlinear bound on the number of distinct values of this map, for point sets with not too many points on a line (this statement is not made explicit in the paper, but is implicit in the proof). Finally, Lund, Sheffer, and De Zeeuw [32] study the rational map $\mathcal{B} : \mathbb{R}^2 \times \mathbb{R}^2 \to \mathbb{R}^2$ that sends a pair of points to the line that is their perpendicular bisector. They prove that $|\mathcal{B}(P \times P)| = \Omega_{M,\varepsilon}(|P|^{8/5-\varepsilon})$ if $P \subset \mathbb{R}^2$ has at most M points on a line or circle.

Elekes and Szabó [19] proved a more general theorem in this vein (their "Main Theorem" [19, Theorem 27]). It is difficult even to state this theorem precisely, so we give only a rough description. Let Y be a D-dimensional irreducible variety over $\mathbb{C}$ of degree at most d, and let $X \subset Y \times Y \times Y$ be an irreducible $2D$-dimensional subvariety of degree at most d with surjective and generically finite projections onto any two of the three factors. Let $A \subset Y$ be a finite set, which is in very general position in the following sense: For any proper subvariety $Z \subset Y$ of degree at most M we have $|A \cap Z| \leq N$. Then

$$|X \cap (A \times A \times A)| = O_{d,D,M,N}(|A|^{2-\eta})$$

with $\eta > 0$ depending on d, D, M, N, unless X is in some specific way related to an algebraic group.

The results of [2, 32, 39] mentioned above fit into this framework: Given a reasonable map $\mathcal{F} : \mathbb{C}^2 \times \mathbb{C}^2 \to \mathbb{C}^2$, its graph $X \subset \mathbb{C}^2 \times \mathbb{C}^2 \times \mathbb{C}^2$ is a variety that satisfies the conditions of the Main Theorem of Elekes and Szabó. Note that [2, 32, 39] not only provide explicit values of η in certain special cases, but they also make the "very general position" condition more precise in those cases. Indeed, they replace the condition that P avoids any algebraic curve (a proper subvariety of $\mathbb{C}^2$) of bounded degree, with the condition that P avoids only lines in the case of [2, 39], or lines and circles in [32].

4.3 Expanding Polynomials over Other Fields

Let us finish by briefly discussing Elekes-Rónyai-type questions over a finite field $\mathbb{F}_q$ and the field $\mathbb{Q}$ of rational numbers. We focus on statements of the form "$|f(A \times A)| = \Omega(|A|^{1+c})$ for a polynomial f that is not of a special form"; there has been much work on conditional expansion bounds (especially over finite fields), of the form "if $|A + A|$ is small, then $|f(A \times A)|$ is large", but we will not discuss these here.

Over finite fields, the question is considerably harder, because there is not yet any finite field equivalent of an incidence bound for algebraic curves like Theorem 1.6. For simplicity, let us restrict to a prime field $\mathbb{F}_p$, and let us ignore small p. There is a dichotomy between *small* subsets of $\mathbb{F}_p$, where very few incidence bounds are known, and *large* subsets of $\mathbb{F}_p$, for which various techniques can be used to obtain incidence bounds.

For instance, Bourgain [3] used the incidence bound of Bourgain, Katz, and Tao [4] for small subsets of $\mathbb{F}_p$ to prove the following expansion bound. For $f(x, y) = x^2 + xy$ and $A \subset \mathbb{F}_p$ with $|A| < p^{c'}$, we have $|f(A \times A)| > |A|^{1+c}$, with $c > 0$ depending on $c' < 1$. This bound has been improved slightly, and generalized to some other polynomials, but the range of polynomials for which such bounds are known remains limited; see Aksoy Yazici et al. [1] for some recent developments.

For large subsets of finite fields (think of $|A| > p^{7/8}$), more comprehensive bounds have been proved. Bukh and Tsimerman [7] proved that $|f(A \times A)| = \Omega_d(|A|^{1+c})$ for $|A| > p^{7/8+c'}$, with $c > 0$ depending on c', if $f \in \mathbb{F}_p[x, y]$ is monic in each variable, not of the form $p(q(x, y))$ with $\deg p \geq 2$, and not of the form $g(x) + h(y)$ or $g(x)h(y)$. The exceptional cases here are close to those in Theorem 1.1. Tao [58] then proved that $|f(A \times A)| = \Omega_d(p)$ for $|A| > p^{15/16}$, unless $f \in \mathbb{F}_p[x, y]$ is additive or multiplicative, matching the condition in Theorem 1.1. The proofs in [7, 58] both used fiber products and Cauchy–Schwarz (somewhat like in Sect. 2.4), and both relied on the bound of Lang and Weil [30, Theorem 1] for points on varieties over finite fields. Some of the ideas in [58] played a role in the proof of Theorem 2.1 in [40]. In [7, Sect. 9], an Elekes-Szabó-type statement over finite fields is conjectured.

One can also ask the Elekes-Rónyai question over $\mathbb{Q}$. Of course, Theorem 1.1 gives a bound there, but it is not clear that every exceptional polynomial from Theorem 1.1 is also exceptional over $\mathbb{Q}$. For instance, although $f(x, y) = x^2 + y^2$ is additive, the construction from Sect. 1.1 would require us to choose A so that A^2 is an arithmetic progression, which is not possible in $\mathbb{Q}$. Solymosi made the following conjecture.

Conjecture 4.7 *Let $f \in \mathbb{Q}[x, y]$ be a polynomial of degree d. For $A \subset \mathbb{Q}$ we have*

$$|f(A \times A)| = \Omega_d(|A|^{1+c}),$$

unless $f(x, y) = g(ax + by)$ or $f(x, y) = g((x + a)^{\alpha}(y + b)^{\beta})$ for $a, b \in \mathbb{Q}$ and positive integers α, β.

References

1. E. Aksoy Yazici, B. Murphy, M. Rudnev, I. Shkredov, Growth estimates in positive characteristic via collisions. International Mathematics Research Notices, **2017**(23), 7148–7189 (1 December 2017), arXiv:1512.06613
2. J. Beck, On the lattice property of the plane and some problems of Dirac, Motzkin, and Erdős in combinatorial geometry. Combinatorica **3**, 281–297 (1983)
3. J. Bourgain, More on the sum-product phenomenon in prime fields and its applications. Int. J. Number Theory **1**, 1–32 (2005)
4. J. Bourgain, N. Katz, T. Tao, A sum-product estimate in finite fields, and applications. Geom. Funct. Anal. **14**, 27–57 (2004)
5. P. Brass, W. Moser, J. Pach, *Research Problems in Discrete Geometry* (Springer, Berlin, 2005)
6. A. Bronner, A. Sheffer, M. Sharir, Distinct distances on a line and a curve (2016) [manuscript]
7. B. Bukh, J. Tsimerman, Sum-product estimates for rational functions. Proc. Lond. Math. Soc. **104**, 1–26 (2012)
8. M. Charalambides, Distinct distances on curves via rigidity. Discret. Comput. Geom. **51**, 666–701 (2014)
9. F. de Zeeuw, A course in algebraic combinatorial geometry, lecture notes available from the author's website http://dcg.epfl.ch/page-84876-en.html (2015)
10. G. Elekes, in *SUMS versus PRODUCTS in Number Theory Algebra and Erdős Geometry*, Paul Erdős and his Mathematics II, Bolyai Society Mathematical Studies, vol. 11 (2002), pp. 241–290
11. G. Elekes, E. Szabó, On triple lines and cubic curves: the orchard problem revisited, arXiv:1302.5777 (2013)
12. G. Elekes, Circle grids and bipartite graphs of distances. Combinatorica **15**, 167–174 (1995)
13. G. Elekes, On the number of sums and products. Acta Arith. **81**, 365–367 (1997)
14. G. Elekes, A combinatorial problem on polynomials. Discret. Comput. Geom. **19**, 383–389 (1998)
15. G. Elekes, A note on the number of distinct distances. Period. Math. Hung. **38**, 173–177 (1999)
16. G. Elekes, On linear combinatorics III, few directions and distorted lattices. Combinatorica **1**, 43–53 (1999)
17. G. Elekes, L. Rónyai, A combinatorial problem on polynomials and rational functions. J. Comb. Theory Ser. A **89**, 1–20 (2000)
18. G. Elekes, M. Sharir, Incidences in three dimensions and distinct distances in the plane. Comb. Probab. Comput. **20**, 571–608 (2011)
19. G. Elekes, E. Szabó, How to find groups? (And how to use them in Erdős geometry?). Combinatorica **32**, 537–571 (2012)
20. G. Elekes, M.B. Nathanson, I.Z. Ruzsa, Convexity and sumsets. J. Number Theory **83**, 194–201 (1999)

21. G. Elekes, M. Simonovits, E. Szabó, A combinatorial distinction between unit circles and straight lines: how many coincidences can they have? Comb. Probab. Comput. **18**, 691–705 (2009)
22. P. Erdős, E. Szemerédi, On sums and products of integers, in *Studies in Pure Mathematics* (Birkhäuser, 1983), pp. 213–218
23. P. Erdős, On sets of distances of *n* points. Am. Math. Mon. **53**, 248–250 (1946)
24. P. Erdős, G. Purdy, Some extremal problems in geometry. J. Comb. Theory **10**, 246–252 (1971)
25. B. Green, T. Tao, On sets defining few ordinary lines. Discret. Comput. Geom. **50**, 409–468 (2013)
26. L. Guth, N.H. Katz, On the Erdős distinct distances problem in the plane. Ann. Math. **181**, 155–190 (2015)
27. R. Hartshorne, *Algebraic Geometry* (Springer, Berlin, 1977)
28. N. Hegyvári, F. Hennecart, Conditional expanding bounds for two-variable functions over prime fields. Eur. J. Comb. **34**, 1365–1382 (2013)
29. S. Konyagin, I. Shkredov, On sum sets of sets, having small product set, in Proceedings of the Steklov Institute of Mathematics, vol. 290, n.1 (August 2015), pp. 288–299, arXiv:1503.05771
30. S. Lang, A. Weil, Number of points of varieties in finite fields. Am. J. Math. **76**, 819–827 (1954)
31. R.J. Lipton, blog post (2009), http://rjlipton.wordpress.com/2009/11/30//the-curious-history-of-the-schwartz-zippel-lemma/
32. B. Lund, A. Sheffer, F. de Zeeuw, Bisector energy and few distinct distances, in *31st International Symposium on Computational Geometry (SoCG 2015)* (2015), pp. 537–552, arXiv:1411.6868
33. J. Matoušek, *Lectures on Discrete Geometry* (Springer, Berlin, 2002)
34. H. Nassajian Mojarrad, T. Pham, C. Valculescu, F. de Zeeuw, *Schwartz–Zippel Bounds for Two-dimensional Products* (2015), arXiv:1507.08181
35. J. Pach, F. de Zeeuw, Distinct distances on algebraic curves in the plane, Comb. Probab. Comput. **26**(1), 99–117 (2017), arXiv:1308.0177; Proceedings of the Thirtieth Annual Symposium on Computational geometry (2014), pp. 549–557
36. J. Pach, M. Sharir, On the number of incidences between points and curves. Comb. Probab. Comput. **7**, 121–127 (1998)
37. O.E. Raz, *A Note on Distinct Distance Subsets* (2016), arXiv:1603.00740
38. O.E. Raz, M. Sharir, *Rigidity of complete bipartite graphs in the plane* (2016) [manuscript]
39. O.E. Raz, M. Sharir, The number of unit-area triangles in the plane: Theme and variations. Combinatorica **37**(6), 1221–1240 (December 2017), arXiv:1501.00379
40. O.E. Raz, M. Sharir, F. de Zeeuw, *Polynomials vanishing on cartesian products: The Elekes–Szabó theorem revisited*, Duke Math. J. **165**(18), 3517–3566 (2016), arXiv:1504.05012; in *31st International Symposium on Computational Geometry (SoCG 2015)* (2015), pp. 522–536
41. O.E. Raz, M. Sharir, F. de Zeeuw, The Elekes–Szabó theorem in four dimensions. Israel Journal of Mathematics **227**(2), 663–690 (August 2018) [manuscript]
42. O.E. Raz, M. Sharir, J. Solymosi, On triple intersections of three families of unit circles, Discret. Comput. Geom. **54**, 930–953 (2015); in *Proceedings of the Thirtieth Annual Symposium on Computational Geometry* (2014), pp. 198–205
43. O.E. Raz, M. Sharir, J. Solymosi, Polynomials vanishing on grids: the Elekes–Rónyai problem revisited. Am. J. Math. arXiv:1401.7419; in *Proceedings of the thirtieth annual symposium on Computational geometry* (2014), pp. 251–260
44. R. Schwartz, J. Solymosi, F. de Zeeuw, Extensions of a result of Elekes and Rónyai. J. Comb. Theory Ser. A **120**, 1695–1713 (2013)
45. M. Sharir, S. Smorodinsky, C. Valculescu, F. de Zeeuw, *Distinct distances between points and lines*. Comput. Geom. **69**, 2–15 (2018), arXiv:1512.09006
46. M. Sharir, J. Solymosi, Distinct distances from three points, Comb. Probab. Comput. **25**(4), 623–632 (2016), arXiv:1308.0814
47. M. Sharir, A. Sheffer, J. Solymosi, Distinct distances on two lines. J. Comb. Theory Ser. A **120**, 1732–1736 (2013)

48. A. Sheffer, The polynomial method, lecture notes from a course at Caltech (2015), http://www.math.caltech.edu/~2014-15/3term/ma191c-sec2/
49. A. Sheffer, E. Szabó, J. Zahl, Point-curve incidences in the complex plane. Combinatorica **38**(2), 487–499 (April 2018), arXiv:1502.0
50. A. Sheffer, J. Zahl, F. de Zeeuw, *Few distinct distances implies no heavy lines or circles*. Combinatorica **36**(3), 349–364 (June 2016), arXiv:1308.5620
51. C.-Y. Shen, Algebraic methods in sum-product phenomena. Isr. J. Math. **188**, 123–130 (2012)
52. J.H. Silverman, J. Tate, *Rational Points on Elliptic Curves* (Springer, Berlin, 1992)
53. J. Solymosi, F. de Zeeuw, *Incidence bounds for complex algebraic curves on cartesian products* (2018) [this volume]
54. J. Solymosi, Bounding multiplicative energy by the sumset. Adv. Math. **222**, 402–408 (2009)
55. J. Spencer, E. Szemerédi, W.T. Trotter, Unit distances in the Euclidean plane, in *Graph Theory and Combinatorics*, (Academic Press, Dublin, 1984), pp. 293–303
56. E. Szemerédi, W.T. Trotter, Extremal problems in discrete geometry. Combinatorica **3**, 381–392 (1983)
57. T. Tao, blog post (2012), http://terrytao.wordpress.com/2012/08/31/the-lang-weil-bound/
58. T. Tao, Expanding polynomials over finite fields of large characteristic, and a regularity lemma for definable sets. Contrib. Discret. Math. **10**, 22–98 (2015)
59. P. Ungar, $2N$ noncollinear points determine at least $2N$ directions. J. Comb. Theory Ser. A **33**, 343–347 (1982)
60. C. Valculescu, F. de Zeeuw, Distinct values of bilinear forms on algebraic curves, Contrib. Discret. Math (2015). arXiv:1403.3867
61. H. Wang, Exposition of Elekes Szabo paper, arXiv:1512.04998 (2015)
62. J. Zahl, A Szemerédi–Trotter type theorem in $\mathbb{R}^4$. Discret. Comput. Geom. **54**, 513–572 (2015)

The Geometry of Abrasion

Gábor Domokos and Gary W. Gibbons

Abstract Our goal is to narrow the gap between the mathematical theory of abrasion and geological data. To this end, we first review existing mean field geometrical theory for the abrasion of a single particle under collisions and extend it to include mutual abrasion of two particles and also frictional abrasion. Next we review the heuristically simplified box model [8], operating with ordinary differential equations, which also describes mutual abrasion and friction. We extend the box model to include an independent physical equation for the evolution of mass and volume. We introduce volume weight functions as multipliers of the geometric equations and use these multipliers to enforce physical volume evolution in the unified equations. The latter predict, in accordance with Sternberg's Law, exponential decay for volume evolution so the extended box model appears to be suitable to match and predict field data. The box model is also suitable for tracking the collective abrasion of large particle populations. The mutual abrasion of identical particles, modeled by the self-dual flows, plays a key role in explaining geological scenarios. We give stability criteria for self-dual flows in terms of the parameters of the physical volume evolution models and show that under reasonable assumptions these criteria can be met by physical systems.

1 Introduction

In geophysics, abrasion is a process where material is removed in small, successive steps from the *abraded particle*. Depending on the environment, this may happen either by colliding with *abrading particles* or by wear due to friction. The former

G. Domokos (✉)
MTA-BME Morphodynamics Research Group and Dept. of Mechanics, Materials and Structures, Budapest University of Technology, Műegyetem rakpart 1-3, Budapest 1111, Hungary
e-mail: domokos@iit.bme.hu

G. W. Gibbons
Dept. of Applied Mathematics and Theoretical Physics, Cambridge University, Wilberforce Road, Cambridge CB3 0WA, UK
e-mail: gwg1@cam.ac.uk

G. Ambrus et al. (eds.), *New Trends in Intuitive Geometry*, Bolyai Society Mathematical Studies 27, https://doi.org/10.1007/978-3-662-57413-3_6

we call *collisional*, the latter *frictional* abrasion. The goal of the geometric theory of abrasion is either to predict, based on current information on the environment, how the shape of the abraded particle will evolve in the future, or, to provide information on past environments based on currently observed particle shape. In this context, 'abraded particle' may refer to anything between the size of a sand-grain and an asteroid and abrasion may occur on the full surface or only on an exposed part thereof.

Not only is the geometric theory yet incomplete, also, there appears to be a substantial gap between existing mathematical models and the field data which could serve their verification. Our goal in this review paper is to narrow this gap by introducing existing models and extending them in a manner which admits direct comparison with field data. Keeping this double objective in mind we will not only describe the mathematical relationship between various models, simultaneously we also provide the explicit formulae which are essential for building computational tools.

1.1 Model Types

As mentioned above, the physical process of abrasion is fundamentally discrete and perhaps closest approximated by *chipping models* [11, 23]. Here the abraded particle is represented by a polyhedron and abrasion is modeled by *chipping events* where the abraded polyhedron is truncated by a plane. While chipping models have been successful in quantitative approximation of abrasion processes [12], their relationship to mean field models remains still unclear. In particular, the limit transition to mean field models appears to be challenging, as essential qualitative differences between smooth surfaces and their fine discretizations have been recently pointed out [10]. In this paper we will not discuss chipping models in detail, however, we will point out some aspects of their relationship to the mean field equations.

The space- and time-averaged mean field equations of abrasion can be formulated as partial differential equations (PDEs). One possible compact way of writing these PDEs is to express the evolution speed v in the direction of the surface normal at each surface point of the abraded particle; in this paper we will adopt this convention. PDE models may describe the shape evolution of a single particle (individual models) or a coupled system of PDEs may describe the mutual abrasion process of two particles (binary models) where the roles of abraded and abrading particle are symmetrical. PDE models may be *local* if they only involve derivatives and may be *nonlocal* if they also involve surface or volume integral terms. As we will show, collisional abrasion of a single particle can be described by local PDEs while binary collisional models as well as the description of frictional abrasion naturally lead to nonlocal equations.

The geometric PDEs describing shape evolution under abrasion are highly nonlinear. Not only is their analytical theory incomplete, even numerical approximations are often nontrivial. The latter involve various types of discretizations which should not be confused with the chipping models - as we pointed out earlier, the latter do not arise immediately as natural discretizations of the PDE models.

Another approach to approximating PDEs is to reduce them to a system of ordinary differential equations (ODEs). There are various ways to perform this reduction, here we will present a heuristic reduction called the *box model* [8] which aims to approximately track the evolution of the orthogonal bounding box of the abrading particle rather than the evolution of the actual shape. While the box model relies on a heuristic simplification of the PDE models, it opens the possibility to study *collective abrasion*, i.e. the simultaneous evolution of many particles under mutual abrasion. The box model reveals many interesting features of collective abrasion and several of these features can be demonstrated within the model [8]. Also, statistical simulations based on the box model have successfully approximated geophysical field data [36]. From the point of view of geophysical applications, a central concept of collisional abrasion is the case where (approximately) identical particles mutually abrade each other and are also being abraded by friction. In the box model we called this case the *s*elf-dual flow [8]. It is of fundamental interest whether self-dual flows are stable or not, i.e. whether the masses of mutually abrading particles converge or diverge. Since the original, geometrical box model in [8] did not discuss physical volume evolution, the stability can not be addressed in that model. In the current paper we complement the geometrical box model with an additional, independent equation modeling physical mass evolution. We will show that in this extended model the self-dual flows are stable under reasonable physical assumptions and we support this claim with field data. These results illustrate that we already have some limited insight into the statistical theory. Nevertheless, the governing 'master' equation for the evolution of statistical properties of abrading particle collectives (a distant analogue of the Boltzmann equation) is still missing.

1.2 Historical Overview

The shape of pebbles has been a matter of discussion since at least the time of Aristotle [24]. In general, the central question is whether particular pebble shapes emerge from the abrasion and transport processes. Aristotle himself claimed that spherical shapes dominate. However, as Aristotle also observed, abrasion is a complex interaction between the abraded particle and the abrading environment represented by 'other objects' (i.e. other pebbles) where not only local properties (e.g. curvatures) but also semi-global effects due to particle shape as well as global effects due to particle transport play an important role. In this process pebbles mutually abrade each other, defining the time evolution both for the abraded pebble and the abrading environment represented by other particles subject to particle transport. In the simplest approach, one neglects the latter effects and regards the abrasion of a single pebble in a constant environment; in our previous classification of models this corresponds to an individual abrasion model. Aristotle's model for individual abrasion may be translated as

$$-v = f(R), \tag{1}$$

where v is the speed with which the pebble's surface moves along the normal (the negative sign indicates that it is moving inward), R is the radial distance from the center of gravity of the abraded pebble, f is a monotonically increasing function of R only, and in particular, f is independent of time. Note that (1) is a partial integro-differential equation, since the location of the center of gravity is determined by time-dependent integrals, so, according to our terminology, (1) is a non-local, individual PDE model of abrasion. The modern theory of individual abrasion, based on local PDE models, appears to start with the pioneering work, both experimental and theoretical, of Lord Rayleigh (son and biographer of the Nobelist) [30–32]. For some earlier work see [28] and a more recent article [7]. Rayleigh mainly considered axisymmetric pebbles and he observed that the ultimate shape was not necessarily, indeed seldom, spherical. He found that some pebbles' shapes are far from ellipsoidal, being much more discoid in shape. He asserted that abrasion cannot be a simple function of the Gaussian curvature K and he proved that ellipsoidal shapes abrade in a self-similar manner under the *Rayleigh flow*

$$-v = \text{constant } K^{\frac{1}{4}}, \tag{2}$$

which is a local, individual PDE model in our classification. As Rayleigh himself pointed out, physical abrasion does not obey (2), so in a physical process ellipsoids do not evolve in a self-similar manner.

Not much later, Firey [15] initiated a rigorous study by adopting an individual, local PDE model rejected by Rayleigh in which the shape evolved according to what Andrews [1] calls the *flow by the Gauss curvature* that is, he studied the PDE

$$-v = cK, \tag{3}$$

where c is a constant and $K = \kappa_1\kappa_2$ is the Gauss curvature. Firey proved that all convex shapes ultimately converge to the sphere under the action of (3). Note that the word *flow* is being used in the sense in dynamical systems theory and it should not be confused with physical fluid flow. Later, this proof was substantially amplified by Andrews [1]. Recently, Durian [14] investigated the statistical distribution of Gaussian curvature on pebble shapes. The physical assumption underlying Firey's model is that the abraded particle (pebble) undergoes a series of small collisions with a very large, smooth abrader, and this might be the case when pebbles are carried by a fast river and collide repeatedly with the riverbed, a process called bedload transport. This concept of *collisional flows* was substantially generalized by Bloore, who postulated that abrasion by a gas of small spherical abraders should be governed by a local equation of the form

$$-v = F(\kappa_1, \kappa_2), \tag{4}$$

where $\kappa_1, \kappa_2 = \frac{1}{R_1}, \frac{1}{R_2}$ and R_1, R_2 are the principal radii of curvatures, v is the speed along the normal at which the local area element dA is being eroded and $F(\kappa_1, \kappa_2)$ is some symmetric function of the principal curvatures κ_1, κ_2. Bloore proved that if

$F(0, 0) \neq 0$ then the only shape abrading under (4) in a self-similar manner is the sphere (Bloore's Similarity Theorem).

Bloore's PDE (5) is the most general form of a *curvature-driven flow*, a broad class of geometric PDEs with physical applications ranging from digital image processing [22, 25] through surface growth phenomena [21] to abrasion models. Beyond being physical models, curvature-driven flows have attracted considerable mathematical attention in their own right. While locally defined, curvature-driven flows have startling global properties, e.g. they can shrink curves and surfaces to round points [1, 16, 19]. These features made these flows powerful tools to prove topological theorems which ultimately led, via their generalizations by Hamilton [18] to Perelman's celebrated proof [29] of the Poincaré conjecture. The global features of curvature-driven flows are mostly related to the monotonic change of quantities, such as the entropy associated with Gaussian curvature [6], or other functionals, such as the Huisken functional [20].

The simplest case of (4) is perhaps abrasion by spherical abraders of radius r (Firey's model corresponds to $r \to \infty$). With brilliant intuition Bloore [4] arrived at the PDE (called Bloore's specific equation)

$$-v = a(1 + 2bH + cK), \tag{5}$$

where a = constant with the dimension of speed, $H = \frac{1}{2}(\kappa_1 + \kappa_2)$ is the mean curvature and b and c are constants with the dimensions of length and length^2 respectively. For spherical abraders of radius r, Bloore gave a statistical argument that

$$b = r, \qquad c = r^2. \tag{6}$$

and for shapes sufficiently close to the sphere, he proved that if the abraded pebble has radius $R > 3r$ then it will become more ellipsoidal whereas for $R \leq 3r$ it will approach the sphere. (Bloore's Loop Theorem). Bloore also proved that if one writes (5) for nearly spherical shapes $r(\theta, \phi, t) = a(t)(1 + \epsilon(\theta, \phi, t))$ and keeps only the lowest (zeroth or first) order terms for ϵ and its derivatives then the equation for ϵ is a modified version of the heat equation on the sphere, where the quantity analogous to heat is the Gaussian curvature (Bloore's Stability Theorem). To obtain the analogy to the heat equation Bloore applies a nonlinear rescaling of time.

In [8] we approximated (5) by a set of ordinary differential equations called the box equations under the assumption that all shapes are ellipsoidal and remain so for all times, i.e. it is sufficient to track the evolution of the orthogonal bounding boxes. We found that both Bloore's Similarity Theorem and Bloore's Loop Theorem remain valid in the collisional box model, the latter could be even generalized for shapes at arbitrary distance from the sphere. The box model was successfully tested against laboratory experiments and recently against a detailed field study along the Williams river, Australia [36]. Since the box model not only captures some basic features of the PDE model but also matches field data, we believe that it is a suitable tool to study statistical properties of abrasion processes and in this paper we will describe the corresponding basic formulae.

1.3 Geometric PDEs and Volume Evolution

The Bloore equation (5) and its box approximations correctly describe the evolution of geometrical shapes, however, these are purely geometrical equations and thus unable to predict the correct time evolution for mass and volume. One important sign of this shortcoming is that the model (5) predicts finite lifetimes for all particles whereas field observations in fluvial environments indicate an exponential volume decay as formulated by Sternberg's empirical formula, also called Sternberg's Law [35]. This indicates that volume evolution has to be derived from physical equations independent of the Bloore model.

Although physically incorrect, the purely geometrical Bloore model (and its box approximations) still predict volume evolution rates depending on the normal speed v from (5) and on the geometry of the surface Σ:

$$\dot{V}^g(v) = \int_\Sigma v dA\,, \tag{7}$$

where the superscript g refers to the geometrical equations and $\dot{()}$ denotes differentiation with respect to time. These rates we call *the geometrical volume evolution* and we derive the exact formulae in Sect. 5. As we can see in (7), $\dot{V}^g$ is a linear function of the normal speed v in (5), i.e.

$$\dot{V}^g(\lambda v) = \lambda \dot{V}^g. \tag{8}$$

Subsequently, in Sect. 6 in the spirit of Firey's work [15] we introduce the *volume weight functions* $f(V(t))$ which depend only on time and do not depend on the location on the surface. These functions enter Bloore's equation instead of the constant a and we also define their analogues in the box equations. If we have an independent physical model for volume evolution predicting volume diminution rate $\dot{V}^p$ (the superscript referring to the independent physical equations) then we can set this equal to the volume diminution predicted by the volume-weighted geometrical equations

$$\dot{V}^g(f(V)v) = \dot{V}^p \tag{9}$$

and this condition yields, via the linear property (8)

$$f(V) = \frac{\dot{V}^p}{\dot{V}^g}. \tag{10}$$

This illustrates that volume weight functions can be used to suppress the geometrical volume evolution rates entirely in favor of the physical ones. After introducing in Sect. 7 the basic equations for the statistical theory of collective abrasion, in Sect. 8 we introduce some models which predict physical volume diminution in accordance with Sternberg's Law, so combining these models with the original geometrical equations via the volume weight functions yields shape and size evolution consistent both with the geometrical Bloore theory as well as Sternberg's empirical formula for volume diminution.

In addition to introducing the volume weight functions and the physical volume evolution into the geometrical model, we also generalize the original Bloore model in other ways. In Sect. 3 we introduce the coupled system of PDEs describing the mutual abrasion of two particles, as well as the box approximations of these equations. All previously mentioned equations deal with collisional abrasion which, as we pointed out in [8], is not capable on its own to adequately describe the collective evolution of pebbles in geological environments. In Sect. 4 we introduce the PDE including friction and also its box approximations. In Sect. 8 we also provide the physical volume evolution model for the frictional case.

Frictional abrasion is particularly significant, because in [8] we showed that in the box flows if identical shapes mutually abrade each other (which we call the self-dual flow) then friction may stabilize nontrivial shapes as global attractors. However, it was not clear whether these shapes are also attractive in size, i.e. whether the self-dual flows are stable with respect to perturbations in size. Earlier we pointed out that global transport resulting in size segregation may stabilize these flows. While that is certainly a valid possibility, in Sect. 9 we show that a potentially more relevant mechanism is defined by the physical models of volume diminution. In the models introduced in our current paper we show the exact condition under which a physical volume diminution model can stabilize the self-dual flows.

2 Collisional Abrasion of an Individual Particle in Constant Environment

2.1 Bloore's Local Equation

The coefficients in Bloore's specific PDE (5) can be also identified for non-spherical abraders. A more sophisticated treatment using Schneider–Weil theory [38] leads to

$$b = \frac{M}{4\pi}, \qquad c = \frac{A}{4\pi}, \tag{11}$$

where

$$M = \int_{\Sigma} H dA, \qquad A = \int_{\Sigma} dA \tag{12}$$

are the integrated mean curvature and area, respectively. Thus one expects on purely dimensional grounds that the first term to be important for pebbles whose linear size is large compared with the size of the abraders while for pebbles whose linear size is comparable with the size of the abraders the second and third terms should be increasingly important. Evidently, when the size of the pebble is comparable with the size of the abraders, a *single pebble* treatment like Bloore's breaks down and the evolution of the abraders must also be considered.

In the mathematics literature the three terms in (5) are often treated separately. The first term in (5)

$$-v = a \tag{13}$$

is called the *Eikonal equation* or the *parallel map* and arises in the study of wave fronts with speed a, satisfying Huygens's principle. Given an initial aspherical surface the Eikonal flow tends to make the surface more aspherical and to develop faces which intersect on edges [13].

The second term in (5)

$$-v = 2abH \tag{14}$$

is called the *mean curvature flow* [5] and often arises in problems where surface tension is important [5, 33]. Given an initial aspherical surface it tends to make the surface more spherical [19].

2.2 Relation to the Kardar–Parisi–Zhang Equation

In soft condensed matter physics, interfaces are often modeled using the *Kardar–Parisi–Zhang equation* for the height function $h = h(x, y)$

$$\frac{\partial h}{\partial t} = \nu\nabla^2 h + \frac{\lambda}{2}(\nabla h)^2 + \eta(x, y, t), \tag{15}$$

where ∇ is with respect to the *flat* metric on $\mathbb{E}^2$ and $\eta(x, y, t)$ is a Langevin-type stochastic Gaussian noise term [21, 27]. It was pointed out in [26] that this was not re-parametrization invariant and is an approximation to a stochastic version of the *mean curvature flow*

$$v = -\nu H + \lambda + \eta(\sigma^A, t). \tag{16}$$

The first term is essentially the functional derivative of surface energy, i.e. a *surface tension term* and the second is the functional derivative of a volume energy i.e. a *pressure* term. In the absence of the stochastic noise, i.e. if $\eta = 0$ and if $\nu, \lambda > 0$, the system should relax to a surface of constant mean curvature $H = \frac{\lambda}{\nu}$. For pebbles $\lambda = a$ and $\nu = -2ab$ and the pressure is negative. In the absence of the noise term, the KPZ equation (15) may, by means of the substitution $w = \exp(\frac{\lambda}{2\nu}h)$, be reduced to the linear diffusion equation for w [3].

2.3 Box Equations

The Bloore equations are partial differential equations and define a flow on the infinite-dimensional space of shapes. In [8] a finite dimensional truncation was introduced which leads to a finite number of ordinary differential equations referred to as the *box equations*. The basic idea is to bound our pebble by a rectangular box of sides $2u_1, 2u_2, 2u_3$ ordered such that $u_1 \leq u_2 \leq u_3$ which defines an inscribed ellipsoid of semi-axes u_1, u_2, u_3. One then writes down three equations

$$- \dot{u}_i = F(\kappa_{1i}, \kappa_{2i}), \tag{17}$$

where $i = 1, 2, 3$ and κ_{1i}, κ_{2i} are now taken to be the curvatures of the inscribed ellipsoid at the ends of the three principal axes $(\pm u_1, 0, 0)$, $(0, \pm u_2, 0)$,$(0, 0, \pm u_3)$. Thus (5) takes the form

$$- \dot{u}_1 = a \left(1 + b \left(\frac{u_1}{u_2^2} + \frac{u_1}{u_3^2} \right) + c \frac{u_1^2}{u_2^2 u_3^2} \right), \quad \text{etc} \tag{18}$$

where etc denotes two further equations obtained by cyclic permutation of the suffices 1, 2, 3.

In [8] it was found convenient to replace the three lengths u_1, u_2, u_3 by two dimensionless ratios and a length $y_1 = \frac{u_1}{u_3}$, $y_2 = \frac{u_2}{u_3}$ and $y_3 = u_3$, yielding

$$\dot{y}_i = aF_i(y_1, y_2, y_3, b, c) = a \left(\frac{F_i^E}{y_3} + 2b \frac{F_i^M}{y_3^2} + c \frac{F_i^G}{y_3^3} \right) \tag{19}$$

$$-\dot{y}_3 = aF_3(y_1, y_2, y_3, b, c) = a \left(1 + \frac{b}{y_3} \frac{y_1^2 + y_2^2}{y_1^2 y_2^2} + \frac{c}{y_3^2} \frac{1}{y_1^2 y_2^2} \right), \tag{20}$$

where

$$F_i^E = y_i - 1, \quad F_i^M = \frac{1 - y_i^2}{2y_i} \quad F_i^G = \frac{1 - y_i^3}{y_i y_j^2}, \quad i, j = 1, 2; i \neq j. \tag{21}$$

By introducing the vector notation $\mathbf{y} = [y_1, y_2, y_3]^T$, $\mathbf{F} = [F_1, F_2, F_3]^T$, (19)–(20) can be rewritten as

$$\dot{\mathbf{y}} = a\mathbf{F}(\mathbf{y}, b, c), \tag{22}$$

which is identical to Eqs. (2.2)–(2.6) of [8].

A special case of the box equations are the *spherical flows* for which $u_1 = u_2 = u_3 = R$, where R is the radius of the sphere. The spherical flows obtained from the box equations in fact coincide with the exact solutions of the full partial differential equations (5) obtained by assuming that Σ is a sphere. Box approximations of other PDE models also yield interesting results. In [34] the authors propose a surface evolution model to describe the spheroidal weathering of rocks. Formula (44) in the paper shows that the box approximation of their model is the $c = 0$ special case of our Eq. (22).

3 Collisional Abrasion of Two, Mutually Colliding Particles

3.1 Binary Bloore Equations

In the Bloore equations the abraders are assumed to be constant in shape and size. It is, however, simple to write down a set of evolution equations for both the abraders and the abraded pebbles as done in [8] for the simplified case, the box equations. In that case we introduced semi-box-lengths v_1, v_2, v_3 for the abrading particles, yielding two dimensionless ratios and one length $z_1 = \frac{v_1}{v_3}$, $z_2 = \frac{v_2}{v_3}$ and $z_3 = v_3$. Retaining the notation of [8] we use the labels y and z for abraded and abrader, by utilising (11), the obvious partial differential equations to consider are

$$-v_y = a\left(1 + 2\frac{M_z}{4\pi}H_y + \frac{A_z}{4\pi}K_y\right), \tag{23}$$

$$-v_z = a\left(1 + 2\frac{M_y}{4\pi}H_z + \frac{A_y}{4\pi}K_z\right). \tag{24}$$

3.2 Binary Box Equations

In the box approximation the mean curvature and surface area integrals in (11) are replaced by the corresponding quantities of the orthogonal bounding box of the incoming particle (which, for simplicity is now taken as the **z** particle):

$$M = 2\pi z_3(z_1 + z_2 + 1), \quad A = 8z_3^2(z_1z_2 + z_1 + z_2). \tag{25}$$

The same quantities can be expressed for the unit cube as $M_1 = 6\pi, \quad A_1 = 24$, so in the box equations we have

$$b(\mathbf{z}) = \frac{M}{M_1} = z_3\frac{z_1 + z_2 + 1}{3} = z_3 f_z^b, \quad c(\mathbf{z}) = \frac{A}{A_1} = z_3^2\frac{z_1 + z_2 + z_1z_2}{3} = z_3^2 f_z^c. \tag{26}$$

The corresponding binary box equations can be written as

$$\dot{\mathbf{y}} = a\mathbf{F}(\mathbf{y}, b(\mathbf{z}), c(\mathbf{z})) = a\mathbf{F}^c(\mathbf{y}, \mathbf{z}), \tag{27}$$

$$\dot{\mathbf{z}} = a\mathbf{F}(\mathbf{z}, b(\mathbf{y}), c(\mathbf{y})) = a\mathbf{F}^c(\mathbf{z}, \mathbf{y}), \tag{28}$$

where superscript c refers to collisional abrasion. Equations (27)–(28) differ from Eqs. (2.13)–(2.14) of [8] by an overall multiplier.

3.3 The Self-dual Flows

As written, the Eqs. (23)–(24) have a solution for which the abraders and abraded have identical forms. This solution we refer to as the *self-dual flow*. For the self-dual flow the labels y and z are redundant and we are left with the single equation

$$-v = a\left(1 + 2\frac{M}{4\pi}H + \frac{A}{4\pi}K\right) \tag{29}$$

which in the box approximation reads

$$\dot{\mathbf{y}} = a\mathbf{F}(\mathbf{y}, b(\mathbf{y}), c(\mathbf{y})) = a\mathbf{F}^c(\mathbf{y}, \mathbf{y}). \tag{30}$$

An important question is whether the self dual flow (29) or its box version (30) are stable within the class of Binary Bloore flows (23)–(24) and Binary Box flows (27)–(28), respectively.

3.4 The Spherical Case

If both particles are spherical (with radii R_y and R_z, respectively), then both the binary Bloore equations (23)–(24) and the binary box Eqs. (27)–(28) collapse to the same two coupled first order ordinary differential equations:

$$-\dot{R}_y = a\left(1 + 2\frac{R_z}{R_y} + \left(\frac{R_z}{R_y}\right)^2\right) \tag{31}$$

$$-\dot{R}_z = a\left(1 + 2\frac{R_y}{R_z} + \left(\frac{R_y}{R_z}\right)^2\right). \tag{32}$$

4 Frictional Abrasion of an Individual Particle: Non-local Theory

4.1 Bloore Equations with Friction

In [8] the effects of mutual friction, both rolling and sliding were incorporated into the box equations. This can be done at the level of the equations describing the complete evolution of the pebble but while the equations remain first order in time they become rather non-local in the coordinates u, v used to parametrize the embedding

$$\mathbf{r} = \mathbf{r}(u, v, t) \tag{33}$$

of the surface Σ into Euclidean space.

We define $R(u, v, t) = |\mathbf{r}(u, v, t) - \bar{\mathbf{r}}(t)|$ to be the distance of the point $\mathbf{r}(u, v, t)$ from the instantaneous centroid $\bar{\mathbf{r}}(t)$ of the pebble. We also define $R_{\max}(t)$ and $R_{\min}(t)$ as the instantaneous maximum and minimum of values of $R(u, v, t)$ over the surface and we postulate that frictional abrasion is governed by

$$\frac{\partial \mathbf{r}(u, v, t)}{\partial t} = -G(R, R_{min}, R_{max})\mathbf{n}(u, v, t), \qquad G > 0. \tag{34}$$

In [8] several constraints on the general form of of the function $G(R, R_{\min}, R_{\max})$ were given and also one example satisfying these constraints was demonstrated, introducing separate terms for sliding and rolling with independent coefficients ν_s, ν_r, respectively and the dimensionless ratios $r_1 = R/R_{min}$, $r_2 = R/R_{max}$:

$$G(R, R_{min}, R_{max}) = \nu_s f_s(r_1, r_2) + \nu_r f_r(r_1, r_2) = \nu_s r_2 r_1^{-n} + \nu_r r_2(1 - r_2^n). \tag{35}$$

According to the arguments discussed in [8], for sufficiently high values of n, this model appears to capture most essential physical features of frictional abrasion. While (35) is clearly just an example ([8] describes also an alternative equation), however, it provides a simple basis for a qualitative analysis.

The geometry of frictional abrasion has not yet been verified in experiments. We also describe an alternative, simplified model based on orthogonal affinity which may be easier to implement numerically. As we will point out, the two models coincide in case of ellipsoidal shapes. We select the planar cross section C_1 with maximal area and from among the planar cross sections orthogonal to C_1 we select C_2 with maximal area. We denote the orthogonal distances to these planes by R_1, R_2, respectively. We also introduce the angles between the tangent (supporting) plane and the aforementioned two planes by γ_1, γ_2, respectively. Using these quantities, we propose to write frictional abrasion as

$$\frac{\partial \mathbf{r}(u, v, t)}{\partial t} = -G(R_1, R_2, \gamma_1, \gamma_2)\mathbf{n}(u, v, t), \qquad G > 0. \tag{36}$$

In the case of affinities orthogonal to the planes C_1 and C_2 we have

$$G(R_1, R_2, \gamma_1, \gamma_2) = \nu_s R_1 \cos\gamma_1 + \nu_r(R_1 \cos\gamma_1 + R_2 \cos\gamma_2). \tag{37}$$

and it can be checked easily that in case of ellipsoidal shapes, (36)–(37) satisfy the constraints given in [8].

Frictional abrasion can be readily introduced into the Bloore equations. As before we use the labels y and z. Since friction is an *additional, independent* mechanism for abrasion it is natural to assume in case of the friction model (35) that

$$-v_y = a\left(1 + 2\frac{M_z}{4\pi}H_y + \frac{A_z}{4\pi}K_y\right) + G(R_y, R_{y\,\min}, R_{y\,\max}), \tag{38}$$

$$-v_z = a\left(1 + 2\frac{M_y}{4\pi}H_z + \frac{A_y}{4\pi}K_z\right) + G(R_z, R_{z\,\min}, R_{z\,\max}), \tag{39}$$

and in case of the friction model (37) we get

$$-v_y = a\left(1 + 2\frac{M_z}{4\pi}H_y + \frac{A_z}{4\pi}K_y\right) + G(R_{y1}, R_{y2}, \gamma_{y1}, \gamma_{y2}), \tag{40}$$

$$-v_z = a\left(1 + 2\frac{M_y}{4\pi}H_z + \frac{A_y}{4\pi}K_z\right) + G(R_{z1}, R_{z2}, \gamma_{z1}, \gamma_{z2}). \tag{41}$$

In the case of spherical flows (35) reduces to a single constant ν_s, so we have

$$-\dot{R}_y = a\left(1 + 2\frac{R_z}{R_y} + \left(\frac{R_z}{R_y}\right)^2\right) + \nu_s, \tag{42}$$

$$-\dot{R}_z = a\left(1 + 2\frac{R_y}{R_z} + \left(\frac{R_y}{R_z}\right)^2\right) + \nu_s. \tag{43}$$

The Eq. (37) can not be interpreted for exact spheres as the direction of the largest cross section is not defined in that case.

4.2 Box Equations with Friction

If we take the $n \to \infty$ limit in the semi-local PDE (35) we obtain for the box variables

$$\dot{u}_1 = -\nu_s y_1 - \nu_r y_1, \quad \dot{u}_2 = -\nu_r y_2, \quad \dot{u}_3 = 0, \tag{44}$$

where ν_s, ν_r are the coefficients for sliding and rolling friction, respectively. We obtain (44) also from the PDE (37). Equation (44) is equivalent to

$$\dot{\mathbf{y}} = \mathbf{F}^f(\mathbf{y}, \nu_s, \nu_r) = \frac{1}{y_3}(\nu_s \mathbf{F}^S + \nu_r \mathbf{F}^R), \tag{45}$$

where

$$\mathbf{F}^S = -[y_1, 0, 0]^T, \qquad \mathbf{F}^R = -[y_1, y_2, 0]^T. \tag{46}$$

We can now simply add collisional and frictional flows (27)–(28) and (45) to obtain the collisional-frictional equations for the two-body problem:

$$\dot{\mathbf{y}} = a\mathbf{F}^c(\mathbf{y}, \mathbf{z}) + \mathbf{F}^f(\mathbf{y}, \nu_s, \nu_r) \tag{47}$$
$$\dot{\mathbf{z}} = a\mathbf{F}^c(\mathbf{z}, \mathbf{y}) + \mathbf{F}^f(\mathbf{z}, \nu_s, \nu_r)\,. \tag{48}$$

5 Volume Evolution in the Geometric Equations

5.1 Geometric Volume Evolution in the Bloore Equations: Spherical Case

The Binary Bloore equations (23)–(24) define the mutual evolution of *observable quantities*, such as maximal width D, surface area A and volume V. In general, we can not obtain closed formulae for their evolution, however, the spherical case admits such computations. In case of spherical particles with radii R_y, R_z volume evolution can be derived by integrating (31)–(32) on the surface, to obtain

$$-\dot{V}_y = -\dot{V}_z = 4a\pi(R_y + R_z)^2 \tag{49}$$

which we call the *geometrical volume evolution* for spheres in the binary Bloore equations.

5.2 Geometric Volume Evolution in the Box Equations

In the box equations we can derive geometric volume evolution for arbitrary shapes. Regardless whether the abrasion is collisional or frictional, the volumes V_y, V_z of the two particles can be expressed as

$$V_y = 8y_1y_2y_3^3\,, \tag{50}$$
$$V_z = 8z_1z_2z_3^3\,. \tag{51}$$

By differentiating (50)–(51) with respect to time we get for the geometric volume evolution:

$$\dot{V}_y^g(\mathbf{y},\dot{\mathbf{y}}) = \frac{d}{dt}(8y_1y_2y_3^3) = 8\left(\dot{y}_1y_2y_3^3 + y_1\dot{y}_2y_3^3 + 3y_1y_2y_3^2\dot{y}_3\right)\,, \tag{52}$$
$$\dot{V}_z^g(\mathbf{z},\dot{\mathbf{z}}) = \frac{d}{dt}(8z_1z_2z_3^3) = 8\left(\dot{z}_1z_2z_3^3 + z_1\dot{z}_2z_3^3 + 3z_1z_2z_3^2\dot{z}_3\right)\,, \tag{53}$$

and we note that $\dot{V}_y^g$, $\dot{V}_z^g$ are linear in $\dot{\mathbf{y}}$, $\dot{\mathbf{z}}$, respectively, i.e.

$$\lambda\dot{V}_y^g(\mathbf{y},\dot{\mathbf{y}}) = \dot{V}_y^g(\mathbf{y}, \lambda\dot{\mathbf{y}})\,, \tag{54}$$

and the same holds for $\dot{V}_z^g$. Now we substitute the collisional Eqs. (27)–(28) into (52)–(53) to obtain the geometric volume evolution specifically for collisional abrasion

$$\dot{V}_y^{g,c}(\mathbf{y},\dot{\mathbf{y}}) = \dot{V}_y^{g,c}(\mathbf{y}, a\mathbf{F}^c(\mathbf{y},\mathbf{z})) = aF^{g,c}(\mathbf{y},\mathbf{z}) \tag{55}$$

$$\dot{V}_z^{g,c}(\mathbf{z},\dot{\mathbf{z}}) = \dot{V}_z^{g,c}(\mathbf{z}, a\mathbf{F}^c(\mathbf{z},\mathbf{y})) = aF^{g,c}(\mathbf{z},\mathbf{y}). \tag{56}$$

The geometric volume evolution under friction can be derived similarly to its collisional counterpart in (55)–(56):

$$\dot{V}_y^{g,f}(\mathbf{y},\dot{\mathbf{y}}) = \dot{V}_y^{g,f}(\mathbf{y}, \mathbf{F}^f(\mathbf{y},\nu_s,\nu_r)) = F^{g,f}(\mathbf{y},\nu_s,\nu_r) \tag{57}$$

$$\dot{V}_z^{g,f}(\mathbf{z},\dot{\mathbf{z}}) = \dot{V}_z^{g,f}(\mathbf{z}, \mathbf{F}^f(\mathbf{z},\nu_s,\nu_r)) = F^{g,f}(\mathbf{z},\nu_s,\nu_r), \tag{58}$$

where $\mathbf{F}^f$ is from (45). We can also compute $F^{g,f}(\mathbf{y},\nu_s,\nu_r)$ explicitly by substituting (45)–(46) into (52):

$$F^{g,f}(\mathbf{y},\nu_s,\nu_r) = \dot{V}_y^{g,f} = \frac{f_1^f}{y_3}y_2y_3^3 + \frac{f_2^f}{y_3}y_1y_3^3 + 3f_3^f y_1y_2y_3^2 = -\frac{V_y}{y_3}(\nu_s + 2\nu_r), \tag{59}$$

where

$$f_1^f(y_1,y_2,\nu_1,\nu_2) = \nu_s F_1^S + \nu_r F_1^R = -\nu_s y_1 - \nu_r y_1 \tag{60}$$

$$f_2^f(y_1,y_2,\nu_1,\nu_2) = \nu_s F_2^S + \nu_r F_2^R = -\nu_r y_2 \tag{61}$$

$$f_3^f(y_1,y_2,\nu_1,\nu_2) = \nu_s F_3^S + \nu_r F_3^R = 0. \tag{62}$$

6 Volume Weighted Individual and Mutual Abrasion

6.1 Volume Weighted Bloore Equations

Bloore's general Eq. (4) and its particular case (5) are local in character and did not take into account the possibility that non-local properties of the pebble might influence the speed of abrasion v. In fact, three years before Bloore, Firey [15] had studied a modification of the Gauss flow (63) of the form

$$-v = \alpha V^p K, \tag{63}$$

where V is the volume of the pebble and α and p are constants. Based on some experimental work [2] consistent with the intuition that more massive pebbles should abrade faster than less massive particles, Firey chose $p = 1$. More generally one might consider replacing (5) by

$$-v = f(V)(1 + 2bH + cK), \tag{64}$$

where $f(V)$ may be considered as a variable speed of attrition for the Eikonal term depending on the mass of equivalently the volume V of the pebble. We can introduce the volume weight functions in the Binary Bloore flows (23)–(24) as:

$$-v_y = f^c(V_y, V_z)\left(1 + 2\frac{M_z}{4\pi}H_y + \frac{A_z}{4\pi}K_y\right) \tag{65}$$

$$-v_z = f^c(V_z, V_y)\left(1 + 2\frac{M_y}{4\pi}H_z + \frac{A_y}{4\pi}K_z\right) \tag{66}$$

and in case of spherical particles, based on (31)–(32), this reduces to

$$-\dot{R}_y = f^c(V_y, V_z)\left(1 + 2\frac{R_z}{R_y} + \left(\frac{R_z}{R_y}\right)^2\right) \tag{67}$$

$$-\dot{R}_z = f^c(V_z, V_y)\left(1 + 2\frac{R_y}{R_z} + \left(\frac{R_y}{R_z}\right)^2\right). \tag{68}$$

In case of both collisional and frictional abrasion we have

$$-v_y = f^c(V_y, V_z)\left(1 + 2\frac{M_z}{4\pi}H_y + \frac{A_z}{4\pi}K_y\right) + f^f(V_y)G(R_y, R_{y\,\min}, R_{y\,\max}) \tag{69}$$

$$-v_z = f^c(V_z, V_y)\left(1 + 2\frac{M_y}{4\pi}H_z + \frac{A_y}{4\pi}K_z\right) + f^f(V_z)G(R_z, R_{z\,\min}, R_{z\,\max}), \tag{70}$$

6.2 Volume Weighted Box Equations

In the box equation approximation one has $V = V(\mathbf{y}) = 8y_1y_2y_3^3$ and (22) becomes

$$\dot{\mathbf{y}} = f(V(\mathbf{y}))\mathbf{F}(\mathbf{y}, b, c). \tag{71}$$

Evidently, the path pursued by a pebble in the space of shapes is unaffected by the prefactor $f(V)$ in (64) merely the speed with which the curve is executed. It is worth noting that Winzer [40], although not stating this explicitly, arrived at a special version of (71) with $b \to \infty$, $c = 0$, $f(V) \equiv V/b$.

We can introduce the volume weight functions in the Binary Box flows (27)–(28) as:

$$\dot{\mathbf{y}} = f^c(V_y(\mathbf{y}), V_z(\mathbf{z}))\mathbf{F}^c(\mathbf{y}, \mathbf{z}) = f^c(\mathbf{y}, \mathbf{z})\mathbf{F}^c(\mathbf{y}, \mathbf{z}) = \hat{\mathbf{F}}^c(\mathbf{y}, \mathbf{z}) \tag{72}$$

$$\dot{\mathbf{z}} = f^c(V_z(\mathbf{z}), V_y(\mathbf{y}))\mathbf{F}^c(\mathbf{z}, \mathbf{y}) = f^c(\mathbf{z}, \mathbf{y})\mathbf{F}^c(\mathbf{z}, \mathbf{y}) = \hat{\mathbf{F}}^c(\mathbf{z}, \mathbf{y}), \tag{73}$$

where ^ indicates that the volume weight is included in the operator. The linear behavior (54) and Eqs. (55)–(56) imply that in the volume weighted box Eqs. (72)–(73) volume evolution will be given by

$$\dot{\hat{V}}_y^{g,c}(\mathbf{y},\dot{\mathbf{y}}) = \dot{\hat{V}}_y^{g,c}(\mathbf{y}, f^c(\mathbf{y},\mathbf{z})\mathbf{F}^c(\mathbf{y},\mathbf{z})) = f^c(\mathbf{y},\mathbf{z})F^{g,c}(\mathbf{y},\mathbf{z}) \tag{74}$$

$$\dot{\hat{V}}_z^{g,c}(\mathbf{z},\dot{\mathbf{z}}) = \dot{\hat{V}}_z^{g,c}(\mathbf{z}, f^c(\mathbf{z},\mathbf{y})\mathbf{F}^c(\mathbf{z},\mathbf{y})) = f^c(\mathbf{z},\mathbf{y})F^{g,c}(\mathbf{z},\mathbf{y}), \tag{75}$$

where ^ refers to the inclusion of the volume weight function and $F^{g,c}$ is given in (55). We introduce the volume weight function in an analogous manner for frictional abrasion based on (45):

$$\dot{\mathbf{y}} = f^f(V_y(\mathbf{y}))\mathbf{F}^f(\mathbf{y},\nu_s,\nu_r) = f^f(\mathbf{y})\mathbf{F}^f(\mathbf{y},\nu_s,\nu_r) = \hat{\mathbf{F}}^f(\mathbf{y},\nu_s,\nu_r) \tag{76}$$

and again ^ indicates that the volume weight is included in the operator. Here again (54) and (57)–(58) imply that in volume weighted frictional box Eq. (76) volume evolution is given by:

$$\dot{\hat{V}}_y^{g,f}(\mathbf{y},\dot{\mathbf{y}}) = \dot{\hat{V}}_y^{g,f}(\mathbf{y}, f^f(\mathbf{y})\mathbf{F}^f(\mathbf{y},\nu_s,\nu_r)) = f^f(\mathbf{y})F^{g,f}(\mathbf{y},\nu_s,\nu_r), \tag{77}$$

where $F^{g,f}$ is given in (59) and $\dot{\hat{V}}_z^{g,f}(\mathbf{z},\dot{\mathbf{z}})$ is defined in the same manner.

Our next goal is to derive the volume weight function $f(V_y, V_z)$ for the Binary Bloore Flows (65)–(66) and Binary Box Flows (72)–(73), based on some physical considerations and to investigate the stability of the volume-weighted self-dual flows. The PDE (65)–(66) only admits the study of the special case where both particles are spherical and we will derive the volume weight function for this case. Subsequently, in an analogous manner, we will identify the volume weight function for general (non-spherical) particle evolution in the box Eqs. (72)–(73).

6.3 Asymmetry of the Volume Weight Function Stabilizing the Self-dual Flows

Before introducing the physical considerations, we point out, purely on geometric grounds, a fundamental property of the volume weight function f: in order to stabilize the self-dual collisional flows, f *needs to be* asymmetrical. It is sufficient to show in the spherical case that the symmetric volume weight function implies instability.

The spherical flow (67)–(68) takes place in the positive quadrant of the $R_y - R_z$ plane with both R_y and R_z decreasing. Defining, as is standard

$$\tan\theta = \frac{R_z}{R_y}, \qquad \tan\psi = \frac{dR_z}{dR_y} \tag{78}$$

we find that the trajectories satisfy

$$\frac{dR_z}{dR_y} = \frac{f(V_z, V_y)}{f(V_y, V_z)} \cot^2\theta\,, \tag{79}$$

or in terms of volumes:

$$\frac{dV_z}{dV_y} = \frac{f(V_z, V_y)}{f(V_y, V_z)}. \tag{80}$$

It is immediately apparent that if f is symmetric, i.e.

$$f(V_z, V_y) = f(V_y, V_z) \tag{81}$$

then we have

$$\frac{dV_z}{dV_y} = 1\,, \tag{82}$$

that is the trajectories are straight lines in the V_y, V_z plane making an angle of $\frac{\pi}{4}$ with the axes. By using $V_y = \frac{4\pi}{3}R_y^3$ and $V_z = \frac{4\pi}{3}R_z^3$, these can be transferred to the $[R_z, R_y]$ plane where straight lines become curves which in the downward direction move away from the straight line $R_z = R_y$. It follows that if the volume weight function $f(V_y, V_z)$ is symmetrical then the self-dual trajectory defined by $R_y = R_z$ is unstable within the class of spherical flows. Beyond showing that asymmetry is a necessary condition for the stability for the self-dual flows, we also show a simple example where it is also sufficient. If we assume that

$$f(V_y, V_z) = \left(\frac{V_y}{V_z}\right)^p \tag{83}$$

then we have

$$\frac{dR_z}{dR_y} = \tan\psi = (\tan\theta)^{2(3p-1)}. \tag{84}$$

If $p < \frac{1}{3}$ and the trajectory lies above the diagonal line $\theta = \frac{\pi}{4}$, then its slope ψ is less than $\frac{\pi}{4}$ and it will move away from the diagonal. If the trajectory lies below the diagonal then its slope ψ is greater than $\frac{\pi}{4}$ and it will again move away from the diagonal. Thus if $p \leq \frac{1}{3}$ the self-dual flow is unstable and if $p > \frac{1}{3}$ it will be stable.

As pointed out in [8], friction can stabilize attractors in the geometric self-dual flows in the $[y_1, y_2]$ space of box ratios. Here we would like to point out that in case of volume-weighted spherical flows, friction also contributes to the relative stabilization of size in the sense that the particle's linear size converges to each other. Since we treat friction as an individual abrasion, any monotonically increasing volume weight function $f^f(V_y)$ associated with friction produces an asymmetry which has an analogous effect to the above-discussed asymmetry of the volume weight function for collisional abrasion.

In the next section we show that asymmetric models (although more complex than (83)) emerge naturally from physical considerations. We will only prove the stabilizing property of the physical volume weight functions for the spherical case, however, they appear to have the same effect for general geometries.

6.4 Derivation of the Volume Weight Function from Physical Models in the Bloore Equations

We assume that volume evolution is given by an independent physical model as

$$\dot{V}_y^p = C_y^c g^c(V_y, V_z) \tag{85}$$

$$\dot{V}_z^p = C_z^c g^c(V_z, V_y), \tag{86}$$

where the superscript p stands for "physical" and the constants C_y^c, C_z^c may differ due to the different hardness of the material of the particles. In the spherical flows we can use (49) to obtain the volume weight function as

$$f(V_y, V_z) = \frac{C_y^c g^c(V_y, V_z)}{4a\pi(R_y + R_z)^2}. \tag{87}$$

Using (87), (67)–(68) can be written as

$$-\dot{R}_y = \frac{C_y^c g^c(V_y, V_z)}{4a\pi(R_y + R_z)^2}\left(1 + 2\frac{R_z}{R_y} + \left(\frac{R_z}{R_y}\right)^2\right) \tag{88}$$

$$-\dot{R}_z = \frac{C_z^c g^c(V_z, V_y)}{4a\pi(R_y + R_z)^2}\left(1 + 2\frac{R_y}{R_z} + \left(\frac{R_y}{R_z}\right)^2\right). \tag{89}$$

Later we give examples for some specific functions $g^c(V_y, V_z)$.

6.5 Derivation of the Volume Weight Function from Physical Models in the Box Equations

Without giving any specific physical abrasion model, in this subsection we show how the volume weight functions f^c, f^f can be formally derived if such models are available. Later on, we give specific examples of some physical models, however, any physical model can be plugged into the equations of this subsection. We only assume that the physical model is defined by volume evolution equations for collisional and frictional abrasion, respectively, as

$$\dot{V}_y^{p,c} = C_y^c g^c(\mathbf{y}, \mathbf{z}) \qquad \dot{V}_y^{p,f} = C_y^f g^f(\mathbf{y}) \tag{90}$$

$$\dot{V}_z^{p,c} = C_z^c g^c(\mathbf{z}, \mathbf{y}) \qquad \dot{V}_z^{p,f} = C_z^f g^f(\mathbf{z}), \tag{91}$$

then by using (74)–(75) and (77) we can set the geometric and physical volume evolution rates to be equal and this condition yields:

$$f^c(\mathbf{y}, \mathbf{z}) = \frac{C_y^c g^c(\mathbf{y}, \mathbf{z})}{F^{g,c}(\mathbf{y}, \mathbf{z})} \tag{92}$$

$$f^f(\mathbf{y}) = \frac{C_y^f g^f(\mathbf{y})}{F^{g,f}(\mathbf{y})} \tag{93}$$

and $F^{g,c}$ and $F^{g,f}$ are given in (55) and (57), respectively. So, based on the above equations and (72)–(73) and (76), the box equations for the combined model (including the physical law for volume evolution) are

$$\dot{\mathbf{y}} = \frac{C_y^c g^c(\mathbf{y}, \mathbf{z})}{F^{g,c}(\mathbf{y}, \mathbf{z})} \mathbf{F}^c(\mathbf{y}, \mathbf{z}) + \frac{C_y^f g^f(\mathbf{y})}{F^{g,f}(\mathbf{y})} \mathbf{F}^f(\mathbf{y}, \nu_s, \nu_r) = \mathbf{F}^u(\mathbf{y}, \mathbf{z}) \tag{94}$$

$$\dot{\mathbf{z}} = \frac{C_z^c g^c(\mathbf{z}, \mathbf{y})}{F^{g,c}(\mathbf{z}, \mathbf{y})} \mathbf{F}^c(\mathbf{z}, \mathbf{y}) + \frac{C_z^f g^f(\mathbf{z})}{F^{g,f}(\mathbf{z})} \mathbf{F}^f(\mathbf{z}, \nu_s, \nu_r) = \mathbf{F}^u(\mathbf{z}, \mathbf{y}), \tag{95}$$

where $\mathbf{F}^c$, $\mathbf{F}^f$ are defined in (22), (27) and (45), respectively and $F^{g,c}$,$F^{g,f}$ are given in (55), (57).

7 Collective Abrasion

Using the above model, a Markov-process can be simulated by regarding $\mathbf{y}$, $\mathbf{z}$ in (94)–(95) as random vectors with *identical* distributions since they represent two random samples of the same pebble population. The evolution of this Markov process (and thus the time evolution of the pebble size and ratio distributions) is of prime interest since it determines the physical relevance of the stable attractors identified in [8]. While the analytical investigation of the Markov process is beyond the scope of this paper, direct simulations are relatively straightforward. We consider N pebbles out of which we randomly draw two with coordinates $\mathbf{y}^0$, $\mathbf{z}^0$ and run Eqs. (94)–(95) for a short time period Δt on these initial conditions to obtain the updated vectors $\mathbf{y}^1$,$\mathbf{z}^1$. In the simplest linear approximation we have the recursive formula

$$\mathbf{y}^{i+1} = \mathbf{y}^i + \Delta t \mathbf{F}^u(\mathbf{y}^i, \mathbf{z}^i) \tag{96}$$

$$\mathbf{z}^{i+1} = \mathbf{z}^i + \Delta t \mathbf{F}^u(\mathbf{z}^i, \mathbf{y}^i) \, . \tag{97}$$

Such an iterative step can be regarded as the cumulative, averaged effect of several collisions between the two selected pebbles. Apparently, the $N = 2$, $\Delta t \to 0$ case is identical to (94)–(95). In [8] we investigated the behavior of the deterministic flows in the special cases of steady state flows (19)–(20) and self-dual flows (30). Multi-body simulations allow the numerical study of the statistical stability of the flows, i.e. one can assess the stability of the above-mentioned special cases.

8 Physical Models of Mass Evolution

It appears to be widely believed that the relationship between volume V and time t follows an exponential law suggested by Sternberg [35]

$$V(t) = V(0)e^{-\frac{t}{t_0}}, \tag{98}$$

where t_0 is a constant. More accurately, Sternberg's Law is usually held to hold for the volume of pebbles as a function of distance along a river or stream. If they are transported along the river at constant speed this is equivalent to (98).

Our goal is to introduce physical collisional models which, on one hand, predict infinite lifetimes (in accordance with Sternberg's law), on the other hand, they can be plugged into the geometric equations via the formulae (94)–(95). We propose first collisional models followed by frictional models.

8.1 Collisional Models

It seems intuitively reasonable that mutual abrasion will be the greater the greater the kinetic energy E_{com} of the colliding particles in their common rest frame. This is given by

$$E_{\text{com}} = \frac{1}{2}\frac{m_y m_z}{m_y + m_z}u^2, \tag{99}$$

where u is the relative velocity of the abrader and the abraded and m_y and m_z are the masses of the pebbles. These will be related to the densities ρ_y and ρ_z and volumes by

$$m_y = \rho_y V_y, \qquad m_z = \rho_z V_z. \tag{100}$$

For a homogeneous ensemble of pebbles it is reasonable to assume $\rho_y = \rho_z$. In binary collisions one might suppose that the rate of reduction of volume is proportional to E_{com} and a power α of the mass. Assuming equal densities and that u^2 is on average a constant, we arrive at the equation for *physical volume evolution*

$$-\dot{V}_y^{c,p} = C_y^c V_y^{\alpha}\frac{V_y V_z}{V_y + V_z} = C_y^c g^c(V_y, V_z) \tag{101}$$

$$-\dot{V}_z^{c,p} = C_z^c V_z^\alpha \frac{V_y V_z}{V_y + V_z} = C_z^c g^c(V_z, V_y), \tag{102}$$

where the superscript p stands for "physical" and the constants C_y^c, C_z^c may differ due to the different hardness of the material of the particles. This results in

$$\frac{dV_z}{dV_y} = \frac{C_z^c}{C_y^c}\left(\frac{V_z}{V_y}\right)^\alpha. \tag{103}$$

We remark that one plausible motivation behind (101)–(102) is Weibull Theory for fragmentation [37, 39] relating the material strength σ_{crit} to the specimen mass m as

$$\sigma_{crit} = \sigma_0 \left(\frac{m}{m_0}\right)^{-\frac{1}{\mu}}, \tag{104}$$

where σ_0 is the strength of the specimen of unit volume m_0 and μ is Weibull's modulus. This formula is based on the statistical distribution of Griffith cracks [17] and $\mu \to \infty$ corresponds to homogeneous material without Griffith cracks. Here we assume that the critical fragmentation energy E_f per fragmented mass m_f, given by

$$\tau_{crit} = \frac{E_f}{m_f}, \tag{105}$$

follows a similar power law

$$\tau_{crit} = \tau_0 \left(\frac{m}{m_0}\right)^{-\frac{1}{\bar{\mu}}}, \tag{106}$$

and similarly to Weibull's modulus, $\bar{\mu} \to \infty$ corresponds to homogeneous material. Using Eqs. (99), (104) and (105) yields (101)–(102) with $\alpha = 1/\bar{\mu}$. Note that $\alpha = 0$ corresponds to homogeneous material. As pointed out in [37], brittle materials are *softening* in fragmentation in the sense that the energy per unit fragmented volume is decreasing with the size of the particle. This behaviour implies in (101)–(102)

$$\alpha \geq 0. \tag{107}$$

In the box equations, via (50)–(51), (101)–(102) is translated into

$$-\dot{V}_y^{c,p} = C_y^c g^c(\mathbf{y}, \mathbf{z}) \tag{108}$$

$$-\dot{V}_z^{c,p} = C_z^c g^c(\mathbf{z}, \mathbf{y}) \tag{109}$$

which can be plugged into (94)–(95). In the spherical case we have

$$g^c(R_y, R_z) = \left(\frac{4\pi}{3}\right)^{1+\alpha} \frac{R_y^{3(1+\alpha)} R_z^3}{R_y^3 + R_z^3} \tag{110}$$

and using (87) this yields for the volume weight function

$$f(V_y, V_z) = \frac{C_y^c}{3a}\left(\frac{4\pi}{3}\right)^{\alpha} \frac{R_y^3 R_z^3}{(R_y^3 + R_z^3)} \frac{R_y^{3\alpha}}{(R_y + R_z)^2}. \tag{111}$$

By substituting (110) into (88) we get the physical evolution equations for spheres.

We also note that (111) is asymmetrical: $f(V_y, V_z) \neq f(V_z, V_y)$. Indeed, in the case of spheres, (101)–(102) yield

$$\frac{dR_z}{dR_z} = \left(\frac{R_z}{R_y}\right)^{3\alpha-2} \tag{112}$$

and we can immediately see that the self-dual trajectory $R_y = R_z$ will therefore be unstable unless $\alpha > \frac{2}{3}$. Recalling that the exponent α was motivated by Weibull theory, this condition suggests that, in the absence of other effects, for nearly homogeneous particles the self-dual flows will be unstable.

8.2 Frictional Models

Here we describe the evolution of mass as a single particle K_y is being abraded by friction and we postulate

$$-\dot{m}_y = \bar{C}_y^f m_y^{\beta}, \qquad \bar{C}_y^f > 0 \tag{113}$$

which, for $\beta = 1$, is essentially a simplified version of Archard's formula [2] by assuming constant velocity and contact area with the abrading surface. If the contact stress approaches the yield stress then higher β values may be appropriate. The case $\beta \geq 1$ corresponds to infinite time horizon and, as we will show in the next subsection, the volume evolution Eqs. (101)–(102) also predict similar behaviour, so for $\beta \geq 1$ the two effects (collisional and frictional abrasion) may compete on the same timescale. In Eq. (113), $\beta \geq 1$ can be motivated by assuming friction caused entirely by the gravity acting on the particle K_y, e.g. the particle is sliding on a free surface. Friction could also occur inside granular assemblies under compressive forces far exceeding the particles own weight; in this case mass will decay in finite time and frictional abrasion will dominate the whole process. However, as we showed in [8], only the continuous interaction of collisional and frictional abrasion can produce the geologically observed dominant pebble box ratios. Based on (113) we have

$$-\dot{V}_y^{p,f} = C_y^f V_y^{\beta} = C_y^f g^f(V_y), \tag{114}$$

where $C_y^f = \bar{C}_y^f/\rho_y$ and again, the superscript p refers to the fact that this evolution is based on physical considerations rather than geometrical ones, superscript f refers to the frictional process. In the box Eq. (114) translates into

$$-\dot{V}_y^{p,f} = C_y^f V_y^\beta = C_y^f (8y_1y_2y_3^3)^\beta = C_y^f g^f(\mathbf{y}) \tag{115}$$

which can be plugged into (94).

8.3 Collective Abrasion: Rescaling of Time

In Sect. 7 we introduced the concept of collective abrasion. In case of two particles under mutual collisions we have assumed that in equal time intervals equal number of collisions occur. If we consider a collection of particles from which we choose random pairs and evolve them under the above-described binary process then the choice of this pairs can follow various rules, in any case, we have to consider that the probability of collision in equal time between two arbitrary particles is not equal. For example, it is a plausible assumption that in the same amount of time a large particle will suffer more collisions than a small particle. We will implement the particular assumption that the number N_y of collisions per unit time suffered by the particle $\mathbf{y}$ is proportional to the ν-power of the relative volumes:

$$N_y \propto \left(\frac{V_y}{V_z}\right)^\nu. \tag{116}$$

Needless to say, this assumption would not make sense in the binary process since from it would follow that the two colliding particles suffer different number of collisions in equal time intervals. Nevertheless, in case of collective abrasion this assumption can be implemented and in essence it requires the rescaling of time. If we denote the time in the collective process by T and time in the original, binary process by t then we have

$$\frac{dT}{dt} = \left(\frac{V_y}{V_z}\right)^\nu. \tag{117}$$

If we study the collective process (96)–(97) process then rescaled time can be implemented by modifying (101)–(102) as

$$-\dot{V}_y^{c,p} = C_y^c \frac{V_y^{(\alpha+\nu+1)} V_z^{(1-\nu)}}{V_y + V_z} = C_y^c \bar{g}^c(V_y, V_z) = C_y^c \bar{g}^c(\mathbf{y}, \mathbf{z}) \tag{118}$$

$$-\dot{V}_z^{c,p} = C_z^c \frac{V_z^{(\alpha+\nu+1)} V_y^{(1-\nu)}}{V_y + V_z} = C_z^c \bar{g}^c(V_z, V_y) = C_z^c \bar{g}^c(\mathbf{z}, \mathbf{y}). \tag{119}$$

As a consequence, if we model collective abrasion then in (94)–(95) $g^c(\mathbf{y}, \mathbf{z})$ has to be replaced by $\bar{g}^c(\mathbf{y}, \mathbf{z})$ and all other formulae remain unchanged.

9 Lifetimes, Sternberg's Law and the Stability of the Self-dual Flows

9.1 Lifetimes, Physical Mass Evolution Models and Sternberg's Law

Bloore's geometric equation apparently predicts finite lifetimes for abrading particles, this is immediately suggested by the constant term on the right hand side of (5). However, not only the constant, but every single term in the geometric equation predicts finite time horizon for the particle and this property is inherited by the box equations (for details see [9]).

Our box model (94)–(95) is constructed in such a way that geometric volume evolution rates $F^{g,c}$,$F^{g,f}$ (given in (55), (57)) are completely suppressed and volume evolution is determined by the physical evolution rates given in (90)–(91). Consequently, the lifetimes for the unified box model (94)–(95) are determined by the lifetimes for the physical volume evolution models (90)–(91) and next we study the latter. As we are about to show, they predict exponential decay for the volume, thus reproducing the empirical law (98) of Sternberg [35]. Needless to say, these models are certainly not unique and others may have similar properties.

We gave two examples of physical evolution models for collisional abrasion in (101)–(102) and (118)–(119). Since the former is just the $\nu = 0$ special case of the latter it suffices to study the latter. We introduce a simple

Lemma *The differential equation* $\dot{f} = -cf^{\gamma}$ *(with* $c = \text{constant} > 0$, $f(t_0) > 0$, $\gamma \neq 1$) *has a solution* $f(t) = \left(f^{1-\gamma}(t_0) - (1-\gamma)(t-t_0)\right)^{1/(1-\gamma)}$ *for* $t \geq t_0$. *Thus if* $\gamma < 1$, $f(t)$ *goes to zero in finite time, whereas if* $\gamma > 1$, *then* $f(t)$ *reaches zero only after an infinite time.*

Similar conclusions could be reached if $c(t)$ varies with time, with $c(t - t_0)$ replaced by $\int_{t_0}^{t} c(t)dt$. In particular, if $c(t) \to 0$ and $\gamma > 1$ then we also have infinite time horizon. Equation (114) describes mass and volume evolution under friction, trivially agree with the equation in the Lemma and for $\beta > 1$ it corresponds to processes with infinite lifetimes. Next we consider Eqs. (118)–(119) for mass evolution under collisional abrasion. We note that in (118)–(119) both variables are strictly monotonically decreasing, regardless of the initial values. This implies that two cases are possible: (I) either V_y or V_z will approach zero while the other volume is still finite or (II) when both volumes approach zero simultaneously at some slope $V_z/V_y = c_0$.

Case (I) Assume V_y approaches zero first and thus we have $Vy << V_z$. Then, if $\nu = 0$, Eq. (118) for $\dot{V}_y$ may be approximated by the equation in the lemma, by setting $f = V_y, \gamma = \alpha + 1, c = -C_y^c$. By assumption (107), $\alpha \geq 0$ and so in all cases $\gamma \geq 1$. It follows that the lifetime for the **y** particle is always infinite, approaching $V_y = 0$ asymptotically. As V_y is asymptotic to zero, based on (119) so is $\dot{V}_z$, so the

z particle will also have infinite time horizon (approaching finite constant mass). If $\nu \geq 0$ then we have $c(t) = (V_y/V_z)^\nu$ and since $V_y \to 0$ we also have $c(t) \to 0$ so this also yields infinite time horizon for both particles.

Case (II) If V_y and V_z vanish together at some slope $V_z/V_y = c_0$ then we can take either to equal f in the lemma and $\gamma = \alpha + 1$, $c = -c_0 C_y^c/(c_0 + 1)$ or $c = -c_0 C_z^c/(c_0 + 1)$. The same conclusion holds.

9.2 Lifetimes and the Volume Weight Functions

Field observations of river pebbles are consistent with Sternberg's Law [35] which predicts that particles live for ever. This gives an important constraint on evolution laws. In our model, the latter determine the volume weight functions and next we shall give some general results on whether or not a volume weight function predicts a finite lifetime by giving a general upper bound on the lifetime.

In the spherical case, based on (67) we can write

$$-\dot{R}_y \geq f(V_y, V_z) \tag{120}$$

and so we have

$$\frac{R_y(t)}{R_y(0)} \leq e^{-\int_0^t \frac{f(V_y, V_z)}{R_y} dt'} \tag{121}$$

which gives exponential decay as long as $\frac{f(V_y, V_z)}{R_y}$ converges for small R_y. In the general case we may obtain volume evolution by integrating (120) over the surface Σ:

$$-\dot{V}_y \geq \int_\Sigma f(V_y, V_z) dA = A_y f(V_y, V_z). \tag{122}$$

Thus we have

$$\frac{V_y(t)}{V_y(0)} \leq e^{-\int_0^t f(V_y, V_z) \frac{A_y}{V_y} dt'} \tag{123}$$

which gives exponential decay as long as $f(V_y, V_z)\frac{A_y}{V_y}$ converges for small V_y.

9.3 Stability of the Self-dual Flows in the Stochastic Process

We can study the evolution of $\rho = V_y/V_z$ under the described process and we can see that $\rho = 1$ is always a solution of (118)–(119). The stability of this solution is of particular interest since it indicates the stability of the self-dual flows in (94)–(95). It is easy to see that the stability of $\rho = 1$ is guaranteed if

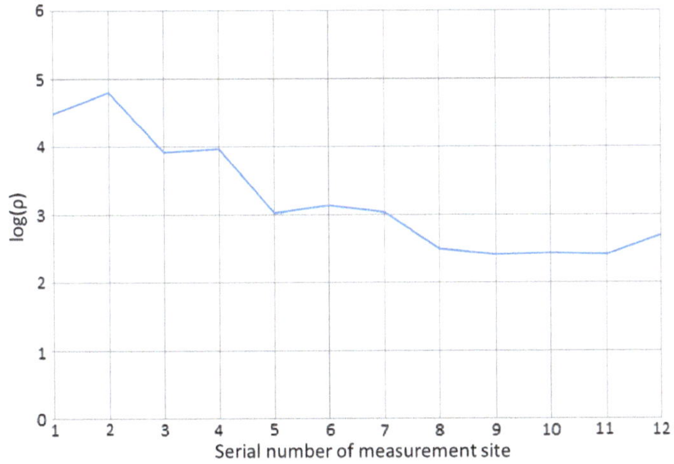

Fig. 1 Field data from the Williams river

$$(1 - \rho)\dot{\rho} > 1 \tag{124}$$

and we can see from (118)–(119) that the condition for stability is

$$\alpha + 2\nu > 1. \tag{125}$$

Now, we expect $\alpha << 1$ if the material is nearly homogeneous; this suggests that the self-dual flows are not stable in the binary process where $\nu = 0$. In other words, our model predicts that the mass ratio of two, mutually abrading particles will diverge if the material is sufficiently homogeneous. We expect to see a similar divergence of mass if we study a smaller assembly of mutually abrading particles. On the other hand, for very large populations, $\nu = 2/3$ is a plausible statistical assumption, relating the number of collisions per unit time to the effective cross section of the particle. So, in a sufficiently large collective process we expect that the self-dual flows will be stable and attractive. This is also confirmed by the field data collected along the Williams river [36] where we measured $\bar{\rho} = V_{max}/V_{min}$ in each sample. Since $\bar{\rho}$ is an upper bound for ρ, its evolution indicates the stability of the $\rho = 1$ solution. In Fig. 1 we plotted $\log(\bar{\rho})$ versus the serial number of the measurement site along the Williams river, the latter can be regarded as an approximate measure of time. As we can see, $\log(\bar{\rho})$ shows a marked decrease along the river thus indicating the stability of the $\rho = 1$ solution.

These considerations also imply that our conclusions regarding the role of segregation in [8] are only valid for the geometric equations. If we study the unified flows then we expect that under the combined effect of collisions and friction, stable attractors in the space $[y_1, y_2]$ of the box ratios will emerge spontaneously and robustly. Also, while segregation by size is catalyzing this process, it is not a pre-condition for the emergence of the attractors. Rather, we expect that abrasion itself will further help to produce pebbles of similar sizes.

Acknowledgements This research was supported by NKFI grant 119245. The comments from Dr Timea Szabó and from Prof. Fred Bloore are greatly appreciated.

References

1. B. Andrews, Guass curvature flow: the fate of rolling stones. Invent. Math. **138**, 151–161 (1999)
2. J.F. Archard, W. Hirst, The wear of metals under unlubricated conditions. Proc. R.Soc. Lond. A **236**, 397–416 (1956)
3. M.T. Batchelora, R.V. Burneb, B.I. Henry, S.D. Watt, Deterministic KPZ model for stromatolite laminae. Phys. A **282**, 123–136 (2000)
4. F.J. Bloore, The shape of pebbles. Math. Geol. **9**, 113–122 (1977)
5. K. Brakke, *The Motion of a Surface by its Mean Curvature* (Princeton University Press, Princeton, 1978)
6. B. Chow, On Harnack's inequailty and entropy for the Gaussian curvature flow. Commun. Pure Appl. Math. **XLIV**, 469–483 (1991)
7. J.E. Dobkins, R.J. Folk, Shape developement on Tahiti-Nui. J. Sediment. Petrol. **40**, 1167–1203 (1970)
8. G. Domokos, G.W. Gibbons, The evolution of pebble shape in space and time. Proc. R. Soc. Lond. **468**(2146), 3059–3079 (2012)
9. G. Domokos, G.W. Gibbons, Geometrical and physical models of abrasion, arXiv preprint (2013), arXiv:1307.5633
10. G. Domokos, Z. Lángi, T. Szabó, On the equilibria of finely discretized curves and surfaces. Monatsh. Math. **168**, 321–345 (2012)
11. G. Domokos, A. Sipos, P. Várkonyi, Continuous and discrete models for abrasion processes. Period. Polytech. Archit. **40**, 3–8 (2009)
12. G. Domokos, D.J. Jerolmack, A.Á. Sipos, Á. Török, How river rocks round: resolving the shape-size paradox. PloS one **9**(2), e88657 (2014). https://doi.org/10.1371/journal.pone.0088657
13. G. Domokos, A. Sipos, G. Szabó, P. Várkonyi, Formation of sharp edges and plane areas of asteroids by polyhedral abrasion. Astrophys. J. **699**, L13–116 (2009)
14. D.J. Durian et al., What is in a Pebble shape? Phys. Rev. Lett. **97**, 028001 (2006). (4 p.)
15. W.J. Firey, The shape of worn stones. Mathematika **21**, 1–11 (1974)
16. M.A. Grayson, The heat equation shrinks embedded plane curves to round points. J. Differ. Geom. **26**, 285–314 (1987)
17. A.A. Griffith, The phenomena of rupture and flow in solids. Philos. Trans. Roy. Soc. A **221**, 163–198 (1921)
18. R. Hamilton, Three-manifolds with positive Ricci curvature. J. Differ. Geom. **17**, 255–306 (1982)
19. G. Huisken, Flow by mean curvature of convex surfaces into spheres. J. Differ. Geom. **20**, 27–266 (1984)
20. G. Huisken, Asymptotic behavior for singularities of the mean curvature flow. J. Differ. Geom **31**, 285–299 (1990)
21. M. Kardar, G. Parisi, Y.C. Zhang, Phys. Rev. Lett. **56**, 889–892 (1986)
22. J.J. Koenderink, The structure of images. Biol. Cybern. **50**, 363–370 (1984)
23. P.L. Krapivsky, S. Redner, Smoothing rock by chipping. Phys. Rev. E **75**(3 Pt 1), 031119 (2006). https://doi.org/10.1103/PhysRevE.75.031119
24. P.D. Krynine, On the antiquity of "sedimentation" and hydrology. Bull. Geol. Soc. Am. **71**, 1721–1726 (1960)

25. C. Lu, Y. Cao, D. Mumford, Surface evolution under curvature flows. J. Vis. Commun. Imgae Represent. **13**, 65–81 (2002)
26. A. Maritan, F. Toigo, J. Koplik, J.R. Banavar, Dynamics of growing interfaces. Phys. Rev. Lett. **69**, 3193–3195 (1992)
27. M. Marsilli, A. Maritan, F. Toigo, J.B. Banavar, Stochastic growth equations and reparameterization invariance. Rev. Mod. Phys. **68**, 963–983 (1996)
28. H.R. Palmer, Observations on the motions of Shingle beaches. Philos. Trans. R. Soc. Lond. **124**, 567–576 (1834)
29. G. Perelman, Ricci flow with surgery on three-manifolds (2003), arXiv:math.DG/0303109v1
30. L. Rayleigh, Pebbles, natural and artificial. Their shape under various conditions of abrasion. Proc. R. Soc. Lond. A **181**, 107–118 (1942)
31. L. Rayleigh, Pebbles, natural and artificial. Their shape under various conditions of abrasion. Proc. R. Soc. Lond. A **182**, 321–334 (1944)
32. L. Rayleigh, Pebbles of regular shape and their production in experiment. Nature **154**, 161–171 (1944)
33. F. Rhines, K. Craig, R. Dehoff, Mechanism of steady-state grain growth in aluminium. Metal. Mater. Trans. **5**, 413–425 (1974)
34. R.C. Sarracino, G. Prasad, M. Hoohlo, A mathematical model of spheroidal weathering. Math. Geol. **19**, 269–289 (1987)
35. H. Sternberg, Untersuchungen uber Langen-und Querprofil geschiebefuhrender Flusse. Z. Bauwes. **25**, 486–506 (1875)
36. T. Szabó, S. Fityus, G. Domokos, Abrasion model of downstream changes in grain shape and size along the Williams River, Australia. J. Geophys. Res. Earth Surf. **118**(4), 2059–2071 (2013). https://doi.org/10.1002/jgrf.20142
37. O. Tsoungui, D. Vallet, J.-C. Charmet, S. Roux, Size effects in single grain fragmentation. Granul. Matter **2**, 19–27 (1999)
38. P.L. Várkonyi, G. Domokos, A general model for collision-based abrasion. IMA J. Appl. Math. **76**, 47–56 (2011)
39. W. Weibull, A statistical theory of the strength of materials. R. Swed. Inst. Eng. Res. **151** (1939)
40. K. Winzer, On the formation of elliptic stones due to periodic water waves. Eur. Phys. J. B **86**, 464 (2013). https://doi.org/10.1140/epjb/e2013-40745-3

Computing Upper Bounds for the Packing Density of Congruent Copies of a Convex Body

Fernando Mário de Oliveira Filho and Frank Vallentin

Abstract In this paper we prove a theorem that provides an upper bound for the density of packings of congruent copies of a given convex body in $\mathbb{R}^n$; this theorem is a generalization of the linear programming bound for sphere packings. We illustrate its use by computing an upper bound for the maximum density of packings of regular pentagons in the plane. Our computational approach is numerical and uses a combination of semidefinite programming, sums of squares, and the harmonic analysis of the Euclidean motion group. We show how, with some extra work, the bounds so obtained can be made rigorous.

1991 Mathematics Subject Classification 52C17 · 90C22

1 Introduction

How much of Euclidean space can be filled with pairwise nonoverlapping congruent (i.e., rotated and translated) copies of a given convex body $\mathcal{K}$?

A union of congruent copies of $\mathcal{K}$ with pairwise disjoint interiors is a *packing* of congruent copies of $\mathcal{K}$, or just a packing of $\mathcal{K}$ for short; below, packings are always packings of congruent copies of the body. The *density* of a packing is the fraction of Euclidean space it covers. Rewritten, the question of the previous paragraph is: What is the maximum density of a packing of congruent copies of $\mathcal{K}$? We call this the *body packing problem.*

The first author was support by Rubicon grant 680-50-1014 from the Netherlands Organization for Scientific Research (NWO). The second author was supported by Vidi grant 639.032.917 from the Netherlands Organization for Scientific Research (NWO).

F. M. de Oliveira Filho (✉)
Faculty of Mathematics and Computer Science, TU Delft,
Van Mourik Broekmanweg 6, 2628, XE Delft, The Netherlands
e-mail: fmario@gmail.com

F. Vallentin
Mathematisches Institut, Universität zu Köln, Weyertal 86–90, 50931 Köln, Germany
e-mail: frank.vallentin@uni-koeln.de

G. Ambrus et al. (eds.), *New Trends in Intuitive Geometry*, Bolyai Society Mathematical Studies 27, https://doi.org/10.1007/978-3-662-57413-3_7

Theorem 1.1, the main theorem of this paper, provides a way to compute upper bounds to the density of any packing of a given convex body $\mathcal{K}$. We then illustrate the use of this theorem by applying computational methods to obtain bounds for packings of regular pentagons in the Euclidean plane — the case when $\mathcal{K} \subseteq \mathbb{R}^2$ is a regular pentagon.

Before presenting our main theorem and its application, let us first survey some of the most interesting cases of the body packing problem. We refer to the following books and surveys for more information: Conway and Sloane [21], Brass, Moser and Pach [11], Bezdek and Kuperberg [8], Fejes Tóth and Kuperberg [29].

Perhaps the most well-known case occurs when $\mathcal{K}$ is a unit ball. We then have the classical *sphere packing problem*. It is easy to find out which sphere packings are optimal (i.e., attain the maximum packing density) in dimensions 1 and 2. In dimension 3, it was conjectured by the German mathematician and astronomer Johannes Kepler (1571–1630) [37] that a certain packing covering $\pi/(3\sqrt{2}) = 0.74048\ldots$ of space is optimal. The Kepler conjecture has been proven by Hales [32], who makes massive use of computers in his proof. On August 10, 2014, the flyspeck project was completed which had the purpose to produce a formal proof of the Kepler conjecture, see [33].

In all other dimensions the best known upper bounds come from the linear programming bound of Cohn and Elkies [15]. For a long time this upper bound was conjectured to provide tight bounds in dimensions 8 and 24 and there was very strong numerical evidence to support this conjecture, see Cohn and Kumar [16] or Cohn and Miller [18]; the only thing missing was a rigorous proof. Very recently, on March 14, 2016 Viazovska [48] announced a proof for dimension 8 and a few days later, on March 21, 2016, building on Viazovska's breakthrough result, Cohn, Kumar, Miller, Radchenko, and Viazovska [17] announced a proof for dimension 24. In dimensions 4, 5, 6, 7, and 9 the linear programming bound of Cohn and Elkies was improved by de Laat, Oliveira, and Vallentin [24], who also provided upper bounds for binary sphere packings, that is, for packings of spheres of two different sizes.

Another case of the body packing problem that has attracted attention happens when $\mathcal{K}$ is a regular tetrahedron in $\mathbb{R}^3$. We called the sphere packing problem "classical", but this adjective most properly applies to the problem of packing tetrahedra, as it was considered by Aristotle (384 BC–322 BC).

In his treatise *De Caelo* (On the Heavens), Aristotle attacks the Platonic theory of assigning geometrical figures, namely the Platonic solids, to the elements, stating (cf. *De Caelo*, Book III, Chapter VIII, in translation by Guthrie [3]):

> This attempt to assign geometrical figures to the simple bodies is on all counts irrational. In the first place, the whole of space will not be filled up. Among surfaces it is agreed that there are three figures which fill the place that contains them — the triangle, the square, and the hexagon: among solids only two, the pyramid and the cube. But they need more than these, since they hold that the elements are more.

Here, the "pyramid" is the regular tetrahedron. Aristotle then thought it to be possible to tile the space with regular tetrahedra. Only much latter, Johannes Müller von Königsberg (1436–1476), commonly known as Regiomontanus, a pioneer of trigonometry, would prove that it is actually not possible to do so — it is amusing to observe that this in fact makes Aristotle's argument stronger.

Regiomontanus' manuscript, titled *De quinque corporibus aequilateris quae vulgo regularis nuncupantur: quae videlicet eorum locum impleant corporalem et quae non, contra commentatorem Aristotelis Averroem,*[1] is lost. The Italian mathematician and astronomer Francesco Maurolico (1494–1575) mentions Regiomontanus' work on a manuscript of very similar title [42]. Considering Regiomontanus' manuscript as lost, he sets out to obtain the same results. He observes (cf. Sect. 2, ibid.) that the angles between the faces of a solid are of importance in determining whether the solid tiles space or not. Nowadays one may easily check that the angle between two faces of a regular tetrahedron is $\arccos \frac{1}{3} \approx 70.52877°$, thus a little less than $360°/5 = 72°$, and one sees that it is therefore impossible to tile $\mathbb{R}^3$ with regular tetrahedra. Maurolico himself did a similar computation (cf. Sect. 73, ibid.):

> (...) Nunc exponam hosce angulos cum suis chordis hic inferius:
>
> Pyramidis angulus – gradus 70. minutiæ 31. secundæ $43\frac{1}{2}$. chorda 1154701.[2]

More on the history of the tetrahedra packing problem can be found in the paper by Lagarias and Zong [39].

In 2006, Conway and Torquato [22] found surprisingly dense packings of tetrahedra. This sparked renewed interest in the problem and a race for the best construction (cf. Lagarias and Zong [39] and Ziegler [50]). The current record is held by Chen, Engel, and Glotzer [13], who found a packing with density ≈ 0.8563, a much larger fraction of space than that which can be covered by spheres. This prompted the quest for upper bounds: We know tetrahedra do not tile space, so the maximum packing density is strictly less than 1. The current record rests with Gravel, Elser, and Kallus [31], who proved an upper bound of $1 - 2.6 \ldots \cdot 10^{-25}$. They are themselves convinced that the bound can be greatly improved:

> In fact, we conjecture that the optimal packing density corresponds to a value of δ [the fraction of empty space] many orders of magnitude larger than the one presented here. We propose as a challenge the task of finding an upper bound with a significantly larger value of δ (e.g., $\delta > 0.01$) and the development of practical computational methods for establishing informative upper bounds.

In 1964, in his famous little book on packing and covering, Rogers noted [44, p. 12]: "Little precise is known about the packing density $\delta(\mathcal{K})$ in three or more dimensions." Since then the general situation did not improve much. In dimension three, the only cases where the optimal packing density is known are the cases when $\mathcal{K}$ is a space filling polytope, or when $\mathcal{K}$ is the unit ball, or when $\mathcal{K}$ is a slight truncation of the rhombic dodecahedron, see [8]. In particular finding good upper bounds for the body packing problem is very difficult. Recently, progress was made

---

[1]On the five equilateral bodies, that are usually called regular, and which of them fill their natural space, and which do not, in contradiction to the commentator of Aristotle, Averroës.

[2]Below I show these angles with their chords:
Angle of the pyramid – 70 degrees. 31 minutes. $43\frac{1}{2}$ seconds. chord 1154701.

by Fejes Tóth, Fodor, and Vígh [28] in the case when $\mathcal{K}$ is the n-dimensional cross polytope.

Our paper can be seen as a step in the search for good upper bounds for the maximum density of body packings in general and tetrahedra packings in particular. Its main theorem is a generalization of the linear programming bound of Cohn and Elkies [15] for the sphere packing density, which provides the best known upper bounds in small and high dimensions (cf. Cohn and Zhao [19]). To specify a sphere packing it suffices to give the centers of the spheres in the packing; this is the reason why the Cohn-Elkies bound is a *linear programming* bound. In our case, to specify a packing of congruent copies of a body, we need also to consider different rotations of the body, and so linear programming is replaced by semidefinite programming.

We apply the theorem to packings of pentagons in the plane because the specific structure of the Euclidean plane simplifies computations and because such packings are interesting in themselves (see e.g. Kuperberg and Kuperberg [38], Casselman [12], and Atkinson, Jiao, and Torquato [4]), obtaining an upper bound of 0.98103 for the density of any such packing. The best known construction is a packing consisting of pentagons placed in two opposite orientations achieving a density of $(5-\sqrt{5})/3 = 0.9213\ldots$ (cf. Kuperberg and Kuperberg, ibid.). Kallus and Kusner [35] showed that the pentagon packing of Kuperberg and Kuperberg cannot be improved by small perturbations.

Using more refined computational tools, for example using a numerically stable complex semidefinite programming solver, it is conceivable that our upper bound could be improved. Our main goal however was to show how the theorem can be applied and that it gives bounds well below the trivial bound of 1. Our long-term goal is to apply the theorem to obtain upper bounds for packings in $\mathbb{R}^3$, in particular tetrahedra packings.

In fact, after the first draft of our paper was finished and submitted, Hales and Kusner [34] improved our upper bound for pentagon packings to 0.9611. For this they used an entirely different method, based on area estimates of Voronoi cells.

1.1 The Main Theorem

We defined the density of a packing informally, as the fraction of space it covers. There are different ways to formalize this definition, and questions arise as to whether every packing has a density and so on. We postpone such discussion to Sect. 2, when we shall prove the main theorem.

Let $\mathrm{SO}(n)$ be the group of rotations of $\mathbb{R}^n$, that is,

$$\mathrm{SO}(n) = \{\, A \in \mathbb{R}^{n\times n} : A^{\mathsf{T}} A = I \text{ and } \det A = 1 \,\}.$$

The set $\mathrm{M}(n) = \mathbb{R}^n \times \mathrm{SO}(n)$ is a group with identity element $(0, I)$, multiplication defined as

$$(x, A)(y, B) = (x + Ay, AB),$$

and inversion given by

$$(x, A)^{-1} = (-A^{-1}x, A^{-1}).$$

The group $\mathrm{M}(n)$, the semidirect product $\mathbb{R}^n \rtimes \mathrm{SO}(n)$, is the *Euclidean motion group* of $\mathbb{R}^n$; it is a noncompact (but locally compact), noncommutative group. When we integrate functions over $\mathrm{M}(n)$, we always use the measure $d(x, A)$, which is the product of the Lebesgue measure dx for $\mathbb{R}^n$ with the Haar measure dA for $\mathrm{SO}(n)$, normalized so that $\mathrm{SO}(n)$ has total measure 1.

A bounded complex-valued function $f \in L^\infty(\mathrm{M}(n))$ is said to be of *positive type* if

$$f(x, A) = \overline{f((x, A)^{-1})} \quad \text{for all } (x, A) \in \mathrm{M}(n)$$

and for all $\rho \in L^1(\mathrm{M}(n))$ we have

$$\int_{\mathrm{M}(n)} \int_{\mathrm{M}(n)} f((y, B)^{-1}(x, A))\rho(x, A)\overline{\rho(y, B)}\, d(y, B)d(x, A) \geq 0.$$

With this we have all we need for presenting the main theorem.

Theorem 1.1 *Let $\mathcal{K} \subseteq \mathbb{R}^n$ be a convex body and let $f \in L^1(\mathrm{M}(n))$ be a bounded real-valued function such that:*

(i) f is continuous and of positive type;
(ii) $f(x, A) \leq 0$ whenever $\mathcal{K}^\circ \cap (x + A\mathcal{K}^\circ) = \emptyset$, where $\mathcal{K}^\circ$ is the interior of $\mathcal{K}$;
(iii) $\lambda = \int_{\mathrm{M}(n)} f(x, A)\, d(x, A) > 0$.

Then the density of any packing of congruent copies of $\mathcal{K}$ is at most

$$\frac{f(0, I)}{\lambda} \operatorname{vol} \mathcal{K},$$

where vol $\mathcal{K}$ *is the volume of* $\mathcal{K}$.

This theorem is a generalization of a theorem of Cohn and Elkies [15] that provides upper bounds for the maximum density of sphere packings, and more generally also for translational packings of convex bodies. The theorem of Cohn and Elkies generalizes Delsarte's linear programming method. Delsarte (see for example the survey of Delsarte and Levenshtein [26]) gave a very general method to determine strong upper bounds for packing problems in compact spaces. The Cohn-Elkies bound deals with the noncompact, commutative group $\mathbb{R}^n$ and our bound deals with the noncompact, noncommutative group $M(n)$.

Our theorem can also be seen as an extension of the Lovász theta number [41], a parameter originally defined for finite graphs, to the infinite packing graph for the body $\mathcal{K}$. Our proof of Theorem 1.1 in Sect. 1 relies on this connection and will make it clear.

Finally, applying Theorem 1.1 to find upper bounds for the densities of packings of a given body $\mathcal{K} \subseteq \mathbb{R}^n$ means finding a good function f satisfying the conditions required in the theorem. In Sect. 5, to find such a function for the case of pentagon packings, we use a computational approach that relies on semidefinite programming (see Sect. 3) and the harmonic analysis of the Euclidean motion group of the plane (see Sect. 4). Here is a place where we see that it is simpler to deal with pentagon packings than with tetrahedra packings, since the formulas describing the harmonic analysis of M(2) are much simpler, specially from a computational perspective, than those describing the harmonic analysis of M(3).

2 Proving the Main Theorem

2.1 *Packing Density and Periodic Packings*

To give a proof of Theorem 1.1 we need to present some technical considerations regarding the density of a packing. Here we follow Appendix A of Cohn and Elkies [15].

Let $\mathcal{K} \subseteq \mathbb{R}^n$ be a convex body and $\mathcal{P}$ be a packing of congruent copies of $\mathcal{K}$. We say that the *density* of $\mathcal{P}$ is Δ if for all $p \in \mathbb{R}^n$ we have

$$\Delta = \lim_{r\to\infty} \frac{\operatorname{vol}(B(p,r) \cap \mathcal{P})}{\operatorname{vol} B(p,r)},$$

where $B(p,r)$ is the ball of radius r centered at p. Not every packing has a density, but every packing has an *upper density* given by

$$\limsup_{r\to\infty} \sup_{p\in\mathbb{R}^n} \frac{\operatorname{vol}(B(p,r) \cap \mathcal{P})}{\operatorname{vol} B(p,r)}.$$

We say that a packing $\mathcal{P}$ is *periodic* if there is a lattice[3] $L \subseteq \mathbb{R}^n$ that leaves $\mathcal{P}$ invariant, i.e., which is such that $\mathcal{P} = x + \mathcal{P}$ for all $x \in L$; here, L is the *periodicity lattice* of $\mathcal{P}$. In other words, an invariant packing consists of some congruent copies of $\mathcal{K}$ arranged inside the fundamental cell of L, and this arrangement repeats itself at each copy of the fundamental cell translated by vectors of the lattice.

Periodic packings have well-defined densities. Moreover, it is not hard to prove that given any packing $\mathcal{P}$, one may define a sequence of periodic packings whose fundamental cells have volumes approaching infinity and whose densities converge to the upper density of $\mathcal{P}$. So in computing bounds for the packing density of a given body, one may restrict oneself to periodic packings. This restriction is particularly interesting because it allows us to compactify the problem, as we will see later on.

[3]A *lattice* is a discrete subgroup of $(\mathbb{R}^n, +)$.

2.2 A Generalization of the Lovász Theta Number

Let $G = (V, E)$ be an undirected graph without loops,[4] finite or infinite. A set $I \subseteq V$ is *independent* if no two vertices in I are adjacent. The *independence number* of G, denoted by $\alpha(G)$, is the maximum cardinality of an independent set of G.

Packings of a given body $\mathcal{K} \subseteq \mathbb{R}^n$ correspond to the independent sets of the *packing graph* of $\mathcal{K}$. This is the graph whose vertices are the elements of $\mathrm{M}(n)$. Here, vertex $(x, A) \in \mathrm{M}(n)$ corresponds to the congruent copy $x + A\mathcal{K}$ of $\mathcal{K}$. Two vertices are adjacent when the corresponding copies of $\mathcal{K}$ cannot both be in the packing at the same time, i.e., when they intersect in their interiors. In other words, distinct vertices (x, A) and (y, B) are adjacent if

$$(x + A\mathcal{K}^\circ) \cap (y + B\mathcal{K}^\circ) \neq \emptyset.$$

Clearly, an independent set of the packing graph corresponds to a packing and vice versa. The packing graph however has infinite independent sets, and so its independence number is also infinite.

If we consider periodic packings we may manage to work with graphs that, though infinite, have a compact vertex set and also a finite independence number. Given a lattice $L \subseteq \mathbb{R}^n$, write $\mathrm{M}(n)/L = (\mathbb{R}^n/L) \times \mathrm{SO}(n)$. Note that this is a compact set. Here, we assume that the fundamental cell of L is big enough so that there exists a nonempty periodic packing with periodicity lattice L.

Consider the graph G_L whose vertex set is $\mathrm{M}(n)/L$ and in which two distinct vertices (x, A) and (y, B) are adjacent if there is $v \in L$ such that

$$(v + x + A\mathcal{K}^\circ) \cap (y + B\mathcal{K}^\circ) \neq \emptyset.$$

In this setting, a vertex $(x, A) \in \mathrm{M}(n)/L$ now represents all bodies $v + x + A\mathcal{K}$ for $v \in L$, and we put an edge between two distinct vertices if any of the corresponding bodies overlap.

So, independent sets of G_L correspond to periodic packings with periodicity lattice L and vice versa. Moreover, G_L has finite independence number, and we actually have that the maximum density of a periodic packing with periodicity lattice L is equal to

$$\frac{\alpha(G_L)}{\mathrm{vol}(\mathbb{R}^n/L)} \mathrm{vol}\, \mathcal{K}. \tag{1}$$

In view of the fact that we may restrict ourselves to periodic packings, as seen in the previous section, if we manage to find an upper bound for $\alpha(G_L)$ for every L, then we obtain an upper bound for the maximum density of any packing of $\mathcal{K}$.

Computing the independence number of a finite graph is an NP-hard problem, figuring in the list of combinatorial problems proven to be NP-hard by Karp [36]. Lovász [41] introduced a graph parameter, the theta number, that provides an upper bound for the independence number of a finite graph and that can be computed in

[4] Like all the graphs considered in this paper.

polynomial time. In Theorem 2.1 we present a generalization of the theta number to graphs defined over certain measure spaces, like the graph G_L.

To present our theorem we need first a few definitions and facts from functional analysis. For background we refer the reader to the book by Conway [20].

Let V be a separable and compact topological space and μ be a finite Borel measure on V which is such that every nonempty open subset of V has nonzero measure. There are many examples of such a space. For instance, any finite set V with the counting measure provides such an example, as does $\mathrm{M}(n)/L$ with its natural measure.

A *Hilbert-Schmidt kernel*, or simply a *kernel*, is a square-integrable function $K\colon V \times V \to \mathbb{C}$. A kernel defines an operator $T_K\colon L^2(V) \to L^2(V)$ as follows: for $f \in L^2(V)$ and $x \in V$ we have

$$(T_K f)(x) = \int_V K(x, y) f(y)\, d\mu(y).$$

An *eigenfunction* of K is a nonzero function $f \in L^2(V)$ such that $T_K f = \lambda f$ for some $\lambda \in \mathbb{C}$. We say that λ is the *eigenvalue* associated with f.

A kernel K is *Hermitian* if $K(x, y) = \overline{K(y, x)}$ for all $x, y \in V$; a Hermitian kernel defines a self-adjoint operator T_K. We say that K is *positive* if it is Hermitian and for all $\rho \in L^2(V)$ we have

$$\int_V \int_V K(x, y)\rho(x)\overline{\rho(y)}\, d\mu(x) d\mu(y) \geq 0.$$

This is equivalent to $\langle T_K \rho, \rho \rangle \geq 0$ for all $\rho \in L^2(V)$, where

$$\langle f, g \rangle = \int_V f(x)\overline{g(x)}\, d\mu(x)$$

is the standard inner product between $f, g \in L^2(V)$. Further still, a Hermitian kernel is positive if and only if all its eigenvalues are nonnegative.

Theorem 2.1 *Let $G = (V, E)$ be a graph, where V is a separable and compact topological space having a finite Borel measure μ such that every nonempty open set of V has nonzero measure. Suppose that kernel $K\colon V \times V \to \mathbb{R}$ satisfies the following conditions:*

(i) K is continuous;
(ii) $K(x, y) \leq 0$ whenever $x \neq y$ are nonadjacent;
(iii) $K - J$ is positive, where J is the constant 1 kernel.

Then, for any number B such that $B \geq K(x, x)$ for all $x \in V$, we have $\alpha(G) \leq B$.

Notice that any kernel K satisfying the conditions of the theorem provides an upper bound for $\alpha(G)$. For a finite graph, the optimal bound given by the theorem is exactly the theta prime number of the graph G, a strengthening of the theta number introduced independently by McEliece, Rodemich, and Rumsey [43] and Schrijver [45].

Such generalizations of the theta number have been considered before by Bachoc, Nebe, Oliveira, and Vallentin [6], where it is proved that the linear programming bound of Delsarte, Goethals, and Seidel [25] for the sizes of spherical codes comes from such a generalization.

To prove the theorem we need the following alternative characterization of continuous and positive kernels: A continuous kernel $K : V \times V \to \mathbb{C}$ is positive if and only if for all m and any choice $x_1,\ldots, x_m$ of points in V the matrix $\big(K(x_i, x_j)\big)_{i,j=1}^{m}$ is positive semidefinite (cf. Lemma 1 in Bochner [9]). In fact, here is where the hypothesis on V is used: To prove this, one needs to use the fact that V is compact, separable, and that the measure μ is finite and nonzero on nonempty open sets.

Proof of Theorem 2.1 Let $I \subseteq V$ be a nonempty finite independent set. Since $K - J$ is a continuous and positive kernel, we have that

$$\sum_{x,y\in I} K(x, y) \geq \sum_{x,y\in I} J(x, y) = |I|^2.$$

Since K satisfies condition (ii), if B is an upper bound on the diagonal entries of K we have that

$$|I|B \geq \sum_{x\in I} K(x, x) \geq \sum_{x,y\in I} K(x, y) \geq |I|^2,$$

and then $B \geq |I|$, as we wanted. □

2.3 A Proof of the Main Theorem

The proof of Theorem 1.1 is similar to the proof of Theorem 3.1 in the paper by de Laat, Oliveira, and Vallentin [24]. We first prove the theorem for functions of bounded support and then extend it to L^1 functions.

Let $f : \mathrm{M}(n) \to \mathbb{R}$ be a function of bounded support satisfying conditions (i)–(iii) in Theorem 1.1. Given a lattice $L \subseteq \mathbb{R}^n$ whose fundamental cell is big enough so that there is a nonempty periodic packing with periodicity lattice L, we use f to define a kernel $K_L : (\mathrm{M}(n)/L) \times (\mathrm{M}(n)/L) \to \mathbb{R}$ satisfying conditions (i)–(iii) of Theorem 2.1 for the graph G_L, defined in the previous section.

In fact, we let

$$\begin{aligned} K_L((x, A), (y, B)) &= \sum_{v\in L} f((y - v, B)^{-1}(x, A)) \\ &= \sum_{v\in L} f(B^{-1}(x - y + v), B^{-1}A) \end{aligned} \tag{2}$$

for every $(x, A), (y, B) \in \mathrm{M}(n)/L$.

Since f has bounded support and $x, y \in \mathbb{R}^n/L$, the sum above is actually a finite sum. This shows not only that K_L is well defined, but also that it is continuous.

We claim that K_L has the following properties:

K1. it is a positive kernel;
K2. the constant 1 function is an eigenfunction of K_L, with eigenvalue λ;
K3. $K_L((x, A), (y, B)) \leq 0$ if $(x, A) \neq (y, B)$ are nonadjacent in G_L;
K4. $f(0, I) \geq K_L((x, A), (x, A))$ for all $(x, A) \in \mathrm{M}(n)/L$.

Once we have established these properties, it becomes clear that the kernel

$$\tilde{K}_L = \frac{\mathrm{vol}(\mathbb{R}^n/L)}{\lambda} K_L$$

satisfies conditions (i)–(iii) of Theorem 2.1 for the graph G_L. In particular, the fact that $\tilde{K}_L - J$ is positive follows directly from K1 and K2 above, because the constant 1 function is an eigenfunction of both $\tilde{K}_L$ and J with associated eigenvalue $\mathrm{vol}(\mathbb{R}^n/L)$ in both cases.

But then from K4 we may take $B = f(0, I)\,\mathrm{vol}(\mathbb{R}^n/L)/\lambda$ in Theorem 2.1 and obtain the bound

$$\alpha(G_L) \leq f(0, I)\frac{\mathrm{vol}(\mathbb{R}^n/L)}{\lambda}.$$

So the maximum density of any periodic packing with periodicity lattice L is bounded from above by (cf. equation (1))

$$\frac{\alpha(G_L)}{\mathrm{vol}(\mathbb{R}^n/L)} \mathrm{vol}\,\mathcal{K} \leq \frac{f(0, I)}{\lambda} \mathrm{vol}\,\mathcal{K},$$

and since L is an arbitrary lattice, we would have a proof of Theorem 1.1.

So we set out to prove K1–K4. Property K1 is implied by the fact that f is of positive type. In fact, since $f(x, A) = \overline{f((x, A)^{-1})}$, kernel K_L is Hermitian by construction. Now take a function $\rho \in L^2(\mathrm{M}(n)/L)$. We also view ρ as the periodic function $\rho\colon \mathrm{M}(n) \to \mathbb{C}$ such that $\rho(x + v, A) = \rho(x, A)$ for all $v \in L$. For $T > 0$, write $\mathrm{M}_T(n) = [-T, T]^n \times \mathrm{SO}(n)$. Then

$$\begin{aligned}
&\int_{\mathrm{M}(n)/L}\int_{\mathrm{M}(n)/L} K_L((x, A), (y, B))\rho(x, A)\overline{\rho(y, B)}\, d(y, B) d(x, A)\\
&= \int_{\mathrm{M}(n)/L}\int_{\mathrm{M}(n)/L} \sum_{v \in L} f((y - v, B)^{-1}(x, A))\rho(x, A)\overline{\rho(y, B)}\, d(y, B) d(x, A).\\
&= \int_{\mathrm{M}(n)/L}\int_{\mathrm{M}(n)} f((y, B)^{-1}(x, A))\overline{\rho(y, B)}\, d(y, B)\rho(x, A)\, d(x, A)\\
&= \lim_{T\to\infty} \frac{\mathrm{vol}(\mathbb{R}^n/L)}{\mathrm{vol}[-T, T]^n} \int_{\mathrm{M}_T(n)}\int_{\mathrm{M}(n)} f((y, B)^{-1}(x, A))\overline{\rho(y, B)}\, d(y, B)\\
&\qquad \cdot \rho(x, A)\, d(x, A)\\
&= \lim_{T\to\infty} \frac{\mathrm{vol}(\mathbb{R}^n/L)}{\mathrm{vol}[-T, T]^n} \int_{\mathrm{M}_T(n)}\int_{\mathrm{M}_T(n)} f((y, B)^{-1}(x, A))\rho(x, A)\overline{\rho(y, B)}\\
&\qquad \cdot d(y, B) d(x, A)\\
&\geq 0.
\end{aligned}$$

Above, from the second to the third line we exchange the sum with the innermost integral and use the fact that, if $h\colon \mathbb{R}^n \to \mathbb{C}$ is an integrable function, then

$$\sum_{v \in L} \int_{\mathbb{R}^n/L} h(x+v)\,dx = \int_{\mathbb{R}^n} h(x)\,dx. \tag{3}$$

To go from the third to the fourth line we notice that the function

$$(x, A) \mapsto \int_{\mathrm{M}(n)} f((y, B)^{-1}(x, A))\overline{\rho(y, B)}\,d(y, B)$$

is periodic with respect to the lattice L. From the fourth to the fifth line we use the fact that f is of bounded support. Finally, from the fifth to the sixth line we apply directly the definition of a function of positive type.

To see K2, we use (3) and notice that for a fixed $(x, A) \in \mathrm{M}(n)/L$ we have

$$\begin{aligned}\int_{\mathrm{M}(n)/L} K_L((x, A), (y, B))\,d(y, B) &= \int_{\mathrm{M}(n)/L} \sum_{v \in L} f((y-v, B)^{-1}(x, A))\,d(y, B)\\ &= \int_{\mathrm{M}(n)} f((y, B)^{-1}(x, A))\,d(y, B)\\ &= \lambda.\end{aligned}$$

To prove K3, recall that $(x, A), (y, B) \in \mathrm{M}(n)/L$ are nonadjacent if for all $v \in L$ we have $(v + x + A\mathcal{K}^\circ) \cap (y + B\mathcal{K}^\circ) = \emptyset$, and this is the case if and only if

$$\mathcal{K}^\circ \cap (B^{-1}(x - y + v) + B^{-1}A\mathcal{K}^\circ) = \emptyset.$$

But then, since f satisfies (ii) in the statement of Theorem 1.1, every summand in (2) will be nonpositive, implying K3.

Property K4 may be similarly proven. In fact, since from start we assumed L has a large enough fundamental cell, for $v \in L$ with $v \neq 0$ we have $\mathcal{K}^\circ \cap (A^{-1}v + \mathcal{K}^\circ) = \emptyset$. But then in expression (2) for $K_L((x, A), (x, A))$, all summands but the one for $v = 0$ will be nonpositive, and the summand for $v = 0$ is exactly $f(0, I)$, proving K4.

So we have K1–K4, and Theorem 1.1 follows for functions f of bounded support. To prove the theorem for a given L^1 function, we approximate it by functions of bounded support as follows.

Let $f \in L^1(\mathrm{M}(n))$ be a real-valued function satisfying conditions (i)–(iii) in Theorem 1.1. For $T > 0$, consider the function $g_T : \mathrm{M}(n) \to \mathbb{R}$ given by

$$g_T(x, A) = \frac{\mathrm{vol}(B(0, T) \cap B(x, T))}{\mathrm{vol}\, B(0, T)} f(x, A),$$

where $B(x, T)$ is the ball of radius T centered at x.

Clearly, g_T is continuous and has bounded support. We claim that it is also a function of positive type.

To see this, we will use a characterization of continuous functions of positive type analogous to the characterization of continuous positive kernels given in Sect. 2.2, namely: A continuous function $f \in L^\infty(\mathrm{M}(n))$ is of positive type if and only if the matrix

$$\big(f((x_j, A_j)^{-1}(x_i, A_i))\big)_{i,j=1}^m$$

is positive semidefinite for any m and any elements $(x_1, A_1),\ldots,(x_m, A_m) \in \mathrm{M}(n)$ (cf. Folland [30], Proposition 3.35).

Let $(x_1, A_1),\ldots, (x_m, A_m) \in \mathrm{M}(n)$ be any given elements. Let $\chi_i : \mathbb{R}^n \to \{0, 1\}$ be the characteristic function of $B(x_i, T)$ and denote by $\langle f, g\rangle$ the standard inner product between functions $f, g \in L^2(\mathbb{R}^n)$. Then

$$\begin{aligned}
g_T((x_j, A_j)^{-1}(x_i, A_i)) &= g_T(A_j^{-1}(x_i - x_j), A_j^{-1}A_i)\\
&= \frac{\mathrm{vol}(B(0,T) \cap B(A_j^{-1}(x_i - x_j), T))}{\mathrm{vol}\, B(0,T)} f((x_j, A_j)^{-1}(x_i, A_i))\\
&= \frac{\mathrm{vol}(B(x_i,T) \cap B(x_j, T))}{\mathrm{vol}\, B(0,T)} f((x_j, A_j)^{-1}(x_i, A_i))\\
&= \frac{\langle \chi_i, \chi_j\rangle}{\mathrm{vol}\, B(0,T)} f((x_j, A_j)^{-1}(x_i, A_i)).
\end{aligned}$$

This shows that the matrix

$$\big(g_T((x_j, A_j)^{-1}(x_i, A_i))\big)_{i,j=1}^m \tag{4}$$

is the Hadamard (entrywise) product of the matrices

$$\big(f((x_j, A_j)^{-1}(x_i, A_i))\big)_{i,j=1}^m \quad \text{and} \quad \frac{1}{\mathrm{vol}\, B(0,T)}\big(\langle \chi_i, \chi_j\rangle\big)_{i,j=1}^m.$$

The first matrix above is positive semidefinite since f is of positive type. The second matrix is positive semidefinite since it is a positive multiple of the Gram matrix of vectors $\chi_1,\ldots,\chi_m$. So we have that (4) is positive semidefinite, and thus g_T is of positive type.

By construction, whenever $f(x, A) \le 0$, also $g_T(x, A) \le 0$. So g_T is a continuous function of bounded support satisfying conditions (i) and (ii) from the statement of Theorem 1.1. This implies immediately that

$$\frac{g_T(0, I)}{\lambda_T} \mathrm{vol}\, \mathcal{K} = \frac{f(0, I)}{\lambda_T} \mathrm{vol}\, \mathcal{K} \tag{5}$$

is an upper bound for the density of any packing of congruent copies of $\mathcal{K}$, where

$$\lambda_T = \int_{\mathrm{M}(n)} g_T(x, A)\, d(x, A).$$

To finish, notice that g_T converges pointwise to f as $T \to \infty$. Moreover, for all T we have $|g_T(x, A)| \leq |f(x, A)|$. So it follows from Lebesgue's dominated convergence theorem that $\lambda_T \to \lambda$ as $T \to \infty$. This together with (5) finishes the proof of Theorem 1.1.

2.4 Using the Symmetry of the Body

Let $\mathcal{K} \subseteq \mathbb{R}^n$ be a convex body. Its *symmetry group* is the subgroup of $\mathrm{SO}(n)$ defined as

$$\mathrm{S}(\mathcal{K}) = \{\, A \in \mathrm{SO}(n) : A\mathcal{K} = \mathcal{K}\, \}.$$

The action by conjugation of an element $B \in \mathrm{SO}(n)$ on a function $f \in L^1(\mathrm{M}(n))$ is given by

$$(B \cdot f)(x, A) = f((0, B)(x, A)(0, B)^{-1}).$$

Suppose now G is a compact subgroup of $\mathrm{S}(\mathcal{K})$. Then in Theorem 1.1 we may restrict ourselves to G-invariant functions $f \in L^1(\mathrm{M}(n))$ without affecting the bound obtained. Here we say that f is *G-invariant* if $B \cdot f = f$ for all $B \in G$.

This restriction to G-invariant functions may make it easier to apply Theorem 1.1. This is actually the case for our application to pentagon packings, as we will see in Sect. 5.

To see that the restriction to G-invariant functions does not affect the bound that can be obtained from Theorem 1.1, notice that, if $f \in L^1(\mathrm{M}(n))$ is a bounded continuous function satisfying conditions (i)–(iii) of Theorem 1.1, then also $B \cdot f$, for $B \in G$, satisfies these conditions.

In fact, to show that $B \cdot f$ is of positive type, let $(x_1, A_1), \ldots, (x_m, A_m) \in \mathrm{M}(n)$. Then

$$\begin{aligned}
&\big((B \cdot f)((x_j, A_j)^{-1}(x_i, A_i))\big)_{i,j=1}^m \\
&= \big(f((0, B)(x_j, A_j)^{-1}(x_i, A_i)(0, B)^{-1})\big)_{i,j=1}^m \\
&= \big(f(((x_j, A_j)(0, B)^{-1})^{-1}((x_i, A_i)(0, B)^{-1}))\big)_{i,j=1}^m,
\end{aligned}$$

and since f is of positive type, $B \cdot f$ is also of positive type (cf. the alternative characterization of continuous functions of positive type in the previous section).

To see that $B \cdot f$ satisfies condition (ii) of Theorem 1.1, notice that, since $B^{-1}\mathcal{K} = \mathcal{K}$, we have $\mathcal{K}^\circ \cap (x + A\mathcal{K}^\circ) = \emptyset$ if and only if $\mathcal{K}^\circ \cap (Bx + BAB^{-1}\mathcal{K}^\circ) = \emptyset$.

Finally, we have

$$\int_{M(n)} (B \cdot f)(x, A)\, d(x, A) = \int_{M(n)} f(x, A)\, d(x, A),$$

and so we see that $B \cdot f$ satisfies the conditions of Theorem 1.1 and provides the same bound as f.

Now, since G is compact, it admits a Haar measure μ which we normalize so that $\mu(G) = 1$. Then it is immediate that the function $\overline{f} \in L^1(M(n))$ such that

$$\overline{f}(x, A) = \int_G (B \cdot f)(x, A)\, d\mu(B)$$

satisfies (i)–(iii) of Theorem 1.1 and provides the same bound as f. Moreover, $\overline{f}$ is G-invariant. So it follows that a restriction to G-invariant functions does not affect the bound of Theorem 1.1.

3 Semidefinite Programming and Sums of Squares

We collect here the basic facts we need from semidefinite programming. For further background we refer to the book by Ben-Tal and Nemirovski [7].

A linear programming problem amounts to maximizing a linear function over a polyhedron, which is the intersection of the nonnegative orthant $\mathbb{R}^n_{\geq 0}$ with an affine subspace. A semidefinite programming problem — a rich generalization of linear programming — amounts to maximizing a linear function over a spectrahedron, the intersection of the cone of positive semidefinite matrices $\mathcal{S}^n_{\succeq 0}$ with an affine subspace. A semidefinite programming problem in primal standard form is

$$\sup \left\{ \langle C, X \rangle : X \in \mathcal{S}^n_{\succeq 0},\ \langle A_j, X \rangle = b_j,\ j = 1, \ldots, m \right\},$$

where C, $A_1, \ldots, A_m$ are given $n \times n$ matrices and where $b_1, \ldots, b_m \in \mathbb{C}$. Here $\langle A, B \rangle = \operatorname{tr}(B^* A)$ denotes the trace inner product between matrices. Matrices C and A_j are usually required to be symmetric (or Hermitian). The seemingly more general setting used here can be easily reduced to this restricted version though.

Semidefinite programming problems are conic optimization problems. Sometimes it is convenient to assume that the variable matrix X has block-diagonal structure, which amounts to changing the cone $\mathcal{S}^n_{\succeq 0}$ to the direct product $\mathcal{S}^{n_1}_{\succeq 0} \times \cdots \times \mathcal{S}^{n_k}_{\succeq 0}$. For solving semidefinite programming problems two types of algorithms are available: the ellipsoid method and interior point methods. The ellipsoid method focuses on the existence of polynomial-time algorithms but no practical implementation is available. In contrast to this there are many very good implementations of interior point methods; De Klerk and Vallentin showed in [23] that a variant of the interior point method for semidefinite programming can run in polynomial time on the Turing machine model.

Semidefinite programming is specially useful for certifying the nonnegativity of polynomials or of trigonometric polynomials via sums of squares. We quickly discuss the univariate case — the multivariate case is a simple extension.

A univariate polynomial $p \in \mathbb{R}[x]$ of degree $2d$ is a sum of squares, i.e., it can be written as

$$p = h_1^2 + \cdots + h_r^2 \quad \text{for some } r \in \mathbb{N} \text{ and } h_1, \ldots, h_r \in \mathbb{R}[x] \text{ of degree at most } d$$

if and only if there is a positive semidefinite matrix Q with

$$p = \langle V, Q \rangle,$$

where V is a matrix of polynomials such that $V_{kl} = P_k(x)P_l(x)$ for some basis P_k of the space of polynomials of degree at most d.

Note $p = \langle V, Q \rangle$ is an identity between polynomials. One can check it by linear equalities — equating the coefficients — once one writes both sides in terms of some basis.

If a polynomial can be written as a sum of squares, then it is clearly nonnegative. For univariate polynomials, the converse is also true. This is not the case in general, however; Laurent [40] presents a survey.

A similar approach can be applied to trigonometric polynomials. Such is an expression of the sort

$$p(\theta) = \sum_{k=-n}^{n} c_k e^{ik\theta},$$

where $c_k = \overline{c_{-k}}$. One way to certify that this trigonometric polynomial is nonnegative for all θ is to write it as a sum of squares, that is, to write it as

$$p(\theta) = |h_1(e^{i\theta})|^2 + \cdots + |h_r(e^{i\theta})|^2$$

for some number r and some univariate polynomials $h_1, \ldots, h_r$. Now, being a sum of squares is equivalent to the existence of an $(n+1) \times (n+1)$ positive semidefinite matrix Q such that

$$p(\theta) = \langle V(e^{i\theta}), Q \rangle,$$

where V is the matrix with $V_{kl}(z) = z^{k-l}$.

4 Harmonic Analysis on M(2)

Our approach to apply Theorem 1.1 in order to obtain upper bounds for the maximum density of pentagon packings is to specify the function f via its Fourier transform. So here we quickly present the facts from the theory of harmonic analysis on M(2) that we will use. We follow Sugiura [47] closely, though we deviate at some points, mainly concerning choices of normalization, as we will see.

For $x, y \in \mathbb{R}^2$, denote by $x \cdot y = x_1y_1 + x_2y_2$ the Euclidean inner product. Let S^1 be the unit circle and for $\varphi, \psi \in L^2(S^1)$ denote by $\langle \varphi, \psi \rangle$ the standard inner product between φ and ψ, i.e.,

$$\langle \varphi, \psi \rangle = \frac{1}{\omega(S^1)} \int_{S^1} \varphi(\xi)\overline{\psi(\xi)}\, d\omega(\xi),$$

where ω is the Lebesgue measure on the unit circle.

For $a \geq 0$ and $(x, A) \in \mathrm{M}(2)$, consider the operator $U^a_{(x,A)} \colon L^2(S^1) \to L^2(S^1)$ defined as follows: For $\varphi \in L^2(S^1)$ we have

$$[U^a_{(x,A)}\varphi](\xi) = e^{2\pi i a x \cdot \xi}\varphi(A^{-1}\xi)$$

for all $\xi \in S^1$. (In the definition of $U^a_{(x,A)}$ we differ from Sugiura [47], who omits the factor 2π, which we include to obtain better formulas — from a computational point of view — later on. Besides changing some normalization parameters, this does not affect the theory as presented by Sugiura.)

Operator $U^a_{(x,A)}$ is a bounded and unitary operator. Moreover, one can easily check that

$$U^a_{(x,A)(y,B)} = U^a_{(x,A)}U^a_{(y,B)}$$

for all $(x, A), (y, B) \in \mathrm{M}(2)$. So the strongly continuous map $\rho_a(x, A) = U^a_{(x,A)}$ provides a representation of M(2) for every $a \geq 0$. Representations ρ_a, for $a > 0$, are all irreducible and pairwise nonequivalent.

Given a function $f \in L^1(\mathrm{M}(2))$, its Fourier transform at $a \geq 0$ is the bounded operator $\widehat{f}(a) \colon L^2(S^1) \to L^2(S^1)$ defined as

$$\widehat{f}(a) = \int_{\mathrm{M}(2)} f(x, A)U^a_{(x,A)^{-1}}\, d(x, A).$$

Having defined the Fourier transform of f, we would like to have an *inversion formula*, that is, a way to compute f back from its Fourier transform. In our case the inversion formula takes the following shape:

$$f(x, A) = 2\pi \int_0^\infty \mathrm{tr}(U^a_{(x,A)}\widehat{f}(a))a\, da, \tag{6}$$

where tr F is the trace of a trace-class operator F. In the following, we only need positive trace-class operators. We define them now briefly, and we refer e.g. to Conway [20] or Folland [30] for further information. A positive bounded operator $F \colon L^2(S^1) \to L^2(S^1)$ is called *trace-class* if there is a complete orthonormal system φ_k consisting of eigenfunctions of F with eigenvalues $\lambda_k \geq 0$ and $\sum_k \lambda_k < \infty$. Then the trace of F is $\mathrm{tr}(F) = \sum_k \lambda_k = \sum_k \langle F\varphi_k, \varphi_k \rangle$.

In (6) we again deviate slightly from the exposition of Sugiura. The extra factor of 2π in the above formula as compared to Theorem 3.1 in his book [47] follows from the different normalization he uses for the measure of M(2).

Of course, it is not always the case that the inversion formula holds or converges everywhere. In the book by Sugiura it is shown that the inversion formula holds for *rapidly decreasing functions* (see Definition 3 in Chapter IV, Sect. 3 in Sugiura [47]).

We will provide explicit formulas for the Fourier transform and hence obtain explicit formulas for f. In this way it will be clear in our application that f is continuous and L^1. To ensure that f is of positive type, the following lemma will be useful. It shows that one can parametrize positive type functions by the positivity of the Fourier transform $\widehat{f}$. Here it shows that the computations needed for applying Theorem 1.1 are much more complicated than those for applying the Cohn-Elkies bound. For the Euclidean motion group $M(n)$ the Fourier transform of a positive type function is positive trace-class-operator-valued, whereas for the translation group $\mathbb{R}^n$ its values are simply nonnegative real numbers.

Lemma 4.1 *Suppose that for each $a \geq 0$ we have that $\widehat{f}(a)$ is a positive, trace-class operator. Then, if the function f defined in* (6) *is bounded and continuous, it is of positive type.*

Proof Take $(x_1, A_1),\ldots, (x_m, A_m) \in \mathrm{M}(2)$. Recalling the alternative characterization of continuous functions of positive type given in Sect. 2.3, we show that the matrix

$$\left(f((x_j, A_j)^{-1}(x_i, A_i))\right)_{i,j=1}^{m}$$

is positive semidefinite.

By construction, this is a Hermitian matrix. From (6), to prove it is positive semidefinite it suffices to show that for all $a \geq 0$ the matrix

$$\left(\mathrm{tr}(U^a_{(x_j,A_j)^{-1}(x_i,A_i)}\widehat{f}(a))\right)_{i,j=1}^{m} \tag{7}$$

is positive semidefinite.

Notice that since each $\widehat{f}(a)$ is trace class, and since $U^a_{(x,A)}$ is a bounded operator, then $U^a_{(x,A)}\widehat{f}(a)$ is trace class for all $(x, A) \in \mathrm{M}(2)$, and so each entry of (7) is well defined.

To see that (7) is positive semidefinite, let $\varphi_1, \varphi_2,\ldots$ be a complete orthonormal system of $L^2(S^1)$. For $i, j = 1,\ldots, m$ we have

$$\begin{aligned}
\mathrm{tr}(U^a_{(x_j,A_j)^{-1}(x_i,A_i)}\widehat{f}(a)) &= \sum_{k=1}^{\infty}\langle U^a_{(x_j,A_j)^{-1}(x_i,A_i)}\widehat{f}(a)\varphi_k, \varphi_k\rangle \\
&= \sum_{k=1}^{\infty}\langle U^a_{(x_j,A_j)^{-1}(x_i,A_i)}\widehat{f}(a)U^a_{(x_j,A_j)^{-1}}\varphi_k, U^a_{(x_j,A_j)^{-1}}\varphi_k\rangle \\
&= \sum_{k=1}^{\infty}\langle U^a_{(x_i,A_i)}\widehat{f}(a)U^a_{(x_j,A_j)^{-1}}\varphi_k, \varphi_k\rangle \\
&= \sum_{k=1}^{\infty}\langle \widehat{f}(a)U^a_{(x_j,A_j)^{-1}}\varphi_k, U^a_{(x_i,A_i)^{-1}}\varphi_k\rangle \\
&= \sum_{k=1}^{\infty}\langle \widehat{f}^{1/2}(a)U^a_{(x_j,A_j)^{-1}}\varphi_k, \widehat{f}^{1/2}(a)U^a_{(x_i,A_i)^{-1}}\varphi_k\rangle.
\end{aligned}$$

Here, we go from the first to the second line by noticing that, since $U^a_{(x_j,A_j)^{-1}}$ is a unitary operator, $U^a_{(x_j,A_j)^{-1}}\varphi_1, U^a_{(x_j,A_j)^{-1}}\varphi_2, \ldots$ is also a complete orthonormal system of $L^2(S^1)$. Finally, from the fourth to the fifth line, we observe that since $\widehat{f}(a)$ is a positive trace-class operator, it has a square-root $\widehat{f}^{1/2}(a)$, a self-adjoint operator such that $\widehat{f}(a) = \widehat{f}^{1/2}(a)\widehat{f}^{1/2}(a)$.

So we see that (7) is a sum of positive semidefinite matrices, the kth summand being the Gram matrix of m vectors, and we are done. □

We finish this section by computing a more explicit formula for the inverse transform. We identify both SO(2) and S^1 with the torus $\mathbb{R}/(2\pi\mathbb{Z})$, and by an abuse of language with the interval $[0, 2\pi]$. We equip $L^2([0, 2\pi])$ with the inner product

$$\langle \varphi, \psi \rangle = \frac{1}{2\pi} \int_0^{2\pi} \varphi(\xi)\overline{\psi(\xi)}\, d\xi$$

for $\varphi, \psi \in L^2([0, 2\pi])$. Then the functions $\chi_r \in L^2([0, 2\pi])$, for $r \in \mathbb{Z}$, defined as $\chi_r(\xi) = e^{ir\xi}$ provide a complete orthonormal system of $L^2([0, 2\pi])$.

We define the *matrix coefficients* of the operator $U^a_{(x,A)}$ on the basis χ_r as

$$u^a_{r,s}(x, A) = \langle U^a_{(x,A)}\chi_s, \chi_r\rangle \quad \text{with } r, s \in \mathbb{Z}.$$

To compute this, we express x in polar coordinates as

$$x = \rho(\cos\theta, \sin\theta)$$

and we see A as the rotation matrix

$$A = A(\alpha) = \begin{pmatrix} \cos\alpha & -\sin\alpha \\ \sin\alpha & \cos\alpha \end{pmatrix}, \tag{8}$$

which rotates vectors counter-clockwise by an angle of α. Then

$$\begin{aligned}
u^a_{r,s}(\rho, \theta, \alpha) &= \langle U^a_{(x,A)}\chi_s, \chi_r\rangle \\
&= \frac{1}{2\pi}\int_0^{2\pi} [U^a_{(x,A)}\chi_s](\xi)\overline{\chi_r(\xi)}\, d\xi \\
&= \frac{1}{2\pi}\int_0^{2\pi} e^{2\pi i a\rho(\cos\theta,\sin\theta)\cdot(\cos\xi,\sin\xi)} e^{is(\xi-\alpha)} e^{-ir\xi}\, d\xi \\
&= \frac{1}{2\pi} e^{-is\alpha}\int_0^{2\pi} e^{2\pi i a\rho\cos(\xi-\theta)} e^{i(s-r)\xi}\, d\xi \\
&= \frac{1}{2\pi} e^{-is\alpha}\int_0^{2\pi} e^{2\pi i a\rho\cos\xi} e^{i(s-r)(\xi+\theta)}\, d\xi \\
&= \frac{1}{2\pi} e^{-i(s\alpha+(r-s)\theta)}\int_0^{2\pi} e^{i(s-r)\xi} e^{2\pi i a\rho\cos\xi}\, d\xi \\
&= i^{s-r} e^{-i(s\alpha+(r-s)\theta)} J_{s-r}(2\pi a\rho).
\end{aligned} \tag{9}$$

Here, $J_n(z)$ is the Bessel function of parameter n. To obtain the last line, we apply Bessel's integral (cf. Watson [49], (1) in Chapter II, Sect. 2.2).

We may then rewrite (6) by expressing the operators $\widehat{f}(a)$ on the basis χ_r, for $r \in \mathbb{Z}$. This gives us

$$\begin{aligned} f(\rho, \theta, \alpha) &= \int_0^\infty \sum_{r,s\in\mathbb{Z}} \widehat{f}(a)_{r,s} u^a_{r,s}(\rho, \theta, \alpha) a\, da \\ &= \int_0^\infty \sum_{r,s\in\mathbb{Z}} \widehat{f}(a)_{r,s} i^{s-r} e^{-i(s\alpha+(r-s)\theta)} J_{s-r}(2\pi a \rho) a\, da. \end{aligned} \tag{10}$$

5 Computations for Pentagon Packings

In this section we present a semidefinite programming problem and show how from its solution a function f can be derived that satisfies conditions (i)–(iii) of Theorem 1.1 when $\mathcal{K}$ is a regular pentagon. We describe the semidefinite programming problem in detail, and then discuss how it can be solved with the computer and how a function f can be obtained from its solution that provides the bound of 0.98103 for the maximum density of packings of regular pentagons in $\mathbb{R}^2$.

Throughout this section, $\mathcal{K}$ will denote the regular pentagon on $\mathbb{R}^2$ whose vertices are the points

$$\frac{1}{2}(\cos(k2\pi/5), \sin(k2\pi/5)) \quad \text{for } k = 0, \ldots, 4.$$

Note the circumscribed circle of $\mathcal{K}$ has radius $1/2$.

The symmetry group of $\mathcal{K}$ is isomorphic to C_5, the cyclic group of order 5. It consists of the rotation matrices $A(k2\pi/5)$, for $k = 0, \ldots, 4$, where $A(\alpha)$ is given in (8).

5.1 *Specifying the Function*

Our approach is to specify the function f required by Theorem 1.1 via its Fourier transform. In this section we discuss our choice for the Fourier transform of f, give explicit formulas for f in terms of its transform, and show which constraints must be imposed on the transform so that f is a real valued, L^1 and continuous function of positive type which is S($\mathcal{K}$)-invariant.

Let $N > 0$ be an integer and $d \geq 1$ be an odd integer. Consider the matrix-valued function φ given by

$$\varphi(a) = \big(\varphi_{r,s}(a)\big)_{r,s=-N}^{N} = \left(\sum_{k=0}^{d} f_{r,s;k} a^{2k}\right)_{r,s=-N}^{N}. \tag{11}$$

Notice that each $\varphi(a)$ is a $(2N+1)\times(2N+1)$ matrix whose entries are even univariate polynomials in the variable a.

We define f as the function whose Fourier transform is

$$\widehat{f}(a) = \varphi(a)e^{-\pi a^2}. \tag{12}$$

Note that we express the operator $\widehat{f}(a)$ in the basis χ_r for $r \in \mathbb{Z}$, as discussed in Sect. 4. Clearly, each $\widehat{f}(a)$ is a trace-class operator. In fact, each $\widehat{f}(a)$ has finite rank.

The reason for our choice for the Fourier transform is that it makes it easy to compute the function f. Let

$$C_{r,s;k}(\rho) = \frac{\Gamma(k+1+|r-s|/2)(\rho\sqrt{\pi})^{|r-s|}}{2\pi^{k+1}\Gamma(|r-s|+1)}.$$

Then using (4.11.24) in Andrews, Askey, and Roy [2], since $J_n(z) = (-1)^n J_{-n}(z)$, we have

$$\begin{aligned}\int_0^\infty a^{2k+1}e^{-\pi a^2} J_{s-r}(2\pi a\rho)\, da &= (-1)^{s-r}\int_0^\infty a^{2k+1}e^{-\pi a^2} J_{|r-s|}(2\pi a\rho)\, da\\ &= (-1)^{s-r} C_{r,s;k}(\rho)\, {}_1F_1\left(\begin{matrix}|r-s|/2-k\\ |r-s|+1\end{matrix}; \pi\rho^2\right)e^{-\pi\rho^2},\end{aligned}$$

where ${}_1F_1$ is the hypergeometric series. Together with (10) and (11), this implies for f the formula

$$\begin{aligned}f(\rho,\theta,\alpha) = \sum_{r,s=-N}^{N}\sum_{k=0}^{d}(-1)^{s-r}i^{s-r}e^{-i(s\alpha+(r-s)\theta)} f_{r,s;k}C_{r,s;k}(\rho)&\\ \cdot\, {}_1F_1\left(\begin{matrix}|r-s|/2-k\\ |r-s|+1\end{matrix}; \pi\rho^2\right)e^{-\pi\rho^2},&\end{aligned} \tag{13}$$

where (ρ,θ,α) parametrizes an element of M(2) as in Sect. 4.

It is immediately clear that, thanks to our choice of Fourier transform, f is an L^1 and continuous function; actually, it is rapidly decreasing. So, by using Lemma 4.1, we see that if $\widehat{f}(a)$ is a positive kernel for each $a \geq 0$, then f is a function of positive type. From the definition of $\widehat{f}(a)$, we see that $\widehat{f}(a)$ is positive for every $a \geq 0$ if and only if the matrices $\varphi(a)$ are positive semidefinite for every a. Notice that requiring $\varphi(a)$ to be positive semidefinite includes requiring $\varphi(a)$ to be Hermitian. This on its turn we achieve by imposing the constraint

$$f_{r,s;k} = \overline{f_{s,r;k}} \quad \text{for all } r,\ s, \text{ and } k.$$

We may further simplify (13) by imposing two extra conditions on the coefficients $f_{r,s;k}$. Namely, when $r-s$ is even and $|r-s|/2-k \leq 0$, the hypergeometric series in (13) becomes a Laguerre polynomial; see also the treatment about the eigenfunction decomposition of the Hankel transform in the book [1, Chapter 9] by Akhiezer. Indeed we have (cf. (6.2.2) in Andrews, Askey, and Roy [2])

$$ {}_1F_1\left(\begin{matrix} |r-s|/2-k \\ |r-s|+1 \end{matrix}; \pi\rho^2\right) = \frac{n!}{(|r-s|+1)_n} L_n^{|r-s|}(\pi\rho^2), $$

where $n = k - |r-s|/2$,

$$ (a)_n = a(a+1)\cdots(a+n-1) \quad \text{for } n > 0 \text{ with } (a)_0 = 1, $$

and L_n^{α} is the Laguerre polynomial of degree n and parameter α.

So we impose on the coefficients $f_{r,s;k}$ the constraints

$$ f_{r,s;k} = 0 \quad \text{if } r-s \text{ is odd or } k < |r-s|/2. \tag{14} $$

Then (13) becomes

$$ f(\rho,\theta,\alpha) = \sum_{\substack{r,s=-N \\ r-s \text{ even}}}^{N} \sum_{k=|r-s|/2}^{d} (-1)^{|r-s|/2} e^{-i(s\alpha+(r-s)\theta)} f_{r,s;k} \cdot D_{r,s;k}(\rho) L_n^{|r-s|}(\pi\rho^2) e^{-\pi\rho^2}, \tag{15} $$

where $D_{r,s;k}(\rho) = C_{r,s;k}(\rho) n!/(|r-s|+1)_n$.

To ensure that f is a real-valued function, we observe from (9) that when $r-s$ is even, $u^a_{r,s}(\rho,\theta,\alpha) = \overline{u^a_{-r,-s}(\rho,\theta,\alpha)}$. Then from (10) it is clear that if $\varphi_{r,s}(a) = \overline{\varphi_{-r,-s}(a)}$ for all $a \geq 0$ and r, s, function f is real valued. So to ensure that f is real valued it suffices to impose the constraint

$$ f_{r,s;k} = \overline{f_{-r,-s;k}} \quad \text{for all } r,\ s, \text{ and } k. \tag{16} $$

Finally, we would like to impose constraints on the coefficients $f_{r,s;k}$ so as to make function f S($\mathcal{K}$)-invariant, that is, so as to have

$$ f(\rho, \theta + l2\pi/5, \alpha) = f(\rho,\theta,\alpha) \quad \text{for } l = 0, \ldots, 4. $$

From (15), it is easy to see that one way of achieving this is to require that

$$ f_{r,s;k} = 0 \quad \text{whenever } r - s \not\equiv 0 \pmod 5. $$

Since we already set $f_{r,s;k} = 0$ when $r-s$ is odd, we end up with the constraint

$$ f_{r,s;k} = 0 \quad \text{whenever } r - s \not\equiv 0 \pmod{10}. \tag{17} $$

To finish, we summarize the constraints imposed on the coefficients $f_{r,s;k}$:

(1) We consider only the pairs r, s such that $r - s \equiv 0 \pmod{10}$ and we set $f_{r,s;k} = 0$ if $k < |r - s|/2$. This has a double effect: It simplifies the hypergeometric series into a Laguerre polynomial and makes the function $\mathrm{S}(\mathcal{K})$-invariant;
(2) We set $f_{r,s;k} = \overline{f_{s,r;k}}$ for all r, s, and k. This makes the matrices $\varphi(a)$ Hermitian. We then require these matrices to be positive semidefinite; this ensures that function f is of positive type;
(3) We set $f_{r,s;k} = \overline{f_{-r,-s;k}}$ for all r, s, and k. This ensures that function f is real-valued.

With these constraints, we obtain the following formula for f:

$$f(\rho, \theta, \alpha) = \sum_{\substack{r,s=-N \\ r-s\equiv 0 \pmod{10}}}^{N} \sum_{k=|r-s|/2}^{d} (-1)^{|r-s|/2} e^{-i(s\alpha+(r-s)\theta)} f_{r,s;k} \cdot D_{r,s;k}(\rho) L_n^{|r-s|}(\pi\rho^2) e^{-\pi\rho^2}. \tag{18}$$

5.2 A Semidefinite Programming Formulation: Basic Setup

Recall our goal is to describe a semidefinite programming problem whose solutions correspond to functions $f \in L^1(\mathrm{M}(n))$ satisfying the conditions of Theorem 1.1. In this section, we take a first step by showing how to formulate the problem of finding a function φ like (11) as a semidefinite programming problem.

We start by making an extra assumption, namely that all coefficients $f_{r,s;k}$ are real. This allows us to work exclusively with real matrices, making the semidefinite programming problem we obtain smaller. Then the optimization problem will be solvable by state-of-the-art semidefinite programming solvers which are numerically stable. In principle, however, everything we describe can be extended to the more general setting of complex coefficients. *It could be*, though we do not know, that such a restriction to real numbers greatly worsens the bound that can be obtained via our approach.

So let φ be given as in (11) with

$$f_{s,r;k} = f_{r,s;k} = f_{-r,-s;k} = f_{-s,-r;k} \quad \text{for all } r, s, \text{ and } k$$

as we require. Write $y = (y_{-N}, \ldots, y_N)$ and consider the polynomial

$$\sigma(a, y) = \sum_{\substack{r,s=-N \\ r-s\equiv 0 \pmod{10}}}^{N} \sum_{k=|r-s|/2}^{d} f_{r,s;k} a^{2k} y_r y_s.$$

Then $\varphi(a)$ is positive semidefinite for all a if and only if σ is a sum of squares (see Sect. 3). (Here, it is easy to see that if σ is a sum of squares, then $\varphi(a)$ is positive semidefinite for all a. The converse is also true; for a proof see Choi, Lam, and Reznick [14]. This fact is related to the Kalman–Yakubovich–Popov lemma in systems and control; see the discussion in Aylward, Itani, and Parrilo [5].)

The constraint that σ is a sum of squares can on its turn be formulated in terms of positive semidefinite matrices. Following the recipe given on Sect. 3, one would obtain a semidefinite programming formulation in terms of a single variable matrix of large size. In our case, however, since σ is an even polynomial in a and since the product $y_r y_s$ only appears when $r - s \equiv 0 \pmod{10}$, we may block-diagonalize the variable matrix, obtaining a formulation in terms of smaller matrices, as we show now.

To this end, let P_0, P_1,... be a sequence of real, even, univariate polynomials such that P_k has degree $2k$. For $j = 0,\ldots, 9$, let

$$\mathcal{I}_j = \{\, r \in \mathbb{Z} : -N \leq r \leq N \quad \text{and} \quad r \equiv j \pmod{10} \,\}.$$

For $i = 0, 1$ and $j = 0,\ldots, 9$, consider the matrix V^{ij} with rows and columns indexed by $\{0, \ldots, \lfloor d/2 \rfloor\} \times \mathcal{I}_j$ such that

$$V^{ij}_{(l,r)(l',s)} = a^{2i} P_l(a) P_{l'}(a) y_r y_s$$

for all $l, l' = 0,\ldots, \lfloor d/2 \rfloor$ and $r, s \in \mathcal{I}_j$. Notice the entries of V^{ij} are even polynomials in a.

Then σ is a sum of squares if and only if there are real, positive semidefinite matrices Q^{ij}, of appropriate dimensions, such that

$$\sigma = \sum_{i=0}^{1} \sum_{j=0}^{9} \langle Q^{ij}, V^{ij} \rangle,$$

where $\langle A, B \rangle = \mathrm{tr}(B^* A)$ denotes the trace inner product between matrices A and B.

Here, it is also important to observe that the symmetry constraints $f_{r,s;k} = f_{s,r;k}$ are implied by the fact that the matrices Q^{ij} are symmetric.

So finding real numbers $f_{r,s;k}$ such that $f_{r,s;k} = f_{s,r;k}$ and such that $\varphi(a)$ is positive semidefinite for all a amounts to finding real positive semidefinite matrices Q^{ij}. Also the other constraints that we imposed on the coefficients $f_{r,s;k}$ can be represented as linear constraints on the entries of the Q^{ij} matrices, as we show now.

For r, s, k with $r - s \equiv 0 \pmod{10}$, let $j \in \{0, \ldots, 9\}$ be such that $r, s \in \mathcal{I}_j$. For $i = 0, 1$, consider the matrix $F^i_{r,s;k}$ with rows and columns indexed by $\{0, \ldots, \lfloor d/2 \rfloor\} \times \mathcal{I}_j$ such that

$$(F^i_{r,s;k})_{(l,r)(l',s)} = \mathrm{coeff}(a^{2k}, a^{2i} P_l(a) P_{l'}(a))$$

for all $l, l' = 0, \ldots, \lfloor d/2 \rfloor$, where for a given polynomial p, $\operatorname{coeff}(a^k, p)$ is the coefficient of monomial a^k in p. Then we obtain the coefficients $f_{r,s;k}$ from the matrices Q^{ij} by the formula

$$f_{r,s;k} = \sum_{i=0}^{1} \langle F^i_{r,s;k}, Q^{ij} \rangle.$$

So constraints (14) and (16) become

$$\sum_{i=0}^{1} \langle F^i_{r,s;k}, Q^{ij} \rangle = 0 \quad \text{if } k < |r-s|/2,$$

$$\sum_{i=0}^{1} (\langle F^i_{r,s;k}, Q^{ij} \rangle - \langle F^i_{-r,-s;k}, Q^{ij'} \rangle) = 0 \quad \text{for all } r, s, \text{ and } k,$$

where $r, s \equiv j \pmod{10}$ and $-r, -s \equiv j' \pmod{10}$. Notice that constraint (17) is already implicit in our formulation, because we enforce by construction that only pairs r, s with $r - s \equiv 0 \pmod{10}$ occur.

Also the function f can be computed from matrices Q^{ij}. To see how, for r, $s = -N, \ldots, N$ such that $r - s \equiv 0 \pmod{10}$ and for $k \geq |r-s|/2$, set

$$[\tau_{r,s}(a^{2k})](\rho, \theta, \alpha) = (-1)^{|r-s|/2} e^{-i(s\alpha + (r-s)\theta)} D_{r,s;k}(\rho) L_n^{|r-s|}(\pi\rho^2), \tag{19}$$

where $n = k - |r-s|/2$. When $k < |r-s|/2$, we set $\tau_{r,s}(a^{2k}) = 0$, and then we extend $\tau_{r,s}$ linearly to all even polynomials in the variable a.

For $i = 0, 1$ and $j = 0, \ldots, 9$, consider the matrix $\mathcal{F}^{ij}$ with rows and columns indexed by $\{0, \ldots, \lfloor d/2 \rfloor\} \times \mathcal{I}_j$ such that

$$[\mathcal{F}^{ij}(\rho, \theta, \alpha)]_{(l,r)(l',s)} = \tau_{r,s}(a^{2i} P_l(a) P_{l'}(a))$$

for all $l, l' = 0, \ldots, \lfloor d/2 \rfloor$ and $r, s \in \mathcal{I}_j$. Then, in view of (18) and since $f_{r,s;k} = 0$ whenever $k < |r-s|/2$, we have

$$f(\rho, \theta, \alpha) = \sum_{i=0}^{1} \sum_{j=0}^{9} \langle \mathcal{F}^{ij}(\rho, \theta, \alpha), Q^{ij} \rangle e^{-\pi\rho^2}. \tag{20}$$

5.3 *Ensuring Nonpositiveness*

How can we ensure that function f, given by (20), satisfies constraint (ii) of Theorem 1.1? This we do also in terms of semidefinite programming constraints.

First, observe that we require $f(x, A) \leq 0$ whenever $\mathcal{K}^\circ \cap (x + A\mathcal{K}^\circ) = \emptyset$. The latter happens if and only if $x \notin (\mathcal{K} - A\mathcal{K})^\circ$, where $\mathcal{K} - A\mathcal{K}$ is the *Minkowski dif-*

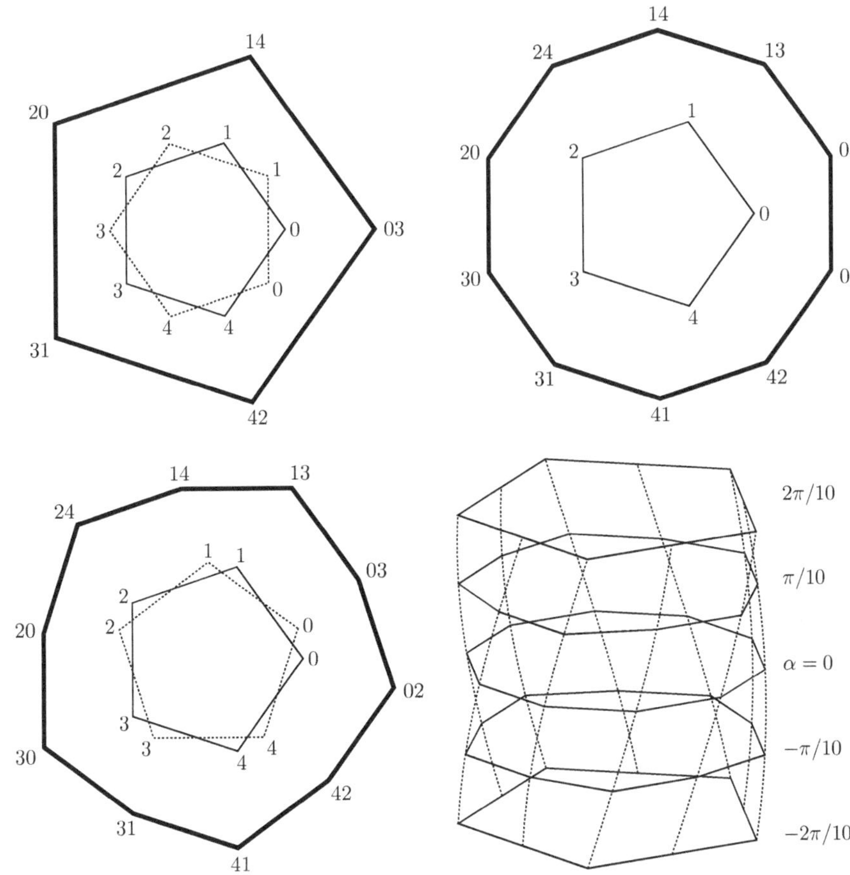

Fig. 1 From left to right, top to bottom. In the first three pictures, we see the Minkowski difference $\mathcal{K} - A(\alpha)\mathcal{K}$ (the outer shape) for $\alpha = -2\pi/10$, 0, and $\pi/10$. The dashed pentagon in the center corresponds to $A(\alpha)\mathcal{K}$. The vertices of the pentagons are numbered from 0 to 4. The vertices of the Minkowski difference are numbered ij, meaning that they correspond to $x - y$, where x is the ith vertex of $\mathcal{K}$ and y is the jth vertex of $A(\alpha)\mathcal{K}$. In the last picture we show the three-dimensional set $\{ (x, \alpha) : x \in \mathcal{K} - A(\alpha)\mathcal{K} \}$. Here, α is on the vertical axis; every section perpendicular to the vertical axis corresponds to a Minkowski difference $\mathcal{K} - A(\alpha)\mathcal{K}$

ference of $\mathcal{K}$ and $A\mathcal{K}$:

$$\mathcal{K} - A\mathcal{K} = \{ y - z : y \in \mathcal{K},\ z \in A\mathcal{K} \}.$$

The Minkowski difference $\mathcal{K} - A\mathcal{K}$ is a polygon for all $A \in \mathrm{SO}(2)$. Its vertices can be explicitly determined; Fig. 1 shows the Minkowski difference when $A = A(\alpha)$ (as defined in (8)) for $\alpha \in [-2\pi/10, 2\pi/10]$. By the symmetry of $\mathcal{K}$, this gives a full characterization of the shape of the Minkowski difference for all α.

Our approach to ensure that f is nonpositive outside of $(\mathcal{K} - A\mathcal{K})^\circ$ consists of two steps. First, we observe that all vertices of $\mathcal{K} - A\mathcal{K}$ have norm at most 1. This implies that we must have $f(x, A) \leq 0$ whenever $\|x\| \geq 1$. This condition on f can be expressed in terms of sums of squares constraints.

Indeed, by writing $z_1 = e^{i\theta}$ and $z_2 = e^{i(\alpha-\theta)}$, we may rewrite (19) as

$$[\tau_{r,s}(a^{2k})](\rho, z_1, z_2) = (-1)^{|r-s|/2} z_1^{-r} z_2^{-s} D_{r,s;k}(\rho) L_n^{|r-s|}(\pi\rho^2).$$

In view of (20), if we then have

$$\sum_{i=0}^{1}\sum_{j=0}^{9}\langle \mathcal{F}^{ij}(\rho, z_1, z_2), Q^{ij}\rangle \leq 0 \quad \text{for all } \rho \geq 1,$$

we have $f(\rho, \theta, \alpha) \leq 0$ whenever $\rho \geq 1$, as we want.

For $j = 0, \ldots, 9$, consider the set

$$\mathcal{P}_j = \{\, (r, s) : 0 \leq r, s \leq N \quad \text{and} \quad r - s \equiv j \pmod{10} \,\}.$$

For $i = 0, 1$, $j = 0, \ldots, 9$, consider the matrix W^{ij} with rows and columns indexed by $\{0, \ldots, \lfloor d/2 \rfloor\} \times \mathcal{P}_j$ such that

$$W^{ij}_{(l,p)(l',p')}(\rho, z_1, z_2) = (\rho^i P_l(\rho) z_1^{-u} z_2^{-v})(\rho^i P_{l'}(\rho) z_1^{u'} z_2^{v'}),$$

where $p = (u, v)$ and $p' = (u', v')$ with $p, p' \in \mathcal{P}_j$, and $l, l' = 0, \ldots, \lfloor d/2 \rfloor$.

If there are real positive semidefinite matrices R^{ij} for $i = 0, 1$ and $j = 0, \ldots, 9$, and S^j for $j = 0, \ldots, 9$, such that

$$\sum_{i=0}^{1}\sum_{j=0}^{9}(\langle \mathcal{F}^{ij}(\rho, z_1, z_2), Q^{ij}\rangle + \langle W^{ij}(\rho, z_1, z_2), R^{ij}\rangle) \\ + \sum_{j=0}^{9}\langle(\rho^2 - 1)W^{0j}(\rho, z_1, z_2), S^j\rangle = 0, \tag{21}$$

then $f(\rho, \theta, \alpha) \leq 0$ for all $\rho \geq 1$. Notice (21) is a polynomial identity on variables ρ, z_1, z_1^{-1}, z_2, and z_2^{-1}. In other words, the left-hand side defines a polynomial and the identity above states that this polynomial must be identically zero. To see that (21) implies that $f(\rho, \theta, \alpha) \leq 0$ whenever $\rho \geq 1$, one only has to notice that, for $\rho \geq 1$ and $\theta, \alpha \in [0, 2\pi]$, the Hermitian matrices

$$W^{ij}(\rho, e^{i\theta}, e^{i(\alpha-\theta)}) \quad \text{and} \quad (\rho^2 - 1)W^{0j}(\rho, e^{i\theta}, e^{i(\alpha-\theta)})$$

are positive semidefinite, and then all inner products in (21) become nonnegative.

Constraint (21) is not enough to ensure, however, that f is nonpositive outside of the Minkowski difference. To ensure nonpositiveness in the remaining region, we use a discretization heuristic: We pick a sample of triples (ρ, θ, α) with $\rho \leq 1$ for which we have to ensure that $f(\rho, \theta, \alpha) \leq 0$ and we do so explicitly for every point of the sample using (20). Afterwards, we have to analyze the solution obtained in order to check that it indeed satisfies condition (ii) of Theorem 1.1. We will give details on this approach in the next section.

One may model the constraint that f is nonpositive outside the Minkowski difference using only sums of squares, without using the discretization approach. The sizes of the matrices get very large, however, making this approach computationally infeasible.

5.4 The Semidefinite Programming Problem and How to Solve It

We now describe the semidefinite programming problem we solve to obtain upper bounds for the pentagon packing density.

Let $N > 0$ be an integer and $d \geq 1$ be an odd integer. Let $\mathcal{S}$ be a finite set of triples (ρ, θ, α) with $\rho \leq 1$ corresponding to elements $(x, A) \in \mathrm{M}(2)$ such that $\mathcal{K}^\circ \cap (x + A\mathcal{K}^\circ) = \emptyset$. We consider the following semidefinite programming problem:

Problem A Find real, positive semidefinite matrices Q^{ij}, R^{ij} for $i = 0, 1$ and $j = 0, \ldots, 9$, and S^j for $j = 0, \ldots, 9$, that minimize

$$\sum_{i=0}^{1}\sum_{j=0}^{9}\langle \mathcal{F}^{ij}(0,0,0), Q^{ij}\rangle$$

subject to the constraints

$$\sum_{i=0}^{1}\langle F^i_{r,s;k}, Q^{ij}\rangle = 0 \quad \text{if } k < |r-s|/2, \text{ where } r, s \equiv j \pmod{10}, \tag{22}$$

$$\sum_{i=0}^{1}(\langle F^i_{r,s;k}, Q^{ij}\rangle - \langle F^i_{-r,-s;k}, Q^{ij'}\rangle) = 0 \quad \text{where } r, s \equiv j \pmod{10} \tag{23}$$
$$\text{and } -r, -s \equiv j' \pmod{10},$$

$$\sum_{i=0}^{1}\sum_{j=0}^{9}(\langle \mathcal{F}^{ij}(\rho, z_1, z_2), Q^{ij}\rangle + \langle W^{ij}(\rho, z_1, z_2), R^{ij}\rangle) \tag{24}$$
$$+\sum_{j=0}^{9}\langle(\rho^2-1)W^{0j}(\rho, z_1, z_2), S^j\rangle = 0,$$

$$\sum_{i=0}^{1}\sum_{j=0}^{9}\langle \mathcal{F}^{ij}(\rho,\theta,\alpha), Q^{ij}\rangle \le 0 \quad \text{for all } (\rho,\theta,\alpha) \in \mathcal{S}, \tag{25}$$

$$\sum_{i=0}^{1}\langle F^{i}_{0,0;0}, Q^{i0}\rangle = 1. \tag{26}$$

Conditions (22)–(25) were already discussed in the previous sections. Notice this is indeed a semidefinite programming problem. In fact, the objective function and all constraints but (24) are clearly linear. As for the polynomial identity (24), one only has to observe that it can be turned into linear constraints by using the fact that a polynomial is identically zero if and only if each monomial has a zero coefficient (cf. Sect. 3).

Of Problem A we have to explain our choice of objective function and also the meaning of constraint (26). To obtain the best possible bound from Theorem 1.1, we wish to minimize $f(0, I)/\lambda$, where

$$\lambda = \int_{\mathrm{M}(2)} f(x, A)\, d(x, A).$$

Constraint (26) is a normalization constraint, setting $\lambda = 1$. Indeed, one has $\lambda = f_{0,0;0}$, since from the definition of $\widehat{f}$ and the inversion formula (cf. Sect. 4) we have

$$\begin{aligned} f_{0,0;0} &= (\widehat{f}(0))_{0,0} = \langle \widehat{f}(0)\mathbf{1}, \mathbf{1}\rangle \\ &= \left\langle \int_{\mathrm{M}(2)} f(x, A) U^{0}_{(x,A)^{-1}}\mathbf{1}\, d(x, A), \mathbf{1}\right\rangle \\ &= \lambda\langle \mathbf{1}, \mathbf{1}\rangle \\ &= \lambda, \end{aligned}$$

where $\mathbf{1} \in L^2(S^1)$ is the constant one function, so that $U^{0}_{(x,A)^{-1}}\mathbf{1} = \mathbf{1}$. Now, the objective function evaluates $f(0, I)$, that we wish to minimize.

To be able to solve Problem A on the computer, the choice of the sequence P_0, P_1, ... of polynomials which we use to define our matrices is essential. A bad choice here can lead to numerical instability that might prevent us from solving the problem.

In particular, we have observed that the monomial basis performs specially badly. A much better choice are normalized Laguerre polynomials, as had been observed in a similar setting by de Laat, Oliveira, and Vallentin [24]. Namely, we set

$$P_k(x) = \mu_k^{-1} L_k^0(2\pi x^2),$$

where μ_k is the absolute value of the coefficient of $L_l^0(2\pi x^2)$ with largest absolute value.

Also essential to the stability of Problem A is the choice of the basis used to express polynomial identity (24). Again, the monomial basis is a poor choice. Instead we use the basis

$$P_k(\rho^2)z_1^{-r}z_2^{-s}$$

for $k = 0, \ldots, d$ and $-N \leq r, s \leq N$ such that $r - s \equiv 0 \pmod{10}$.

This means that in order to express constraint (24), we expand the corresponding polynomial in the above basis, and then require each coefficient of the expansion to be zero.

In preliminary tests with reasonably dense samples for constraint (25), we observed that most variables in Problem A did not seem to play a role, at least for the values of d and N that we considered. So we decided to discard all variable matrices except for

$$Q^{00},\ Q^{05},\ Q^{10},\ Q^{15},\ R^{00},\ R^{05},\ S^0,\ \text{and } S^5,$$

and we observed that this did not have much effect on the optimal value of the problem, while providing for simpler and more stable problems. From now on, when we refer to Problem A it should be understood that we only use the variables listed above.

We now have a complete description of the semidefinite programming problem to be solved, let us sketch how we obtained the bound of 0.98103 for the pentagon packing density.

We first solve Problem A (with less variables, as explained above) for $d = 11$ and $N = 5$, using a sample with 537 points. This sample we pick as follows. We first pick 5 uniformly spaced values for α in $[-2\pi/10, 0]$, starting with $-2\pi/10$ and ending with 0. For each such value of α, we pick in the square $[-1, 1]^2$ a uniformly spaced grid of 50×50 points, and add to the sample all triples (ρ, θ, α), where (ρ, θ) corresponds to a grid point outside of the Minkowski difference $(\mathcal{K} - A(\alpha)\mathcal{K})^\circ$ and such that $\rho \leq 1$. Moreover, the symmetry of $\mathcal{K}$ allows us to restrict our sample considerably — Fig. 2 has an example.

We observed, by evaluating the function f obtained via this approach, that this small sample is already enough to enforce condition (ii) of Theorem 1.1 on most of the required domain. To really obtain a function f satisfying the conditions of Theorem 1.1, however, we have to work a bit more.

Since we use a numerical solver for semidefinite programming, the solutions we obtain for Problem A are not really feasible, but almost feasible. So we cannot be *a priori* certain that the bound given by Problem A is really an upper bound.

To deal with this issue, we use the same approach outlined by de Laat, Oliveira, and Vallentin [24], which we briefly explain here. First, we solve Problem A in order to get an estimate of its optimal value; say z^* is the numerical optimal value obtained. Then, we solve a version of Problem A in which the objective function is removed but a constraint

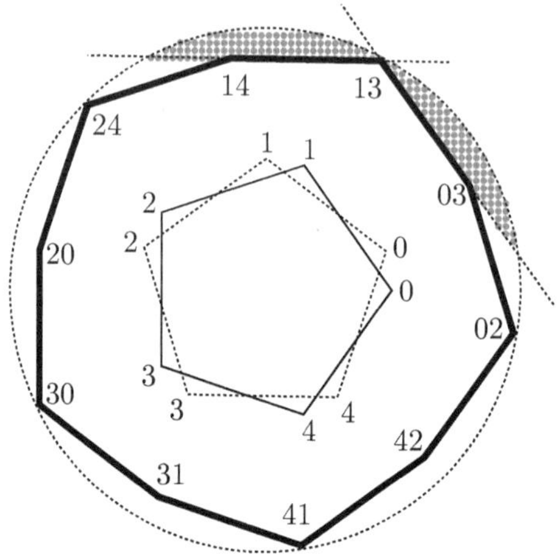

Fig. 2 The points in gray are an example of a sample used in Problem A; here we show the points in the sample for $\alpha = \pi/10$. Each facet F of the Minkowski difference defines a line l_F, its supporting hyperplane, and for the sample we would then pick all points in the grid that are inside the circle of radius 1 and that lie, for some facet F of the Minkowski difference, on the side of l_F that does not contain the origin. Since we work with S($\mathcal{K}$)-invariant functions, however, we need not choose all these points: It suffices to consider only two adjacent facets of the Minkowski difference, instead of all the facets

$$\sum_{i=0}^{1}\sum_{j=0}^{9}\langle \mathcal{F}^{ij}(0,0,0), Q^{ij}\rangle \le z^* + 10^{-5}$$

is added.

This problem is a feasibility problem, and for this reason the solver will return a solution that is strictly feasible, i.e., a solution in which the solution matrices are *positive definite*, if one can be found.

In this way, we manage to obtain a solution of Problem A having objective value close to what the optimal value is supposed to be, in which each matrix has a minimum eigenvalue around 10^{-6}, whereas the constraints are satisfied up to an absolute error of 10^{-9}. By projecting the solution obtained onto the affine subspace generated by constraints (22), (23), (24), and (26), using double-precision floating point arithmetic, we manage to drop the absolute error to 10^{-22}, while not changing much the minimum eigenvalues of the solution matrices. We give the matrices Q^{00}, Q^{10}, Q^{05}, Q^{15} parametrizing function f in Fig. 3.

So the approach detailed by de Laat, Oliveira, and Vallentin [24] applies. Namely, since the minimum eigenvalues of the solution matrices are big compared to the absolute errors, we may be sure that by changing the solution matrices slightly, we may ensure that the constraints are satisfied, thus obtaining a truly feasible solution, without significantly changing the objective value. Notice that we do not need to carry out this change in practice, it suffices to know that it can be done.

$$
Q^{00} = \begin{pmatrix}
0.471618 & 0.260411 & -0.236109 & -0.377199 & -0.180939 & -0.030009 \\
0.260411 & 7.592679 & 10.207992 & 7.759719 & 2.888178 & 0.397423 \\
-0.236109 & 10.207992 & 16.826377 & 15.319807 & 6.525211 & 1.046526 \\
-0.377199 & 7.759719 & 15.319807 & 15.894902 & 7.419904 & 1.313099 \\
-0.180939 & 2.888178 & 6.525211 & 7.419904 & 3.894405 & 0.800412 \\
-0.030009 & 0.397423 & 1.046526 & 1.313099 & 0.800412 & 0.192548
\end{pmatrix}
\quad
Q^{10} = \begin{pmatrix}
5.699753 & 10.718186 & 7.500505 & 3.282071 & 0.724819 & 0.038715 \\
10.718186 & 28.673593 & 23.889232 & 11.778999 & 2.842942 & 0.168553 \\
7.500505 & 23.889232 & 21.623810 & 11.602148 & 3.011834 & 0.198710 \\
3.282071 & 11.778999 & 11.602148 & 6.963134 & 1.985246 & 0.150341 \\
0.724819 & 2.842942 & 3.011834 & 1.985246 & 0.620808 & 0.057626 \\
0.038715 & 0.168553 & 0.198710 & 0.150341 & 0.057626 & 0.008410
\end{pmatrix}
$$

$$
Q^{05} = \begin{pmatrix}
0.031237 & 0.000014 & 0.073890 & 0.000014 & 0.066751 & -0.000043 & 0.042216 & -0.000070 & 0.015181 & -0.000041 & 0.001998 & -0.000008 \\
0.000014 & 0.031237 & 0.000014 & 0.073890 & -0.000043 & 0.066751 & -0.000070 & 0.042216 & -0.000041 & 0.015181 & -0.000008 & 0.001998 \\
0.073890 & 0.000014 & 0.187082 & -0.000040 & 0.206253 & -0.000252 & 0.157013 & -0.000336 & 0.063369 & -0.000200 & 0.008807 & -0.000043 \\
0.000014 & 0.073890 & -0.000040 & 0.187082 & -0.000252 & 0.206253 & -0.000336 & 0.157013 & -0.000200 & 0.063369 & -0.000043 & 0.008807 \\
0.066751 & -0.000043 & 0.206253 & -0.000252 & 0.332785 & -0.000662 & 0.314941 & -0.000794 & 0.140408 & -0.000475 & 0.020309 & -0.000106 \\
-0.000043 & 0.066751 & -0.000252 & 0.206253 & -0.000662 & 0.332785 & -0.000794 & 0.314941 & -0.000475 & 0.140408 & -0.000106 & 0.020309 \\
0.042216 & -0.000070 & 0.157013 & -0.000336 & 0.314941 & -0.000794 & 0.322670 & -0.000932 & 0.148145 & -0.000561 & 0.021668 & -0.000129 \\
-0.000070 & 0.042216 & -0.000336 & 0.157013 & -0.000794 & 0.314941 & -0.000932 & 0.322670 & -0.000561 & 0.148145 & -0.000129 & 0.021668 \\
0.015181 & -0.000041 & 0.063369 & -0.000200 & 0.140408 & -0.000475 & 0.148145 & -0.000561 & 0.068732 & -0.000343 & 0.010103 & -0.000082 \\
-0.000041 & 0.015181 & -0.000200 & 0.063369 & -0.000475 & 0.140408 & -0.000561 & 0.148145 & -0.000343 & 0.068732 & -0.000082 & 0.010103 \\
0.001998 & -0.000008 & 0.008807 & -0.000043 & 0.020309 & -0.000106 & 0.021668 & -0.000129 & 0.010103 & -0.000082 & 0.001494 & -0.000022 \\
-0.000008 & 0.001998 & -0.000043 & 0.008807 & -0.000106 & 0.020309 & -0.000129 & 0.021668 & -0.000082 & 0.010103 & -0.000022 & 0.001494
\end{pmatrix}
$$

$$
Q^{15} = \begin{pmatrix}
0.142874 & -0.000133 & 0.319521 & -0.000347 & 0.232797 & -0.000415 & 0.107381 & -0.000323 & 0.028240 & -0.000116 & 0.003013 & -0.000008 \\
-0.000133 & 0.142874 & -0.000347 & 0.319521 & -0.000415 & 0.232797 & -0.000323 & 0.107381 & -0.000116 & 0.028240 & -0.000008 & 0.003013 \\
0.319521 & -0.000347 & 0.716870 & -0.000892 & 0.529669 & -0.001028 & 0.250830 & -0.000783 & 0.068280 & -0.000279 & 0.007494 & -0.000021 \\
-0.000347 & 0.319521 & -0.000892 & 0.716870 & -0.001028 & 0.529669 & -0.000783 & 0.250830 & -0.000279 & 0.068280 & -0.000021 & 0.007494 \\
0.232797 & -0.000415 & 0.529669 & -0.001028 & 0.414894 & -0.001077 & 0.216990 & -0.000767 & 0.066166 & -0.000268 & 0.007884 & -0.000021 \\
-0.000415 & 0.232797 & -0.001028 & 0.529669 & -0.001077 & 0.414894 & -0.000767 & 0.216990 & -0.000268 & 0.066166 & -0.000021 & 0.007884 \\
0.107381 & -0.000323 & 0.250830 & -0.000783 & 0.216990 & -0.000767 & 0.130349 & -0.000518 & 0.045032 & -0.000179 & 0.005780 & -0.000015 \\
-0.000323 & 0.107381 & -0.000783 & 0.250830 & -0.000767 & 0.216990 & -0.000518 & 0.130349 & -0.000179 & 0.045032 & -0.000015 & 0.005780 \\
0.028240 & -0.000116 & 0.068280 & -0.000279 & 0.066166 & -0.000268 & 0.045032 & -0.000179 & 0.017003 & -0.000065 & 0.002284 & -0.000007 \\
-0.000116 & 0.028240 & -0.000279 & 0.068280 & -0.000268 & 0.066166 & -0.000179 & 0.045032 & -0.000065 & 0.017003 & -0.000007 & 0.002284 \\
0.003013 & -0.000008 & 0.007494 & -0.000021 & 0.007884 & -0.000021 & 0.005780 & -0.000015 & 0.002284 & -0.000007 & 0.000314 & -0.000001 \\
-0.000008 & 0.003013 & -0.000021 & 0.007494 & -0.000021 & 0.007884 & -0.000015 & 0.005780 & -0.000007 & 0.002284 & -0.000001 & 0.000314
\end{pmatrix}
$$

Fig. 3 Matrices Q^{00}, Q^{10}, Q^{05}, Q^{15} parametrizing the function f after the projection

Finally, we still have to show that the function f thus obtained satisfies condition (ii) of Theorem 1.1. We have said that f satisfies condition (ii) for most of the points on the required domain. For instance, since we have constraint (24), we know that $f(\rho, \theta, \alpha) \leq 0$ for all $\rho \geq 1$. There are, however, points (ρ, θ, α) with $\rho \leq 1$ for which we have $f(\rho, \theta, \alpha)$ positive, while (ii) would require this value to be nonpositive.

Though f does not satisfy condition (ii) of Theorem 1.1 for the pentagon $\mathcal{K}$, it satisfies this condition once we enlarge $\mathcal{K}$ slightly. Indeed, f satisfies condition (ii) for the pentagon $1.02\mathcal{K}$. This we may verify by picking a fine enough sample of points (ρ, θ, α) with $\rho \leq 1$ for which f has to be nonpositive, and computing the minimum value of f on this sample using 256-bit-precision floating point. By computing the derivatives of f, we may estimate how fine the sample has to be and how large the absolute value of the minimum of f on the sample has to be, in order for us to be sure that f is nonpositive in the whole required region.

A side effect of our restriction of the variables is that the function f we obtain is by construction such that

$$f(\rho, \theta, \alpha + l2\pi/5) = f(\rho, \theta, \alpha)$$

for all integer l (cf. (18)). This and the symmetry of $\mathcal{K}$ helps us restrict the sample to points (ρ, θ, α) with $\alpha \in [-2\pi/10, 2\pi/10]$. To obtain our bound, we had to use a sample of about 6.5 million points to check that f satisfies condition (ii) of Theorem 1.1. Details of this procedure can be found in the paper [27] by Dostert, Guzmán, Oliveira, and Vallentin.

Enlarging the body $\mathcal{K}$ worsens the bound given by Theorem 1.1, but since we consider a small enlargement of $\mathcal{K}$, we still manage to obtain the bound of 0.98103.

Finally, we mention some of the computational tools used to generate the semidefinite programming problem and solve it. To generate the problem, we use a C++ program with a custom-made C++ library for generating semidefinite programming problems, in particular dealing with sums of squares constraints. As a solver we used CSDP [10], and to analyze the resulting solution and check that it is feasible[5] we used a mix of SAGE [46] and C++.

Acknowledgements We are thankful to Pier Daniele Napolitani and Claudia Addabbo from the Maurolico Project, who provided us with a transcript of Maurolico's manuscript. In particular, Claudia Addabbo provided us with a draft of her commented Italian translation of the manuscript.

[5] http://maurolico.free.fr.

References

1. N.I. Akhiezer, Lectures on Integral Transforms, in *Translations of Mathematical Monographs 70* (American Mathematical Society, 1988)
2. G.E. Andrews, R. Askey, R. Roy, in *Special Functions, Encyclopedia of Mathematics and its Applications 71* (Cambridge University Press, Cambridge, 1999)
3. Aristotle, *On the Heavens*, translation by W.K.C. Guthrie (Harvard University Press, Cambridge, 2006)
4. S. Atkinson, Y. Jiao, S. Torquato, Maximally dense packings of two-dimensional convex and concave noncircular particles. Phys. Rev. E **86**, 031302 (2012)
5. E. Aylward, S. Itani, P.A. Parrilo, Explicit SOS decompositions of univariate polynomial matrices and the Kalman-Yakubovich-Popov lemma, in *Proceedings of the 46th IEEE Conference on Decision and Control* (2007), pp. 5660–5665
6. C. Bachoc, G. Nebe, F.M. de Oliveira Filho, F. Vallentin, Lower bounds for measurable chromatic numbers. Geom. Funct. Anal. **19**, 645–661 (2009)
7. A. Ben-Tal, A. Nemirovski, *Lectures on Modern Convex Optimization: Analysis, Algorithms, and Engineering Applications* (SIAM, Philadelphia, 2001)
8. A. Bezdek, W. Kuperberg, Dense packing of space with various convex solids, in *Geometry — Intuitive, Discrete, and Convex, A Tribute to László Fejes Tóth*, Bolyai Society Mathematical Studies, ed. by I. Bárány, K.J. Böröczky, G. Fejes Tóth, J. Pach (Springer, Berlin, 2013), pp. 66–90
9. S. Bochner, Hilbert distances and positive definite functions. Ann. Math. **42**, 647–656 (1941)
10. B. Borchers, CSDP, A C library for semidefinite programming. Optim. Methods Softw. **11**, 613–623 (1999)
11. P. Brass, W. Moser, J. Pach, *Research Problems in Discrete Geometry* (Springer, Berlin, 2005)
12. B. Casselman, Can you do better? in *Feature Column of the AMS*, http://www.ams.org/samplings/feature-column/fc-2012-11 (2012)
13. E.R. Chen, M. Engel, S.C. Glotzer, Dense crystalline dimer packings of regular tetrahedra. Discrete Comput. Geom. **44**, 253–280 (2010)
14. M.D. Choi, T.Y. Lam, B. Reznick, Real zeros of positive semidefinite forms I. Mathematische Zeitschrift **171**, 1–26 (1980)
15. H. Cohn, N.D. Elkies, New upper bounds on sphere packings I. Ann. Math. **157**, 689–714 (2003)
16. H. Cohn, A. Kumar, Optimality and uniqueness of the Leech lattice among lattices. Ann. Math. **170**, 1003–1050 (2009)
17. H. Cohn, A. Kumar, S.D. Miller, D. Radchenko, and M.S. Viazovska, The sphere packing problem in dimension 24. Ann. Math. (2) **185**(3), 1017–1033 (2017). arXiv:1603.06518 [math.NT]
18. H. Cohn, S.D. Miller, *Some properties of optimal functions for sphere packing in dimensions 8 and 24* (2016) 23p. arXiv:1603.04759 [math.MG]
19. H. Cohn, Y. Zhao, Sphere packing bounds via spherical codes. Duke Math. J. **163**, 1965–2002 (2014)
20. J.B. Conway, *A Course in Functional Analysis, Graduate Texts in Mathematics 96* (Springer, New York, 1985)
21. J.H. Conway, N.J.A. Sloane, *Sphere packings, lattices and groups (Grundlehren der mathematischen Wissenschaften)*, vol. 290, 3rd edn. (Springer, New York, 1999)
22. J.H. Conway, S. Torquato, Packing, tiling, and covering with tetrahedra. Proc. Natl. Acad. Sci. USA **103**, 10612–10617 (2006)
23. E. de Klerk, F. Vallentin, On the Turing model complexity of interior point methods for semidefinite programming. SIAM J. Optim. **26**(3), 1944–1961 (2016). arXiv:1507.03549 [math.OC]
24. D. de Laat, F.M. de Oliveira Filho, F. Vallentin, Upper bounds for packings of spheres of several radii. Forum Math. Sigma **2**, e23 (42 pages) (2014)
25. P. Delsarte, J.M. Goethals, J.J. Seidel, Spherical codes and designs. Geom. Dedic. **6**, 363–388 (1977)

26. P. Delsarte, V.I. Levensthein, Association schemes and coding theory. IEEE Trans. Inf. Theory **IT–44**, 2477–2504 (1988)
27. M. Dostert, C. Guzmán, F.M. de Oliveira Filho, F. Vallentin, New upper bounds for the density of translative packings of three-dimensional convex bodies with tetrahedral symmetry. Discrete Comput. Geom. **58**, 449–481 (2017). arXiv:1510.02331 [math.MG]
28. G. Fejes Tóth, F. Fodor, V. Vígh, The packing density of the n-dimensional cross-polytope. Discrete Comput. Geom. **54**, 182–194 (2015)
29. G. Fejes Tóth, W. Kuperberg, Packing and covering with convex sets, in *Handbook of Convex Geometry*, ed. by P.M. Gruber, J.M. Wills (North-Holland, Amsterdam, 1993), pp. 799–860
30. G.B. Folland, *A Course in Abstract Harmonic Analysis* (Studies in Advanced Mathematics, CRC Press, Boca Raton, 1995)
31. S. Gravel, V. Elser, Y. Kallus, Upper bound on the packing density of regular tetrahedra and octahedra. Discrete Comput. Geom. **46**, 799–818 (2011)
32. T.C. Hales, A proof of the Kepler conjecture. Ann. Math. **162**, 1065–1185 (2005)
33. T.C. Hales, M. Adams, G. Bauer, D. Tat Dang, J. Harrison, T. Le Hoang, C. Kaliszyk, V. Magron, S. McLaughlin, T. Tat Nguyen, T. Quang Nguyen, T. Nipkow, S. Obua, J. Pleso, J. Rute, A. Solovyev, A. Hoai Thi Ta, T. Nam Tran, D. Thi Trieu, J. Urban, K. Khac Vu, R. Zumkeller, *A formal proof of the Kepler conjecture* (2015) 21p. arXiv:1501.02155 [math.MG]
34. T.C. Hales, W. Kusner, *Packings of regular Pentagons in the plane* (2016) 26p. arXiv:1602.07220 [math.MG]
35. Y. Kallus, W. Kusner, *The local optimality of the double lattice packing* (2015) 23p. arXiv:1509.02241 [math.MG]
36. R.M. Karp, Reducibility among combinatorial problems, in: *Complexity of Computer Computations*, ed. by R.E. Miller, J.W. Thatcher. Proceedings of a symposium on the Complexity of Computer Computations, (IBM Thomas J. Watson Research Center, Yorktown Heights, Plenum Press, New York, 1972), pp. 85–103
37. J. Kepler, Vom sechseckigen Schnee (*Strena seu de Nive sexangula*, published in 1611), translation with introduction and notes by Dorothea Goetz, *Ostwalds Klassiker der exakten Wissenschaften* 273, (Akademische Verlagsgesellschaft Geest u. Portig K.-G, Leipzig, 1987)
38. G. Kuperberg, W. Kuperberg, Double-lattice packings of convex bodies in the place. Discrete Comput. Geom. **5**, 389–397 (1990)
39. J.C. Lagarias, C. Zong, Mysteries in packing regular tetrahedra. Notices Amer. Math. Soc. **59**, 1540–1549 (2012)
40. M. Laurent, Sums of squares, moment matrices and optimization, in *Emerging Applications of Algebraic Geometry*, IMA Volumes in Mathematics and its Applications, ed. by M. Putinar, S. Sullivant (Springer, Berlin, 2009), pp. 157–270
41. L. Lovász, On the Shannon capacity of a graph. IEEE Trans. Inf. Theory **IT–25**, 1–7 (1979)
42. F. Maurolico, De quinque solidis, quae vulgo regularia dicuntur, quae videlicet eorum locum impleant, et quae non, contra commentatorem Aristotelis, Averroem, 1529
43. R.J. McEliece, E.R. Rodemich, H.C. Rumsey Jr., The Lovász bound and some generalizations. J. Comb. Inf. Syst. Sci. **3**, 134–152 (1978)
44. C.A. Rogers, *Packing and Covering* (Cambridge University Press, 1964)
45. A. Schrijver, A comparison of the Delsarte and Lovász bounds. IEEE Trans. Inf. Theory **IT–25**, 425–429 (1979)
46. W.A. Stein et al. *Sage Mathematics Software (Version 4.8)*. The Sage Development Team (2012). http://www.sagemath.org
47. M. Sugiura, *Unitary Representations and Harmonic Analysis: An Introduction* (Kodansha Scientific Books, Tokyo, 1990)
48. M.S. Viazovska, The sphere packing problem in dimension 8. Ann. Math. (2) **185**(3), 991–1015 (2017). arXiv:1603.04246 [math.NT]
49. G.N. Watson, *A Treatise on the Theory of Bessel Functions* (Cambridge University Press, 1922)
50. G.M. Ziegler, Three mathematics competitions, in *An Invitation to Mathematics: From Competitions to Research*, ed. by D. Schleicher, M. Lackmann (Springer, Berlin, 2011), pp. 195–206

Two Geometrical Applications of the Semi-random Method

Péter Hajnal and Endre Szemerédi

Abstract The semi-random method was introduced in the early eighties. In its first form of the method lower bounds were given for the size of the largest independent set in hypergraphs with certain uncrowdedness properties. The first geometrical application was a major achievement in the history of Heilbronn's triangle problem. It proved that the original conjecture of Heilbronn was false. The semi-random method was extended and applied to other problems. In this paper we give two further geometrical applications of it. First, we give a slight improvement on Payne and Wood's upper bounds on a Ramsey-type parameter, introduced by Gowers. We prove that any planar point set of size $\Omega\left(\frac{n^2 \log n}{\log\log n}\right)$ contains n points on a line or n independent points. Second, we give a slight improvement on Schmidt's bound on Heilbronn's quadrangle problem. We prove that there exists a point set of size n in the unit square that doesn't contain four points with convex hull of area $\mathcal{O}(n^{-3/2}(\log n)^{1/2})$.

1 Introduction

The semi-random method was introduced for graphs in [1]. Later it was extended to 3-uniform hypergraphs in [7]. The method was further extended in [2, 4].

A hypergraph $\mathcal{H}$ on the vertex set V is a subset of $\mathcal{P}(V)$, the power set of V. I.e. $\mathcal{H}$ is a collection of certain subsets of V, called edges. If the edges have a common size,

Partially supported by TÉT_12_MX-1-2013-0006 and by National Research, Development and Innovation Office – NKFIH Fund No. SNN-117879. Supported by ERC-AdG. 321104, and OTKA Grant NK104186.

P. Hajnal (✉)
Bolyai Institute, University of Szeged, Szeged, Hungary
e-mail: hajnal@math.u-szeged.hu

P. Hajnal · E. Szemerédi
Alfréd Rényi Institute of Mathematics, Hungarian Academy of Sciences, Budapest, Hungary
e-mail: szemered@renyi.hu

E. Szemerédi
Department of Computer Science, Rutgers University, New Brunswick, NJ, USA

G. Ambrus et al. (eds.), *New Trends in Intuitive Geometry*, Bolyai Society Mathematical Studies 27, https://doi.org/10.1007/978-3-662-57413-3_8

say k, then we say that $\mathcal{H}$ is k-uniform. In a hypergraph $\mathcal{H}$ a vertex set $I \subset V$ is called an independent set iff it doesn't contain any edge as a subset. The maximum size of the independent sets of $\mathcal{H}$ is denoted by $\alpha(\mathcal{H})$. There are several results concerning independent sets in 3-uniform uncrowded hypergraphs. From hypergraph theory we recall that the degree of a vertex x $(deg(x))$ is the number of edges, containing x. Also a k-cycle $(k \geq 2)$ in $\mathcal{H}$ is a sequence of k different vertices: $x_1, \ldots, x_{k-1}, x_k = x_0$ and a sequence of k different edges: $E_1, \ldots, E_k$ such that $x_{i-1}, x_i \in E_i$ for $i = 1, 2 \ldots, k$. The cycle above is called a simple cycle iff $E_i \cap \left(\cup_{j:j \neq i} E_j\right) = \{x_{i-1}, x_i\}$ for $i = 1, 2 \ldots, k$. We quote the earliest result on hypergraphs using the semi-random method.

Theorem 1 ([7], Lemma 1) *Let $\mathcal{H}$ be a 3-uniform hypergraph on v vertices. Let $\overline{d}$ denote the average degree of $\mathcal{H}$. Assume that $\overline{d} \leq t^2$ and $1 \ll t \ll v^{1/10}$.*

If $\mathcal{H}$ doesn't contain simple cycles of length at most 4, then

$$\alpha(\mathcal{H}) = \Omega\left(\frac{v}{t}\sqrt{\log t}\right).$$

In our applications we might have many simple cycles of length 3 and 4. We need the following strengthening of the basic bound:

Theorem 2 ([4], Theorem 2) *Let $\mathcal{H}$ be a k-uniform hypergraph on v vertices. Let Δ be the maximum degree of $\mathcal{H}$. Assume that $\Delta \leq t^{k-1}$ and $1 \ll t$. If $\mathcal{H}$ doesn't contain a 2-cycle (two edges with at least two common vertices), then*

$$\alpha(\mathcal{H}) = \Omega\left(\frac{v}{t}(\log t)^{\frac{1}{k-1}}\right).$$

We give two new geometrical applications of the above bound.

In the first application we consider a question asked by Gowers [5]. Given a planar point set $\mathcal{P}$, what is the minimal size of $\mathcal{P}$ that guarantees that one can find n points on a line or n independent points (no three on a line) in it? He noted that the grid shows that $\Omega(n^2)$ many points are necessary, and in the case of $2n^3$ many points without n points on a line a simple greedy algorithm finds n independent points. Payne and Wood [9] improved the upper bound to $\mathcal{O}(n^2 \log n)$. They also considered an arbitrary point set with much fewer points than n^3 and without n points on a line. But instead of the greedy algorithm they used Spencer's lemma, which is based on a simple probabilistic sparsification. They also used the Szemerédi–Trotter theorem in order to bound the number of edges of their hypergraph.

We improve the previous upper bound methods. We also start with a random sparsification. After some additional preparation (we get rid of 2-cycles) we are able to use a semi-random method (see [4]) to find a large independent set.

Theorem 3 *Let $\mathcal{P}$ be an arbitrary planar point set of size $\Omega\left(\frac{n^2 \log n}{\log \log n}\right)$. Then we can find n points in $\mathcal{P}$, that are incident to a line or independent.*

Our second application is closely related to Heilbronn's triangle problem [8, 10–15]. Take a "nice" unit area domain D (usually a square, disc or a regular triangle). Place n points into D and find the smallest area among the triangles determined by the chosen points. Let $H_\triangle(n)$ denote the maximum of this parameter over all possible choices of n points.

Instead of triangles we can take k-tuples of our point set and consider the area of the convex hull of the k chosen points. We denote the corresponding parameter by $H_k(n)$ (so $H_3(n) = H_\triangle(n)$). The best lower bound on $H_\triangle(n)$ [7], and some trivial observations are summarized in the next line:

$$\Omega\left(\frac{\sqrt{\log n}}{n^2}\right) = H_\triangle(n) \le H_4(n) \le H_5(n) \le \cdots = \mathcal{O}\left(\frac{1}{n}\right).$$

We mention two major open problems: Is it true that $H_\triangle(n) = O(1/n^{2-o(1)})$ and $H_4(n) = o(1/n)$?

Our interest is in the lower bound on $H_4(n)$. Schmidt [15] proved that $H_4(n) = \Omega(n^{-3/2})$. The proof is a construction of a point set by a simple greedy algorithm. In [3] the authors provide a new proof, and extensions of this result. They also proposed an open question, which they have not yet been able to resolve: that is whether Schmidt's bound can be improved by a logarithmic factor. With the help of the semi-random method we are able to improve Schmidt's bound and settle the problem of [3].

Theorem 4 *There exists a point set of size n in the unit square that doesn't contain four points with convex hull of area* $\mathcal{O}(n^{-3/2}(\log n)^{1/2})$.

In some cases we closely follow the preceding papers. Since it is hard to refer technical details we repeat the necessary arguments. This way our paper is self-contained. Throughout the paper we will use log to denote the logarithm of base 2, we omit all floor and ceiling signs, whenever these are not essential, and assume that n is large enough.

2 The Proof of Theorem 3

Let $\mathcal{P}$ be a planar point set of size $N = \frac{n^2 \log n}{\log\log n}$, not containing n points on a line. We must show that it contains a large independent set. The collinear triples of $\mathcal{P}$ form a 3-uniform hypergraph $\mathcal{H}_3$. Independent subsets of the point set are the independent sets of the hypergraph. The collinear quadruples of $\mathcal{P}$ form a 4-uniform $\mathcal{H}_4$ hypergraph. Edges correspond to $K_4^{(3)}$'s (four vertices with all four triples as edges) in $\mathcal{H}_3$.

First we consider the size of $\mathcal{H}_3$, and $\mathcal{H}_4$. A line with i incident points determines $\binom{i}{3}$ many edges of $\mathcal{H}_3$ and $\binom{i}{4}$ many edges of $\mathcal{H}_4$. Let t_i denote the number of lines that contain exactly i points of $\mathcal{P}$. Our assumption gives that $0 = t_n = t_{n+1} = \cdots$. Similarly let $t_{\ge i}$ denote the number of lines that contain at least i points of $\mathcal{P}$ ($t_{\ge i} = t_i + t_{i+1} + \cdots + t_{n-1}$). Then

$$|\mathcal{H}_3| = \sum_{i=2}^{n-1} \binom{i}{3} t_i \le \sum_{i=2}^{n-1} \left(i^2 \sum_{j=i}^{n-1} t_j \right) = \sum_{i=2}^{n-1} i^2 t_{\ge i}.$$

The Szemerédi–Trotter Theorem [16] says that $t_{\ge i} = \mathcal{O}(|\mathcal{P}|^2/i^3 + |\mathcal{P}|/i) = \mathcal{O}(N^2/i^3)$ (in our case $N^2/i^3 \gg N/i$). For a suitable constant see [6] (Theorems 18.6 and 18.7): $t_{\ge i} \le 1000N^2/i^3$. Thus,

$$|\mathcal{H}_3| \le \sum_{i=2}^{n-1} i^2 t_{\ge i} \le \sum_{i=2}^{n-1} i^2 1000 \frac{N^2}{i^3} = 1000N^2 \sum_{i=2}^{n-1} \frac{1}{i} \le 2000N^2 \log n = 2000 \frac{n^4 (\log n)^3}{(\log\log n)^2}.$$

Similarly

$$|\mathcal{H}_4| \le \sum_{i=2}^{n-1} i^3 t_{\ge i} \le \sum_{i=2}^{n-1} i^3 1000 \frac{N^2}{i^3} = 1000N^2 \sum_{i=2}^{n-1} 1 \le 1000N^2 n = 1000 \frac{n^5 (\log n)^2}{(\log\log n)^2}.$$

Consider a random subset of $\mathcal{P}$, that we obtain keeping each point with probability p, and throwing away with probability $1 - p$ (and doing this independently).

Let

$$p = \frac{1}{100} \left(\frac{1}{n^{1/3} N^{1/3}} \right) = \frac{(\log\log n)^{1/3}}{100n (\log n)^{1/3}}.$$

After the random sparsification let $\boldsymbol{\mathcal{P}}$, $\boldsymbol{\mathcal{H}}_3$, and $\boldsymbol{\mathcal{H}}_4$ be the set of the surviving points, collinear triples, and collinear quadruples (these are random sets, throughout the paper we use bold face to denote random variables). It is obvious that

$$\mathbb{E}(|\boldsymbol{\mathcal{P}}|) = pN = \frac{n(\log n)^{2/3}}{100(\log\log n)^{2/3}},$$

$$\mathbb{E}(|\boldsymbol{\mathcal{H}}_3|) = p^3 |\mathcal{H}_3| \le \frac{2n(\log n)^2}{1000 \log\log n},$$

$$\mathbb{E}(|\boldsymbol{\mathcal{H}}_4|) = p^4 |\mathcal{H}_4| \le \frac{n(\log n)^{2/3}}{100000(\log\log n)^{2/3}}.$$

We have chosen the probability so that the number of the surviving edges of $\mathcal{H}_4$ will be negligible compared to the number of surviving vertices.

Using elementary probability theory the following events will hold at the same time with high probability:

$$\frac{n(\log n)^{2/3}}{1000(\log\log n)^{2/3}} < |\boldsymbol{\mathcal{P}}| < \frac{n(\log n)^{2/3}}{10(\log\log n)^{2/3}},$$

$$|\mathcal{H}_4| \leq \frac{n(\log n)^{2/3}}{10000(\log\log n)^{2/3}},$$

$$|\mathcal{H}_3| < \frac{n(\log n)^2}{100\log\log n}.$$

Hence $\overline{d}(\mathcal{H}_3) < \frac{30(\log n)^{4/3}}{(\log\log n)^{1/3}}$, where $\overline{d}$ denotes the average degree.

Now choose one of the good outcomes of the above probabilistic process such that all the above events hold. Let $\mathcal{H}_3^{(0)}$, and $\mathcal{H}_4^{(0)}$ the corresponding 3- and 4-uniform hypergraphs.

Consider $\mathcal{H}_3^{(0)}$, and throw away all points of surviving quadruples of $\mathcal{H}_4^{(0)}$, and throw away each point that has degree higher than $\frac{100(\log n)^{4/3}}{(\log\log n)^{1/3}}$. Let $\mathcal{L}$ denote the "leftover" points with the "leftover" triples.

We are still left with at least one third of the points. Hence the leftover hypergraph has at least $\frac{n(\log n)^{2/3}}{3000(\log\log n)^{2/3}}$ many vertices. By throwing away the high degree vertices the maximal degree of $\mathcal{L}$ at most is $\frac{100(\log n)^{4/3}}{(\log\log n)^{1/3}}$. Furthermore $\mathcal{L}$ is very "uncrowded": we cannot have 2-cycles in $\mathcal{L}$. Indeed, two edges along the same pair of vertices would give a quadruple that is an edges in $\mathcal{H}_4$. Our process eliminated all of them, hence there are no 2-cycles in $\mathcal{L}$. The conditions in Theorem 2 are satisfied with $v = \Theta(\frac{n(\log n)^{2/3}}{(\log\log n)^{2/3}})$, $t = \frac{10(\log n)^{2/3}}{(\log\log n)^{1/6}}$, and $\log t = \Theta(\log\log n)$.

The rest of the proof is the application of the Theorem 2 and simple arithmetic:

$$\alpha(\mathcal{H}_3) \geq \alpha(\mathcal{L}) \geq c\frac{v}{t}\sqrt{\log t} \geq c'n,$$

where c is the constant in Theorem 2 and c' can be determined from c based on our previuos calculation.

Theorem 3 can be easily deduced from this: Let $\nu = \min\{n, c'n\}$. Our point set contains $\Theta(\frac{\nu^2\log\nu}{\log\log\nu})$ points and our argument guarantees that it contains ν points on a line or ν independent points. The proof is complete.

3 The Proof of Theorem 4

Let $S := \{(x, y) \in \mathbb{R}^2 : |x|, |y| \leq 1/2\}$ be a unit square on the plane. Choose N (a parameter that will be chosen later) random points (independently with uniform distribution) from $(1/2)S = \{(x/2, y/2); (x, y) \in S\}$. Let $\mathcal{P}$ be the point set $\{P_1, P_2, \ldots, P_N\}$ we obtain this way. $\mathcal{P}$ is a random point set. The reason we place our points into $(1/2)S$ is technical. This way we know that any connecting line of two points from $\mathcal{P}$ has an intersection with S of length $\Theta(1)$, furthermore any distance determined by points of $\mathcal{P}$ is smaller than 0.9.

Consider the following 4-uniform hypergraph $\mathcal{Q}$ on the vertex set $\mathcal{P}$: A point set of size 4, $\{P, Q, R, S\}$ forms an edge iff $Area(PQRS) < \tau$, where $Area(PQRS)$ is the area of the convex hull of $\{P, Q, R, S\}$, and τ is a threshold, to be determined

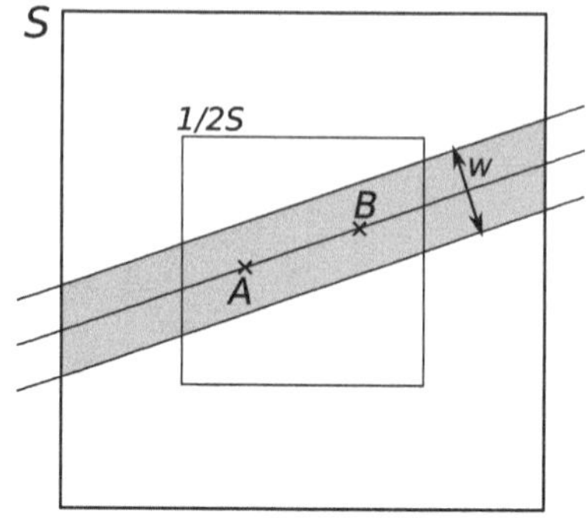

Fig. 1 The shaded region is $strip(AB, w)$. Its area is $\Theta(w)$

later. (Similarly $Area(PQR)$ is the area of the PQR triangle.) $\mathcal{Q}$ is a random 4-uniform hypergraph.

The major part of the proof is bounding the expected values of combinatorial parameters of $\mathcal{Q}$.

Let $A, B \in \mathcal{P}$ be two different points and

$$\mathbf{deg}(A, B) = |\{\{C, D\} : \{A, B, C, D\} \in \mathcal{Q}\}| \leq |\{(C, D) : \{A, B, C, D\} \in \mathcal{Q}\}| \tag{1}$$

i.e. denotes the number of edges of $\mathcal{Q}$, that contains both A and B. The upper bound counts ordered pairs (hence it is an overcounting by a factor of 2). Our goal is to give an upper bound for this parameter. We will count how many ordered pairs of points C, D are considered when $deg(A, B)$ is determined.

Let $strip(AB, w)$ denote the set of points from S, that are in the strip of width w with midline AB (see Fig. 1). I.e. $strip(AB, w)$ contains those points of S that have distance at most $w/2$ from the line AB.

Fix A and B, and let $d = dist(A, B)(< 1)$. $deg(A, B)$ counts certain C, D pairs of points, see (1). We distinguish cases according to the position of C, an arbitrary point from $\mathcal{P} - \{A, B\}$ and we bound the possible positions of the D's that contributes to $deg(A, B)$ with the current C.

Case 1: $C \notin strip(AB, 4\tau/d)$.

In this case the area of $ABC\triangle$ is at least τ, hence this C doesn't contributes to $deg(A, B)$.

Case 2: $C \in strip(AB, 4\tau/\sqrt{d})$. Note that $strip(AB, 4\tau/\sqrt{d})$ has area $\Theta(\tau/\sqrt{d})$. We distinguish two subcases:

Case 2a: $D \notin strip(AB, 4\tau/d)$. Similarly to Case 1 no D contributes to $deg(A, B)$.

Case 2b: $D \in strip(AB, 4\tau/d)$. Note that this strip has area $\Theta(\tau/d)$.

Case3: $C \in strip(AB, 4\tau/d) - strip(AB, 4\tau/\sqrt{d})$ (note that $d < \sqrt{d} < 1$). $strip(AB, 4\tau/d) - strip(AB, 4\tau/\sqrt{d})$ has area $\Theta(\tau/d)$.

The contributing D's must come from $strip(AB, 4\tau/d) \cap strip(AC, 4\tau/dist(A, C))$.

Elementary geometry gives that the above region has area $\Theta(\tau^2/Area(ABC\triangle))$ (see the green parallelogram on Fig. 2), bounding the possible positions of

Fig. 2 The shaded region is the space for those C's where Case 3 applies. The green region contains those D's, that can form an edge of $\mathcal{Q}$ with A, B and C

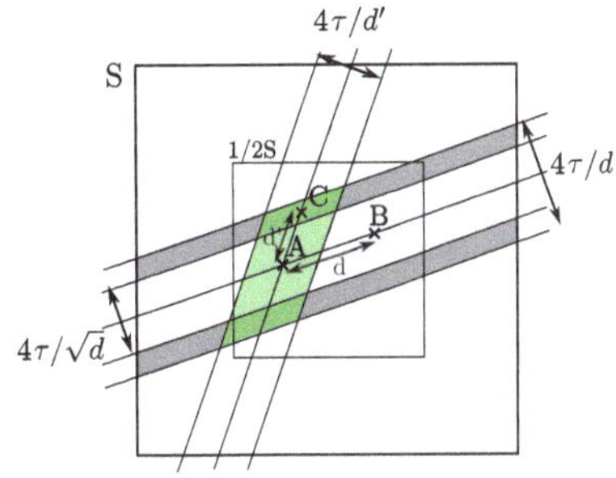

contributing D's. Since we are in Case 3 we have $Area(ABC\triangle) = \Omega(d \cdot \tau/\sqrt{d}) = \Omega(\tau\sqrt{d})$, hence the parallelogram has area $\mathcal{O}(\tau/\sqrt{d})$.

The expected value of $\mathbf{deg}(A, B)$ can be bounded easily. The contributing C,D's are covered by Case 2b and Case 3. In both cases the contributing C's and D's are coming from a restricted domain with known area. Since the choice of C and D are independent the number of contributing (C, D)'s in expectation is a product of the two expectations, that we can bound. In each of the three cases this product is $\mathcal{O}(\tau^2 d^{-3/2} N^2)$. Hence

$$\mathbb{E}(\mathbf{deg}(A, B)) = \mathcal{O}(\tau^2 (dist(A, B))^{-\frac{3}{2}} N^2).$$

With a similar argument we obtain bound on the number of 2-cycles through $A, B \in \mathcal{P}$. In a 4-uniform hypergraph there are two types of 2-cycles: (I): $\{A, B, C, D\}$, $\{A, B, C', D'\}$ and (II): $\{A, B, C, D\}$, $\{A, B, C, D'\}$ (now different symbols denote different points). Let $\mathcal{C}_I(A, B)$, resp. $\mathcal{C}_{II}(A, B)$ denote the number of 2-cyles of type (I), resp. type (II) for given points, A and B. Bounding the expected value of $\mathcal{C}_I(A, B)$ is easy, based on the previous calculation

$$\mathbb{E}(\mathcal{C}_I(A, B)) = \mathcal{O}(\tau^4 (dist(A, B))^{-3} N^4).$$

Bounding the expected value of $\mathcal{C}_{II}(A, B)$ is a little bit more technical. We distinguish the contribution of C's that satisfy Case 2 and those that satisfy Case 3:

$$\mathbb{E}(\mathcal{C}_{II}(A, B)) = \mathcal{O}((N\tau/\sqrt{d})(N\tau/d)^2 + (N\tau/d)(N\tau/\sqrt{d})^2) = \mathcal{O}(\tau^3 N^3 d^{-2.5}).$$

Now we sparsify our point set a little bit in order to have a lower bound on the minimal distance determined by our points.

Let $\delta = \frac{1}{100} N^{-1/2}$. We count the number of pairs in $\mathcal{P}$ that are closer than δ. Let $C(\mathcal{P})$ be the set of these pairs (this is a random set). Let $C_A(\mathcal{P}) = \mathcal{P} \cap Disc(A; \delta)$, where $Disc(A; \delta)$ denote the disc of radius δ centered at A. It is clear that $|C(\mathcal{P})| = 1/2 \sum_{A \in \mathcal{P}} |C_A(\mathcal{P})|$ and $\mathbb{E}(|C_A(\mathcal{P})|) \le (N-1) Area(Disc(A; \delta)) = 1/2\pi \cdot \delta^2 N < 1/1000$. Hence

$$\mathbb{E}(|C(\mathcal{P})|) \le N/1000.$$

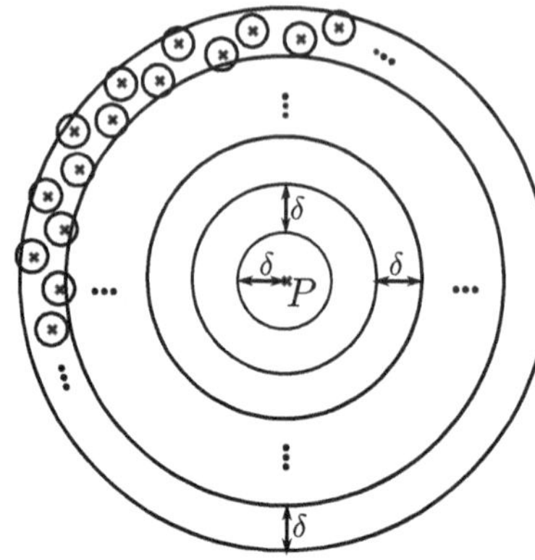

Fig. 3 The annuli around P, and the elementary volume argument in the proof

With high probability $|C(\mathcal{P})| \leq N/4$. After deleting these pairs we obtain $\mathcal{P}_0$, our new point set. $\mathcal{P}_0$ has size at least $N/2$ with high probability, and the distance of any two points of it is at least δ.

Let $\mathcal{Q}_0$ be the restriction of $\mathcal{Q}$ to $\mathcal{P}_0$. From now on we will work with $\mathcal{Q}_0$.

Lemma 5 *Let $\mathcal{M}$ be a set of M points from S so that the minimal distance among them is at least δ. Let $P \in S$. Let $Ann_i(P, \delta)$ be the annulus*

$$Ann_i(P; \delta) = \{X \in \mathbb{R}^2 : (i-1)\delta < dist(P, X) \leq i\delta\}.$$

$Ann_1(P; \delta)$, $Ann_2(P; \delta)$, ..., $Ann_{\mathcal{O}(\delta^{-1})}(P; \delta)$ are disjoint and cover S (hence they cover our point set). Furthermore at most $\mathcal{O}(i)$ of our M points can be covered by $Ann_i(P, \delta)$ (Fig. 3).

Proof The covering property is obvious. The bound on the number of points in the annulus is a simple volume argument: Draw $Disc(A, \delta/3)$ for all points $A \in \mathcal{M} \cap Ann_i(P; \delta)$. These discs are disjoint subsets of $\{X \in \mathbb{R}^2 : (i-4/3)\delta < dist(P, X) \leq (i+1/3)\delta\}$. The claim follows immediately. □

Using Lemma 5 and the earlier estimation for $deg(A, B)$'s it is easy to bound the expected number of edges, in $\mathcal{Q}_0$:

$$\mathbb{E}(deg_{\mathcal{Q}_0}(A)) \leq \sum_{B \in \mathcal{P}_0} \mathbb{E}(\mathbf{deg}(A, B)) = \sum_{i=1}^{\mathcal{O}(N^{1/2})} \sum_{B \in Ann_i(A,\delta) \cap \mathcal{P}_0} \mathbb{E}(\mathbf{deg}(A, B))$$
$$\leq \sum_{i=1}^{\mathcal{O}(N^{1/2})} \sum_{B \in Ann_i(A,\delta) \cap \mathcal{P}_0} \mathcal{O}(\tau^2 N^2 (i/\sqrt{N})^{-3/2})$$

$$\leq \sum_{i=1}^{\mathcal{O}(N^{1/2})} i \cdot \mathcal{O}(\tau^2 N^{2.75} i^{-3/2}) = \mathcal{O}(\tau^2 N^{2.75}) \sum_{i=1}^{\mathcal{O}(N^{1/2})} i \cdot i^{-3/2}$$
$$= \mathcal{O}(\tau^2 N^{2.75})\mathcal{O}(N^{0.25}) = \mathcal{O}(\tau^2 N^3).$$

Hence

$$\mathbb{E}(|\mathcal{Q}_0|) = \mathcal{O}(\tau^2 N^4).$$

The bound of the expected number of 2-cycles ($\mathcal{C} = \mathcal{C}_I + \mathcal{C}_{II}$) is similar:

$$\mathbb{E}(\mathcal{C}_I) \leq \sum_{A,B\in\mathcal{P}_0} \mathbb{E}(\mathcal{C}_I(A, B)) = \sum_{A\in\mathcal{P}_0} \sum_{i=1}^{\mathcal{O}(N^{1/2})} \sum_{B\in Ann_i(A,\delta)\cap\mathcal{P}_0} \mathbb{E}(\mathcal{C}_I(A, B))$$
$$= \sum_{A\in\mathcal{P}_0} \sum_{i=1}^{\mathcal{O}(N^{1/2})} \sum_{B\in Ann_i(A,\delta)\cap\mathcal{P}_0} \mathcal{O}(\tau^4 \cdot i^{-3} N^{1.5} \cdot N^4)$$
$$= \sum_{A\in\mathcal{P}_0} \sum_{i=1}^{\mathcal{O}(N^{1/2})} \mathcal{O}(\tau^4 \cdot i^{-2} \cdot N^{5.5}) = \sum_{A\in\mathcal{P}_0} \mathcal{O}(\tau^4 N^{5.5}) \sum_{i=1}^{\mathcal{O}(N^{1/2})} i^{-2} =$$
$$= \mathcal{O}(\tau^4 N^{6.5}).$$

$$\mathbb{E}(\mathcal{C}_{II}) \leq \sum_{A,B\in\mathcal{P}_0} \mathbb{E}(\mathcal{C}_{II}(A, B)) = \sum_{A\in\mathcal{P}_0} \sum_{i=1}^{\mathcal{O}(N^{1/2})} \sum_{B\in Ann_i(A,\delta)\cap\mathcal{P}_0} \mathbb{E}(\mathcal{C}_{II}(A, B))$$
$$= \sum_{A\in\mathcal{P}_0} \sum_{i=1}^{\mathcal{O}(N^{1/2})} \sum_{B\in Ann_i(A,\delta)\cap\mathcal{P}_0} \mathcal{O}(\tau^3 \cdot i^{-2.5} N^{1.25} \cdot N^3)$$
$$= \sum_{A\in\mathcal{P}_0} \sum_{i=1}^{\mathcal{O}(N^{1/2})} \mathcal{O}(\tau^3 \cdot i^{-1.5} \cdot N^{4.25}) = \sum_{A\in\mathcal{P}_0} \mathcal{O}(\tau^3 N^{4.25}) \sum_{i=1}^{\mathcal{O}(N^{1/2})} i^{-1.5} =$$
$$= \mathcal{O}(\tau^3 N^{5.25}).$$

$$\mathbb{E}(\mathcal{C}) = \mathbb{E}(\mathcal{C}_I + \mathcal{C}_{II}) = \mathcal{O}(\tau^4 N^{6.5}) + \mathcal{O}(\tau^3 N^{5.25}).$$

Now choose one of the good outcomes of the above probabilistic process so that $\mathcal{P}_0$ and $\mathcal{Q}_0$ satisfies the following properties:: the number of points is $N/2$, the number of quadruples is $\mathcal{O}(\tau^2 N^4)$, the number of 2-cycles is $\mathcal{O}(\tau^4 N^{6.5}) + \mathcal{O}(\tau^3 N^{5.25})$ (the two terms correspond to the two types of 2-cycles: the first term to cycles on six points, the second term to cycles on five points). Let $\mathcal{Q}_1$ be the 4-uniform hypergraph we obtained this way.

In order to get rid of the 2-cycles we need a random sparsification (as in the previous section): with probability p keep a point and with probability $1-p$ throw it away, and do this independently for all points. Let $\mathcal{Q}_1$ be the random 4-uniform hypergraph we obtain this way. Its parameters can easily be bounded:

$$\mathbb{E}(|V(\mathcal{Q}_1)|) = \Theta(pN),$$

$$\mathbb{E}(|\mathcal{Q}_1|) = \mathcal{O}(p^4\tau^2N^4),$$

$$\mathbb{E}(\mathcal{C}) = \mathcal{O}(p^6\tau^4N^{6.5}) + \mathcal{O}(p^5\tau^3N^{5.25}).$$

The end of the proof is straightforward: We choose p so that

$$\mathcal{C} \ll |V(\mathcal{Q}_1)|. \tag{2}$$

Choose one of the good outcomes of the above probabilistic process such that we obtain a 4-uniform hypergraph with the property that after deleting the points of the 2-cycles we obtain a leftover hypergraph with $\Theta(pN)$ points, and $\mathcal{O}(p^4\tau^2N^4)$ edges, and without 2-cycles. Let $\overline{d}$ denote the average degree. Throw away the points with degree at least $10\overline{d}$. The leftover hypergraph (without 2-cycles) is denoted by $\mathcal{L}$ and its parameters are:

$$|V(\mathcal{L})| = \Theta(pN),$$

$$|\mathcal{L}| = \mathcal{O}(p^4\tau^2N^4),$$

$$\Delta(\mathcal{L}) = \mathcal{O}(p^3\tau^2N^3).$$

Now we can apply Theorem 2. We choose N, τ such that $\alpha(\mathcal{L}) \geq n$ will hold. The n points forming an independent set will prove Theorem 4.

Set the parameters as follows:

$$p := n^{-0.001}, \quad N := n^{1.01}, \quad \tau := n^{-3/2}\sqrt{\log n}.$$

Now we are going to check that with this choice of parameters (2) is satisfied:

$$\begin{aligned}\mathbb{E}(\mathcal{C}) &= \mathcal{O}(p^6\tau^4N^{6.5}) + \mathcal{O}(p^5\tau^3N^{5.25})\\ &= \mathcal{O}(n^{0.006}(n^{-6}\log^2 n)\cdot n^{6.565}) + \mathcal{O}(n^{0.005}(n^{-4.5}\log n)n^{5.3025}) = o(n),\end{aligned}$$

and at the same time

$$\mathbb{E}(|V(\mathcal{Q}_1)|) = \Theta(pN) = \Theta(n^{1.009}).$$

Hence getting rid of 2-cycles is easy.

In order to apply Theorem 2 we introduce a parameter t, such that $\Delta(\mathcal{L}) \le t^3$. Based on our previous estimate $\Delta(\mathcal{L}) = \mathcal{O}(p^3\tau^2N^3)$, the right choice for t is

$$t = \Theta(p\tau^{2/3}N) = \Theta(n^{0.001}(n^{-1}\log^{1/3} n)n^{1.01}) = \Theta(n^{0.009}\log^{1/3} n).$$

Hence Theorem 2 is applicable and it provides the following bound:

$$\alpha(\mathcal{L}) \ge \frac{\Omega(pN)}{t}\log^{1/3} t = \Omega(n).$$

Thus, $\alpha(\mathcal{L}) \ge cn$ for certain constant c, and the theorem is proven by the same scaling argument as Theorem 3.

References

1. M. Ajtai, J. Komlós, E. Szemerédi, A dense infinite Sidon sequence. Eur. J. Comb. **2**(1), 1–11 (1981)
2. M. Ajtai, J. Komlós, J. Pintz, J. Spencer, E. Szemerédi, Extremal uncrowded hypergraphs. J. Comb. Theory Ser. A **32**(3), 321–335 (1982)
3. C. Bertram-Kretzberg, T. Hofmeister, H. Lefmann, An algorithm for Heilbronn's problem. SIAM J. Comput. **30**(2), 383–390 (2000)
4. R. Duke, H. Lefmann, V. Rödl, On uncrowded hypergraphs. Random Struct. Algorithms **6**(2–3), 209–212 (1995)
5. T. Gowers, A Geometric Ramsey Problem, http://mathoverflow.net/questions/50928/a-geometric-ramsey-problem. Accessed May 2016
6. S. Jukna, *Extremal Combinatorics, With Applications in Computer Science*, 2nd edn., Texts in Theoretical Computer Science (Springer, Heidelberg, 2011)
7. J. Komlós, J. Pintz, E. Szemerédi, A lower bound for Heilbronn's problem. J. Lond. Math. Soc **25(2)**(1), 13–24 (1982)
8. J. Komlós, J. Pintz, E. Szemerédi, On Heilbronn's triangle problem. J. Lond. Math. Soc. **24(2)**(2), 385–396 (1981)
9. M.S. Payne, D.R. Wood, On the general position subset selection problem. SIAM J. Discret. Math. **27**(4), 1727–1733 (2013)
10. K.F. Roth, Estimation of the area of the smallest triangle obtained by selecting three out of n points in a disc of unit area, in *Analytic Number Theory (Proc. Sympos. Pure Math., Vol. XXIV, St. Louis Univ., St. Louis, Mo., 1972)* (Amer. Math. Soc, Providence, 1973), pp. 251–262
11. K.F. Roth, On a problem of Heilbronn. J. Lond. Math. Soc. **26**, 198–204 (1951)
12. K.F. Roth, On a problem of Heilbronn II. Proc. Lond. Math. Soc. **25**(3), 193–212 (1972)
13. K.F. Roth, On a problem of Heilbronn III. Proc. Lond. Math. Soc. **25**(3), 543–549 (1972)
14. K.F. Roth, Developments in Heilbronn's triangle problem. Adv. Math. **22**(3), 364–385 (1976)
15. W. Schmidt, On a problem of Heilbronn. J. Lond. Math. Soc. **4**(2), 545–550 (1971/72)
16. E. Szemerédi, W. Trotter, Extremal problems in discrete geometry. Combinatorica **3**(3–4), 381–392 (1983)

Erdős–Szekeres Theorems for Families of Convex Sets

Andreas F. Holmsen

Abstract The well-known Erdős–Szekeres theorem states that every sufficiently large set of points in the plane containing no three points on a line, has a large subset in convex position. This classical result has been generalized in several directions. In this article we review recent progress related to one such direction, initiated by Bisztriczky and Fejes Tóth, in which the points are replaced by convex sets.

1 Introduction

1.1 The Erdős–Szekeres Theorem

One of the foundational results of combinatorial geometry and Ramsey theory is the famous theorem of Erdős and Szekeres from 1935 on points in convex position.

Theorem 1.1 (Erdős–Szekeres [11]) *For every integer $n \geq 3$ there exists a minimal integer $f(n)$ with the following property. Any set of $f(n)$ points in the plane such that no three points are on a line, has n points in convex position.*

Here, a set of points is in *convex position* if no point is in the convex hull of the others, or in other words, they form the vertices of a convex polygon. The condition that no three points are on a line is referred to as being in *general position*. (Being in general position usually depends on the dimension of the ambient space, but here we will only be working in two dimensions.)

Theorem 1.1 asserts the *existence* of a certain integer function $f(n)$, which leads to the problem of determining its precise values. This is one of the longest-standing open problems in combinatorial geometry. In their original paper [11], Erdős and Szekeres gave the upper bound of $f(n) \leq \binom{2n-4}{n-2} + 1 \approx 4^n/\sqrt{n}$, and around 30 years later, Erdős and Szekeres [12] gave, for every $n \geq 3$, a construction of a set of 2^{n-2} points in general position with no n points in convex position.

A. F. Holmsen (✉)
Department of Mathematical Sciences, KAIST Daejeon, Daejeon, South Korea
e-mail: andreash@kaist.edu

G. Ambrus et al. (eds.), *New Trends in Intuitive Geometry*, Bolyai Society Mathematical Studies 27, https://doi.org/10.1007/978-3-662-57413-3_9

Conjecture 1.2 (Erdős – Szekeres [12]) *For every integer $n \geq 3$, we have*

$$f(n) = 2^{n-2} + 1.$$

This conjecture has been verified for $n \leq 6$ [11, 12, 35]. The values $f(4) = 5$ and $f(5) = 9$ can be proven "by hand", but $f(6) = 17$ is based on a computer-aided proof. For general n, the upper bound has been improved little by little over the years, until Suk [34], in late April 2016, announced a major breakthrough on Conjecture 1.2, showing that $f(n) \leq 2^{n+o(n)}$. In this article, however, our main focus will be on a generalization of Theorem 1.1 which we describe next.

1.2 The Erdős–Szekeres Theorem for Convex Sets

In 1989, Bisztriczky and Fejes Tóth gave the following generalization of the Erdős–Szekeres theorem.

Theorem 1.3 (Bisztriczky–Fejes Tóth [3]) *For any integer $n \geq 3$ there exists a minimal positive integer $g(n)$ with the following property. Any disjoint family of $g(n)$ compact convex sets in the plane such that any three members are in convex position, has n members in convex position.*

Here, a *disjoint family* is one where its members are pairwise disjoint, and a family is in *convex position* if no member is contained in the convex hull of the union of the other members. We obtain Theorem 1.1 when each member of the family is a point, and so clearly we have

$$f(n) \leq g(n)$$

for every n.

The proof in [3] relies on repeated applications of Ramsey's theorem and gives an upper bound $g(n) \leq t_n(t_{n-1}(\ldots(t_1(n))\ldots))$, where $t_k(x)$ denotes a tower of exponents of height $k-1$ with x at the top, that is, $t_1(x) = x$ and $t_{i+1}(x) = 2^{t_i(x)}$. However, Bisztriczky and Fejes Tóth conjectured that the two functions, $f(n)$ and $g(n)$, are in fact equal (which has been verified for $n \leq 6$ [3, 4, 8]).

Conjecture 1.4 (Bisztriczky–Fejes Tóth [3]) *For every integer $n \geq 3$, we have $f(n) = g(n)$.*

In this article we will survey various results related to Theorem 1.3, that is, the generalization of Theorem 1.1 to *families of convex sets*. For the many other generalizations and extensions of Theorem 1.1 we refer the reader to the surveys [1, 26]. Throughout, $f(n)$ will always refer to the function whose existence is guaranteed by Theorem 1.1, while $g(n)$ refers to the function of Theorem 1.3.

In Sect. 2 we briefly review three standard proofs of Theorem 1.1. These are probably known to anyone familiar with the original Erdős–Szekeres theorem.

In Sect. 3 we focus on Theorem 1.3, reviewing several proofs and improvements on the upper bound on $g(n)$. Some of these improvements lead to an interesting Ramsey-type problem concerning *monotone paths* in uniform hypergraphs which will be reviewed. We also take a look at a generalization of Theorem 1.1 in the abstract setting of *oriented matroids*, which happens to be very closely related to Theorem 1.3.

Finally, in Sect. 4 we review the recent proof of a conjecture by Pach and Tóth, which concerns relaxing the disjointness hypothesis in Theorem 1.3.

2 Three Proofs of Theorem 1.1

Here we take a look at the three standard methods for bounding the function $f(n)$ in Theorem 1.1.

The first two use the Ramsey number $R_k(s, t)$, that is, the minimal integer m such that for any red-blue coloring of the k-element subsets of an m-element set, there exists an s-element subset such that all its k-element subsets are red, or a t-element subset such that all its k-element subsets are blue.

The use of Ramsey numbers does not give particularly good bounds, but such arguments are nevertheless instructive when considering generalizations such as Theorem 1.3.

The last proof is the well-known "cup-cap" argument which gives an exponential upper bound. Ideas from this argument are also important in Suk's recent record-breaking bound.

2.1 *First Argument* [11]

This is based on the following observations.

- Carathéodory's theorem. Let P be a set of points in the plane. If every four points in P are in convex position, then P is in convex position.
- Any set of five points in the plane such that no three are on a line, has four points in convex position. In other words, $f(4) = 5$.

Suppose we have $R_4(n, 5)$ points in the plane in general position. Color a subset of size 4 *red* if it is in convex position, and *blue* if it is not. By Ramsey's theorem there are n points such that every subset of size 4 is red, or 5 points such that every subset of size 4 is blue. In the first case, the n points are in convex position by Carathéodory's theorem. The second case can not occur since $f(4) = 5$. It follows that $f(n) \leq R_4(n, 5)$.

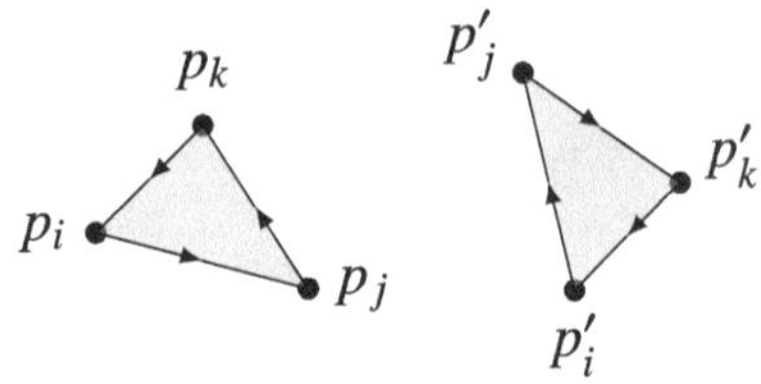

Fig. 1 On the left, (p_i, p_j, p_k) has positive orientation. On the right, (p'_i, p'_j, p'_k) has negative orientation

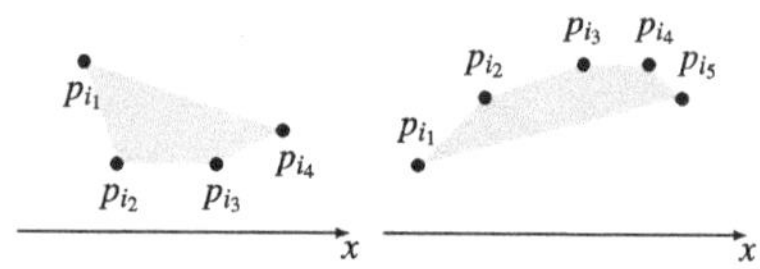

Fig. 2 On the left, $(p_{i_1}, p_{i_2}, p_{i_3}, p_{i_4})$ form a 4-cup. On the right, $(p_{i_1}, p_{i_2}, p_{i_3}, p_{i_4}, p_{i_5})$ form a 5-cap

2.2 *Second Argument* [22]

This is based on the *order-type* of a set of points in the plane. Let p_i, p_j, p_k be points in the plane. We say that the ordered triple (p_i, p_j, p_k) has *positive (negative)* orientation if they are in general position and appear in counter-clockwise (clockwise) order on $\text{conv}(\{p_i, p_j, p_k\})$. (See Fig. 1.)

We have the following observation.

- Let $P = \{p_1, p_2, \dots, p_n\}$ be a set of points in the plane. If every ordered triple (p_i, p_j, p_k), with $i < j < k$, has the same orientation, then P is in convex position.

Suppose we have $N = R_3(n, n)$ points in the plane in general position. Order the points arbitrarily $p_1, p_2, \dots, p_N$. For every $i < j < k$, color the the subset $\{i, j, k\}$ *red* if the ordered triple (p_i, p_j, p_k) has positive orientation, and *blue* if (p_i, p_j, p_k) has negative orientation. By Ramsey's theorem there is a subset of size n in which every ordered triple has the same orientation, and consequently, by the observation, the corresponding n points are in convex position. It follows that $f(n) \leq R_3(n, n)$.

2.3 *The "Cup-Cap" Argument* [11]

This is based on the observation from the second argument, but the twist is choose a specific ordering of the points to obtain more structure. For any finite set of points in the plane we may choose a coordinate system in such a way that the points have distinct x-coordinates, and we order the points in the order of their x-coordinates.

With respect to this fixed ordering, the ordered set of points $(p_{i_1}, p_{i_2}, \dots, p_{i_k})$ is called a *k-cup (k-cap)* if every ordered triple has positive (negative) orientation. (See Fig. 2.)

The advantage of this ordering is that it allows us to define the following "critical configuration".

- Let $A = (a_1, \ldots, a_k)$ be a k-cup and $B = (b_1, \ldots, b_l)$ an l-cap, and suppose $a_k = b_1$. Then $A \cup B$ contains a $(k+1)$-cup or an $(l+1)$-cap.

Without going into details, this critical configuration can be used to establish a recurrence relation which shows that any set in the plane in general position with more than $\binom{k+l-4}{k-2}$ points, contains a k-cup or an l-cap. Since cups and caps are in convex position (by the observation in the second argument), this results in the bound $f(n) \leq \binom{2n-4}{n-2} + 1$.

The bound above was the best known for more than 60 years, until a series of improvements [7, 20, 37, 38], which all rely on clever modifications of the "cup-cap" argument.

Recently, Vlachos [39] discovered a new critical configuration consisting of two cups and one cap. The analysis and resulting inductive argument is more complicated, but yields a slight improvement on the upper bound on $f(n)$, which was further improved by Mojarrad [24]. Mojarrad and Vlachos combined their results [25] to obtain the bound, $f(n) \leq \binom{2n-5}{n-2} - \binom{2n-8}{n-3} + 2$. Asymptotically, this is an improvement of the original Erdős and Szekeres bound by a factor of $\frac{7}{16}$.

Let us also mention that the "cup-cap" argument typically uses the critical configuration to set up an inductive argument. But recently, a different approach was discovered in which such critical configurations are used to give combinatorial proofs using the pigeon-hole principle. In [27], Moshkovitz and Shapira generalized the original "cup-cap" argument to a combinatorial setting and gave a bijective proof of the original Erdős–Szekeres bound. Norin and Yuditsky [28] gave a bijective proof based on the critical configuration of Vlachos to obtain a bound on $f(n)$ which is slightly weaker than the bound in [25], by a lower order term.

The current best bound on $f(n)$ is due to Suk [34], which shows

$$f(n) \leq 2^{n+4n^{4/5}}.$$

His proof also uses the "cup-cap" argument in an essential way, but the most important additional ingredient is a *positive fraction* version of the Erdős–Szekeres theorem originally discovered by Bárány and Valtr [2]. This essentially says that, for a point set X in general position in the plane, and any $k \geq 3$, it is possible to find disjoint subsets $X_1, \ldots, X_k$ each containing a positive fraction of the points of X, such that *every* system of representatives $x_1 \in X_1, \ldots, x_k \in X_k$ are in convex position in the given cyclic order. Based on this structure theorem (more precisely a sharper quantitative version due to Pór and Valtr [32]), Suk establishes the stated upper bound by ingenious combinations of Dilworth's theorem and the pigeon-hole principle.

3 Disjoint and Non-crossing Families

3.1 Disjoint Families

For integers $n \geq k \geq 3$, let $h(n, k)$ be the smallest integer with the following property. Any disjoint family of $h(n, k)$ compact convex sets in the plane such that every k members are in convex position, has n members in convex position.

The numbers $h(n, k)$ were introduced by Bisztriczky and Fejes Tóth [5] as a tool to improve their bound on $g(n)$ from [3]. Note that we have

$$h(n, k) \leq h(n, 3) = g(n).$$

By applying the first argument of Sect. 2, bounds on $h(n, 4)$ and $g(4)$ will immediately imply a bound on $h(n, 3)$. Consider a disjoint family F of convex sets in the plane such that every three members of F are in convex position. By coloring the subfamilies of size four *red* and *blue* according to whether or not they are in convex position, we obtain the bound

$$g(n) \leq R_4(h(n, 4), g(4)).$$

It is reasonable to expect that $h(n, k)$ is considerably smaller than $g(n)$ when $k \geq 4$, and in [5] an exponential upper bound was given on $h(n, 4)$, and quadratic upper bounds on $h(n, k)$ for all $k \geq 5$. Some improvements were given in [30], and the current best bounds are given in the following.

Theorem 3.1 *For $h(n, k)$ defined as above we have*

(1) $$2\lfloor\frac{n+1}{4}\rfloor^2 < h(n, 4) \leq n^3 \qquad [30].$$

(2) $$n + \lfloor\frac{n-1}{k-2}\rfloor \leq h(n, 5) \leq 6n - 11 \qquad [36].$$

(3) $$n + \lfloor\frac{n-1}{k-2}\rfloor \leq h(n, k) \leq n + \lfloor\frac{n}{k-5}\rfloor + 1 \quad (k \geq 6) \qquad [36].$$

In [30], Pach and Tóth also show that $g(n) \leq \binom{2n-4}{n-2}^2 + 1$. Their argument first applies Ramsey's theorem on interval graphs to obtain a large subfamily in which every member has a line transversal or each pair can be separated by a vertical line. They then show that in either case one can apply a modification of the "cup-cap" argument.

3.2 Non-crossing Families and Monotone Paths

A compact convex set with non-empty interior is called a *convex body*. Two convex bodies are *non-crossing* if they have at most two boundary points in common, and a family of convex bodies is called non-crossing if its members are pairwise non-crossing. The following generalization of Theorem 1.3 is due to Pach and Tóth.

Theorem 3.2 (Pach – Tóth [31]) *For every integer $n \geq 3$ there exists a minimal integer $g_1(n)$ with the following property. Any non-crossing family of $g_1(n)$ convex bodies in the plane such that every three members are in convex position, has n members in convex position.*

For any disjoint family F of compact convex sets in the plane it is easy to see that if we add a sufficiently small ϵ-disk to each member, then the resulting family F' is a disjoint family of convex bodies, and therefore, also non-crossing. Moreover, a subfamily of F is in convex position if and only if the corresponding subfamily of F' is in convex position. Thus Theorem 3.2 is indeed a generalization of Theorem 1.3, and clearly we have

$$g(n) \leq g_1(n).$$

Pach and Tóth did not calculate an explicit bound on $g_1(n)$, but their argument relies on repeated applications of Ramsey's theorem.

The first explicit bound on $g_1(n)$ was given by Hubard et al. [19], who showed that $g_1(n)$ grows at most double exponentially in n. The key to their argument was to introduce the notion of *orientations* of triples of convex bodies, which generalizes the notion of order-types for point sets in the plane.

Let C_1, C_2, C_3 be non-crossing convex bodies in general position the plane. We say that the ordered triple (C_1, C_2, C_3) has *positive (negative) orientation* if there exists points $p_1 \in C_1$, $p_2 \in C_2$, $p_3 \in C_3$ which lie on the boundary of $\text{conv}(C_1 \cup C_2 \cup C_3)$, such that the ordered triple (p_1, p_2, p_3) has positive (negative) orientation (as defined in Sect. 2). It is clear that if the convex bodies are in convex position, then the triple has at least one orientation, but it is also possible that it has *both* orientations. (See Fig. 3.)

Lemma 3.3 (Hubard et al. [19]) *If a non-crossing family of convex bodies in the plane can be ordered in such a way that every triple has clockwise orientation (or every triple has counter-clockwise orientation), then the family is in convex position.*

This is a generalization of the observation from the second argument of Sect. 2, and results in the bound $g(n) \leq R_3(n, n) = N$. Order the convex bodies arbitrarily $C_1, C_2, \ldots, C_N$. For every $i < j < k$ color the subset $\{i, j, k\}$ *red* if the ordered triple

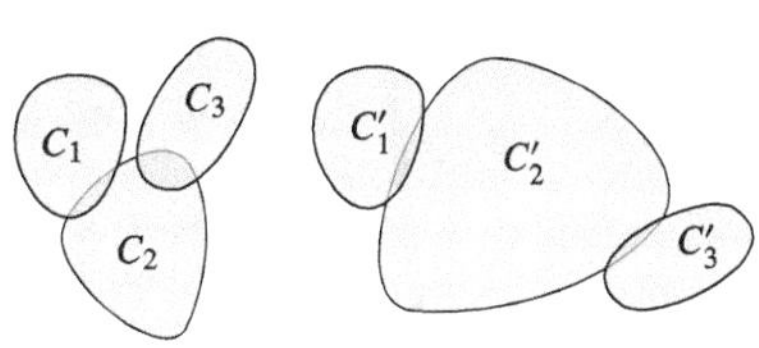

Fig. 3 Two non-crossing triples of convex bodies. On the left, the ordered triple (C_1, C_2, C_3) has positive orientation. On the right, the ordered triple (C_1', C_2', C_3') has both orientations

(C_i, C_j, C_k) has positive orientation, and *blue* if the ordered triple (C_i, C_j, C_k) does *not* have positive orientation. By Ramsey's theorem there exists a subfamily of size n satisfying the hypothesis of Lemma 3.3.

The idea of using orientations of triples of convex bodies was further exploited by Fox et al. [14], by ordering the family in a particular way as in the "cup-cap" argument. This lead them to introduce the notion of Ramsey numbers for *monotone paths* in *ordered* hypergraphs, which we now describe.[1]

For integers k and N, let K_N^k denote the complete k-uniform hypergraph with vertex set $\{1, 2, \dots, N\}$. For any n integers $1 \leq j_1 < j_2 < \cdots < j_n \leq N$ we call the set of edges

$$\{\{j_i, j_{i+1}, \dots, j_{i+k-1}\} \ : \ i = 1, 2, \dots, n - k + 1\}$$

a *monotone path* of length n. Let $M_r^k(n)$ denote the smallest integer N such that for any coloring of the edges of K_N^k with r colors, there exists a monotone path of length n whose edges all have the same color.

The concept of monotone paths was applied by Fox et al. [14] to obtain the bound

$$g_1(n) \leq M_3^3(n).$$

Their idea is to order the family of convex bodies according to the order in which we meet the members of the family as we sweep a vertical line from left to right across the plane (which we may assume induces a unique linear ordering on the family). They then color the ordered triples according to whether they have only positive orientation, only negative orientation, or both orientations.

By a detailed geometric argument it is then shown that a monochromatic monotone path of length n corresponds to a subfamily of size n in which each ordered triple has the same orientation. The result then follows by applying Lemma 3.3.

The general problem of bounding $M_r^k(n)$ was also treated in [14]. They showed that $M_2^3(n) = \binom{2n-4}{n-2} + 1$ by a simple generalization of the "cup-cap" argument, and for $r \geq 3$ they proved lower and upper bounds bounds for $M_r^3(n)$ which are tight apart from a logarithmic factor in the exponent. By a "stepping-up" approach, developed by Erdős and Hajnal, they also obtain bounds on $M_r^k(n)$ for arbitrary $k \geq 3$.

The bounds on $M_r^3(n)$ were investigated further Moshkovitz and Shapira [27] who established a remarkable correspondence between the numbers $M_r^3(n)$ and the enumeration of *high-dimensional partitions*.

A *d-dimensional partition* is a d-dimensional (hyper)matrix with non-negative integer entries which is weakly decreasing in every line, that is, $A_{i_1,\dots,i_t,\dots,i_d} \geq A_{i_1,\dots,i_{t+1},\dots,i_d}$ for all $i_1, \dots, i_d$ and $1 \leq t \leq d$. (See Fig. 4.)

Let $P_d(n)$ denote the number of $n \times \cdots \times n$ d-dimensional partitions with entries from $\{0, 1, \dots, n\}$. The surprising correspondence between Ramsey numbers for monotone paths and integer partitions is given by the following.

[1] A closely related concept is *transitive colorings* of ordered hypergraphs which was introduced by Eliáš and Matoušek [10], and was used for a different generalization of the "cup-cap" argument.

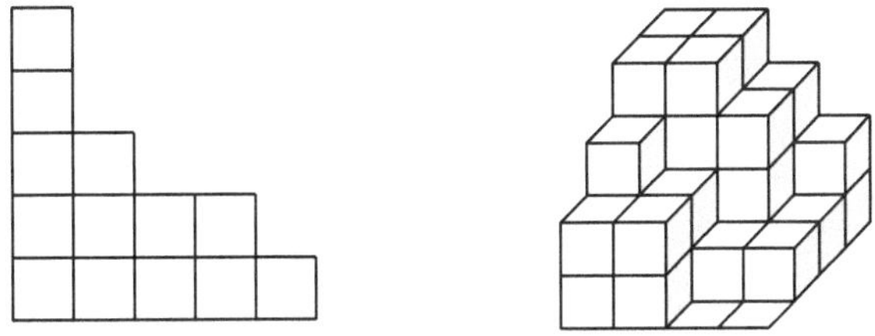

5	3	2	2	1

4	4	3	2
4	4	3	1
3	2	1	1
2	2	0	0

Fig. 4 On the left, a one dimensional partition. On the right, a two dimensional partition. Above are graphical representations as "stacks of boxes"

Theorem 3.4 (Moshkovitz–Shapira [27]) *For every $r \geq 2$ and $n \geq 4$ we have*

$$M_r^3(n) = P_{r-1}(n-2) + 1$$

It is simple to show that $P_1(n) = \binom{2n}{n}$, and a celebrated result of MacMahon [23] states that

$$P_2(n) = \prod_{1 \leq i,j,k \leq n} \frac{i+j+k-1}{i+j+k-2}.$$

For $d \geq 3$, the numbers $P_d(n)$ are not known, but in general we have the following bounds on the Ramsey numbers for monotone paths.

Theorem 3.5 (Moshkovitz–Shapira [27]) *For every $k \geq 3$, $r \geq 2$, and sufficiently large n we have*

$$t_{k-1}(n^{r-1}/2\sqrt{r}) \leq M_r^k(n) \leq t_{k-1}(2n^{r-1}).$$

By the argument of Fox et al. [14], this implies

$$g_1(n) \leq 2^{cn^2}.$$

3.3 Generalized Configurations

Loosely speaking, a *pseudoline arrangement* is a finite collection of curves in the plane which behave like an arrangement of lines. More specifically, each pseudoline is the image of a line under a homeomorphism of the plane (and therefore partitions the plane into two unbounded regions), and each pair of pseudolines intersect exactly

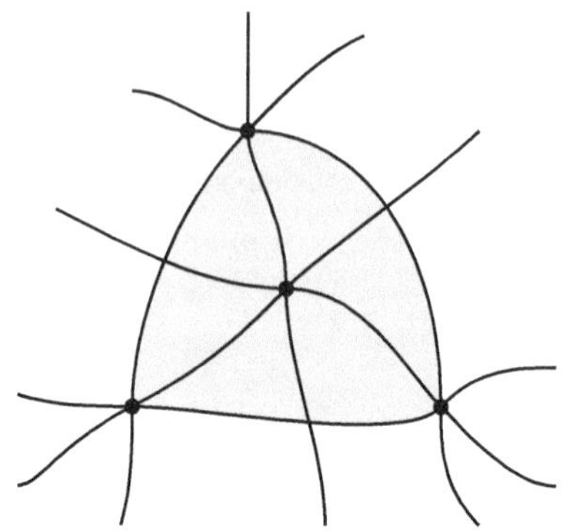

Fig. 5 A generalized configuration with four points. The shaded region is the convex hull

once, where they are required to cross. A *generalized configuration* is a finite set of points in the plane where each pair of points is contained in a unique pseudoline in such a way that the resulting set of pseudolines form a pseudoline arrangement. Although we have defined generalized configurations in a somewhat "geometric" manner, it should be noted that generalized configurations can be characterized by purely combinatorial axioms.[2]

Given a generalized configuration, the underlying pseudoline arrangement induces a convexity structure on the point configuration in a natural way: Each pair of points of the configuration bound a unique *pseudosegment* contained in the associated pseudoline. The complement of the set of pseudosegments determined by the pairs of the configuration is a collection of open connected regions in the plane, and the *convex hull* of the configuration is the complement of the unique unbounded region. (See Fig. 5.) We say that a generalized configuration is in *convex position* if no point of the configuration is contained in the convex hull of the remaining points.

Many of the basic theorems of convexity extend to generalized configurations, for instance, a set of points is in convex position if and only if every four of its points are in convex position (which generalizes Carathéodory's theorem). It is also easy to see that every generalized configuration on 5 points has four points in convex position. Just draw the pseudosegments connecting every pair. Since K_5 is non-planar, two of the segments must cross and their endpoints must be in convex position. By the first argument of Sect. 2 we get the following extension of Theorem 1.1.

Theorem 3.6 (Goodman–Pollack [17]) *For every integer $n \geq 3$ there exists a minimal integer $c(n)$ with the following property. Any generalized configuration of size $c(n)$ contains n points in convex position.*

Every finite set of points in general position in the plane gives rise to a generalized configuration (when we connect each pair by the straight line determined by them), however, there exists generalized configurations which are not obtained in

[2]Other names for generalized configurations found in the literature are *uniform rank 3 acyclic oriented matroids* [6] or *CC-systems* [21]. See also the survey [15] for further information.

this way. In fact, most generalized configurations are *non-realizable* [13, 18]. Therefore Theorem 3.6 is a proper generalization of Theorem 1.1, and we have the trivial relation

$$f(n) \leq c(n).$$

Conjecture 3.7 (Goodman–Pollack [17]) *For every integer $n \geq 3$ we have*

$$f(n) = c(n).$$

It is known that Conjecture 3.7 holds for all $n \leq 6$ [35]. Moreover, it is easily seen that Suk's recent upper bound for $f(n)$ does not actually depend on the straightness of the lines, and his proof translates more or less word for word to generalized configurations.[3] In particular, we have the following.

Theorem 3.8 (Suk [34]) *For every $n \geq n_0$, where n_0 is a sufficiently large absolute constant, we have $c(n) \leq 2^{n+4n^{4/5}}$.*

Recently, Dobbins et al. established the following correspondence between Theorems 3.2 and 3.6.

Theorem 3.9 (Dobbins et al. [8]) *The Erdős–Szekeres problems for non-crossing families of convex bodies in the plane and for generalized configurations are equivalent. In other words, we have $g_1(n) = c(n)$.*

It should be remarked that the non-crossing condition in Theorem 3.9 can not be dropped. This follows from a construction of Pach and Tóth, and will be discussed further in the next section.

In summary, we have the following bounds for the Erdős–Szekeres functions discussed so far.

$$2^{n-2} + 1 \leq f(n) \leq g(n) \leq g_1(n) = c(n) \leq 2^{n+4n^{4/5}} \qquad \text{(for } n \geq 7\text{)}$$

$$2^{n-2} + 1 = f(n) = g(n) = g_1(n) = c(n) \qquad \text{(for } n \leq 6\text{)}$$

We would like to point out the two main steps of the proof of Theorem 3.9. First, let us call a family of convex bodies *orientable* if the family is non-crossing and each ordered triple of its members has a unique orientation. This means that every three members are in convex position, and admit either positive or negative orientations, but not both.

---

[3] A slight warning may be in order: The basis for Suk's argument is the *positive fraction* Erdős–Szekeres theorem, originally due to Bárány and Valtr [2]. Suk's proof uses a quantitatively improved version due to Pór and Valtr [32], but their proof is essentially a counting argument and works just as well for generalized configurations!

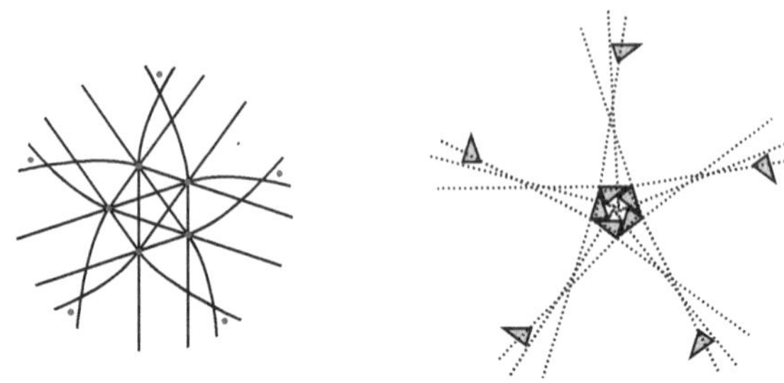

Fig. 6 On the left, a "non-realizable" generalized configuration based on Goodman and Pollack's "bad pentagon" [16]. On the right, a realization of the same generalized configuration by convex bodies

The following lemma asserts that generalized configurations and orientable families of convex bodies are the same as far as their "convexity structures" are concerned. (See Fig. 6.)

Lemma 3.10 (Dobbins et al. [8])

(1) For every generalized configuration C there exists an orientable family F of convex bodies and a bijection $\alpha : C \to F$ such that for any subset $C' \subset C$, we have that C' is in convex position if and only if the subfamily $\alpha(C') \subset F$ is in convex position.

(2) For every orientable family F of convex bodies there exists a generalized configuration C and a bijection $\beta : F \to C$ such that for any subfamily $F' \subset F$, we have that F' is in convex position if and only if the subset $\beta(F') \subset C$ is in convex position.

As a consequence, the Erdős–Szekeres problems for generalized configurations and for *orientable* families of convex bodies are equivalent, and in particular

$$c(n) \leq g_1(n).$$

The second step is to show the reverse inequality, and here we must deal with non-crossing families of convex bodies which are not orientable. (These were precisely the reason why Fox et al. [14] considered the Ramsey numbers $M_3^3(n)$ rather than $M_2^3(n)$.)

Using an elementary homotopy argument it is shown that any non-crossing family of convex bodies can be "mutated" into an orientable family of convex bodies while maintaining control of the subfamilies which are in convex position. In particular, we have the following

Lemma 3.11 (Dobbins et al. [8]) *For every non-crossing family F such that every three members are in convex position, there exists an orientable family G and a bijection $\varphi : F \to G$ such that, if a subfamily $H \subset G$ is in convex position, then the subfamily $\varphi^{-1}(H) \subset F$ is in convex position.*

As a consequence of Lemma 3.10 *(2)* and Lemma 3.11 we get

$$g_1(n) \leq c(n).$$

Lemmas 3.10 and 3.11 provide a general method for reducing non-crossing families of convex bodies to generalized configurations. This method can be applied to other generalizations of Theorem 1.1 previously proven separately for point sets, then for families of convex bodies. For instance, we get the following common generalization of the results from the papers [2, 29, 32, 33].[4]

Theorem 3.12 *For every integer $n \geq 3$ there exists integers $k = k(n)$ and $r = r(n)$ with the following property. Let F be a non-crossing family of convex bodies in the plane such that every three members of F are in convex position. There exists a subfamily $G \subset F$ with $|G| \leq r$, such that $F \setminus G$ can be partitioned into $n \cdot k$ parts*

$$F_{(i,j)}\,,\ 1 \leq i \leq n, 1 \leq j \leq k$$

which satisfy

(1) $|F_{(i,j)}| = |F_{(i',j')}|$ for all i, j, i', j'.
(2) For every fixed j and arbitrary choice of sets $C_1 \in F_{(1,j)}, \ldots, C_n \in F_{(n,j)}$, the sets $C_1, \ldots, C_n$ are in convex position.

4 The Conjecture of Pach and Tóth

Pach and Tóth [31] investigated the possibility of generalizing Theorem 3.2 even further by weakening the non-crossing condition. However, they gave a construction of an infinite family of straight-line segments such that any three are in convex position, but no four are.

On the other hand, they were able to show that for a family of segments such that every *four* members are in convex position, there exists a large subfamily in convex position, provided that the family is sufficiently large. Based on this result, they conjectured the following generalization of Theorem 3.2.

Conjecture 4.1 (Pach–Tóth [31]) *For any integer $k > 2$, there exists a constant s_k and a function $t_k(n)$ with the following property. Any family of $t_k(n)$ convex bodies in the plane such that any two share at most k common boundary points and any s_k are in convex position, has n members in convex position.*

At first sight, it seems natural to guess that the constant s_k should tend to infinity as k tends to infinity. However, Dobbins et al. [9] recently gave a proof of Conjecture 4.1 which shows that $s_k \leq 5$ for every k.

The proof of Conjecture 4.1 is somewhat long and technical, and this section is devoted to outline the main ideas of the proof.

[4]Theorem 3.12 was announced by Pór and Valtr in [32] for the case of pairwise disjoint bodies, but their proof is complicated and appears only in an unpublished manuscript.

4.1 Systems of x-Monotone Curves

By an *x-monotone curve* we mean the graph of a continuous function $f : [0, 1] \to \mathbb{R}$ drawn in the vertical strip $[0, 1] \times \mathbb{R}$. A *system* (of x-monotone curves) is a finite collection S of at least three x-monotone curves which satisfy the following conditions.

(1) The members of S have distinct endpoints.
(2) The number of intersections of any pair of members of S is finite.
(3) No two members of S are tangent, that is, any two members of S cross at every intersection.
(4) The intersection of any three members of S is empty.

Here are some further definitions. A system S is called a *k-system*, if every pair of members of S cross at least once and at most k times. By taking a subset of at least three members of S we obtain a *subsystem*. The system S is called *upper (lower) convex* if every member of S appears at least once on the upper (lower) envelope. Note that a 1-system is equivalent to a (marked) arrangement of pseudolines, and we may think of k-systems as generalizations of collections of real polynomials of degree k.

The main results of [9] are following Ramsey-type results regarding k-systems.

Theorem 4.2 (Dobbins et al. [9]) *For all integers $k \geq 1$ and $n \geq 3$ there exists an integer $u_k(n)$ with the following properties. Let S be a k-system with $u_k(n)$ members.*

(1) If $k \leq 2$ and every subsystem of size 3 is upper convex, then S has a subsystem of size n which is upper convex.
(2) If $k \leq 4$ and every subsystem of size 4 is upper convex, then S has a subsystem of size n which is upper convex.
(3) If $k \geq 5$ and every subsystem of size 5 is upper convex, then S has a subsystem of size n which is upper convex.

The bounds on $u_k(n)$ are in terms of certain Ramsey numbers and are probably very far from the truth.

We will soon explain how Theorem 4.2 can be used to prove Conjecture 4.1, but first we mention the following "symmetric" variant.

Theorem 4.3 (Dobbins et al. [9]) *For all integers $k \geq 1$ and $n \geq 3$ there exists an integer $v_k(n)$ with the following property. Any k-system with $v_k(n)$ members has a subsystem of size n which is upper convex or lower convex.*

For $k = 1$, Theorem 4.3 is a dual version of the Erdős–Szekeres "cup-cap" theorem. Thus, the precise value of $v_1(n)$ is known and equals $\binom{2n-4}{n-2} + 1$. It is easy to see that $v_k(3) = 3$ and $v_k(n) \leq v_{k+1}(n)$ for every k, but it is not known whether the last inequality is strict for some $n > 3$.

Let us now show how to prove Conjecture 4.1. Consider a family F of convex bodies such that any two members share at most k common boundary points and that every 5 members are in convex position. It is no loss in generality to assume that the members of F are in a generic position, which means we may assume that the boundaries cross at each intersection point and that no three members share a common supporting tangent line. Note that under these assumptions, the number of common boundary points between two members of F is equal to the number of common supporting tangents, and therefore k must be a positive even number.

For a convex body C, recall that its support function $\sigma_C : \mathbb{S}^1 \to \mathbb{R}$ is defined as

$$\sigma_C(\theta) = \max_{p \in C} \langle \theta, p \rangle,$$

where $\langle \cdot, \cdot \rangle$ denotes the standard Euclidean inner product. Using this map we can associate each member $C \in F$ with an x-monotone curve

$$C^* = \{(t, \sigma_C(2\pi t)) \ : \ 0 \leq t \leq 1\}.$$

Since the members of F were assumed to be in generic position, the corresponding collection $S = \{C^*\}_{C \in F}$ is a k-system. The crucial observation is that a subfamily $F' \subset F$ is in convex position if and only if the corresponding subsystem $S' = \{C^*\}_{C \in F'}$ is upper convex. By Theorem 4.2(3) we get that $t_k(n) \leq u_k(n)$. (Note that when $k = 4$ it is sufficient to assume that every *four* members of F are in convex position.)

4.2 Uniform Systems

Here we discuss some aspects of the proof of Theorem 4.2. By Ramsey's theorem, every sufficiently large k-system must contain a subsystem which is highly uniform. The proof of Theorem 4.2 given in [9] boils down to a characterization of such uniform systems, which may be thought of as "higher order" analogues of cups and caps.

Consider a system consisting of precisely three curves. In this case the crossings are linearly ordered, and a crossing between two of the curves occurs either above or below the remaining curve. We can therefore encode the "crossing pattern" of the system as a word on a two letter alphabet, $\{a, b\}$, where the letter "a" corresponds to a crossing occurring *above* the remaining curve, and the letter "b" corresponds to a crossing occurring *below* the remaining curve. We call this word the *signature* of the system.[5] (See Fig. 7.)

A system is called *uniform* if it contains at least 4 curves and every subsystem of size 3 has the same signature, and a system S containing 3 curves is called *extendable* if there exists a uniform system such that every subsystem of size 3 has the same signature as S. It turns out that if a system of 3 curves is extendable, then it can be

[5] In [9] they use a different (but equivalent) encoding of the signature which is more useful for the proof. Since we do not discuss the details of the proof here, the current encoding will suffice.

Fig. 7 The system on the left has signature $aabaab$. The system on the right has signature $bababa$

Fig. 8 On the right, a uniform 2-system where each subsystem of size 3 has signature $bbabba$. On the left, the 3-system with signature $baaabaabb$ is not extendable

extended to arbitrarily large uniform systems. Observe that in a uniform system each pair of curves cross the same number of times. (See Fig. 8.)

Using a characterization of extendable systems of size 3, given in [9], one can establish the following results which are easily seen to imply Theorems 4.2 and 4.3.

Theorem 4.4 (Dobbins et al. [9]) *Every uniform system is upper convex or lower convex.*

Theorem 4.5 (Dobbins et al. [9]) *Let S be a uniform system where each pair of curves cross precisely k times.*

(1) If $k \leq 2$, then S is upper convex or only 2 paths appear on the upper envelope of S.
(2) If $k \leq 4$, then S is upper convex or at most 3 paths appear on the upper envelope of S.
(3) If $k \geq 5$, then S is upper convex or at most 4 distinct paths appear on the upper envelope of S.

5 Final Remarks

Now that the original Erdős–Szekeres conjecture has nearly been settled, the main remaining open problem in this area is to determine the precise values of the function $f(n)$. This will certainly be quite challenging.

As for disjoint and non-crossing families of convex bodies, it seems likely that the methods from the proof of Theorem 3.9 could be used to generalize Theorem 3.1 to non-crossing families, and that one could possibly even close the gap on the bounds on $h(n, 4)$.

It would also be interesting to obtain better bounds for the function $t_k(n)$ conjectured by Pach and Tóth. The argument given by Dobbins et al. which shows that $t_k(n) \leq u_k(n)$, can actually be reversed to show that any k-system can be "dualized" to give produce a family of convex bodies such that any two share at most k common boundary points (for even k). This suggest that further study of k-systems is necessary to obtain better estimates on $t_k(n)$. One of the main problems in this respect is to give good lower bound constructions. This is because the condition that every s_k members are in convex position changes the nature of the problem significantly from the original Erdős–Szekeres problem for points when $s_k > 3$. Already for families of convex bodies such that any two share at most 4 common boundary points and any four are in convex position we do not even have a superpolynomial lower bound. (Recall from Theorem 3.1 that if the bodies are *pairwise disjoint*, then the best known lower bound is quadratic in n.)

Finally, let us also mention the problem of improving the bounds on the functions $v_k(n)$ from Theorem 4.3. It is known that $v_1(n) = \binom{2n-4}{n-2} + 1$ (by the "cup-cap" theorem). However, we do not know of any better lower bound for $k > 1$, and it is unclear whether there is actually a dependency on the number of crossings. We conclude with the following (rather bold) conjecture.

Conjecture 5.1 (Dobbins et al. [9]) *For every integer $n \geq 3$ there exists a minimal integer $S(n)$ with the following property. Any system of x-monotone curves of size $S(n)$ such that each pair of curves cross at least once, has a subsystem of size n which is upper convex or lower convex.*

References

1. I. Bárány, G. Károlyi, Problems and results around the Erdős–Szekeres convex polygon theorem, in *Discrete and computational geometry (Tokyo, 2000)*, Lecture Notes in Comput. Sci. (Springer, Berlin, 2001), pp. 91–105
2. I. Bárány, P. Valtr, A positive fraction Erdős-Szekeres theorem. Discret. Comput. Geom. **19**, 335–342 (1998)
3. T. Bisztriczky, G. Fejes, Tóth, A generalization of the Erdős-Szekeres convex n-gon theorem. J. Reine Angew. Math. **395**, 167–170 (1989)
4. T. Bisztriczky, G. Fejes, Tóth, Nine convex sets determine a pentagon with convex sets as vertices. Geom. Dedicata **31**, 89–104 (1989)
5. T. Bisztriczky, G. Fejes Tóth, Convexly independent sets. Combinatorica **10**, 195–202 (1990)
6. A. Björner, M.L. Vergnas, B. Sturmfels, N. White, G.M. Ziegler, *Oriented Matroids*, 2nd edn., Encyclopedia of Mathematics and its Applications (Cambridge University Press, Cambridge, 1999)
7. F.R.K. Chung, R.L. Graham, Forced convex n-gons in the plane. Discret. Comput. Geom. **19**, 367–371 (1998)
8. M.G. Dobbins, A.F. Holmsen, A. Hubard, The Erdős-Szekeres problem for non-crossing convex sets. Mathematika **60**, 463–484 (2014)
9. M.G. Dobbins, A.F. Holmsen, A. Hubard, Regular systems of paths and families of convex sets in convex position. Trans. Am. Math. Soc. **368**, 3271–3303 (2016)
10. M. Eliáš, J. Matoušek, Higher-order Erdős-Szekeres theorems. Adv. Math. **244**, 1–15 (2013)
11. P. Erdős, G. Szekeres, A combinatorial problem in geometry. Compositio Math. **2**, 463–470 (1935)
12. P. Erdős, G. Szekeres, On some extremum problems in elementary geometry. Ann. Univ. Sci. Budapest. Eötvös Sect. Math. **3–4**, 53–62 (1960/1961)
13. S. Felsner, P. Valtr, Coding and counting arrangements of pseudolines. Discret. Comput. Geom. **46**, 405–416 (2011)
14. J. Fox, J. Pach, B. Sudakov, A. Suk, Erdős-Szekeres-type theorems for monotone paths and convex bodies. Proc. Lond. Math. Soc. **105**, 953–982 (2012)

15. J.E. Goodman, Pseudoline arrangements, in *Handbook of Discrete and Computational Geometry*, 2nd edn. (CRC Press Ser. Discrete Math. Appl., FL, 2004), pp. 97–128
16. J.E. Goodman, R. Pollack, On the combinatorial classification of nondegenerate configurations in the plane. J. Comb. Theory Ser. A **29**, 220–235 (1980)
17. J.E. Goodman, R. Pollack, A combinatorial perspective on some problems in geometry, *Proceedings of the Twelfth Southeastern Conference on Combinatorics, Graph Theory and Computing, Vol. I (Baton Rouge, La., 1981)*,vol. 32 (1981), pp. 383–394
18. J.E. Goodman, R. Pollack, Upper bounds for configurations and polytopes in R^d. Discret. Comput. Geom. **1**, 219–227 (1986)
19. A. Hubard, L. Montejano, E. Mora, A. Suk, Order types of convex bodies. Order **28**, 121–130 (2011)
20. D.J. Kleitman, L. Pachter, Finding convex sets among points in the plane. Discret. Comput. Geom. **19**, 405–410 (1998)
21. D.E. Knuth, *Axioms and Hulls*, Lecture Notes in Computer Science (Springer, Berlin, 1992)
22. M. Lewin, A new proof of the Erdős-Szekeres theorem. Math. Gaz. **60**, 136–138 (1976)
23. P.A. MacMahon, *Combinatorial Analysis, Two volumes (bound as one)* (Chelsea Publishing Co., New York, 1960)
24. H.N. Mojarrad, *On the Erdős–Szekeres Conjecture* (2015). arXiv:1510.06255
25. H.N. Mojarrad, G. Vlachos, An improved upper bound for the Erdős-Szekeres Conjecture. Discret. Comput. Geom. **56**, 165–180 (2016)
26. W. Morris, V. Soltan, The Erdős-Szekeres problem on points in convex position–a survey. Bull. Am. Math. Soc. **37**, 437–458 (2000)
27. G. Moshkovitz, A. Shapira, Ramsey Theory, integer partitions and a new proof of the Erdős-Szekeres Theorem. Adv. Math. **262**, 1107–1129 (2014)
28. S. Norin, Y. Yuditsky, Erdős-Szekeres without induction. Discret. Comput. Geom. **55**, 963–971 (2016)
29. J. Pach, J. Solymosi, Canonical theorems for convex sets. Discret. Comput. Geom. **19**, 427–435 (1998)
30. J. Pach, G. Tóth, A generalization of the Erdős-Szekeres theorem to disjoint convex sets. Discret. Comput. Geom. **19**, 437–445 (1998)
31. J. Pach, G. Tóth, Erdős-Szekeres-type theorems for segments and noncrossing convex sets. Geom. Dedicata **81**, 1–12 (2000)
32. A. Pór, P. Valtr, The partitioned version of the Erdős-Szekeres theorem. Discret. Comput. Geom. **28**, 625–637 (2002)
33. A. Pór, P. Valtr, On the positive fraction Erdős-Szekeres theorem for convex sets. Eur. J. Comb. **27**, 1199–1205 (2006)
34. A. Suk, On the Erdős–Szekeres convex polygon problem. J. Am. Math. Soc. **30**, 1047–1053 (2017). arXiv:1604.08657
35. G. Szekeres, L. Peters, Computer solution to the 17-point Erdős-Szekeres problem. ANZIAM J. **48**, 151–164 (2006)
36. G. Tóth, Finding convex sets in convex position. Combinatorica **20**, 589–596 (2000)
37. G. Tóth, P. Valtr, Note on the Erdős-Szekeres theorem. Discret. Comput. Geom. **19**, 457–459 (1998)
38. G. Tóth, P. Valtr, The Erdős-Szekeres theorem: upper bounds and related results, *Combinatorial and Computational Geometry*, vol. 52 (Cambridge Univ. Press, Cambridge, 2005), pp. 557–568
39. G. Vlachos, *On a conjecture of Erdős and Szekeres* (2015). arXiv:1505.07549

Configuration Spaces of Equal Spheres Touching a Given Sphere: The Twelve Spheres Problem

Rob Kusner, Wöden Kusner, Jeffrey C. Lagarias and Senya Shlosman

Abstract The problem of twelve spheres is to understand, as a function of $r \in (0, r_{max}(12)]$, the configuration space of 12 non-overlapping equal spheres of radius r touching a central unit sphere. It considers to what extent, and in what fashion, touching spheres can be varied, subject to the constraint of always touching the central sphere. Such constrained motion problems are of interest in physics and materials science, and the problem involves topology and geometry. This paper reviews the history of work on this problem, presents some new results, and formulates some conjectures. It also presents general results on configuration spaces of N spheres of radius r touching a central unit sphere, with emphasis on $3 \leq N \leq 14$. The problem of determining the maximal radius $r_{max}(N)$ is a version of the Tammes problem, to which László Fejes Tóth made significant contributions.

R. Kusner (✉)
Department of Mathematics & Statistics, University of Massachusetts,
Amherst, MA 01003, USA
e-mail: profkusner@gmail.com

W. Kusner
Institute of Analysis and Number Theory, Graz University of Technology,
8010 Graz, Austria
e-mail: wkusner@tugraz.at; wkusner@gmail.com

W. Kusner
Department of Mathematics, Vanderbilt University , Nashville, TN 37240, USA
e-mail: w.kusner@vanderbilt.edu

J. C. Lagarias
Department of Mathematics, University of Michigan, Ann Arbor, MI 48109-1043, USA
e-mail: lagarias@umich.edu

S. Shlosman
Skolkovo Institute of Science and Technology, Moscow, Russia
e-mail: senya.shlosman@univ-amu.fr; shlos@iitp.ru

S. Shlosman
Aix Marseille Université, Université de Toulon, CNRS, CPT, UMR 7332,
13288 Marseille, France

S. Shlosman
Institute for Information Transmission Problems, RAS, Moscow, Russia

G. Ambrus et al. (eds.), *New Trends in Intuitive Geometry*, Bolyai Society Mathematical Studies 27, https://doi.org/10.1007/978-3-662-57413-3_10

2010 Mathematics Subject Classification 11H31 · 49K35 · 52C17 52C25 · 53C22 · 55R80 · 57R70 · 58E05 · 58K05 · 70G10 · 82B05

1 Introduction

This paper studies constrained configuration spaces of N equal spheres of radius r touching a central sphere of radius 1, with emphasis on small values of N, particularly $N = 12$.

The "problem of the 13 spheres," so named by Schütte and van der Waerden [88] and Leech [66], asks whether there exists any configuration of 13 non-overlapping unit spheres that all touch a central unit sphere. It was raised in the time of Newton by David Gregory, and eventually resolved mathematically as impossible. Its resolution established that the "kissing number" of equal 3-dimensional spheres is 12. Quantitatively, one can ask what is the maximum radius for 13 spheres all touching a central sphere of unit radius. It is less than 1 and its exact value was determined by Musin and Tarasov [77] in 2013.

This paper treats a related problem: *How can N spheres of equal radius r touch a given central sphere of radius* 1*, in what patterns, and how are these patterns related? What is the topology of the corresponding constrained configuration space of such spheres?* One may also ask how the topology changes as the radius r varies. In this paper we review the remarkable history of this problem for radius $r = 1$, the sphere packing case, and $N = 12$, the kissing number. We prove results on the structure of this configuration space and formulate several conjectures.

This problem has come up in physics and materials science. Many atoms and molecules are roughly spherical, and their local interactions are governed by how many of them can get close to a single atom. The arrangements possible for 13 nearby spheres, and allowable motions between them, are relevant to the nature of local interactions, to measuring the entropy of local configurations, and to phase changes in certain materials. We are especially motivated by a statement that Frank [40] made in the context of supercooling of fluids, given in Sect. 2.7. By insisting that exactly 12 equal spheres touch a 13-th central sphere, possibly of a different radius, we obtain a mathematical toy problem that can be subjected to careful analysis.

As a mathematical problem, the twelve spheres problem has both a metric geometry aspect and a topology aspect. László Fejes Tóth made major contributions to the metric geometry of the problem, which concerns extremal questions, formulated as densest packing problems. In connection with the Tammes problem, described in Sect. 3, he found the largest radius of 12 spheres that can touch a central sphere of radius 1, realized by the dodecahedral configuration DOD, and found other extremal configurations of touching spheres for smaller N. He posed the Dodecahedral Conjecture concerning the minimal volume Voronoi cell in a unit sphere packing, and posed another conjecture characterizing all configurations that pack space with every sphere touching exactly 12 neighboring spheres. Both of these conjectures are now proved. This paper focuses on the topological aspect of the twelve spheres problem; more generally we discuss the topology of configuration spaces for general N, and the allowable motions and rearrangements of such configurations. We survey what is known for small N, in the range $3 \leq N \leq 14$.

1.1 Configuration Spaces

The arrangements of the N touching spheres are encoded in the associated *configuration space* of N-tuples of points on the surface of a unit sphere that remain at a suitable distance from each other. This space has nontrivial topology and geometry. In topology the general subject of configuration spaces started in the 1960s with the consideration of topological spaces whose points denote configurations of a fixed number N of labeled points on a manifold.

This paper considers the *constrained* configuration space $\mathrm{Conf}(N)[r]$ of N non-overlapping spheres of radius r which touch a central sphere $\mathbb{S}^2$ of radius 1, centered at the origin. (Here "non-overlapping" means the spheres have disjoint interiors.) It can also be visualized as the space of N spherical caps on the sphere, which are obtained as the radial projection of the external spheres onto the surface of the central sphere, whose *angular diameter* $\theta = \theta(r)$ is a known function of r.

The centers of these caps define a constrained N-configuration on $\mathbb{S}^2$ where no pair of points can approach closer than angular separation θ. For generic ("non-critical") values of r for a range of values $0 < r < r_{max}(N)$, this space is a compact $2N$-dimensional manifold with boundary, not necessarily connected.

The group $SO(3)$ acts as global symmetries of $\mathrm{Conf}(N)[r]$ by rigidly rotating the N-configuration of spheres touching the central sphere. The *reduced constrained configuration space* $\mathrm{BConf}(N)[r] = \mathrm{Conf}(N)[r]/SO(3)$ is obtained by identifying rotationally equivalent configurations. For generic values of r it is a compact $(2N-3)$-dimensional manifold with boundary; for the case of 12 spheres this is a 21-dimensional manifold. The subject of constrained configuration spaces has in part been developed for applications to fields such as robotics. For an introduction to the robotics aspect, see generally Abrams and Ghrist [1] or Farber [31].

This paper surveys results for small N on the metric geometry problem of determining the maximum allowable radius $r_{max}(N)$ for $\mathrm{Conf}(N)[r]$ (equivalently $\mathrm{BConf}(N)[r]$) to be nonempty; this is a variant of the Tammes problem, also treated in the literature under the name *optimal spherical codes* (see Sect. 3).

This paper also studies the topology of configuration spaces of a fixed radius r, and the changes in topology in such spaces as the radius r is varied. In the latter case the configuration space changes topology at a set of *critical radius values*. Associated to these special values are *critical configurations*, which are extremal in a suitable sense. The change in topology is described by a generalization of Morse theory applicable to the radius function r, which we discuss in Sect. 4. To determine these changes one studies the occurrence and structure of the critical configurations. The simplest example of such topology change concerns the connectivity of the space of configurations as a function of r, reported by the rank of the 0-th homology group of the configuration space.

The 12 spheres problem includes as its most important special case that of unit spheres, where the sphere radius $r = 1$. This special case is the one relevant to sphere packing in dimension 3. We treat the topological space BConf(12)[1] in Sects. 5 and 6, and formulate several conjectures related to it. The radius $r = 1$ is a critical radius, and two configurations FCC and HCP on the boundary of the space BConf(12)[1] are critical configurations. The topology of BConf(12)[1] appears to be very complicated, and its cohomology groups have not been determined. In Sect. 6 we describe how it is possible to move in the space BConf(12)[1] to deform any dodecahedral configuration DOD of 12 labeled spheres to any other labeled DOD configuration, permuting the 12 spheres arbitrarily, a result due to Conway and Sloane. This suggests the (folklore) conjecture asserting that $r = 1$ is the largest radius value for which the configuration space BConf(12)[r] is connected, i.e. it is the largest r for which the 0-th cohomology group of BConf(12)[r] has rank 1.

This paper establishes some new results. It makes the observation (in Sect. 4.4) that the family of 5-configurations of spheres achieving $r_{max}(5)$ (see Fig. 6) is topologically complex. It completely determines (in Sect. 4.6) the cohomology of BConf($\mathbb{S}^2$, 4)[r] for allowable r. It makes precise the notion of N-configurations being *critical for maximizing* the injectivity radius on BConf($\mathbb{S}^2$, N), and provides a necessary and sufficient *balancing condition* (Theorem 4.11) for criticality, prefatory to a "Morse theory" for such min-type functions [64]. And it formulates several new conjectures in Sects. 6.5 and 6.6.

1.2 Physics and Materials Science

Configuration spaces are of interest in physics and materials science. Jammed configurations are a granular materials criterion for a stable packing. According to Torquato and Stillinger [95, p. 2634] they are: "particle configurations in which each particle is in contact with its nearest neighbors in such a way that mechanical stability of a specific type is conferred to the packing." Packings of rigid disks and spheres have been studied extensively by simulation (Lubachevsky and Stillinger [70], Donev et al. [25]). It has been empirically discovered that randomly ordered hard spheres achieve in random close packing a density around 66 percent [90], and pass through a jamming transition around 64 percent [67, p. 355]. The appearance of a jamming phase transition, signaled by a change in shear modulus, and the formation of a glass state, is relevant in studying the behavior of colloidal suspensions and granular materials. The large rearrangement of structure required in making a phase transition is relevant in the phenomenon of supercooling of liquids (see Sect. 2.7). The nature of glass transitions has been called "the deepest and most interesting unsolved problem in solid state theory" (Anderson [3]). For articles and reviews of these topics, see generally Ediger et al. [26], O'Hern et al. [79], and Liu and

Nagel [67]. For a survey of hard sphere models, including the idea of a liquid-solid phase transition in packings, see generally Löwen [68].

One may make an analogy between the configuration spaces $\mathrm{BConf}(N)[r]$ treated here and a sphere packing model for jamming studied in [79], which treats spheres having repulsive local potential at zero density and zero applied stress, and includes hard spheres for one model parameter value. In the latter model, the order parameter is the packing fraction of the spheres. In the configuration space model, a proxy value for the packing fraction is the radius parameter r, which determines the fraction of surface area of $\mathbb{S}^2$ covered by the N spherical caps. An analogue of the jamming transition value in the configuration space model is then the maximal radius $r_{conn}(N)$ at which the constrained configuration space $\mathrm{BConf}(N)[r]$ remains connected; this property is detected by the 0-th cohomology group. Finer topological invariants of this kind are then supplied by the various critical values r_j at which the ranks of the individual cohomology groups $H^k(\mathrm{BConf}(N)[r], \mathbb{Q})$ change. Our configuration model is simplified in being 2-dimensional, with constrained configurations on the surface of a 2-sphere $\mathbb{S}^2$, which, however, has the new feature of positive curvature, giving a compact constrained configuration space. For the jamming problem itself, the space of (constrained) configurations of hard spheres in a large 3-dimensional box seems a more appropriate space. The general direction of inquiry investigating the transition of topological invariants (like Betti numbers) of configuration spaces as the radius parameter is varying could shed new light on the nature of jamming transitions. For further remarks, see Sect. 7.

1.3 Roadmap

The sections of the paper have been written with the aim to be independently readable. We prove some results for general N, but Sects. 2, 5 and 6 focus on the case $N = 12$. Sect. 2 gives a brief history of results on the twelve spheres problem, stemming from its special role in connection with sphere packing in dimension 3. Section 3 surveys results on the maximal radius $r_{\max}(N)$ possible for a configuration of N equal spheres touching a central sphere of radius 1, for small N. This problem is a version of the Tammes problem. Section 4 begins with the topology of configuration spaces of N points in $\mathbb{R}^2$ and on the 2-sphere $\mathbb{S}^2$, corresponding to radius $r = 0$. It then considers spaces of configurations of equal spheres of radius r touching a sphere of radius 1 for variable $0 < r \leq r_{\max}(N)$. It defines a notion of critical configuration in the spirit of min-type Morse theory. Section 5 discusses the special configuration space of 12 unit spheres touching a 13-th central sphere, i.e. the case $r = 1$. It focuses on properties of the FCC configuration, the HCP configuration and the dodecahedral configuration DOD. It shows that the FCC and HCP configurations are critical (in the sense of Sect. 4.2) in the reduced configuration spaces $\mathrm{BConf}(12)[1]$. It also shows that there are continuous deformations in $\mathrm{BConf}(12)[1]$ moving a dodecahedral configuration to an FCC configuration, resp. moving it to an HCP configuration. Section 6 considers the problem of permutability of the spheres of the dodecahedral configuration for $r = 1$, conjecturing that the space BConf(12)[1] is connected, and that this is the largest value of r where connectedness holds. It also considers the $r > 1$ case and formulates several conjectures about disconnectedness. Section 7 makes some concluding remarks.

2 The Twelve Spheres Problem: History

We begin with some historical vignettes concerning configurations of 12 spheres touching a central sphere, as they have come up in physics, astronomy, biology and materials science.

2.1 *Kepler (1611)*

Johannes Kepler (1571–1630) studied packings and crystals in his 1611 pamphlet "The Six-cornered Snowflake" [61]. In it he asserts that the densest sphere packing of equal spheres is the FCC packing, or "cannonball packing." He states that this packing has 12 unit spheres touching each central sphere:

> In the second mode, not only is every pellet touched by its four neighbors in the same plane, but also by four in the plane above and four below, so throughout one will be touched by twelve, and under pressure spherical pellets will become rhomboid. This arrangement will be more compatible to the octahedron and the pyramid. The packing will be the tightest possible, so that in no other arrangement could more pellets be stuffed into the same container.[1]

He expands on the construction as follows:

> Thus, let B be a group of three balls; set one A, on it as apex; let there be also another group C, of six balls, and another D, of ten, and another E, of fifteen. Regularly superpose the narrower on the wider to produce the shape of a pyramid. Now, although in this construction each one in the upper layer is seated between three in the lower, yet if you turn the figure round so that not the apex but the whole side of the pyramid is uppermost, you will find, whenever you peel off one ball from the top, four lying below it in square pattern. Again as before, one ball will be touched by twelve others, to with, by six neighbors in the same plane, and by three above and three below. Thus in the closest pack in three dimensions, the

[1] "Iam si ad structuram solidorum quam potest fieri arctissimam progredaris, ordinesque ordinibus superponas, in plano prius coaptatos aut ii erunt quadrati A aut trigonici B: si quadrati aut singuli globi ordinis superioris singulis superstabunt ordinis inferioris aut contra singuli ordinis superioris sedebunt inter quaternos ordinis inferioris. Priori modo tangitur quilibit globis a quattuour cirucmstantibus in eodem plano, ab uno supra se, et ab uno infra se: et sic in universum a six aliis, eritque ordo cubicus, et compressione facta fient cubi: sed no erit arctissima coaptatio. Posteriori modo praeterquam quod quileibet globus a quattuor circumstantibus in eodem plano tangitur etiam a quattuor infra se, et a quattuor supra se, et sic in universum a duodecim tangetur; fientque compressione ex globosis rhombica. Ordo hic magis assimilabitur octahedro et pyramidi. Coaptatio fiet arctissima, ut nullo praetera ordine plures globuli in idem vas compingi queant." [Translation: Colin Hardie [61, p. 15]]

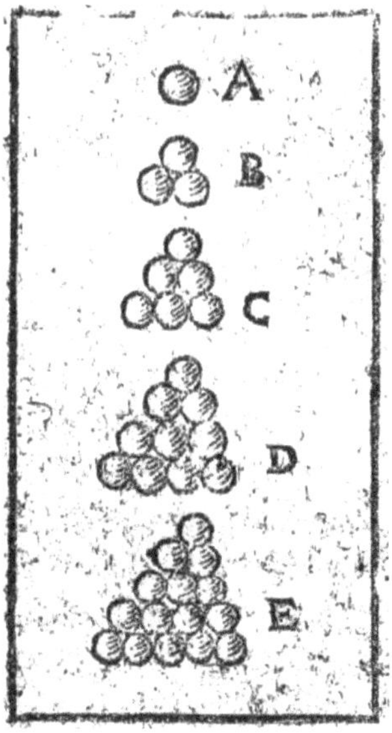

Fig. 1 Woodcut of Kepler sphere arrangements [61]

triangular pattern cannot exist without the square, and vice versa. It is therefore obvious that the loculi of the pomegranate are squeezed into the shape of a solid rhomboid....[2]

The cannonball packing had been studied earlier by the English mathematician Thomas Hariot [Harriot] (1560–1621). Hariot was mathematics tutor to Sir Walter Raleigh, designed some of his ships, wrote a treatise on navigation, and went on an expedition to Virginia in 1585–1587 as surveyor, reporting on it in 1590 in [53], his only published book. He computed a chart in 1591 on how to most efficiently stack cannonballs using the FCC packing, and computed a table of the number of cannonballs in such stacks (see Shirley [91, pp. 242–243]). Hariot supported the atomic theory of matter, in which case macroscopic objects may be packed in arrangements of very tiny spherical objects, i.e. atoms [60, Chap. III]. He corresponded with Kepler in 1606–1608 on optics, and mentioned the atomic theory in a December 1606 letter as a possible way of explaining why some light is reflected, and some refracted, at the surface of a liquid. Kepler replied in 1607, not supporting the atomic theory.

[2]"Esto enim *B* copula trium globorum. Ei superpone *A* unum pro apice; esto et alia copula senum globorum *C*, et alia denum *D* et alia quindenum *E*. Impone semper angustiorem latiori, ut fiat figura pyramidis. Etsi igitur per hanc impositionem singuli superiores sederunt into trinos inferiores: tamen iam versa figura, ut non apex sed integrum latus pyramidis sit loc superiori, quoties unum globulum deglberis e summis, infra stabunt quattuor ordine quadrato. Et rursum tangetur unus globus ut prius, et duodecim aliis, a sex nempe circumstantibus in eodem plano tribus supra et tribus infra. Ita in solida coaptatione arctissima non potest ess ordo triangularis sine quadrangulari, nec vicissim. Patet igitur, acinos punici mali, materiali necessitate concurrente cum rationaibus incrementi acinorum, exprimi in figuri rhombici corporis..." [Translation by Colin Hardie [61, p. 17]].

The known correspondence of Hariot with Kepler does not deal directly with sphere packing (Fig. 1).

The statement that the maximal density of a sphere packing in 3-dimensional space equals $\frac{\pi}{\sqrt{18}} \approx 0.74048$, which is attained by the FCC packing, is called the *Kepler Conjecture*. It was settled affirmatively in the period 1998–2004 by Hales with Ferguson [65]. A second generation proof, which is a formal proof checked entirely by computer, was recently completed in a project led by Hales [51].

2.2 *Newton and Gregory (1694)*

In 1694 Isaac Newton (1642–1727) and David Gregory (1659–1708) had a discussion of touching spheres related to preparing a second edition of Newton's *Principia*. It concerned the question whether the "fixed stars" are subject to gravitational attraction. What force is "balancing" their apparent fixed positions?

Gregory [80, Vol III, p. 317] summarized in a memorandum a conversation with Newton on 4 May 1694 concerning the brightest stars as:

> To discover how many stars there are of a given magnitude, he [Newton] considers how many spheres, nearest, second from them, third etc. surround a sphere in a space of three dimensions, there will be 13 of first magnitude, 4 × 13 of second, 9 × 4 × 13 of third.[3]

Newton's own star table "A Table of ye fixed Starrs for ye yeare 1671" records 13 first magnitude stars, 43 of the second magnitude, 174 of third magnitude (see [80, Vol II, p. 394]).

Newton drafted a new Proposition to be included in a second edition of the *Principia*, stating [59, p. 81, in translation]:

> Proposition XV. Theorem XV. The fixed stars are at rest in the heavens and are separated by enormous distances from our Sun and from each other.

In a draft proof he wrote [59, p. 85, in translation]:

> That the stars are at huge distances from our Sun is clear enough from the absence of parallax; and that they lie at no less distances from each other may be inferred from their differing apparent magnitudes. For there are 13 stars of the first magnitude and roughly the same number of equal spheres can be arranged about a central sphere equal to them.

and:

> For if around some sphere there are arranged more spheres of about the same size, the number of spheres which surround it closely will be 12 or 13; at the second stage about 50; at the third about 110 [roughly 9 × 12]; at the fourth, 200 [16 × $12\frac{1}{2}$], ...

---

[3]"Ut noscatur quot sunt stellae magnitudinis 1 ae, 2 dae, 3 ae & c. considerando quot spherae proximae, seundae ab his 3 ae & c. spheram in spatio trium dimensionis circumstent: erunt 13 primae, 4 × 13 2-dae, 9 × 4 × 13 3 ae."

This argument is similar to one of Kepler [62, Liber I, Pars II, p. 138] (translation in Koyré [63, p. 80]), with roots in the claim of Giordano Bruno, that all stars are suns.

After further work, through several drafts, Newton abandoned this Proposition (according to Hoskin [59]). It was not included in the second edition of the *Principia*, when it was later published in 1713.

Gregory continued study of the geometric problem underlying the spacing of stars. In an (unpublished) notebook[4] he considered the packing problem of 2-dimensional disks in concentric rings and, in 3 dimensions, that of equal spheres, noting that 13 spheres might touch a given equal sphere [80, Vol III, Letter 441, Note (10), p. 321]. He considered the 13 sphere question in later years, making the following memorandum in 1704 [56, p. 21]:

> *Oxon. 23 Novr 1704.* Mr. Kyl[5] said that if 13 equal spheres touch an equal inmost sphere, 9×13 must touch one that include these former 14, because there is nine times as much surface to stand on. I told him that we must reckon by the surface passing through their centers.

A manuscript of Gregory on Astronomy, translated into English and posthumously published in 1715, states [46, p. 289, sic]:

> For if every Fix'd Star did the office of a Sun, to a portion of the Mundane space nearly equal to this that our Sun commands, there will be as many Fix'd Stars of the first Magnitude, as there can be Systems of this sort touching and surrounding ours; that is, as many equal Spheres as can touch an equal one in the middle of them. Now, 'tis certain from Geometry, that thirteen Spheres can touch and surround one in the middle equal to them, (for *Kepler* is wrong in asserting, in *B.* I of the *Epit.*[6] that there may be twelve such, according to the number of Angles of an *Icosaedrum*,)

Thus Gregory expressed a definite opinion that 13 spheres might touch.

2.3 *Bender, Hoppe, Günther (1874)*

The issue of whether 13 equal spheres might touch a central equal sphere was discussed in the physics literature in the period 1874–1875, with contributions by C. Bender [10], Reinhold Hoppe [58] and Siegmund Günther [47]. Hoppe noted a mathematical gap in the argument of Bender. Günther offered a physical intuition, but no proof. They all concluded that at most 12 unit spheres could touch a central unit sphere. In 1994 Hales [49] noted a mathematical gap in the argument of Hoppe.

---

[4] This notebook is at Christ Church, Oxford, according to J. Leech [66].

[5] John Keill (1671–1721) succeeded Gregory as Savilian Professor.

[6] This is Kepler [62].

2.4 Barlow (1883)

In another context the crystallographer William Barlow (1845–1934) noted another optimal sphere packing, the *Hexagonal Close Packing* (HCP). In a paper "Probable nature of the internal symmetry of crystals" [8, p. 186] he considered five symmetry types for crystal structure. The third kind of symmetry he describes is the FCC packing (FIG. 4*a*). He then stated:

> A fourth kind of symmetry, which resembles the third in that each point is equidistant from the twelve nearest points, but which is of a widely different character than the three former kinds, is depicted if layers of spheres in contact arranged in the triangular pattern (plan d) are so placed that the sphere centers of the third layer are over those of the first, those of the fourth layer over those of the second, and so on. The symmetry produced is hexagonal in structure and uniaxial (FIG. 5*a*).

Here "Plan *d*" is the two-dimensional hexagonal packing, and FIG. 5*a* depicts the HCP packing. He suggested that the atoms in a crystal of quartz ($Si\,O_2$) occur with the fourth kind of symmetry (see Fig. 2).

Barlow also stated later in the paper [8, Figs. 7 and 8, p.207] the following about twinned crystal arrays with a connecting layer (Fig. 3):

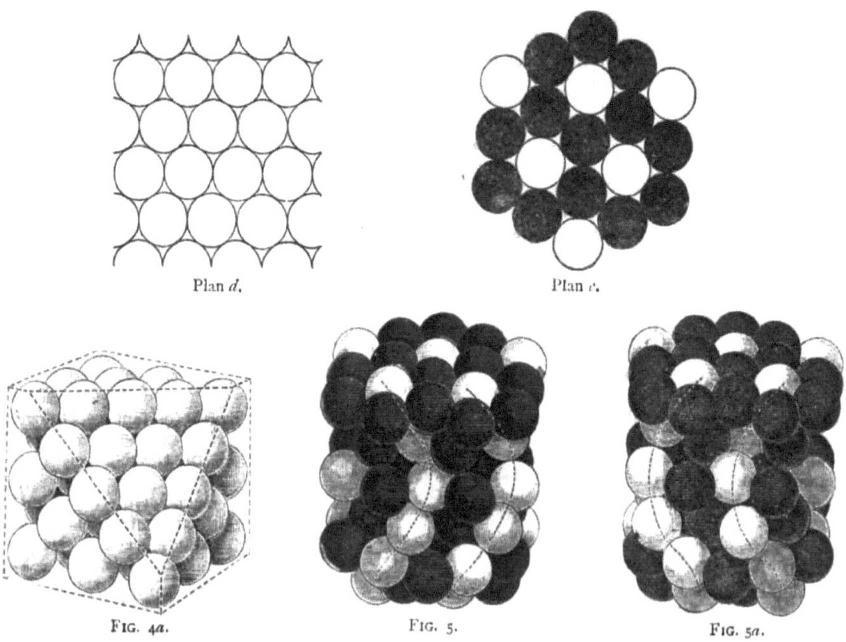

Fig. 2 Barlow FCC and HCP packings [8]

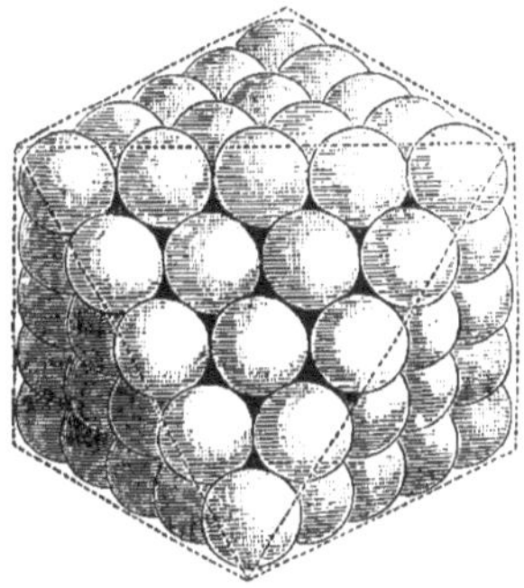

FIG. 7.

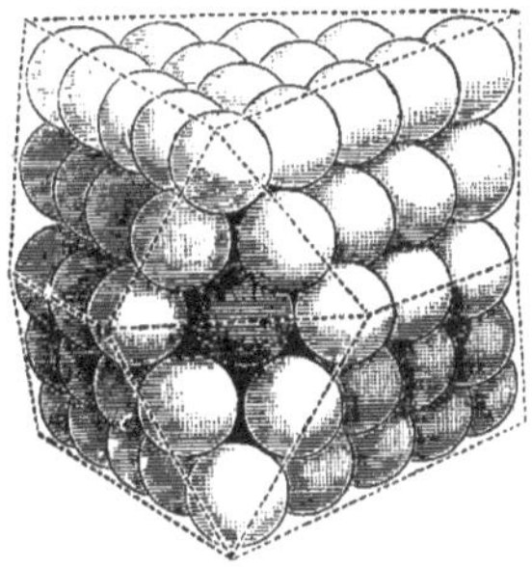

FIG. 8.

Fig. 3 Barlow Twinned Crystal Packing [8]

> The peculiarities of *crystal-grouping* displayed in twin crystals can be shown to favour the supposition that we have in crystals symmetrical arrangement rather than symmetrical shape of atoms or small particles. Thus if an octahedron be cut in half by a plane parallel to two opposite faces, and the hexagonal faces of separation, while kept in contact and their centres coincident, are turned one upon the other through 60°, we know that we get a familiar example of a form found in some twin crystals. And a stack can be made of layers of spheres placed triangularly in contact to depict this form as readily as to depict a regular octahedron, the only modification necessary being for the layers above the centre layer to be placed as though turned bodily through 60°, from the position necessary to depict an octahedron (compare (FIG. 7 and FIG. 8). The modification, as we see, involves *no departure from the condition that each particle is equidistant from the twelve nearest particles.*

2.5 Tammes (1930)

The Dutch botanist Pieter Merkus Lambertus Tammes made in 1930 a study of the equidistribution of pores on pollen grains [92]. He asked the question: What is the maximum number of circular caps $N(\theta)$ of angular diameter θ that can be placed without overlap on a unit sphere? Here θ is measured from the center of the unit sphere $\mathbb{S}^2$ in $\mathbb{R}^3$. Tammes [92, Chap. 3] empirically determined that $N(\frac{\pi}{2}) = 6$, while $N(\theta) \leq 4$ for $\theta > \frac{\pi}{2}$. Let $\theta = \theta(N)$ denote the maximal value of θ having $N(\theta) = N$. He concluded that $\theta(5) = \theta(6) = \frac{\pi}{2}$.

The problem of determining various values of $N(\theta)$ is now called the *Tammes problem*. It is related to a dual question of determining the maximal radius $r(N)$ possible for N equal spheres all touching a central sphere of radius 1. Namely, the maximal value of $\theta := \theta(N)$ having $N(\theta) = N$, determines the maximal allowable radius $r(N)$ of N spheres touching a central unit sphere by a formula given in Lemma 3.1.

2.6 Fejes Tóth (1943)

In 1943 László Fejes Tóth [34] conjectured that the volume of any Voronoi cell of any sphere packing of $\mathbb{R}^3$ by unit spheres is minimized by the dodecahedral configuration of 12 unit spheres touching a central sphere. The Voronoi cell of the central sphere is then a regular dodecahedron circumscribed about the sphere. The packing density of the dodecahedron is approximately 0.7546, which is larger than the density of the known FCC packing of $\mathbb{R}^3$. This conjecture became known as the *Dodecahedral Conjecture* and was settled affirmatively in 2010 (see Sect. 5.4).

2.7 Frank (1952)

The problem of molecular rearrangement in the liquid-solid phase transition is relevant in materials. The structure of ordinary ice, the H_2O phase labeled ice I_h, has an HCP packing of its oxygen atoms, as observed in 1921 by Dennison [24]. Note that the hydrogen atoms are free to change their orientations to some extent (Pauling [81]). Water exhibits a phenomenon of supercooling at standard pressure down to -48°C; under special rapid cooling it can avoid freezing down to -137°C, and enter a glassy phase (Angell [5]).

In 1952 Frederick Charles Frank [40] argued that supercooling can occur because the common arrangements of molecules in liquids assume configurations far from what they would assume if frozen. He wrote:

> Consider the question of how many different ways one can put twelve billiard balls in simultaneous contact with another one, counting as different the arrangements which cannot be transformed into each other without breaking contact with the centre ball?' The answer is *three*. Two which come to the mind of any crystallographer occur in the face-centred cubic and hexagonal close packed lattices. The third comes to the mind of any good schoolboy, and it is to put one at the centre of each face of a regular dodecahedron. That body has five-fold axes, which are abhorrent to crystal symmetry: unlike the other two packings, this one cannot be continuously extended in three dimensions. You will find that the outer twelve in this packing do not touch each other. If we have mutually interacting deformable spheres, like atoms, they will be a little closer to the centre in this third kind of packing; and if one assumes they are argon atoms (interacting in pairs with attractive and repulsive potentials proportional to r^{-6} and r^{-12}) one may calculate that the binding energy of the group of thirteen is 8.4% greater than for the other two packings. This is 40% of the lattice energy per atom in the crystal. I infer that this will be a very common grouping in liquids, that most of the groups of twelve atoms around one will be of this form, that freezing involves a substantial rearrangement, and not merely an extension of the same kind of order from short distances to long ones; a rearrangement which is quite costly of energy in small localities, and which only becomes economical when extended over a considerable volume, because unlike the other packing it can be so extended without discontinuities.

The three local arrangements Frank specifies we shall label as FCC (face-centered-cubic), HCP (hexagonal close packing) and DOD (dodecahedral), for convenience. The crystalline arrangements of FCC and HCP are "extremal" (i.e. on the boundary of the configuration space), while the balls in DOD configuration are free to move independently.

Frank's assertion that there are exactly three possible arrangements is *false* if taken literally. There are continuous deformations between any arrangement of types FCC, HCP and DOD and any of the other types (see Sect. 5.4). There is however an important kernel of truth in Frank's statement, which buttresses his argument made concerning the existence of supercooling: each of the three arrangements above is "remarkable" in some sense (see Sect. 5). To move from a large arrangement of spheres having many DOD configurations to one frozen in the HCP packing requires substantial motion of the spheres.

2.8 Schütte and van der Waerden (1953)

In a paper titled "Das Problem der dreizehn Kugeln" ["The problem of the thirteen balls"] Kurt Schütte and Bartel Leendert van der Waerden [88] gave a rigorous proof that one cannot have 13 unit spheres touching a given central sphere.

There has been much further work on this problem. In his 1956 paper titled "The problem of 13 spheres" John Leech [66] gave a two page proof of the impossibility of 13 unit spheres touching a unit sphere. More accurately he stated: "In the present paper I outline an independent proof of this impossibility, certain details which are tedious rather than difficult have been omitted." Various authors have written to fill in such details, which balloon the length of the proof. These include work of Maehara [71] in 2001, who gave in 2007 a simplified proof [72]. Other proofs of the thirteen spheres problem were given by Anstreicher [4] in 2004 and Musin [75] in 2006.

2.9 Fejes Tóth (1969)

In 1969 László Fejes Tóth [39] discussed the problem of characterizing those sphere packings in space that have the property that every sphere in the packing touches exactly 12 neighboring spheres. The FCC and HCP packing both have this property, as already noted by Barlow [8]. There are in addition uncountably many other packings, obtained by stacking plane layers of hexagonally packed spheres ("penny packing"), where there are two choices at each level of how to pack the next level. Fejes Tóth conjectured that all such packings are obtained in this way.

This conjecture of Fejes Tóth was settled affirmatively by Hales [50] in 2013.

2.10 Conway and Sloane (1988)

In their book *Sphere Packings, Lattices and Groups,* John H. Conway and Neil J. A. Sloane considered the question: *What rearrangements of the* 12 *unit spheres are possible using motions that maintain contact with the central unit sphere at all times?* In [20, Chap. 1, Appendix: Planetary Perturbations] they sketch a resultasserting: The configuration space of 12 unit spheres touching a 13-th allows arbitrary permutations of all 12 touching spheres in the configuration. That is, if the spheres are labeled and in the DOD configuration, it is possible, by moving them on the surface of the central sphere, to arbitrarily permute the spheres in a DOD configuration.

We will describe the motions in detail to obtain such permutations in Sect. 6.

3 Maximal Radius Configurations of N Spheres: The Tammes Problem

What is the maximal radius $r(N)$ possible for N equal spheres all touching a central sphere of radius 1? This problem is closely related to the *Tammes problem* discussed above, which concerns instead the maximum number of circular caps $N(\theta)$ of angular diameter θ that can be placed without overlap on a sphere. The latter problem is also the problem of constructing good spherical codes (see [20, Chap. 1, Sect. 2.3]).

3.1 Radius Versus Angular Diameter Parameterization

One can convert the angular measure θ into the radius of touching spheres; for a sphere touching a central unit sphere, its associated spherical cap on the central sphere is the radial projection of its points onto the surface of the central sphere.

Lemma 3.1 *For a fixed $N > 2$, the maximal value of $\theta := \theta_{max}(N)$ having $N(\theta) = N$ determines the maximal allowable radius $r_{max}(N)$ of N spheres touching a central unit sphere, using the formula*

$$r_{max}(N) = \frac{\sin\left(\frac{\theta(N)}{2}\right)}{1 - \sin\left(\frac{\theta(N)}{2}\right)}.$$

Conversely, given $r = r_{max}(N)$, we obtain

$$\theta_{max}(N) = 2\arcsin\left(\frac{r}{1+r}\right),$$

choosing $0 < \theta(N) < \pi$.

Proof From the right triangle in Fig. 4 we have

$$\sin\frac{\theta}{2} = \frac{r}{1+r}.$$

This relation gives a bijection of the interval $0 \le \theta < \pi$ to the interval $0 \le r < \infty$. □

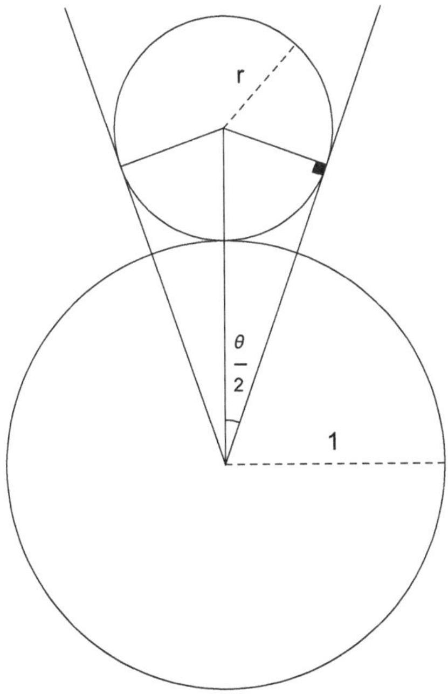

Fig. 4 Angular measure θ related to radius r

3.2 *Rigorous Results for Small N*

The Tammes problem has been solved exactly for only a few values of N, including $3 \le N \le 14$ and $N = 24$.

3.2.1 Fejes Tóth: $N = 3, 4, 6, 12$

The Tammes problem was solved for $N = 3, 4, 6$ and 12 by László Fejes Tóth [33] in 1943, where extremal configurations of touching points for $N = 3$ are attained by vertices of an equilateral triangle arranged around the equator, and for $N = 4, 6, 12$ by vertices of regular polyhedra (tetrahedron, octahedron and icosahedron) inscribed in the unit sphere. Fejes Tóth proved the following inequality: For N points on the surface of the unit sphere, at least two points can always be found with spherical distance

$$d \le \arccos\Big(\frac{(\cot\omega)^2 - 1}{2}\Big), \quad \text{with} \quad \omega = \Big(\frac{N}{N-2}\Big)\frac{\pi}{6}.$$

Note that d is the edge-length of a spherical equilateral triangle with the expected area for an element of an N-vertex triangulation of $\mathbb{S}^2$. The inequality is sharp for $N = 3, 4, 6$ and 12 for the specified configurations above.

In 1949 Fejes Tóth [35] gave another proof of his inequality. His result was re-proved by Habicht and van der Waerden [48] in 1951. After converting this result to the r-parameter using Lemma 3.1, we may re-state his result for $N = 12$ as follows.

Theorem 3.2 (Fejes Tóth (1943))

(1) *The maximum radius of* 12 *equal spheres touching a central sphere of radius* 1 *is:*

$$r_{max}(12) = \frac{1}{\sqrt{\frac{5+\sqrt{5}}{2}} - 1} \approx 1.1085085.$$

Here $r_{max}(12)$ *is a real root of the fourth degree equation* $x^4 - 6x^3 + x^2 + 4x + 1 = 0$.

(2) *An extremal configuration achieving this radius is the* 12 *vertices of an inscribed regular icosahedron (equivalently, face-centers of a circumscribed regular dodecahedron).*

3.2.2 Schütte and van der Waerden: $N = 5, 7, 8, 9$

The Tammes problem was solved for $N = 5$ in 1950 by van der Waerden, building on work of Habicht and van der Waerden [48]. It was solved for $N = 7$ by Schütte. These solutions, plus those of van der Waerden for $N = 8$ and Schütte for $N = 9$ appear in Schütte and van der Waerden [87]. They give a history of these developments in [87, p. 97].

Their paper used geometric methods, introducing and studying the allowed structure of the graphs describing the touching patterns of arrangements of N equal circles on $\mathbb{S}^2$. These graphs are now called *contact graphs*, and Schütte and van der Waerden credit their introduction to Habicht. Schütte also conjectured candidates for optimal configurations for $N = 10, 13, 14, 15, 16$ and van der Waerden conjectured candidates for $N = 11, 24, 32$ (see [87]).

L. Fejes Tóth presented the work of Schütte and van der Waerden in his 1953 book on sphere-packing [36, Chap. VI]. This book uses the terminology of *maximal graph* for the graph of a configuration achieving the maximal radius for N. In 1959 Fejes Tóth [37] noted that the set of vertices of a square antiprism gave an extremal $N = 8$ configuration on the 2-sphere.

3.2.3 Danzer: $N = 6, 7, 8, 9, 10, 11$

In his 1963 Habilitationsschrift [22] (see the 1986 English translation [23]), Ludwig Danzer made a geometric study of the contact graph for a configuration of N circles on the surface of a sphere. This graph has a vertex for each circle and an edge for each pair of touching circles. A contact graph is called *maximal* if it occurs for a set of circles achieving the maximal radius $r_{max}(N)$. It is called *optimal* if it has the minimum number of edges among all maximal contact graphs. A contact graph is called *irreducible* if the radius cannot be improved by altering a single vertex. For each small N, Danzer found a complete list of irreducible contact graphs. He used this analysis to prove the conjectures of Schütte and van der Waerden [87] above for the cases $N = 10, 11$.

Theorem 3.3 (Danzer (1963))
(1) *For $7 \leq N \leq 12$ there is, up to isometry, a unique θ-maximizing unlabeled configuration of spheres with $N(\theta) = N$.*
(2) *For $N = 12$, the vertices of a regular icosahedron form the unique θ-maximizing configuration. The θ-maximizing configuration for $N = 11$ is a regular icosahedron with one vertex removed.*

In [23, Theorem II] Danzer classified irreducible sets for $72° = \frac{2\pi}{5} < \theta < 90° = \frac{2\pi}{4}$. There are additional N-irreducible graphs for $N = 7, 8, 9, 10$ in these cases. For $N = 6, 7, 8, 9$ he finds one optimal set and one irreducible set with one degree of freedom. He also finds for $N = 8$ an irreducible set with two degrees of freedom. For $N = 10$ he finds one optimal set, two irreducible sets with no degrees of freedom, and five with at least one degree of freedom. Danzer states that the irreducible sets with no degrees of freedom (presumably) give relative optima. An irreducible graph having a degree of freedom fails to be relatively optimal, since deforming along its degree of freedom leads to a boundary graph with an additional edge, where the extrema is reached.

Danzer's work was not published in a journal until the 1980s. In the interm, Böröczky [11] gave another solution for $N = 11$, and Hárs [54] for $N = 10$.

3.2.4 Musin and Tarasov: $N = 13, 14$

Very recently the Tammes problem was solved for the cases $N = 13$ and $N = 14$ by Oleg Musin and Alexey Tarasov [76, 78]. Their proofs were computer-assisted, and made use of an enumeration of all irreducible configuration contact graphs (see [77]). Earlier work on configurations of up to 17 points was done by Böröczky and Szabó [12, 13]).

3.2.5 Robinson: $N = 24$

The case $N = 24$ was solved in 1961 by Raphael M. Robinson [85]. He proved a 1959 conjecture of Fejes Tóth [38], asserting that the extremal $\theta(24) \approx 43° 42'$ and that the extremal configuration of 24 sphere centers are the vertices of a snub cube. Coxeter [21, p. 326] describes the snub cube.

Table 1 The tammes problem for small N

N	$\theta_{max}(N)$	$r_{max}(N)$	Configuration	Source
3	$\frac{2\pi}{3} = 120°$	$3 + 2\sqrt{3} \approx 6.4641$	Equilateral Triangle	Fejes Tóth [34]
4	$\cos^{-1}(-\frac{1}{3}) \approx 109.4712°$	$2 + \sqrt{6} \approx 4.4495$	Regular Tetrahedron	Fejes Tóth [34]
5	$\frac{2\pi}{4} = 90°$	$1 + \sqrt{2} \approx 2.4142$	Triangular Bipyramid	Fejes Tóth [34]
6	$\frac{2\pi}{4} = 90°$	$1 + \sqrt{2} \approx 2.4142$	Regular Octahedron	Fejes Tóth [34]
7	$\approx 77.8695°$	≈ 1.6913	[No name]	Schutte and van der Waerden [87]
8	$\approx 74.8585°$	≈ 1.5496	Square Antiprism	Schutte and van der Waerden [87]
9	$\approx 70.5288°$	$\frac{1+\sqrt{3}}{2} \approx 1.3660$	[No name]	Schutte and van der Waerden [87]
10	$\approx 66.1468°$	≈ 1.2013	[No name]	Danzer [22]
11	$\approx 63.4349°$	$\frac{1}{\sqrt{\frac{5+\sqrt{5}}{2}} - 1} \approx 1.1085$	Regular Icosahedron minus one vertex	Danzer [22]
12	$\approx 63.4349°$	$\frac{1}{\sqrt{\frac{5+\sqrt{5}}{2}} - 1} \approx 1.1085$	Regular Icosahedron	Fejes Tóth [34]
13	$\approx 57.1367°$	≈ 0.9165	[No name]	Musin and Tarasov [77]
14	$\approx 55.6706°$	≈ 0.8759	[No name]	Musin and Tarasov [78]
24	$\approx 43.6908°$	≈ 0.5926	Snub Cube	Robinson [85]

3.3 *The Tammes Problem: Optimal Contact Graphs and Optimal Parameters*

Table 1 summarizes optimal angular parameters and radius parameters on the Tammes problem for $3 \leq N \leq 14$ and $N = 24$ (see Aste and Weaire [7, Sect. 11.6]). The configuration name given is associated to the vertices in the corresponding polyhedron being inscribed in a sphere, e.g. an icosahedron has $N = 12$ vertices (see Melnyk et al. [73, Table 2]). In the case $N = 5$ the polyhedron is any from a family of trigonal bipyramids, including the square pyramid as a degenerate case. For $N = 11$ the polyhedron is a singly capped pentagonal antiprism, i.e. the icosahedron with one vertex deleted. The cases $N = 7, 9$ and 10 are described in [73, pp. 1747–1749]. Figure 5 shows schematically the optimal contact graphs for $3 \leq N \leq 14$.

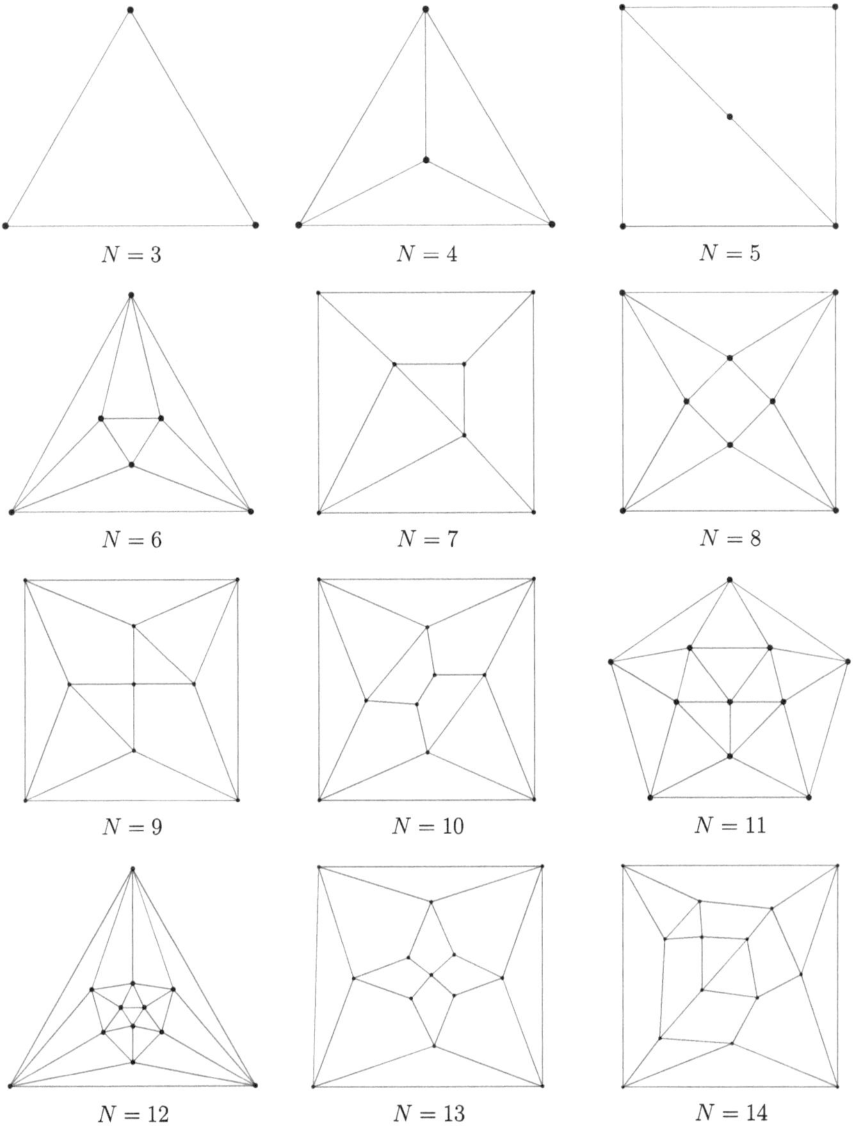

Fig. 5 Optimal contact graphs associated to the Tammes configurations for $N = 3$ to $N = 14$

3.4 *The Tammes Problem Maximal Radius: General N*

The most basic question about the maximal radius in the Tammes problem concerns the distinctness of maximal values. The following conjecture was proposed by R. M. Robinson [86, p. 297].

Table 2 The Tammes problem radii given as algebraic numbers

N	$r_{max}(N)$	Minimal equation	Figure
3	$3+2\sqrt{3} \approx 6.4641$	$X^2 - 6X - 3$	Equilateral Triangle
4	$2+\sqrt{6} \approx 4.4495$	$X^2 - 4X - 2$	Regular Tetrahedron
6	$1+\sqrt{2} \approx 2.4142$	$X^2 - 2X - 1$	Regular Octahedron
7	≈ 1.6913	$X^6 - 6X^5 - 3X^4 + 8X^3 + 12X^2 + 6X + 1$	[No name]
8	≈ 1.5496	$X^4 - 8X^3 + 4X^2 + 8X + 2$	Square antiprism
9	≈ 1.3660	$2X^2 - 2X - 1$	[No name]
10	≈ 1.2013	$4X^6 - 30X^5 + 17X^4 + 24X^3 - 4X^2 - 6X - 1$	[No name]
12	$\frac{1}{\sqrt{\frac{5+\sqrt{5}}{2}}-1} \approx 1.10851$	$X^4 - 6X^3 + X^2 + 4X + 1$	Regular Icosahedron
24	≈ 0.59262	$X^6 - 10X^5 + 23X^4 + 20X^3 - 5X^2 - 6X - 1$	Snub Cube

Conjecture 3.4 (Robinson (1969)) For all $N \geq 4$ the maximal radius satisfies

$$r_{max}(N) < r_{max}(N-1)$$

except possibly for $N = 6, 12, 24, 48, 60$ and 120.

One has $r_{max}(N) = r_{max}(N-1)$ for $N = 6$ and $N = 12$ by results already given above. In 1991 Tarnai and Gáspár [94] established that $r_{max}(24) < r_{max}(23)$. The remaining cases $N = 48, 60$ and 120 are open, but since strict inequality holds for $N = 24$ we expect strict inequality to hold for these values too. However up to now it has been computationally difficult to determine $r_{max}(N)$ for such large N.

We turn to a potentially easier question. The known exact values of $r_{max}(N)$ are algebraic numbers, i.e. roots of some univariate polynomial having integer coefficients. Table 2 below presents algebraicity data for $3 \leq N \leq 14$.
We ask: *Is the optimal radius $r_{max}(N)$ an algebraic number for each $N \geq 3$?* The reason to expect such an algebraicity result to hold is that such a radius should be specified by at least one optimal graph that is *rigid*, i.e. it permits no local deformations preserving optimality, up to isometry. Danzer showed rigidity to be the case for $7 \leq N \leq 12$. The equal length constraints of the edges of the contact graph give a system of polynomial equations with integer coefficients that the coordinates of the sphere centers must satisfy. One may expect that its real solution locus will include some real algebraic solutions for the sphere centers, leading to algebraicity of the radius. Even if the rigidity result fails and deformations occur (as happens for $N = 5$), it could still be the case that the optimal radius is algebraic.

4 Configuration Spaces of N Spheres Touching a Central Sphere

We start with the classical *configuration space* $\mathrm{Conf}(N) := \mathrm{Conf}(\mathbb{S}^2, N)$ of N distinct labeled points on the unit 2-sphere $\mathbb{S}^2$. One may regard these as the points where N surrounding spheres touch a central sphere. Note that $\mathrm{Conf}(N)$ is an open submanifold of the N-fold product $(\mathbb{S}^2)^N$ of unit spheres. We will also consider the *reduced configuration space*

$$\mathrm{BConf}(N) := \mathrm{Conf}(N)/SO(3),$$

which divides out the space $\mathrm{Conf}(N)$ by the orientation-preserving isometry group $SO(3)$ of the unit 2-sphere $\mathbb{S}^2$ in $\mathbb{R}^3$. The elements of $SO(3)$ move all configurations to isometric configurations, and these moves are permitted on any configuration. The space $\mathrm{Conf}(N)$ is a non-compact $(2N)$-dimensional manifold and the space $\mathrm{BConf}(N)$ is a non-compact $(2N-3)$-dimensional manifold. We assume $N \geq 3$ to avoid degenerate cases.

We denote a configuration $\mathbf{U} := (\mathbf{u}_1, \mathbf{u}_2, ..., \mathbf{u}_N)$, where the $\mathbf{u}_j \in \mathbb{S}^2$ are distinct points. The *angular distance* between points $\mathbf{u}_1, \mathbf{u}_2 \in \mathbb{S}^2$ is the angle $\mathbf{u}_1, 0, \mathbf{u}_2$ subtended at the center of the unit sphere that the N spheres all touch; its value is at most π.

Definition 4.1 The *injectivity radius function* $\rho : \mathrm{Conf}(\mathbb{S}^2, N) \to \mathbb{R}_+$ assigns to a configuration $\mathbf{U} := (\mathbf{u}_1, \mathbf{u}_2, \ldots, \mathbf{u}_N) \in (\mathbb{S}^2)^N$ the value

$$\rho(\mathbf{U}) := \frac{1}{2}\big(\min_{i \neq j} \theta(\mathbf{u}_i, \mathbf{u}_j)\big),$$

where $\theta(\mathbf{u}_i, \mathbf{u}_j)$ denotes the angular distance between $\mathbf{u}_i$ and $\mathbf{u}_j$. In particular $0 < \rho(\mathbf{U}) \leq \frac{\pi}{2}$. Since the function ρ is invariant under the action of $SO(3)$, it yields a well-defined function on $\mathrm{BConf}(N; \theta)$, which we also denote ρ.

Our main topic in this section is the study of spaces which are *superlevel sets* for the injectivity radius function ρ, and how these change at configurations which are critical for maximizing ρ.

Definition 4.2 We define the *(constrained) angular configuration spaces*

$$\mathrm{Conf}(N; \theta) = \mathrm{Conf}(\mathbb{S}^2, N; \theta) := \{\mathbf{U} = (\mathbf{u}_1, ..., \mathbf{u}_N) : \theta(\mathbf{u}_i, \mathbf{u}_j) \geq \theta \text{ for } 1 \leq i < j \leq N\}$$

for angles $0 < \theta \leq \pi$, whose points label configurations of N distinct marked labeled points $\mathbf{u}_i$ which are pairwise at *angular distance* at least θ from each other. Equivalently

$$\mathrm{Conf}(N; \theta) := \{\mathbf{U} = (\mathbf{u}_1, ..., \mathbf{u}_N) : \rho(\mathbf{U}) \geq \frac{\theta}{2}\}.$$

The *reduced (constrained) angular configuration spaces* $\mathrm{BConf}(N;\theta)$ are

$$\mathrm{BConf}(N;\theta) := \mathrm{Conf}(N;\theta)/SO(3).$$

This space is well-defined since rotations preserve angular distance.

The spaces $\mathrm{Conf}(N;\theta)$ and $\mathrm{BConf}(N;\theta)$ are compact topological spaces. Away from critical values θ the spaces $\mathrm{Conf}(N;\theta)$ are closed manifolds with boundary; at critical points they need not be manifolds. The descriptions as superlevel sets show that the spaces $\mathrm{Conf}(N;\theta)$ are ordered by set inclusion as decreasing functions of θ. If $\theta_1 > \theta_2$ then we have

$$\mathrm{Conf}(N;\theta_1) \subset \mathrm{Conf}(N;\theta_2) \subset \mathrm{Conf}(N).$$

We have similar inclusions for $\mathrm{BConf}(N;\theta)$. The interiors of these spaces are

$$\mathrm{Conf}^+(N;\theta) := \left\{\mathbf{U} = (\mathbf{u}_1, \ldots, \mathbf{u}_N) : \rho(\mathbf{U}) > \frac{\theta}{2}\right\}$$

and

$$\mathrm{BConf}^+(N;\theta) := \mathrm{Conf}^+(N;\theta)/SO(3),$$

respectively. They are open manifolds for all values of the parameter θ.

The (constrained) angular configuration space $\mathrm{Conf}(N;\theta)$ can be reparametrized as the *(constrained) radial configuration space*

$$\mathrm{Conf}(N)[r] := \mathrm{Conf}(\mathbb{S}^2, N)[r]$$

which consists of N marked labeled spheres of equal radius $r = r(\theta)$ in $\mathbb{R}^3$ which all touch a given unit radius central sphere $\mathbb{S}^2$, with the N touching points being the labeled points of the configuration. The radius $r(\theta)$ is determined by the condition that the spherical cap on the central sphere $\mathbb{S}^2$ obtained by radial projection of all the points of the given touching sphere has angular diameter exactly θ. A configuration belongs to $\mathrm{Conf}(N;\theta)$ exactly when the N spheres of radius $r(\theta)$ have disjoint interiors. There is a maximal angle $\theta_{max} := \theta_{max}(N)$ for which N spherical caps of that angular measure can fit on the surface of $\mathbb{S}^2$ without overlap of their interiors.

Recall from the proof of Lemma 3.1 that the value $r = r(\theta)$ is implicitly given by the equation

$$\sin\frac{\theta}{2} = \frac{r}{1+r}.$$

For $N \geq 3$ the function $r(\theta)$ is monotone increasing in θ up to a maximal value $r_{max}(\theta)$, so we may use either r or θ to parametrize the family of all spaces $\mathrm{Conf}(N;\theta)$ or $\mathrm{Conf}(N)[r]$. Note that we can identify the configuration spaces $\mathrm{Conf}(N)$ with $\mathrm{Conf}(N;0)$, as well as with $\cup_{\theta>0}\mathrm{Conf}(N;\theta)$ and $\cup_{r>0}\mathrm{Conf}(N)[r]$.

In the following subsections we review the topology and geometry of the spaces $\mathrm{Conf}(N;\theta)$ and $\mathrm{BConf}(N;\theta)$:

- In Sect. 4.1 we describe results on the homotopy type and cohomology of the configuration spaces $\mathrm{Conf}(N)$ and reduced configuration spaces $\mathrm{BConf}(N)$, which are well studied.
- In Sect. 4.2 we use ideas from Morse theory, applied to the injectivity radius function ρ, to study general features of the change in topology for fixed N as the angular parameter θ (or radius parameter r) is increased. The topology changes at certain *critical values* of θ. Since the injectivity radius function ρ is semi-algebraic, for each N we expect there to be a finite set of critical values of ρ on $\mathrm{BConf}(N)$. We give a balancing criterion for a configuration in $\mathrm{BConf}(N)$ to be critical.
- In Sect. 4.3 we show that for small enough θ, the angular configuration space $\mathrm{Conf}(N;\theta)$ has the same homotopy type as $\mathrm{Conf}(N)$ and hence the same cohomology.
- In Sect. 4.4 we treat large θ near θ_{max}, in terms of the radius parameter r. For larger angular diameter θ, the topology of $\mathrm{Conf}(N;\theta)$ may differ drastically from that of $\mathrm{Conf}(N)$. For example, in Table 1 for the Tammes problem many of the maximal configurations are isolated, and the associated labeled spheres cannot be continuously interchanged for large θ near $\theta_{max}(N)$. In such situations, the space $\mathrm{Conf}(N;\theta)$ is disconnected for large θ, while the space $\mathrm{Conf}(N)$ is connected. We conjecture that near $\theta_{max}(N)$ the cohomology of $\mathrm{BConf}(N)[r]$ is concentrated in dimension 0 and discuss the associated Betti number.
- In Sect. 4.5 we show by examples that the set of critical configurations at a critical value can have many connected components and can have variable dimension.
- In Sect. 4.6 we treat the case of topology change as θ varies for the simplest case $N=4$. We determine all the critical values for the injectivity radius function ρ on $\mathrm{BConf}(N)$.
- In Sect. 4.7 we briefly discuss topology change in cases $N\geq 5$. The study of properties of the $N=12$ case is deferred to Sects. 5 and 6.

4.1 Topology of Configuration Spaces

Configuration spaces $\mathrm{Conf}(\mathbb{X},N)$ of N distinct, labeled points on a d-dimensional manifold $\mathbb{X}$ have been studied as fundamental spaces in topology. Recall that the *configuration space* of labeled N-tuples on a manifold $\mathbb{X}$ is

$$\mathrm{Conf}(\mathbb{X},N) := \{(x_1,x_2,...,x_N)\in\mathbb{X}^N : x_i\neq x_j \text{ if } i\neq j\}.$$

The symmetric group Σ_N acts freely on the space $\mathrm{Conf}(\mathbb{X},N)$ to permute the points, and

$$B(\mathbb{X},N) := \mathrm{Conf}(\mathbb{X},N)/\Sigma_N$$

is the configuration space of *unlabeled* (i.e. unordered) N-tuples of points on $\mathbb{X}$. Configuration spaces of this type were first considered in the 1960s by Fadell and Neuwirth [28, 30]. The state of the art for $\mathbb{X} = \mathbb{R}^d$ and $\mathbb{S}^d$ as of 2000 is given in Fadell and Husseini [29]. Other useful references are Totaro [96] and Cohen [17].

We begin with the most well-known of these spaces, the configuration space $\mathrm{Conf}(\mathbb{R}^2, N)$ of labeled N-tuples of points on $\mathbb{X} = \mathbb{R}^2$, the plane. Fadell and Neuwirth [30] showed that the unlabeled configuration space $B(\mathbb{R}^2, N)$ is a classifying space (Eilenberg-Maclane space $K(\pi, 1)$), with fundamental group $\pi_1 = \pi$ isomorphic to Artin's *braid group* B_N on N strings. Thus the cohomology of $B(\mathbb{R}^2, N)$ is just the cohomology of B_N; it was computed by Fuks [41] and Cohen [16].

The labeled configuration space $\mathrm{Conf}(\mathbb{R}^2, N)$ is by definition the complement of a finite set of complex hyperplanes given by $x_i = x_j$ for $i \neq j$ in $\mathbb{C}^N \cong (\mathbb{R}^2)^N$. This arrangement is sometimes called the (complexified) A_{N-1}-arrangement of hyperplanes (see Postnikov and Stanley [84]), where A_{N-1} refers to a Coxeter group. The space $\mathrm{Conf}(\mathbb{R}^2, N)$ is also a classifying space with fundamental group equal to the *pure braid group* PB_N, the subgroup of B_N consisting of all N-strand braids which induce the identity permutation. It sits in a short exact sequence $0 \to PB_N \to B_N \to \Sigma_N \to 0$. The rational cohomology of $\mathrm{Conf}(\mathbb{R}^2, N)$ is then the cohomology of the pure braid group; the cohomology ring structure of PB_N was determined by Arnold [6] in 1969.

The *Betti numbers* of a topological space are the ranks of its homology groups (which equal the ranks of its cohomology groups, with coefficients in a field, here $\mathbb{Q}$ or $\mathbb{C}$.) The generating function for this sequence of ranks is called the *Poincaré polynomial.* Arnold [6, Corollary 2] determined the Poincaré polynomial for the pure braid group PB_N on N strands to be

$$P_N(t) = (1+t)(1+2t)\cdots(1+(N-1)t). \tag{4.1}$$

Table 3 gives these Betti numbers for small N. They are of combinatorial interest, being unsigned Stirling numbers of the first kind,

$$\dim H^k(PB_N, \mathbb{Q}) = \left[{N \atop N-k} \right], \quad \text{for} \quad 0 \le k \le N-1$$

(see [45, Sect. 6.1]).

Our interest here is configuration spaces on

$$\mathbb{X} = \mathbb{S}^2 = \{\mathbf{u} = (x, y, z) : \mathbf{u} \cdot \mathbf{u} = x^2 + y^2 + z^2 = 1\},$$

the unit 2-sphere embedded in $\mathbb{R}^3$. The configuration spaces of $\mathbb{S}^2$ have a close relationship to those of $\mathbb{R}^2$, since $\mathbb{S}^2$ is the one-point compactification of $\mathbb{R}^2$. Note also that $\mathbb{S}^2$ is homeomorphic to $\mathbb{CP}^1$, the complex projective line. Tautologically the configuration space $\mathrm{Conf}(\mathbb{S}^2, 1) \cong \mathbb{S}^2$; and the space $\mathrm{Conf}(\mathbb{S}^2, 2)$ is homeomorphic to an $\mathbb{R}^2$-bundle over $\mathbb{S}^2$, hence homotopy equivalent to $\mathbb{S}^2$ (see [17, Example 2.4]).

Table 3 Betti numbers of pure braid group cohomology $H^k(PB_N, \mathbb{Q}) \cong H^k(\mathrm{BConf}(\mathbb{R}^2, N), \mathbb{Q})$

	k								
N	0	1	2	3	4	5	6	7	8
1	1	0	0	0	0	0	0	0	0
2	1	1	0	0	0	0	0	0	0
3	1	3	2	0	0	0	0	0	0
4	1	6	11	6	0	0	0	0	0
5	1	10	35	50	24	0	0	0	0
6	1	15	85	225	274	120	0	0	0
7	1	21	175	735	1624	1764	720	0	0
8	1	28	322	1960	6769	13132	13068	5040	0
9	1	36	546	4536	22449	67284	118124	109584	40320

For $N \geq 3$ points we have a well-known result, given as follows in Feichtner and Ziegler [32, Theorem 1].

Theorem 4.3 *For $N \geq 3$ the configuration space $\mathrm{Conf}(\mathbb{S}^2, N)$ of N distinct labeled points on the 2-sphere is the total space of a trivial $PSL(2, \mathbb{C})$-bundle over $\mathcal{M}_{0,N}$, the moduli space of conformal structures on the N-punctured complex projective line, modulo conformal automorphisms. Hence there is a homeomorphism*

$$\mathrm{Conf}(\mathbb{S}^2, N) \cong PSL(2, \mathbb{C}) \times \mathcal{M}_{0,N}.$$

Note that $PSL(2, \mathbb{C})$ is homotopy equivalent to its maximal compact subgroup $SO(3)$, the group of orientation preserving isometries of $\mathbb{S}^2$.

The $SO(3)$-action on $\mathrm{Conf}(\mathbb{S}^2, N)$ permits us to rotate the first point to the north pole, from which we stereographically project the rest of the unit sphere to the plane $\mathbb{R}^2$. Since we are still free to rotate about the north pole, which corresponds to $SO(2)$ rotations in the plane, we can identify $\mathrm{Conf}(\mathbb{R}^2, N-1)/SO(2)$ with the reduced N-configuration space $\mathrm{BConf}(\mathbb{S}^2, N)$. The action of $SO(2)$ on $\mathrm{Conf}(\mathbb{R}^2, N-1)$ is free if $N \geq 3$, so we can regard $\mathrm{Conf}(\mathbb{R}^2, N-1)$ as a principal $SO(2)$-bundle over $\mathrm{BConf}(\mathbb{S}^2, N)$.

This principal bundle has a section, and thus is a product bundle, so the Poincaré polynomial $\tilde{P}_N(t)$ of the base $\mathrm{BConf}(\mathbb{S}^2, N)$ may be computed as the quotient of the well-known Poincaré polynomials

$$P(t) = P_{N-1}(t) = \big(1+t\big)\big(1+2t\big)\cdots\big(1+(N-2)t\big) \tag{4.2}$$

for $\mathrm{Conf}(\mathbb{R}^2, N-1)$ from Eq. (4.1) and $p(t) = (1+t)$ for $SO(2)$. It follows that $\mathrm{BConf}(\mathbb{S}^2, N)$ and $\mathcal{M}_{0,N}$ both have Poincaré polynomial

$$\tilde{P}_N(t) = P(t)/p(t) = \big(1+2t\big)\cdots\big(1+(N-2)t\big). \tag{4.3}$$

By taking the alternating sum of each row, or more directly by evaluating $P_N(-1)$, we can compute the Euler characteristic

$$\chi(\mathrm{BConf}(N)) = (-1)^{N-3}(N-3)! \tag{4.4}$$

of $\mathrm{BConf}(N)$ for $N \geq 3$.

We also have [32, Proposition 2.3]:

Theorem 4.4 (Feichtner–Ziegler (2000)) *For $N \geq 3$ the moduli space $\mathcal{M}_{0,N}$ is homotopy equivalent to the complement of the affine complex braid arrangement of hyperplanes $\mathcal{M}({}^{aff}\mathcal{A}^{\mathbb{C}}_{N-2})$ of rank $N-2$, since*

$$\mathcal{M}_{0,N} \times \mathbb{C} \simeq \mathcal{M}({}^{aff}\mathcal{A}^{\mathbb{C}}_{N-2}).$$

Its integer cohomology algebra is torsion-free. It is generated by 1-dimensional classes $e_{i,j}$ with $1 \leq i < j \leq N-1$ with $(i,j) \neq (1,2)$ and has a presentation as an exterior algebra

$$H^*(\mathcal{M}({}^{aff}\mathcal{A}^{\mathbb{C}}_{N-2})) \cong \Lambda^* \mathbb{Z}^{\binom{N-1}{2}-1}/\mathcal{I},$$

where the ideal $\mathcal{I}$ is generated by elements

$$e_{1,i} \wedge e_{2,i}, \quad 2 < i \leq N-1$$

and

$$e_{i,\ell} \wedge e_{j,\ell} - e_{i,j} \wedge e_{j,\ell} + e_{i,j} \wedge e_{i,\ell} \quad 1 \leq i < j < \ell \leq N-1,\ (i,j) \neq (1,2).$$

Here the complexified A_{N-2}-arrangement of hyperplanes $\mathcal{A}^{\mathbb{C}}_{N-2}$ of rank $N-2$ is cut out by the hyperplanes

$$z_i - z_j = 0 \quad 1 \leq i < j \leq N-1.$$

Its complement $\mathcal{M}(\mathcal{A}^{\mathbb{C}}_{N-2}) := \mathbb{C}^{N-1} \setminus \bigcup \mathcal{A}^{\mathbb{C}}_{N-2}$ is homeomorphic to $\mathrm{Conf}(\mathbb{C}, N-1)$. The associated affine arrangement is:

$${}^{aff}\mathcal{A}^{\mathbb{C}}_{N-2} := \{(z_1, z_2, \cdots, z_{N-1} \in \mathcal{A}^{\mathbb{C}}_N : z_2 - z_1 = 1\}.$$

Treating $\mathbb{C}^{N-2} \cong \{(z_1, z_2, \cdots, z_{N-1}) : z_2 - z_1 = 1\}$, we set

$$\mathcal{M}({}^{aff}\mathcal{A}^{\mathbb{C}}_{N-2}) := \mathbb{C}^{N-2} \setminus {}^{aff}\mathcal{A}^{\mathbb{C}}_{N-2}.$$

A more refined result determines the integral cohomology ring for the configuration spaces of spheres, which includes torsion elements. It was determined by Feichtner and Ziegler, who obtained in the special case of $\mathrm{Conf}(\mathbb{S}^2, N)$ the following result (see [32, Theorem 2.4]).

Theorem 4.5 (Feichtner–Ziegler (2000)) *For $N \geq 3$, the integer cohomology ring $H^*(\mathrm{Conf}(\mathbb{S}^2, N), \mathbb{Z})$ has only 2-torsion. It is given as*

$$H^*(\mathrm{Conf}(\mathbb{S}^2, N), \mathbb{Z}) \cong (\mathbb{Z}(0) \oplus \mathbb{Z}/2\mathbb{Z}(2) \oplus \mathbb{Z}(3))) \otimes \Lambda^*(\oplus_{i=1}^{\binom{N-1}{2}-1} \mathbb{Z}(1))/\mathcal{I},$$

in which $\mathcal{I}$ is the ideal of relations given in Theorem 4.4.

In this result the expression $G(i)$ denotes a direct summand of G in cohomology of degree i, e.g. there is a $\mathbb{Z}/2\mathbb{Z}$ direct summand in $H^2(\mathrm{Conf}(\mathbb{S}^2, N), \mathbb{Z})$.

4.2 Generalized Morse Theory and Topology Change

Morse theory, as treated in Milnor [74], concerns how topology changes for the *sublevel sets*

$$\mathbb{U}^t := \{u \in \mathbb{U} : f(u) \leq t\}$$

of a given, sufficiently nice, real-valued function f on a manifold $\mathbb{U}$, as the level set parameter t varies. At the *critical values* of the function, where its gradient vanishes, the topology changes. This change can be described by adding up the contributions of individual *critical points* of the function that occur at the critical values. More precisely, a *Morse function* is a smooth enough function that has only isolated *critical points*, each of which is non-degenerate, and arranged so that only one critical point occurs at each critical level $f(u) = t$. Here *non-degenerate* means that the function f is twice-differentiable and its Hessian matrix $[\frac{\partial^2 f}{\partial u_i \partial u_j}]$ is nonsingular at the critical point. The topology of a sublevel set $\mathbb{U}^t$ is changed as t ascends past a critical value, up to homotopy, by attaching a cell of dimension equal to the *index* of the critical point: the number of negative eigenvalues of the Hessian.

Our interest here will be in *superlevel sets*

$$\mathbb{U}_s := \{u \in \mathbb{U} : f(u) \geq s\},$$

whose topology changes as s descends past a critical value by attaching a cell of dimension equal to the *co-index* of the critical point: the number of positive eigenvalues of the Hessian.

In the 1980s Goresky and MacPherson [44] developed Morse theory on more general topological spaces than manifolds, namely *stratified spaces* in the sense of Whitney [98], and applicable to a wider class of real-valued functions. The configuration spaces such as $\mathrm{Conf}(N; \theta)$ studied here are in general stratified spaces in Whitney's sense, because viewed using the r-parameter they are real semi-algebraic varieties.

For the case at hand of $\mathbb{U} = \mathrm{Conf}(\mathbb{S}^2, N)$ and the injectivity radius function ρ, we have a further problem that ρ is not a Morse function. Its critical points are degenerate and non-isolated, and even the notion of "critical" needs care in defining, since ρ is a min-function of a finite number of smooth functions (see Definition 4.1). Technically, the angular distance function from $\mathbf{u}$ is not smooth at the antipodal point $-\mathbf{u}$, with angular distance π on $\mathbb{S}^2$; however we can treat these functions as if they were smooth using the following trick, valid for the nontrivial cases $N \geq 3$ where $\rho_{\max} \leq \frac{\pi}{3}$: simply include the constant function $\frac{\pi}{3}$ among those functions over which we take the min, and smoothly cut off the other pairwise angular distance functions $\theta(\mathbf{u}_i, \mathbf{u}_j)$ if they exceed $\frac{2\pi}{3}$.

An appropriate version of Morse theory that applies in this context, called *min-type Morse theory*, has only recently been sketched by Gershkovich and Rubinstein [43] (see also Baryshnikov et al. [9]). Related work includes Carlsson et al. [15] and Alpert [2]. The treatment of [9] studies a notion of topologically critical value.

In what follows we develop an alternative max-min approach to criticality and a Morse theory for the injectivity radius function ρ on configurations that is in the spirit of the criticality theory for maximizing *thickness* or normal injectivity radius (also known as *reach*) on configurations of curves subject to a length constraint (or in a compact domain of $\mathbb{R}^3$, or in $\mathbb{S}^3$) studied earlier in optimal ropelength and rope-packing problems by Cantarella et al. [14]. This approach provides a notion of critical configuration, refining the notion of a critical value. The Farkas Lemma (and its infinite-dimensional generalizations in the case of the ropelength problem) is a key tool used in these works that relates criticality to the existence of a balanced system of forces on the configuration. A more detailed treatment is planned in [64].

To understand criticality for the injectivity radius function ρ on $\mathrm{Conf}(N) = \mathrm{Conf}(\mathbb{S}^2; N)$, we first need to make sense of varying a configuration $\mathbf{U} = (\mathbf{u}_1, ..., \mathbf{u}_N) \in \mathrm{Conf}(N) \subset (\mathbb{S}^2)^N$ along a tangent vector $\mathbf{V} = (\mathbf{v}_1, ..., \mathbf{v}_N)$ to $\mathrm{Conf}(N)$ at $\mathbf{U}$; here $\mathbf{v}_i$ is a tangent vector to $\mathbb{S}^2$ at $\mathbf{u}_i$, for $i = 1, 2, ..., N$. For sufficiently small t we can define a nearby configuration

$$\mathbf{U}\#t\mathbf{V} := \left(\frac{\mathbf{u}_1 + t\mathbf{v}_1}{|\mathbf{u}_1 + t\mathbf{v}_1|}, ..., \frac{\mathbf{u}_N + t\mathbf{v}_N}{|\mathbf{u}_N + t\mathbf{v}_N|}\right) \in \mathrm{Conf}(N) \subset (\mathbb{S}^2)^N$$

by translating and projecting each factor back to $\mathbb{S}^2$. In particular, the $\mathbf{V}$-directional derivative f' of a smooth function f on $\mathrm{Conf}(N)$ at $\mathbf{U}$ is simply $f' = \frac{d}{dt}|_{t=0} f(\mathbf{U}\#t\mathbf{V})$, so $\mathbf{U}$ is a critical point for smooth f provided all its $\mathbf{V}$-directional derivatives vanish at $\mathbf{U}$; this means that the increment $f(\mathbf{U}\#t\mathbf{V}) - f(\mathbf{U}) = o(t)$, where $o(t)$ is a function which tends to 0 faster than linearly.

Remark The operation taking $\mathbf{U}$ to $\mathbf{U}\#t\mathbf{V}$ can be thought of as the spherical analog of translating $\mathbf{U}$ by $t\mathbf{V}$ via vector addition in the linear case, hence the suggestive sum notation. The map taking $t\mathbf{V}$ to $\mathbf{U}\#t\mathbf{V}$ approximates (to within $o(t^2)$) the exponential map at $\mathbf{U}$.

Now we make precise "max-min criticality" for the injectivity radius function ρ.

Definition 4.6 A configuration $\mathbf{U} = (\mathbf{u}_1, ..., \mathbf{u}_N) \in \mathrm{Conf}(N)$ is *critical for maximizing* ρ provided for every tangent vector $\mathbf{V} = (\mathbf{v}_1, ..., \mathbf{v}_N)$ to $\mathrm{Conf}(N)$ at $\mathbf{U}$ we have, as $t \to 0$,

$$[\rho(\mathbf{U}\#t\mathbf{V}) - \rho(\mathbf{U})]_+ = o(t),$$

where $[g]_+ = \max\{g, 0\}$ denotes the positive part of g. Equivalently, a configuration $\mathbf{U}$ is critical if *no* variation $\mathbf{V}$ can *increase* ρ to first order.

Otherwise, a configuration $\mathbf{U}$ is *regular*, that is, there exists a variation $\mathbf{V}$ which *does* increase ρ to first order, and so, by the definition of ρ as a min-function, this means that for all pairs $(\mathbf{u}_i, \mathbf{u}_j)$ realizing the minimal angular distance $\theta(\mathbf{u}_i, \mathbf{u}_j) = \theta_o$, their distances increase to first order under the variation $\mathbf{V}$ as well. Note that the set of regular configurations is open. If each configuration in this $\{\rho = \frac{\theta_o}{2}\}$-level set is regular, then this level is *topologically regular:* that is, there a deformation retraction from $\mathrm{Conf}(N; \frac{\theta_o}{2} - \varepsilon)$ to $\mathrm{Conf}(N; \frac{\theta_o}{2} + \varepsilon)$ for some $\varepsilon > 0$ (see [9, Lemmas 3.2, 3.3 and Corollary 3.4]).

Definition 4.7 For $\mathbf{U} \in \mathrm{Conf}(N; \theta)$, the *contact graph* of $\mathbf{U}$ is the graph embedded in $\mathbb{S}^2$ with vertices given by points $\mathbf{u}_i$ in $\mathbf{U}$ and edges given by the geodesic segments $[\mathbf{u}_i, \mathbf{u}_j]$ when $\theta(\mathbf{u}_i, \mathbf{u}_j) = \theta$.

Examples of contact graphs for extremal values of the Tammes problem were given in Fig. 5 of Sect. 3.

Definition 4.8 A *stress graph* for $\mathbf{U} \in \mathrm{Conf}(N; \theta)$ is a contact graph with nonnegative weights w_e on each geodesic edge $e = [\mathbf{u}_i, \mathbf{u}_j]$.

A stress graph gives rise to a system of *tangential forces* associated to each geodesic edge $e = [\mathbf{u}_i, \mathbf{u}_j]$ of the contact graph. These forces have magnitude w_e, are tangent to $\mathbb{S}^2$ at each point $\mathbf{u}_i$ of $\mathbf{U}$, and are directed along the outward unit tangent vectors $T_e|_{\mathbf{u}_i}, T_e|_{\mathbf{u}_j}$ to the edge e at its endpoints $\mathbf{u}_i, \mathbf{u}_j$, respectively.

Definition 4.9 A stress graph is *balanced* if the vector sum of the forces in the tangent space of $\mathbb{S}^2$ at $\mathbf{u}_i$ is zero for all points of $\mathbf{U}$. A configuration $\mathbf{U}$ is *balanced* if its underlying contact graph has a balanced stress graph for some choice of nonnegative, not-everywhere-zero weights on its edges.

Theorem 4.10 *To each critical value $\frac{\theta}{2}$ for the injectivity radius ρ, there exists a balanced configuration $\mathbf{U}$ with $\rho(\mathbf{U}) = \frac{\theta}{2}$. The vertices of the contact graph are a subset of the points in $\mathbf{U}$ and the geodesic edges of the contact graph all have length θ.*

Proof As in [9, Corollary 3.4 and Eq. 2], since ρ is a min-function on $\mathrm{Conf}(N) \subset (\mathbb{S}^2)^N$, if $\frac{\theta}{2}$ is not a topologically regular value of ρ, then some configuration $\mathbf{U} \in \rho^{-1}(\frac{\theta}{2})$ is balanced. Because $\rho(\mathbf{U}) = \frac{\theta}{2}$, the conditions on the vertices and edge lengths are clearly met. □

We now prove a converse result.

Theorem 4.11 *If a configuration* $\mathbf{U}$ *on* $\mathbb{S}^2$ *is balanced, then* $\mathbf{U}$ *is critical for maximizing the injectivity radius* ρ.

We will need a preliminary lemma. Consider a planar graph G embedded on the unit sphere $\mathbb{S}^2$ via a map $\mathbf{u} : G \to \mathbb{S}^2$ which is C^2 on the edges of G. (By slight abuse of notation, a point on its image in $\mathbb{S}^2$ may also be denoted by $\mathbf{u}$.) Suppose each edge e of G is assigned a *nonzero* weight $w_e \in \mathbb{R}$. Let $L_e(\mathbf{u})$ denote the length of edge $\mathbf{u}(e)$ induced by the map $\mathbf{u}$, and let $\mathcal{L}(\mathbf{u}) = \sum w_e L_e(\mathbf{u})$ be the total *weighted length* of the embedded graph $\mathbf{u}(G)$. We can vary the map $\mathbf{u}$ using a C^2 vector field $\mathbf{v}$, just as we varied a configuration: for sufficiently small t, each point $\mathbf{u} \in \mathbb{S}^2$ on the image of the graph is moved to $\mathbf{u}\#t\mathbf{v} = \frac{\mathbf{u}+t\mathbf{v}}{|\mathbf{u}+t\mathbf{v}|}$. Let $\mathcal{L}'(\mathbf{v})$ denote the first derivative at $t = 0$ of weighted length for this varied graph, i.e. the *first variation* of $\mathcal{L}(\mathbf{u})$ along $\mathbf{v}$.

Lemma 4.12 *The first variation* $\mathcal{L}'(\mathbf{v})$ *of the weighted lengh* $\mathcal{L}(\mathbf{u})$ *for the embedded graph* $\mathbf{u}(G)$ *vanishes for every vector field* $\mathbf{v}$ *on* $\mathbb{S}^2$ *if and only if the following two conditions hold:*

(1) *each edge* e *joining a pair of vertices* e^-, e^+ *of* G *maps to a geodesic arc* $\mathbf{u}(e) = [\mathbf{u}(e^-), \mathbf{u}(e^+)] = [\mathbf{u}^-, \mathbf{u}^+]$ *in the embedded graph* $\mathbf{u}(G)$*;*

(2) *at any vertex* $\mathbf{u}$ *of the embedded graph* $\mathbf{u}(G)$*, the weighted sum* $\sum w_{e^*} T_{e^*}|_{\mathbf{u}} = 0$*, where the sum is taken over the subset of edges* $\mathbf{u}(e^*)$ *incident to* $\mathbf{u}$*, and where* $T_{e^*}|_{\mathbf{u}}$ *is the outer unit tangent vector of* $\mathbf{u}(e^*)$ *at* $\mathbf{u}$.

Proof This lemma is a direct consequence of the first variation of length formula

$$L_e'(\mathbf{v}) = \mathbf{v} \cdot T|_{\mathbf{u}(e^-)}^{\mathbf{u}(e^+)} - \int_{\mathbf{u}(e)} \mathbf{v} \cdot \mathbf{k}$$

(see, for example, Hicks [55, Chap. 10, Theorem 7, p. 148]). Here T is the unit tangent vector field of the edge $\mathbf{u}(e)$, and $\mathbf{k}$ is the geodesic curvature vector of $\mathbf{u}(e)$; with respect to any local arclength parameter on $\mathbf{u}(e)$, the geodesic curvature vector is the projection to $\mathbb{S}^2$ of the acceleration: $\mathbf{k} = \ddot{\mathbf{u}} + \mathbf{u}$, which is tangent to $\mathbb{S}^2$ and normal to $\mathbf{u}(e)$, and which vanishes iff $\mathbf{u}(e)$ is a geodesic arc.

Now express $\mathcal{L}'(\mathbf{v}) = \sum w_e L_e'(\mathbf{v})$ as a sum of edge terms and vertex terms. The geodesic arc condition (1) – that $\mathbf{k} = 0$ along every edge – implies the edge terms in $\mathcal{L}'(\mathbf{v})$ all vanish for any variation $\mathbf{v}$ of the map $\mathbf{u}$; and the force balancing condition (2) implies all vertex terms vanish for any variation $\mathbf{v}$.

Conversely, given any interior image point $\mathbf{u}$ of an edge, take a variation $\mathbf{v}$ supported in an arbitrarily small neighborhood of $\mathbf{u}$, and orthogonal to $\mathbf{u}(e)$ at $\mathbf{u}$: the vanishing of $\mathcal{L}'(\mathbf{v})$ implies condition (1) that $\mathbf{k} = 0$; similarly, at any given vertex $\mathbf{u}$, consider a pair of variations $\mathbf{v}_1, \mathbf{v}_2$ supported in an arbitrarily small neighborhood of $\mathbf{u}$ which approximate an orthogonal pair of translations of the tangent space to $\mathbb{S}^2$ at $\mathbf{u}$: the vanishing of $\mathcal{L}'(\mathbf{v})$ for both of these $\mathbf{v}_1, \mathbf{v}_2$ implies the forces balance (2). □

Remark In case $w_e = 0$, vanishing for the first variation of $\mathcal{L}$ does not imply $\mathbf{u}(e)$ is a geodesic arc: instead, the edges with nonzero weights form a balanced geodesic subgraph of the original embedded graph $\mathbf{u}(G)$.

Lemma 4.12 suggests the following definition.

Definition 4.13 An embedded graph satisfying properties (1) and (2) is called a *balanced geodesic graph.* (Note that there is no requirement here that the geodesic edge lengths are integer multiples of some basic length, as would be the case for a contact graph.)

Lemma 4.12 shows that a balanced geodesic graph has vanishing first variation of weighted length $\mathcal{L}$, even if some of its edge weights w_e are zero.

Proof of Theorem 4.11 By hypothesis, there are non-negative edge weights (not all zero) so that the resulting stress graph $\mathbf{u}(G)$ for the configuration $\mathbf{U}$ is balanced. By Lemma 4.12 the first variation $\mathcal{L}'(\mathbf{v})$ of weighted length for $\mathbf{u}(G)$ vanishes for all variation vector fields $\mathbf{v}$ on $\mathbb{S}^2$.

Suppose (to the contrary) that $\mathbf{U}$ were *not* critical for maximizing the injectivity radius ρ. Then there would be a variation $\mathbf{V}$ of $\mathbf{U}$ so that every geodesic edge of the stress graph has length increasing at least linearly in $\mathbf{V}$. Extend $\mathbf{V}$ to an ambient C^2 variation vector field $\mathbf{v}$ on $\mathbb{S}^2$. Since the edge weights are *nonnegative*, and not all zero, that implies the weighted length of the stress graph also increases at least linearly in $\mathbf{v}$, a contradiction. □

Remark A key property of balanced configurations is that for each $N \geq 3$ the set of radii r such that BConf(N) contains a balanced configuration of injectivity radius $\frac{\theta(r)}{2}$ is finite. It follows that *the set of critical radius values for* BConf(N) *is finite.*

This finiteness result can be proved using the structure of the spaces BConf$(N)[r]$ as real semi-algebraic sets, which we consider in [64]. We will assume this finiteness result holds in the discussions in Sect. 4.4; it can be directly verified for small N.

4.3 *Small Radius Case*

For small radii, it is convenient to state results for $r = r(\theta)$ in terms of the angle parameter θ. For sufficiently small angles, the superlevel sets Conf$(N; \theta)$ will have the same homotopy type as the full configuration space Conf(N). In terms of the radius function, the conclusion of this result applies for $0 \leq r < r_1(N)$, where $r_1(N) = \sin\left(\frac{\pi}{N}\right) / \left(1 - \sin\left(\frac{\pi}{N}\right)\right)$ is the smallest critical value for Conf$(N)[r]$.

Theorem 4.14 *Suppose* $N \geq 3$. *The smallest critical value for maximizing* ρ *on* BConf(N) *is* $\frac{\pi}{N}$, *achieved uniquely by the* N*-Ring configuration of equally spaced points along a great circle. Moreover, for angular diameter* $0 \leq \theta < \frac{2\pi}{N}$ *the following hold.*

(1) *The space* Conf$(N; \theta)$ *is a strong deformation retract of the full configuration space* Conf$(N) =$ Conf$(N; 0)$.

(2) *The reduced space* BConf$(N; \theta) =$ Conf$(N; \theta)/SO(3)$ *is a strong deformation retract of the full reduced configuration space* BConf(N).

Consequently each has, respectively, the same homotopy type and cohomology groups as the corresponding full configuration space.

Proof This result corresponds to [9, Theorem 5.1]. First note that by using equal weights on each of its edges, the N-Ring is balanced and hence a critical configuration by Theorem 4.11. The balanced contact graph on $\mathbb{S}^2$ of a $\frac{\theta}{2}$-critical N-configuration has geodesic edges with angular length θ. In order to balance, its total angular length must be at least 2π, the length of a complete great circle. Thus if $\theta < \frac{2\pi}{N}$, then the total length $N\theta < 2\pi$ and there is no balanced N-configuration in $\mathrm{Conf}(N;\theta)$ and $\frac{\theta}{2}$ is not a critical value for ρ. In this case, a weighted ρ-subgradient-flow provides the strong deformation retraction of $\mathrm{Conf}(N)$ to $\mathrm{Conf}(N;\theta)$. □

Corollary 4.15 *For $\theta < \frac{2\pi}{N}$ and $N \geq 4$ the configuration spaces* $\mathrm{Conf}(N;\theta)$ *and* $\mathrm{BConf}(N;\theta)$ *are path-connected, but not simply-connected.*

Proof These spaces have the same homotopy type as $\mathrm{Conf}(N)$ (resp. $\mathrm{BConf}(N)$), which is connected since $H^0(\mathrm{Conf}(N), \mathbb{Q}) = \mathbb{Q}$ (resp. $H^0(\mathrm{BConf}(N), \mathbb{Q}) = \mathbb{Q}$). They each are closures of open manifolds and are connected, so are path-connected. We have $H^1(\mathrm{BConf}(n), \mathbb{Q}) = \mathbb{Q}^k$ for some $k = k(N) \geq 2$, using the formula (4.3) applied for $N \geq 4$, so $\mathrm{BConf}(N)$ is not simply-connected. Finally, $\mathrm{Conf}(N)$ is not simply connected via the product decomposition in Theorem 4.3. □

4.4 Large Radius Case

We consider reduced configuration spaces $\mathrm{BConf}(N)[r]$ having radius parameter r sufficiently close to $r_{max}(N)$, depending on N. Using the finiteness of the set of critical values, there is an $\epsilon(N) > 0$ such that the upward "gradient flow" of the injectivity radius function ρ (or of the corresponding touching-sphere radius function r) defines a deformation retraction from $\mathrm{BConf}(N)[r]$ to $\mathrm{BConf}(N)[r_{max}(N)]$ for the range $r_{\max}(N) - \epsilon(N) < r < r_{max}(N)$.

The simplest topology that may occur at $r_{max}(N)$ is where $\mathrm{BConf}(N)[r_{\max}(N)]$ has all its connected components contractible; the property holds for most small N – in fact, for all $N \leq 14$ except $N = 5$. When it holds, the cohomology groups for $\mathrm{BConf}(N)[r]$ in this range of r will have the following very simple form:

Purity Property. *There is some $\epsilon = \epsilon(N) > 0$ such that for*

$$r_{\max}(N) - \epsilon(N) < r < r_{max}(N)$$

there is an integer $s = s(N) \geq 1$ such that the cohomology groups of the reduced configuration space are

$$H^k(\mathrm{BConf}(N)[r], \mathbb{Z}) = \begin{cases} \mathbb{Z}^{s(N)} & \text{if } k = 0 \\ 0 & \text{if } k \geq 1. \end{cases}$$

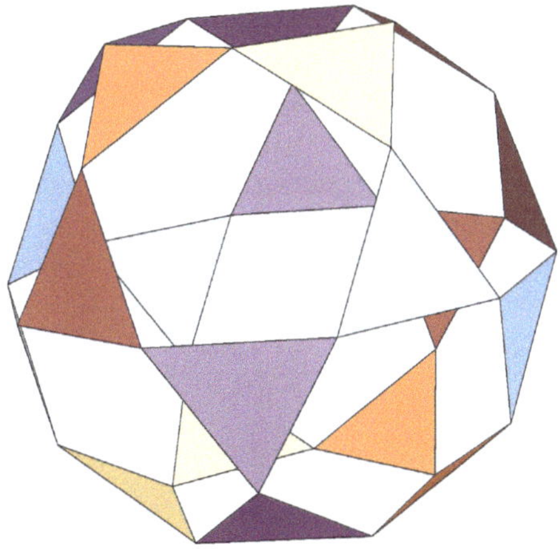

Fig. 6 The r-maximal stratified set for $N = 5$

For $N = 5$ the cohomology does *not* have the Purity Property. The reduced configuration space $\mathrm{BConf}(5)[r]$ is 7-dimensional for $r < r_{max}(5)$ but becomes 2-dimensional at $r = r_{max}(5)$. Some optimal maximum radius configurations at $r_{max}(5) = 1 + \sqrt{2}$ have room for an extra sphere (giving $N = 6$): the sphere centers form five vertices of an octahedron, and either vertex in an antipodal pair of vertices can freely and independently move towards the unoccupied sixth vertex of the octahedron. The resulting reduced configuration space $\mathrm{BConf}(5)[1 + \sqrt{2}] = \mathrm{BConf}(5; \frac{\pi}{2})$ is a simplicial 2-complex which is not contractible; it is pictured schematically in Fig. 6. It has a single connected component having Euler characteristic $\chi(\mathrm{BConf}(5; \frac{\pi}{2})) = -10$. For further discussion of this space as a critical stratified set, see Sect. 4.5.

Does the purity property hold for all or most large N? We do not know. One might expect that extremal configurations for high values of N at $r = r_{max}(N)$ will have most spheres are held in a rigid structure, and for r near it all individual spheres will only be able to move in a tiny area around them, each contributing a connected component to the reduced configuration space. Against this expectation, computer experiments packing N equal-radius two-dimensional disks confined to a unit disk suggest the possibility for some N that extremal configurations could have *rattlers*, which are loose disks that have motion permitted even at $r = r_{max}$ (Lubachevsky and Graham [69]). However, even with rattlers one could still have contractibility of individual connected components. The hypothesis of extremal configurations being rigid (and unique) is known to hold for $6 \leq N \leq 12$.

When the purity property holds one can (in principle) determine the number of connected components for the set of near-maximal configurations; call it $s = s(N)$. This value depends on the symmetries of each maximal configuration under the $SO(3)$ action. Denoting the isomorphism types of the connected components of maximal rigid (labeled) configurations of N points at $r = r_{max}(N)$ by $C_{i,N}$ for $1 \leq i \leq e(N)$, one would have

$$s(N) = \sum_i \frac{N!}{|Aut(C_{i,N})|}.$$

For $3 \le N \le 12$, excluding $N \ne 5$, the extremal configurations for the Tammes problem are known to be unique up to isometry; call them $C_{1,N}$. The analysis of Danzer given in Theorem 3.3 covers the cases $7 \le N \le 10$. For the case $N = 12$, the unique extremal configuration DOD of vertices of an icosahedron has $Aut(C_{1,12}) = A_5$, the alternating group, of order 60, whence $s(12) = \frac{12!}{60} = 7983360$.

4.5 Structure of Critical Strata

Connected components of critical strata necessarily have dimension at least three from the $SO(3)$-action. In what follows we consider reduced critical strata that are quotients by this action. At a critical value ρ there can be several disconnected reduced critical strata, and such strata can have positive dimension. We give examples of each.

For $N = 5$ a positive dimensional reduced critical stratified set occurs at the maximal radius value $r_{max}(5) = 1 + \sqrt{2}$. The set of (reduced) critical configurations forms a family, which is two-dimensional, containing multiple strata. A generic contact graph at the maximal injectivity radius $\rho = \frac{\pi}{4}$ is a Θ-graph having 2 polar vertices and 3 equatorial vertices. This contact graph, depicted in Fig. 5, has 3 faces and 6 edges and is optimal. The three angles between equatorial vertices can range between $\frac{\pi}{2}$ and π, with the condition that their sum is 2π, defining a 2-simplex. As long as none of the equatorial angles is π, criticality is achieved using weights that are non-zero on all the edges. When an equatorial angle is π, corresponding to a corner of the 2-simplex, these equatorial vertices may be regarded as a new pair of polar vertices. In this configuration, as the angles between equatorial angles go to π, some weights of the stress graph can go to 0 and the support of the weights degenerates to a 4-Ring. The limit contact graph consists of the edges of a square pyramid whose base is that 4-Ring. This gives a non-optimal contact graph with 5 faces and 8 edges.

For $N = 12$ there are several distinct reduced critical strata at the critical value $\rho = 1$, two of which correspond to the FCC-configuration and HCP-configurations, singled out in Frank's discussion in Sect. 2.7. These configurations are defined in Sect. 5.2 below, and their criticality is shown in Theorem 5.3.

4.6 Topology Change for Variable Radius: $N = 4$

For very small N it is possible to completely work out all the critical points and the changes in topology. We illustrate such an analysis on the simplest nontrivial example $N = 4$ (see Fig. 7, explained below).

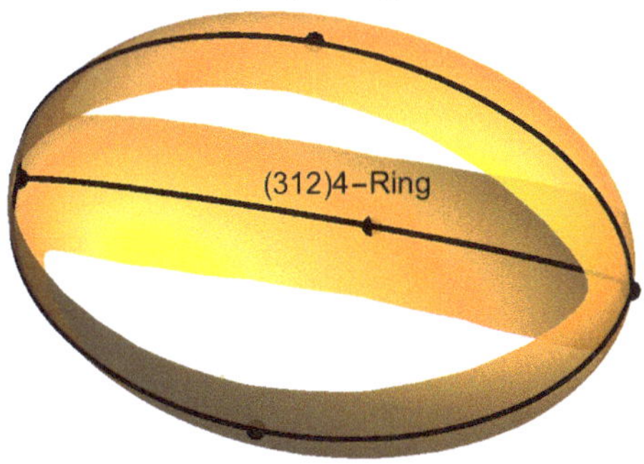

Fig. 7 Part of the configuration space for $N = 4$

We consider the reduced superlevel sets BConf$(4)[r]$. Since Conf(4) is 8-dimensional, away from the critical values these spaces are 5-dimensional manifolds with boundary.

If we ignore the labelling of points and classify the contact graphs for four vertices, there are exactly two geometrically distinct ρ-critical 4-configurations in BConf(4):

(1) The 4-Ring of four equally spaced points around a great circle on $\mathbb{S}^2$ with $\theta = \frac{\pi}{2}$ which is a saddle configuration for ρ. There is a 1-dimensional subspace of the tangent space to BConf(4) at the 4-Ring along which ρ increases to second order, i.e. the co-index is $k = 1$. The critical value for r for the 4-Ring is $r_1 = 1 + 2\sqrt{2} \approx 3.8284$.

(2) The vertices of the regular tetrahedron TET with $\theta = \cos^{-1}(-\frac{1}{3})$, which is the maximizing configuration for ρ on BConf(4), i.e the co-index $k = 0$. The critical value for r for TET is $r_2 = r_{max}(4) = 2 + \sqrt{6} \approx 4.4495$.

There are two intervals $(0, r_1)$ and $(r_1, r_2]$ on which the topology of BConf$(4)[r]$ remains constant. From Theorem 4.3, it can be seen that on the interval $(0, r_1)$, BConf$(4)[r]$ is homeomorphic to BConf(4). This has the homotopy type of $\mathbb{R}^2$ punctured at two points, hence

$$H^0(\text{BConf}(4)[r], \mathbb{Z}) = \mathbb{Z}, \quad H^1(\text{BConf}(4)[r], \mathbb{Z}) = \mathbb{Z}^2,$$
$$H^k(\text{BConf}(4)[r], \mathbb{Z}) = 0, \text{ for } k \geq 2.$$

On the open interval (r_1, r_2), the manifold BConf$(4)[r]$ has two connected components, each diffeomorphic to a 5-ball, which can be seen from the strong deformation retraction to BConf$(4)[r_2]$ consisting of the two points associated to the orientated labelings of TET configurations, TET$^+$ and TET$^-$. Hence

$$H^0(\text{BConf}(4)[r], \mathbb{Z}) = \mathbb{Z}^2, \quad H^k(\text{BConf}(4)[r], \mathbb{Z}) = 0, \text{ for } k \geq 1.$$

Table 4 Betti numbers for reduced configuration space cohomology $H^k(\mathrm{BConf}(\mathbb{S}^2, N); \mathbb{Q})$

	k									
N	0	1	2	3	4	5	6	7	8	9
3	1	0	0	0	0	0	0	0	0	0
4	1	2	0	0	0	0	0	0	0	0
5	1	5	6	0	0	0	0	0	0	0
6	1	9	26	24	0	0	0	0	0	0
7	1	14	71	154	120	0	0	0	0	0
8	1	20	155	580	1044	720	0	0	0	0
9	1	27	295	1665	5104	8028	5040	0	0	0
10	1	35	511	4025	18424	48860	69264	40320	0	0
11	1	44	826	8624	54649	214676	509004	663696	362880	0
12	1	54	1266	16884	140889	761166	2655764	5753736	6999840	3628800

Figure 7 above is only a schematic picture, since we cannot draw a 5-dimensional manifold. It compresses four of the dimensions. The visible points take r-values with $r_1 \le r \le r_2$. The value $r = r_1$ is a circular vertical ring in the middle, and the values of r increase as one moves to the left or right, reaching a maximum at TET^+ and at TET^-.

From Table 4, we can easily compute the Euler characteristic $\chi(\mathrm{BConf}(4)) = -1$. The indexed sum of critical points of the function $\rho : \mathrm{BConf}(4) \to \mathbb{R}$ gives an alternative computation of the Euler characteristic as

$$\chi(\mathrm{BConf}(4)) = \sum_k (-1)^k \#(\text{ critical points of co-index} = k\ ).$$

We count the *labeled* configurations in BConf(4): since the 4-Ring has symmetry group D_4 of order 8 in $SO(3)$, there are $3 = |\Sigma_4/D_4|$ critical points of this type with co-index 1; and since TET has symmetry group A_4 of order 12 in $SO(3)$, there are really $2 = |\Sigma_4/A_4|$ critical points of this type with co-index 0; and so we obtain

$$\chi(\mathrm{BConf}(4)) = 2 - 3 = -1,$$

as predicted. In fact, the Morse complex for ρ captures the fact that BConf(4) itself has the homotopy type of the Θ-graph: there are 2 vertices (0-cells) in the complex corresponding to the maxima (co-index 0) TET^+ and TET^- configurations; there are 3 edges (1-cells) corresponding to the 3 saddle (co-index 1) 4-Ring configurations.

4.7 *Topology Change for Variable Radius: $N \geq 5$*

The complexity of the changes in topology of the configuration space grows rapidly with N. For larger values of N there are many ρ-critical configurations which are not maximal.

The value $N = 12$ is large enough to be extremely challenging to obtain a complete analysis of the critical configurations of the configuration space, and to analyze the variation of the topology as a function of the radius r. The Betti numbers for $N = 12$ for radius $r = 0$ given in Table 4 differ greatly from those at $r = r_{max}(12)$ where the cohomology of $\mathrm{BConf}(12)[r_{max}(12)]$ is entirely in dimension 0, according to the Purity Property, which holds for $N = 12$ by results in Sect. 3. This topology change involves millions of (labeled) critical points. Its full investigation remains a task for the future.

5 Unit Radius Configuration Space for 12 Spheres

In this section, we discuss $\mathrm{Conf}(12)[1]$ and $\mathrm{BConf}(12)[1]$, the configuration spaces of 12 unit spheres touching a central unit sphere $\mathbb{S}^2$. These configuration spaces are remarkable and have some special properties. The value $r = 1$ is a critical value and that $\mathrm{BConf}(12)[1]$ has (at least) two geometrically distinct critical points, the FCC and HCP configurations. We believe that $r = 1$ is the maximal radius where the spheres $\mathrm{BConf}(12)[r]$ are arbitrarily permutable with motions remaining on $\mathbb{S}^2$ (see Sect. 6.5).

The case where all spheres are unit spheres has been extensively studied in connection with sphere packing. The value $r = 1$ is a critical value of the radius function r, and we will see that the associated configuration spaces $\mathrm{Conf}(12)[1]$ and $\mathrm{BConf}(12)[1]$ are not manifolds. To better understand their topology, it is useful to consider the spaces $\mathrm{Conf}(12)[r]$ and $\mathrm{BConf}(12)[r]$ for r in a neighborhood of 1. These are stratified spaces naturally embedded in $(\mathbb{S}^2)^{12}$ and filtered by r. For non-critical values of r, the spaces $\mathrm{Conf}(12)[r]$ and $\mathrm{BConf}(12)[r]$ are submanifolds with boundary. For all $r < r_{max}(12)$, the space $\mathrm{Conf}(12)[r]$ has top dimension 24. After factoring out the ambient $SO(3)$-action, the space $\mathrm{BConf}(12)[r]$ has top dimension 21.

5.1 *Three Remarkable Configurations of Unit Spheres:* DOD, FCP, HCC

We now consider the three configurations of 12 touching spheres singled out by Frank [40]. In Fig. 8, the three polyhedra have vertices located at the 12 touching sphere centers of these configurations and centroids located at the center of the central sphere. The edges of these polyhedra specify the contact graphs of these configurations, also pictured schematically in Fig. 8. The DOD configuration realizes the optimal contact graph for $N = 12$ given in Fig. 5 in Sect. 3.3.

- The DOD configuration is obtained by placing 12 spheres touching a central 13-th sphere at the vertices of an inscribed icosahedron; such touching points are also the centers of the faces of a circumscribed dodecahedron. It has oriented symmetry group A_5, the icosahedral group, of order 60 and in BConf(12)[1] there are $\frac{|\Sigma_{12}|}{|A_5|} = \frac{12!}{60} = 7983360$ of these.
- The FCC configuration is obtained by stacking three layers of the hexagonal lattice, with the third layer not lying over the first layer. The inscribed polyhedron formed by the convex hull of the 12 points of the FCC configuration where the spheres touch the central sphere is a cuboctahedron. The circumscribed dual polyhedron which has the 12 points as the center of its faces is a rhombic dodecahedron. It has oriented symmetry group Σ_4, the octahedral group, of order 24 and in BConf(12)[1] there are $\frac{|\Sigma_{12}|}{|\Sigma_4|} = \frac{12!}{24} = 19958400$ of these.
- The HCP configuration is obtained by stacking three layers of the hexagonal lattice, with the third layer lying directly over the first layer. The inscribed polyhedron formed by the convex hull of the 12 points of HCP where the spheres touch the central sphere is a triangular orthobicupola. This polyhedron is the Johnson solid J_{27}. The circumscribed dual polyhedron which has the 12 points as the center of its faces is a trapezoidal rhombic dodecahedron. It has oriented symmetry group D_3, the dihedral group of order 6, and in BConf(12)[1] there are $\frac{|\Sigma_{12}|}{|D_3|} = \frac{12!}{6} = 79833600$ of these.

The DOD configuration is an interior point of BConf(12)[1], while the FCC and HCP configurations are elements of $\partial\,$BConf(12)[1].

5.2 Three Remarkable Configurations of Unit Spheres: Rigidity Properties

We first consider rigidity properties of these packings. In the following definition we identify a sphere tangent to the 2-sphere with the circular disk (i.e. spherical cap) on $\mathbb{S}^2$ that it produces by radial projection.

Definition 5.1 (cf. Connelly [19, p. 1863]) A packing of disks on $\mathbb{S}^2$ is *locally jammed* if each disk is held fixed by its neighbors. That is, no disk in the packing can be moved if all the other disks are held fixed. We say a configuration of disks is *jammed* if it can only be moved by rigid motions. We call it *completely unjammed* if each disk can be moved slightly while holding all the other disks fixed.

Theorem 5.2 *The* DOD *configuration in* Conf(12)[1] *is completely unjammed. Its space of (infinitesimal) deformations has dimension* 24. *The deformation space is* 21 *dimensional if viewed in* BConf(12)[1].

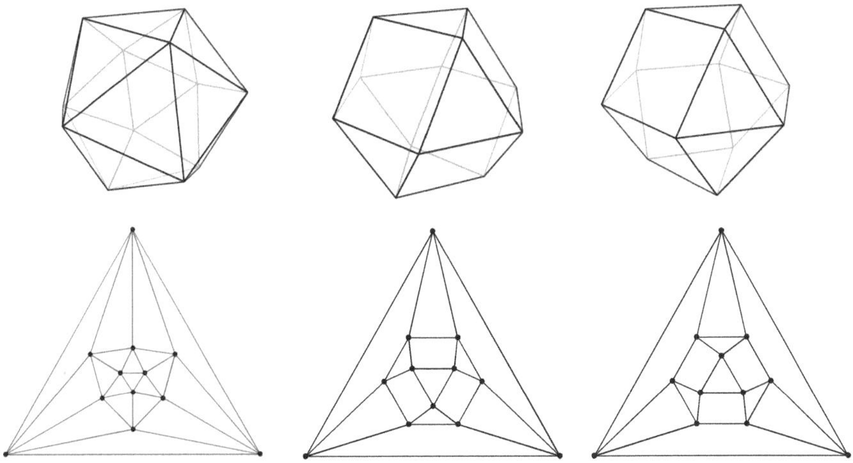

Fig. 8 The DOD, FCC, and HCP configurations with their corresponding contact graphs

Proof The maximal radius for 12 spheres is $r_{\max}(12) > 1$ and is achieved in the DOD configuration. Therefore the deformation space at DOD is full dimensional. □

In contrast, both the FCC and HCP configurations are *locally jammed*, i.e. they are rigid against motion of any one disk while holding all the other disks fixed; each of their infinitesimal deformation spaces has codimension at least 2.

In Sect. 5.4, we will describe a deformation of the DOD packing to the FCC packing. This deformation, properly adjusted, has 6 moving balls during its final phase arriving at FCC. (The 6 fixed balls form an antipodal pair of "triangles".) We believe this value 6 to be the smallest number of moving balls needed to unjam the FCC configuration. For a manual on how to unlock FCC, see Sect. 8.

A deformation of the DOD configuration to the HCP configuration, also described in Sect. 5.4, requires 9 moving balls at the instant of arrival at HCP. We believe this value 9 to be the smallest number of moving balls needed to unjam the HCP configuration. For a manual on how to unlock HCP, see Sect. 8. A possible reason for the larger number of moving balls needed to unjam the HCP configuration compared with that of the FCC configuration is that the HCP configuration has fewer local symmetries.

5.3 Three Remarkable Configurations of Unit Spheres: Criticality Properties

The FCC and HCP configurations are critical configurations for $r = 1$. According to Theorem 4.11 it suffices to show that these configurations carry balanced contact graph structures.

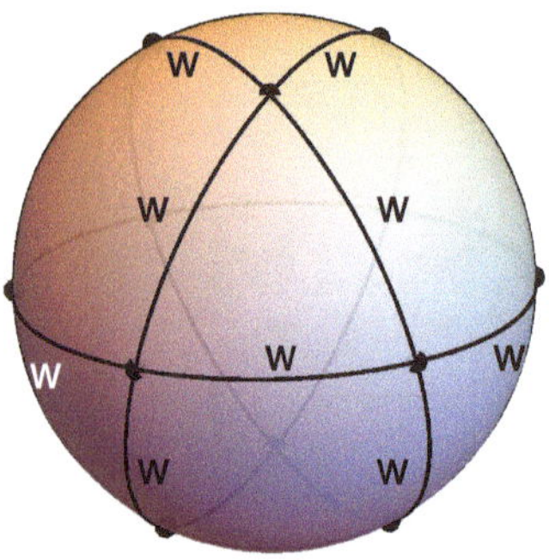

(a) FCC Configuration

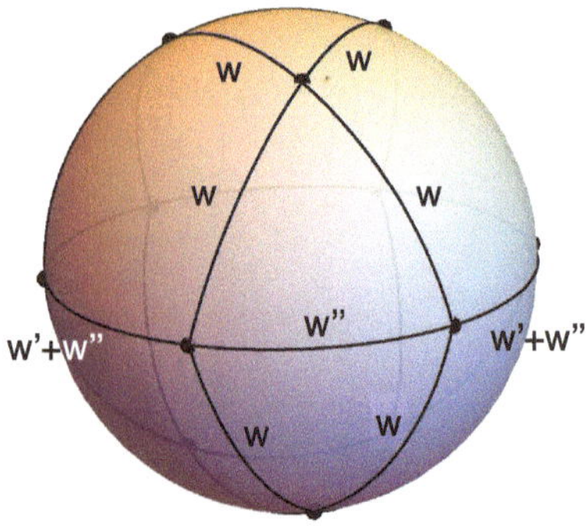

(b) HCP Configuration

Fig. 9 Stress graphs for FCC and HCP configurations

Theorem 5.3 *The* FCC *configuration and* HCP *configuration for* $r = 1$ *carry balanced contact graph structures. Consequently,* $r = 1$ *is a critical value of the radius function on* BConf(12).

Proof By Theorem 4.11, a sufficient condition for the criticality of a configuration for maximizing injectivity radius is that its contact graph can be balanced. That is, a set of positive weights may be assigned to the edges of the contact graph so that at each vertex the weighted vector sum, defined by the outward tangent vectors to the incident edges, vanishes.

We now indicate weight values for the FCC and HCP configurations (see Fig. 9).

(1) At radius $r = 1$ for the FCC configuration, the stress graph is balanced when all the weights are equal. This can be seen from the cubic- or Σ_4-symmetry of the contact graph.

(2) At radius $r = 1$ for the HCP configuration, consider a weight w_1 on edges between triangular faces and square faces, a weight w_2 on edges between pairs of square faces, and a weight w_3 on edges between pairs of triangular faces. From the structure of the contact graph, it is possible to choose a constant $w_1 > 0$ and find a weight $w_2 = w' > 0$ which balances the associated stress graph. This suffices to balance this configuration with some zero weights. However, it is also possible to add a uniform constant weight $w'' = w_3 > 0$ to the equatorial great circle, giving a balanced stress graph with positive weights on all edges of w_1, $w_2 := w' + w''$, $w_3 := w''$. □

The DOD configuration is not a critical configuration for $r = 1$; instead it is a critical configuration at the maximal radius $r = r_{max}(12) \approx 1.10851$. As noted in Sect. 2, Fejes Tóth [34] conjectured that this configuration does have a certain extremality property for local packing by equal spheres, that it gives a minimizer for a single Voronoi cell of a unit sphere packing. This statement, the Dodecahedral Conjecture, was proved in 2010 by Hales and McLaughlin [52].

Theorem 5.4 (Hales and McLaughlin (2010)) *A* DOD *configuration of unit spheres minimizes the volume of a Voronoi cell of a unit sphere with center at the origin of* $\mathbb{R}^3$ *over all sphere packing configurations of unit spheres containing that sphere.*

The volume of this Voronoi cell gives a local sphere packing density of approximately 0.7546, which exceeds the sphere packing density $\frac{\pi}{\sqrt{18}} \approx 0.74048$ in 3-dimensional space.

5.4 Three Remarkable Configurations of Unit Spheres: Deformation Properties

We now show that the DOD configuration can be continuously deformed inside BConf(12)[1] to the FCC configuration and to the HCP configuration.

Theorem 5.5

(1) *On the space* Conf(12)[1] *there is a continuous deformation of the* DOD *configuration to the* FCC *configuration that remains in the interior* $\mathrm{Conf}^+(12)[1]$ *of* Conf(12)[1] *till the final instant.*

(2) *There is also a continuous deformation of the* DOD *configuration to the* HCP *configuration that remains in the interior* $\mathrm{Conf}^+(12)[1]$ *of* Conf(12)[1] *till the final instant.*

The motions of these two deformations, measured from the touching points of the 12 spheres to the central sphere, can be given by piecewise analytic functions on the 2-sphere. The proof of Theorem 5.5 is given in Sects. 5.4.1–5.4.6. An unlocking manual for doing it is given in the Appendix (Sect. 8).

5.4.1 Coordinates for the Icosahedral Configuration DOD

To describe the deformations we will need coordinates. In Sects. 5.4.1–5.4.5 we suppose a ball with radius 1 centered at 0 touches all the 12 balls of the same given radius r. We initially allow all values of $r \in (0, r_{max}(12)]$, but in the move M_6 described in later subsections we will necessarily restrict to $r \leq 1$. We take DOD to consist of 12 equal balls of radius r, touching a central unit sphere at 12 vertices of an inscribed icosahedron I. We view this icosahedron I as embedded in $\mathbb{R}^3$ with Cartesian coordinates so that we have (Fig. 10):

- Its centroid is at the center (0, 0, 0) of the unit sphere.
- It has two opposite faces parallel to the xy-plane. In other words, for some $z = h$, the intersections $I \cap \{z = \pm h\}$ are triangular faces of I. Here $h = \frac{\phi^2}{\sqrt{3\phi^2+3}} \approx 0.79659$, where $\phi = \frac{1+\sqrt{5}}{2}$ is the golden ratio.

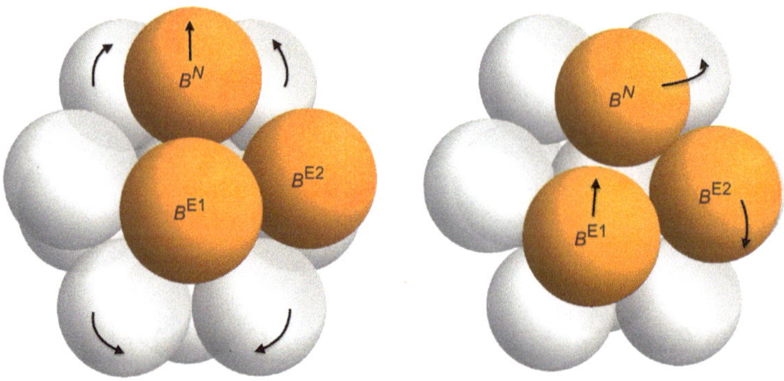

(a) Phase 1 (b) Phase 2

Fig. 10 The 6-move M_6

The *north balls* are those which have their centers in the plane $\{z = h\}$, while the *south balls* have their centers in the plane $\{z = -h\}$. The three north balls form a *north triangle* which is centrally symmetric to the *south triangle* formed by the three south balls, as in FCC. Together with north and south triangles, we have 6 remaining balls, which will be called *equatorial*, even though in the initial configuration they do not have their centers on the equator. The equator lies in the plane $z = 0$.

To fix their positions, let the *Greenwich meridian* be defined as $\mathbb{S}^2 \cap \{x \geq 0, y = 0\}$, and the longitude $\varphi \in [0, 2\pi)$ be measured from it in the counterclockwise direction viewed from the north pole (0, 0, 1). We require:

- The center of one of the north balls is in the half-plane of the Greenwich meridian, i.e. this ball touches the Greenwich meridian. Let us call this ball B^N .
- The center of one of the equatorial balls is in the half-plane of the Greenwich meridian. Let us call it B^{E1} . It will necessarily be in the southern hemisphere.

This fixes the location of all 12 balls. With this orientation of the icosahedron, the meridians of the north triangle are spaced by $\frac{2\pi}{3}$. Furthermore the meridians of the other three balls in the northern hemisphere are also spaced by $\frac{2\pi}{3}$ and the meridians combined are spaced by $\frac{\pi}{3}$ as in FCC. The same holds for the six balls in the southern hemisphere.

5.4.2 The M_6-move: Two Variants

We now define the "6-move" deformation M_6, which has two variants, one leading from DOD to the FCC configuration, and the other leading to the HCP configuration. This move proceeds in two phases.

The first phase is the same for both variants. It moves the 6 balls that are not equatorial at constant speed along meridians towards the poles, until they form north and south triangles of three mutually touching balls. The 6 equatorial balls do not move.

In the second phase, all 12 balls are moving. In both variants, the 6 equatorial balls, initially not on the equator, move towards the equator along their meridians at constant speed, to arrive on the equator at the end of the move, forming a ring of six balls on the equator. This ring is an allowed configuration only if $r \leq 1$. They do not touch during this move, until the last moment, and then all touch if $r = 1$. At the same time, the north and south triangles will rotate about the polar axis at a variable speed, the same for all six, in such a way as to avoid the equatorial balls. They will rotate by $\frac{\pi}{3}$ to their final position. For the FCC move, the north triangle and south triangle rotate in the same direction, while for the HCP move they rotate in opposite directions. A key issue is to suitably specify the variable speed of rotation.

5.4.3 First Phase of the M_6-move: Two Triangles Move away from the Equator

Denote by P the parallel $z = z_1$ where the three centers of the north triangle stop. Let $-P$ be the parallel $z = -z_1$ where the south triangle stops.

5.4.4 Second Phase of the M_6-move: Rotating the Two Triangles

Each of the six centers of the equatorial balls, initially not on the equator, will move at constant speed along their respective meridians towards their final positions on the equator. Parametrize this motion so that at $t = 0$, the 6 balls are at their initial positions, while at $t = 1$, the 6 balls are at their final positions on the equator.

The centers of the north triangle are on P at $t = 0$ and will remain on P throughout the move. Similarly, the centers of the south triangle are on $-P$ at $t = 0$ and will remain on $-P$ throughout the move. The triangles simply rotate.

It now suffices to specify functions $\varphi^N(t)$, which describe the *increment* of the longitude of the north triangle during the time $[0, t]$ and $\varphi^S(t)$, the *increment* of the longitude of the south triangle during the time $[0, t]$. We will take $\varphi^N(t)$ to be a continuous, non-decreasing function, with $\varphi^N(0) = 0$, $\varphi^N(1) = \frac{\pi}{6}$.

We get two different moves to FCC and HCP depending on which direction the south triangle rotates. The motion $\varphi^S = \varphi^N$ will take us to FCC, and choosing the opposite rotation $\varphi^S = -\varphi^N$ will take us to HCP.

5.4.5 The Choice of φ^N

The function φ^N, defined so that no ball from the two triangles hits any equatorial one, is certainly not unique.

Here is a minimal definition of φ^N, beginning at the second phase. Recall that our balls are open, and that:

- the center of one of the three north balls, B^N, is on the half-plane of the Greenwich meridian.
- the center of one of the equatorial balls, B^{E1}, is also on the half-plane of the Greenwich meridian.
- there is an equatorial ball with the longitude $\frac{\pi}{3}$; call it B^{E2}.

Note that the center of B^{E1} is south of the plane $z = 0$, while that of B^{E2} is north of the plane $z = 0$.

Throughout the second phase of M_6 the ball B^{E1} will move north while B^{E2} moves south. Let $\mathrm{B}^{E1}(t)$, $\mathrm{B}^{E2}(t)$ denote their positions, $t \in [0, 1]$. Then for every $\varphi \in [0, 2\pi)$ define $\mathrm{B}^N(\varphi)$ to be the ball with the center at P and with the longitude φ. For example, $\mathrm{B}^N(0) = \mathrm{B}^N$.

Now let us define the function φ^N as follows:

$$\varphi^N(t) = \inf\left\{\varphi \geq 0 : \mathrm{B}^N(\varphi) \cap \mathrm{B}^{E1}(t) = \varnothing\right\}. \tag{5.1}$$

Clearly, $\varphi^N(t) = 0$ for all t small enough. The only thing one needs to check is that

$$\mathrm{B}^N\big(\varphi^N(t)\big) \cap \mathrm{B}^{E2}(t) = \varnothing \tag{5.2}$$

holds for all $t \in [0, 1]$.

Lemma 5.6 *An increment function $\varphi^N(\cdot)$ exists: $\varphi^N(t)$ as defined by* (5.1) *satisfies* (5.2).

Proof We use Euler coordinates on the sphere. The latitude ϑ of the parallel P is $(\pi - \theta)$, where θ satisfies

$$\sin(\theta) = \frac{1}{\sqrt{3}}, \quad \cos(\theta) = \sqrt{\frac{2}{3}}, \quad \tan(\theta) = \frac{1}{\sqrt{2}}.$$

By symmetry, it is enough to consider the movement of three balls:

- B^N on the parallel P. Its initial angle $\varphi(t = 0) = 0$. The latitude ϑ of B^N is constant.
- B^{E1} on the Greenwich meridian $\varphi = 0$. Its initial latitude is $\vartheta(t = 0) = \left(\frac{2\pi}{3} - \theta\right) < 0$ and final latitude is $\vartheta(t = 1) = 0$. On the interval, its latitude is given by $\vartheta(t) = \left(\frac{2\pi}{3} - \theta\right)t$.
- B^{E2} on the meridian $\varphi = \frac{\pi}{3}$. Its initial latitude is $\vartheta(t = 0) = -\left(\frac{2\pi}{3} - \theta\right) > 0$ and final latitude is $\vartheta(t = 1) = 0$. On the interval, its latitude is given by $\vartheta(t) = \left(\theta - \frac{2\pi}{3}\right)t$.

The function $\varphi^N(t)$ is uniquely defined by:

- $\varphi^N(t) \geq 0$.
- At every time $t < 1$ and after initial contact, the ball $\mathrm{B}^N(t)$ touches the ball $\mathrm{B}^{E1}(t)$.

To complete the proof of Lemma 5.6 it remains to check that the balls $\mathrm{B}^N(t)$ and $\mathrm{B}^{E2}(t)$ of radius $r \leq 1$ are disjoint for $0 < t < 1$. This fact will follow from the next lemma. □

Lemma 5.7 *Let $0 < t < 1$. Consider an isosceles spherical triangle ABO, where A has $\varphi = 0$, $\vartheta(t) = \left(\frac{2\pi}{3} - \theta\right)t$, B has $\varphi = \frac{\pi}{3}$, $\vartheta(t) = -\left(\frac{2\pi}{3} - \theta\right)t$, and $O = O(t)$ is defined by $\vartheta(O) = \pi - \theta$ and the touching condition. Then $|AO| = |BO| > \frac{\pi}{3}$.*

Proof Let D be the middle point of the arc AB. It does not depend on t and is given by $\vartheta(D) = 0$, $\varphi(D) = \frac{\pi}{6}$. Let $\varkappa(t)$ be the arc perpendicular to the arc AB at D. Then $O(t)$ is simply the intersection of $\varkappa(t)$ and the parallel P.

The triangle $O(t)\,DA(t)$ is a right triangle. Evidently, the legs $O(t)\,D$ and $A(t)\,D$ become shorter as t increases. Hence the hypotenuse $O(t)\,A(t)$ becomes shorter as well. Since $O(1)\,A(1) = \frac{\pi}{3}$, the proof follows. □

5.4.6 Completion of Proof of Theorem 5.5

Proof of Theorem 5.5 Lemmas 5.6 and 5.7 complete a proof that there exists a deformation path from the DOD configuration to a FCC configuration and to a HCP configuration, respectively. However, the deformation path obtained does not satisfy one required condition of the theorem: remaining in the interior of the configuration space. It exits from the interior of Conf(12)[1] at the end of the first phase and remains on the boundary during the second phase: the three north balls are touching and the three south balls are touching.

We can modify the construction above so that no balls touch throughout the deformation until the final instant. To do this we halt the first phase just short of the three balls touching, at $z = 1 - \epsilon_1$. Then in the second phase, we allow z to increase monotonically in the north triangle at some variable speed $\psi(t)$ as the rotation proceeds, in such a way as to avoid contact between the three north balls and the equatorial balls. The south triangle z variable is to decrease monotonically in the reflected motion of $-z$ at the same time. Lemma 5.7 implies that if z approaches 1 rapidly enough in the motion that we can again avoid contact; this is an open condition at each point t, so by compactness of the motion interval we have a finite subcover to attain it. □

Remark

(1) This motion process can be continued by concatenation with an inverse M_6 using $-\varphi(1-t)$, in such a way as to arrive back at a DOD configuration, differently labeled. This is possible because there are two exit directions (tangent vectors) from the FCC configuration and two exit directions from the HCP configuration in BConf(12)[1]. Section 6.1 studies the group of permutations of the 12 labels obtainable by such deformations.

(2) Starting from the FCC or HCP configuration, there is a reference frame in which the north triangle remains fixed. The inverse of the second phase of M_6 describes a move which unlocks the FCC configuration with 6 moving balls and 6 fixed

balls, and which unlocks the HCP configuration with 9 moving balls and 3 fixed balls (see Sect. 8).

5.5 *Buckminister Fuller's "Jitterbug"*

According to his recollection, on 25 April 1948 Buckminster Fuller found a "jitterbug" construction given by a jointed framework motion that, among other things, permits an FCC configuration, given as the vertices of a cuboctahedron, to be continuously deformed into a DOD configuration, given as the vertices of an icosahedron (see [89, p. 273]).

In Buckminster Fuller's construction, the joint distances remain constant during the motion, so that they can be rigid bars, while the radii of the associated touching spheres continuously contract during the deformation. At each instant during the motion the central sphere and the 12 touching spheres can all have equal radii without overlapping, and this radius varies monotonically in time.

In retrospect one may see that it is possible to rescale space during the motion via homotheties varying in time such that all spheres retain the fixed radius 1 throughout the deformation. In this case the joint lengths will change continuously in the motion. The rescaled motion no longer corresponds to a physical object with rigid bars, but it does give a continuous motion in the configuration space of 12 equal spheres touching a 13-th central sphere that continuously deforms the FCC configuration to the DOD configuration.

The work of Buckminster Fuller on the "jitterbug" movable jointed framework is described in Schwabe [89]. Fuller described it in his book *Synergetics* [42, Sect. 460.00-463.00]. The construction is also described in Edmondson [27, Chap. 11], with a detailed analysis in Verheyen [97].

Remark The "jitterbug" motion immediately enters the interior $\mathrm{BConf}^+(12)[1]$ after the initial instant, in contrast to the "unlockings" described in Appendix 8, which adhere to its boundary.

6 Permutability of the DOD Configuration: Connectedness Conjectures

The number of connected components of the configuration space $\mathrm{Conf}(12)[r]$ is related to the ability to permute labeled spheres by deformations within $\mathrm{Conf}(12)[r]$. The possible permutability of the (labeled) spheres in the DOD configuration in $\mathrm{Conf}(12)[r]$ depends on the radius r of the touching spheres.

6.1 *Permutations of* DOD *Configurations for Radius* $r = 1$

Conway and Sloane [20, Chap. 1, Appendix, pp. 29–30] give a terse proof that for radius 1 the labels on labeled spheres in DOD configurations can be arbitrarily permuted using continuous deformations inside the space Conf(12)[1].

Theorem 6.1 (Permutability at radius $r = 1$) *For the radius parameter* $r = 1$*, each labeled* DOD *configuration can be continuously deformed in the configuration space* BConf(12)[1] *to a* DOD *configuration at the same* 12 *touching points with any permutation of the labeling.*

We follow the outline in Conway and Sloane [20, Chap. 1, Appendix, pp. 29–30]. A main ingredient is an additional set of permutation moves which we call M_5-moves, detailed next.

6.2 *The* M_5*-move*

Beginning from the DOD configuration centered at the origin, we rotate it so that two opposite balls have their centers on the z axis. Call these balls N and S. Note that the centers of 5 of the 10 remaining balls are in the northern half-space, while the remaining 5 centers are in the southern half-space. Call these balls $\mathrm{U}_1 \ldots \mathrm{U}_5$ and $\mathrm{V}_1 \ldots \mathrm{V}_5$, respectively.

First phase. Move the 5 northern U_j balls towards N, in such a way that their centers remain on their corresponding meridians, until each of them touches N. Note that these 5 balls do not touch each other, only N. Indeed, because their centers are located at the latitude $\vartheta = \frac{\pi}{6}$, when viewed from the z-axis each of the 5 balls subtends the dihedral angle $\arccos\left(\frac{1}{3}\right)$ of a regular tetrahedron; but the longitude difference between the neighboring ball centers is $\frac{2\pi}{5}$, so there remains a tiny longitude gap

$$\zeta = \frac{2\pi}{5} - \arccos\left(\frac{1}{3}\right) \approx 0.025 > 0.$$

The 5 southern V_j balls may be moved into the southern hemisphere in the same manner.

Second phase. Note that for $r = 1$, the 6 northern balls fit into the northern half-space, while the 6 southern balls fit into the southern half-space. The union of all the 6 balls in the northern hemisphere may be rotated by $\frac{2\pi}{5}$ as a rigid body, keeping the remaining balls fixed.

Third phase. Reverse the first phase.

The net result of an M_5-move is a cyclic permutation σ_2 of DOD of length 5, which is an even permutation.

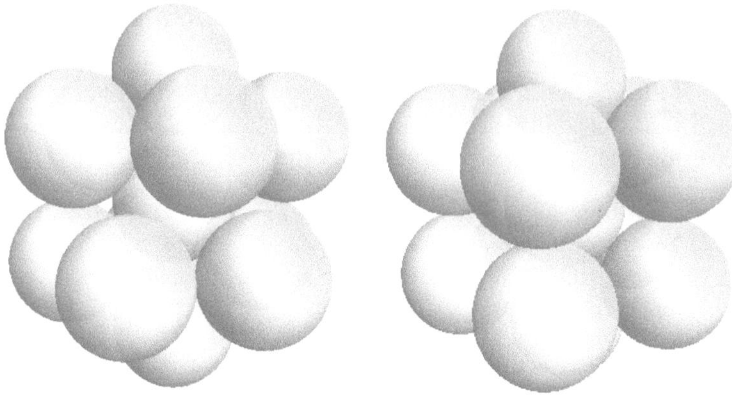

Fig. 11 The 5-move M_5

Remark The halfway point of the M_5-move, as illustrated in the right side of Fig. 11, is itself a "remarkable" configuration: the set of critical configurations for $r = 1$ contains a 4-simplex, which has this halfway point at its center. The family of such configurations is in many ways similar to the maximal stratified set that occurs for $N = 5$ (see Fig. 6).

6.3 *Proof of Permutability Theorem at Radius* $r = 1$

Proof of Theorem 6.1 The first step is to show that there is a continuous deformation of DOD to itself, which permutes the labels by an odd permutation. To exhibit it, we use the move M_6, defined before, and deform DOD into an FCC configuration (note that we can do it for all $r \leq 1$, but not for $r > 1$). Note that the FCC configuration has three axes of 4-fold symmetry passing through the opposite squares of four balls. By rotating $\frac{\pi}{2}$ around any such an axis and then deforming our configuration back to DOD via M_6^{-1}, we induce a a permutation σ_1 of 12 balls, which is a product of three (disjoint) cyclic permutations, each of length 4. Every such cycle is an odd permutation, hence their product σ_1 is also odd.

The second step uses M_5-moves. Each such move gives a cyclic permutation of order 5. Since there are 12 options for choosing N, we get 12 such 5-cycles $\sigma_2^{(i)}$. It is shown in Conway and Sloane (using an elegant argument about the Mathieu group M_{12}, see [20, pp. 328–330]) that all such $\sigma_2^{(i)}$ generate the alternating group A_{12}, the subgroup of even permutations of Σ_{12}.

Combined with any odd permutation σ_1, the full permutation group Σ_{12} is generated. □

6.4 Persistence of M_5-moves to Some $r > 1$

The move M_5 can be modified in such a way that it continues to work for all values $r \leq r_1$, for some r_1 slightly bigger than 1. We first explain the modification and then propose the value of r_1. The modification deals only with the second phase of M_5. In order to explain it, it is enough to follow the 10 longitude values of the touching points of our balls, which may be considered as points on the equator. For $r = 1$, the northern 5 balls U_j correspond to the longitude values u_j, for $j = 1, ..., 5$, and we can suppose that at the initial moment these values are $u_j(t = 0) = (j - 1)\frac{2\pi}{5}$. The longitude values v_j are defined similarly, corresponding to the southern balls V_j, and $v_j(t = 0) = (j - 1)\frac{2\pi}{5} + \frac{\pi}{5}$. Our initial move looks now as follows:

$$u_j(t) = (j - 1 + t)\frac{2\pi}{5},\ v_j(t) = v_j(0)\,.$$

Of course there is no need for all the u_j to move with the same speed; the only constraint is that the difference between consecutive u_j should equal or exceed $\arccos\left(\frac{1}{3}\right)$ at all times. In particular, we can modify the speeds in such a way that at any time t, we have $u_j(t) = v_j(t)$ for at most one value of j.

Now let the radius r be slightly bigger than 1. Then, at the moment t when $u_j(t) = v_j(t)$, the corresponding balls U_j, V_j will overlap.

This, however, can be remedied by making the following small deformation of our 12-configuration:

- the ball U_j moves up along its meridian, by the distance $(r - 1)$.
- the ball N moves along the same meridian in the same direction by the distance $2(r - 1)$.
- the ball V_j moves down along its meridian, by the distance $(r - 1)$.
- the ball S moves along the same meridian in the same direction by the distance $2(r - 1)$.
- other balls may be rearranged in such a way that they do not intersect.

The non-overlap condition can be satisfied when $(r - 1) > 0$ is small enough, since there were no other collisions.

Below we will show there will be 5 *bottleneck configurations* that one encounters on the way to perform the modified M_5 move. Each one defines a value $r_1^{(j)} > 1$, for $1 \leq j \leq 5$ which is the maximal radius for which this configuration is allowed. We set $r_1 := \min_{j=1,\ldots,5} r_1^{(j)} > 1$.

Theorem 6.2 *For every $r \leq r_1$ the move M_5 can be modified in such a way that one can reach from an initial labeled* DOD *configuration any labeled* DOD *configuration whose labels are an even permutation of the initial labels. That is, the alternating group A_{12} is generated by the compositions of different M_5 moves.*

Proof There will occur 5 bottleneck 12-configurations of the r-balls touching the unit central ball, described by certain touching patterns that correspond to the configurations appearing during the move M_5 at the moment when the ball U_j passes due north of the ball V_j.

The 5 bottleneck configurations have a common pattern: 4 touching balls centered on the same meridian, two in the northern half-space, and the remaining two in the southern half-space. We denote them by N, U_j, V_j, S. This set of 4 balls is symmetric with respect to the plane $z = 0$. Strictly speaking, as r is slightly bigger than 1, the balls N and S are now centered on the meridian *opposite* the one containing U_j and V_j.

The eight other balls are the remaining ones from $U_1, \dots, U_5, V_1, \dots, V_5$. Each pair $\{U_i, V_i\}$ touches, as do the pairs $\{U_i, N\}$, as well as the pairs $\{V_i, S\}$. The 5 bottleneck configurations differ in how the additional pairs of balls touch. Each of the rows in the following list completes a different touching pattern that occurs as the move M_5 is performed:

$$\{U_1, U_2\}, \{V_2, U_3\}, \{V_3, U_4\}, \{V_4, U_5\}, \{V_5, V_1\}, \qquad (j = 1)$$

$$\{U_2, U_3\}, \{V_3, U_4\}, \{V_4, U_5\}, \{V_5, V_1\}, \{U_1, U_2\}, \qquad (j = 2)$$

$$\{U_3, U_4\}, \{V_4, U_5\}, \{V_5, V_1\}, \{U_1, V_2\}, \{U_2, U_3\}, \qquad (j = 3)$$

$$\{U_4, U_5\}, \{V_5, V_1\}, \{U_1, V_2\}, \{U_2, V_3\}, \{U_3, U_4\}, \qquad (j = 4)$$

$$\{V_5, V_1\}, \{U_1, V_2\}, \{U_2, V_3\}, \{U_3, V_4\}, \{U_4, U_5\}. \qquad (j = 5)$$

Observe that for any $r > 1$, and for any of the 5 touching patterns, such a configuration is unique if it exists, and that it *does* exist for $(r - 1)$ sufficiently small. We define $r_1^{(j)}$ as the maximal values for which the above configurations exist. □

We are not asserting that the value $r_1 > 1$ defined above is the true critical value above which the (small perturbation of the) move M_5 cannot be performed. Indeed, we imposed some a priori constraints in making our construction of the modified M_5, and did not rule out the possibility of a more "optimal" modification of M_5 (Fig. 12).

Definition 6.3 *Let R_1 be the maximal value of the radius r for which there exists some modified move M_5. We call it the upper critical radius.*

From the previous theorem we know that $R_1 \geq r_1 > 1$. We expect R_1 to be a critical value for maximizing the radius function.

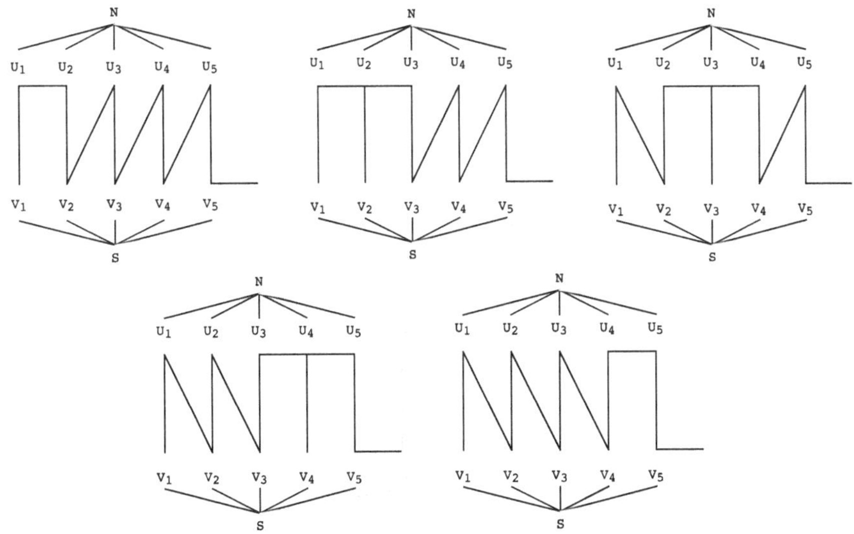

Fig. 12 Contact graphs for the bottleneck configurations $j = 1$ to $j = 5$

6.5 Connectedness Conjectures for **Conf(12)[1]**

Based on Theorems 5.5 and 6.1 about deformation and permutability of labeled DOD configurations, it is natural to propose the following statement.

Conjecture 6.4 (Connectedness Conjecture) The configuration space Conf(12)[1] is connected. That is, every set of 12 distinct labeled points on the 2-sphere pairwise separated by spherical angle at least $\frac{\pi}{3}$ can be deformed into 12 other distinct labeled points, with all points maintaining a spherical angle at least $\frac{\pi}{3}$ apart during the deformation.

This problem appears to be approachable but difficult to prove, despite the supporting evidence of permutability in Theorem 6.1. One may approach it by cutting the space $\mathrm{BConf}(12)[r_0]$ into many small path-connected "convex" pieces and gluing them together in some fashion. The computational size of the problem, since the dimension of the space is 21, and has a complicated boundary, is daunting.

We also propose a stronger statement.

Conjecture 6.5 (Strong Connectedness Conjecture) The radius $r = 1$ is the largest radius value at which configuration space $\mathrm{Conf}(12)[r]$ is connected.

In support of Conjecture 6.5, the M_6-move appears to be possible only when $r \leq 1$. At one time instant it has 6 spheres fitting in a ring around the equator, a condition which is allowed only for $r \leq 1$. We also know that the $r = 1$ satisfies the necessary condition of being a "critical value" for the r-parameter.

We formulate one further conjecture concerning the connectivity structure of BConf(12)[1]. It is based on a further analysis not included here (see [64]), which indicates that BConf(12)[1] has at each of the $\frac{12!}{24}$ FCC-configurations, and at each of the $\frac{12!}{6}$ HCP-configurations, a unique tangent line along which it can be approached from the interior $\mathrm{BConf}^+(12)[1]$. In addition, this analysis shows that each of these is a *local cut point*, a point of a space which when removed, disconnects a small open neighborhood of the point. That is, these configurations are points at which the space BConf(12)[r] locally disconnects as r increases past 1. The following conjecture asserts that these local cut points form unavoidable bottlenecks in BConf(12)[1] in making certain rearrangements of spheres in the configuration space.

Conjecture 6.6 (FCC and HCP Bottlenecks) Any piecewise smooth continuous curve in the reduced configuration space BConf(12)[1] which starts at a labeled DOD configuration and ends at another labeled DOD configuration with the labels permuted by an odd permutation must necessarily pass through either an FCC configuration or a HCP configuration.

This conjecture asserts a specific way in which the FCC and HCP configurations may play a remarkable role in rearrangements of 12-configurations, illuminating the assertion of Frank [40] in Sect. 2.7.

6.6 Disconnectedness Conjectures for $r > 1$

For the region $1 < r \leq R_1$ we propose the following conjecture.

Conjecture 6.7 (Two Connected Components) Let R_1 be the upper critical radius defined after Theorem 6.2. Then the space Conf(12)[r] for $1 < r < R_1$ has exactly two connected components. Two labeled configurations, DOD and σ(DOD), where $\sigma \in \Sigma_{12}$ is a permutation of twelve labels, belong to different connected components of Conf(12)[r] if and only if the permutation σ is odd.

In view of the existence of the M_5-move, the argument in Theorem 6.1 indicates that there can be at most 2 connected components containing DOD configurations in this region. Conjecture 6.7 asserts there are exactly these two, and no other, connected components.

We next note that the five "bottlenecks" in the 5-move lead to the possibility of connected components not containing any DOD configuration, for certain ranges of r. During the M_5 move joining two DOD configurations, there are 5 bottlenecks all through which one can pass at least up to a radius $r_1 > 1$. There is however room for configurations of spheres of larger radius occurring between the bottlenecks. If we increase the radius above the smallest two of the bottleneck radii, it may be possible for a sphere to get stuck in the middle of one of these regions, so it can neither go backwards nor forwards via the M_5-move to a DOD configuration. Assuming that "trapped" configurations (from blocking of the M_5 move) exist containing no

DOD configuration, eventually as we increase r some "trapped" configuration must become a critical configuration. It would then be a local maximum in the configuration space, an isolated point in some BConf(12)[r]. The critical value at which this occurs would necessarily be strictly smaller than $r_{max}(12)$, using the result of Danzer (in Theorem 3.3) that the extremal configuration for $N = 12$ is unique.

Conjecture 6.8 (Non-DOD Components) There is a nonempty interval of values of $r > 1$ such that the reduced configuration space BConf(12)[r] has connected components that do not contain any copy of a DOD configuration.

A positive answer to this question raises the possibility of a value of r for which the number of connected components of BConf(12)[r] exceeds the number of labeled DOD configurations, which is $\frac{12!}{12\times 5} = 7983360$. To obtain the latter number, let us take the DOD configuration and label its 12 balls. There are 12! such labelings. Two labeled configurations are equivalent (i.e. the same in BConf(12)[r]) iff one can be obtained from the other by the $SO(3)$ rotation action. Clearly, there are 12×5 labelings in every equivalence class.

Finally, for the region of r-values very close to $r_{max}(12)$, we assert that each of the spaces Conf(12)[r] and BConf(12)[r] has exactly $\frac{12!}{12\times 5} = 7983360$ connected components, with each component containing a DOD configuration. This fact follows assuming the finiteness of the set of critical radius values r for BConf(12), since at the point $r_{max}(12)$ only the DOD configurations survive, according to the uniqueness result of Danzer (Theorem 3.3), and the topology of BConf(12)[r] does not change above the next largest critical value of r below $r_{max}(12)$.

7 Concluding Remarks

This paper treats configuration spaces of touching spheres for very small values of N. We have shown that the configuration space of 12 equal spheres touching a central 13-th sphere is already large enough to exhibit interesting behavior in its critical points. Concerning 12-sphere configurations in the equal radius case $r = 1$ we have made the following observations.

- We have clarified an assertion of Frank [40] given in Sect. 2.7, showing that in the space BConf(12)[1] there are deformations interconnecting all FCC, HCP and DOD configurations.
- We have given evidence suggesting that BConf(12)[1] is a connected space, and conjectured that $r = 1$ is the largest parameter value where BConf(12)[r] is connected.
- We have shown that all elements of the finite set of FCC and HCP configurations lie on the boundary of the topological space BConf(12)[1] and are critical points for maximizing the radius parameter.
- We have conjectured that a continuous deformation of 12 spheres in a DOD configuration to a permutation of itself that is an odd permutation of its elements,

then the deformation must pass through one of the (finite set of) FCC and HCP configurations in BConf(12)[1]; they are "unavoidable" points.

Many challenging and computationally difficult problems remain to better understand the constrained configuration space BConf(12)[1].

As mentioned in the introduction, configuration spaces are of interest in physics and materials science, particularly in connection with jamming in materials. Hard sphere models which view spheres packed inside a box have been extensively studied for jamming. Materials scientists have studied configuration spaces of small numbers of hard spheres by simulation in connection with nanomaterials. Recently, Holmes-Cerfon [57] developed an algorithm that enumerates rigid sphere clusters and has determined those with up to 16 spheres. The cases of small numbers (but larger than the N treated here) of spheres were studied in Phillips et al. [82] and Glotzer et al. [83], giving estimates for extremal configurations at values of N larger than can be currently treated mathematically. We note that simulations of phase space can sample only a small part of it. In the simulation experiments reported in [82] for $N = 12$ equal spheres, the experimenters were unable to detect that the radii at which the M_5-move and the M_6-move permutation cease being feasible are in fact different (as discussed in Sect. 6.4).

Study of the jamming problem leads to the sub-problem concerning what is a good notion of rigidity for such configurations. There is a notion of "locally jammed configuration" in which no particle can move if its neighbors are fixed. The Tammes problem — also called the (extremal) spherical codes problem — of determining $r_{max}(N)$ is analogous to the problem of determining maximally dense jammed configurations of spheres in a box. Various notions of rigidity for spherical codes were formulated in Tarnai and Gáspár [93]. More recently Cohn et al. [18] give a mathematical treatment of rigidity of extremal n-dimensional spherical codes.

In configuration theory models like BConf$(N)[r]$ of this paper, critical configurations at critical values of the radius parameter might serve as a proxy for locally jammed configurations. One can view the balancing condition in Theorem 4.11 as a weak form of the locally jammed condition. However only a subclass of critical configurations will be locally jammed in the sense above.

8 Appendix: Unlocking Manual for the FCC and HCP Configurations

8.1 *The* FCC *Configuration*

To unlock the FCC configuration, a good way is to do it with the help of a friend, hereafter called Charles.[7] Please follow these steps:

[7] Aprés Charles Radin

(1) Ask Charles to hold the 3 north balls and the 3 south balls firmly in their positions. These 6 polar balls remain fixed during the whole process. As a result, the 13-th central ball stays fixed as well.
(2) Roll the remaining 6 equatorial balls in a direction roughly parallel to the equator. If properly lubricated, this does not require a big effort.
(3) The equatorial balls can all be pushed either to the east or to the west, in a coordinated way.
(4) At all times you must ensure the 6 rolling balls touch the central ball. This requires some practice, but it is possible and not terribly hard.
(5) Observe that the 6 balls roll around the central ball along the equatorial "valley" between the polar balls kept fixed by Charles. These rolling balls cannot always move equatorially, but instead move north and south slightly, in an alternating manner, as you roll them.
(6) Because the 6 rolling balls move north and south, some of them do not touch each other any more: free space may appear between them. Also, some space can be created between them and the 6 balls kept fixed by Charles. This is normal.
(7) As you proceed by $\frac{\pi}{3}$, the 6 rolling balls realign in the equatorial plane, touching each other and the polar balls. Note that at this moment the configuration is locked back into FCC. Each of the 6 rolled balls is touching two of its equatorial neighbors, one ball to the north and one ball to the south.

8.2 *The* HCP *Configuration*

Unlocking the HCP configuration is similar to the FCC configuration, except that Charles has somewhat more to do. Please follow these steps:

(1) Ask Charles to hold firmly the three north balls and the three south balls. The three south balls will remain fixed during the whole process. But the north triangle has to be rotated as a whole in its plane, at some constant speed, which can be either eastward or westward (there are two choices). It will move through an angle $\frac{2\pi}{3}$. The 13-th central ball stays fixed as before.
(2) Roll all the remaining 6 balls in the (roughly same) equatorial direction as the north triangle is rotating. This movement direction is forced on all six equatorial balls by the motion of the north triangle.
(3) The rest of the process goes basically in the same way as for the FCC configuration.
(4) As Charles proceeds to rotate the north triangle by $\frac{2\pi}{3}$, you proceed by $\frac{\pi}{3}$, the six middle balls align back into the equatorial plane, touching each other and the six polar balls. Note that at this moment the configuration is locked back into the HCP configuration. Each of the 6 rolled balls is touching two of its equatorial neighbors, one ball to the north and one ball to the south.

Note that for FCC, the equatorial balls underwent cyclic permutation of length 6. For HCP, the equatorial balls underwent a cyclic permutation of length 6 and the 3 northern balls a cyclic permutation of length 3. These give odd permutations of FCC and HCP.

Acknowledgements The authors were each supported by ICERM in the Spring 2015 program on "Phase Transitions and Emergent Properties." R. K. was also supported by the University of Pennsylvania Mathematics Department sabbatical visitor fund and by MSRI via NSF grant DMS-1440140. W. K. was also supported by Austrian Science Fund (FWF) Project 5503. J. L. was supported by NSF grant DMS-1401224 and by a Clay Senior Fellowship at ICERM. Part of the work of S. S. has been carried out in the framework of the Labex Archimede (ANR-11-LABX-0033) and of the A*MIDEX project (ANR-11-IDEX-0001-02), funded by the "Investissements d'Avenir" French Government programme managed by the French National Research Agency (ANR). Part of the work of S. S. has been carried out at IITP RAS. The support of Russian Foundation for Sciences (Project No. 14-50-00150) is gratefully acknowledged. The authors thank Bob Connelly, Sharon Glotzer, Mark Goresky, Tom Hales and Oleg Musin for helpful comments. Parts of Sect. 4.1 are adapted from unpublished notes by R. K. and John Sullivan (MSRI, 1994) about critical configurations of "electrons" on the sphere.

References

1. A. Abrams, R. Ghrist, Finding topology in a factory: configuration spaces. Am. Math. Mon. **109**(2), 140–150 (2002)
2. H. Alpert, Restricting cohomology classes to disk and segment configuration spaces. Topol. Appl. **230**, 51–76 (2017)
3. P.W. Anderson, Through the glass lightly. Science **267**, 1615 (1995)
4. K. Anstreicher, The thirteen spheres: a new proof. Discret. Comput. Geom. **31**, 613–625 (2004)
5. C. Austin, Angell, Insights into phases of liquid water from study of its unusual glass-forming properties. Science **1**(319), 582–587 (2008)
6. V.I. Arnold, The cohomology ring of dyed braids, (Russian) Mat. Zametki **5**, 227–231 (1969)
7. T. Aste, D. Weaire, *The Pursuit of Perfect Packing* (Institute of Physics Publishing, London, 2000)
8. W. Barlow, Probable nature of the internal symmetry of crystals. Nature **29**(186–188), 205–207 (1883)
9. Y. Baryshnikov, P. Bubenik, M. Kahle, Min-type Morse theory for configuration spaces of hard spheres. Int. Math. Res. Not. IMRN **2014**(9), 2577–2592 (2014)
10. C. Bender, Bestimmung der grössten Anzahl gleich grosser Kugeln, welche sich auf eine Kugel von demselben Radius, wie die übrigen, auflegen lassen. Acrhiv der Mathematik und Physik **56**, 302–306 (1874)
11. K. Böröczky, The problem of Tammes for $n = 11$. Studia Sci. Math. Hung. **18**(2–4), 165–171 (1983)
12. K. Böröczky, L. Szabó, Arrangements of 13 Points on a Sphere, in *by A*, ed. by Discrete Geometry (Marcel Dekker, Bezdek (New York, 2003), pp. 111–184
13. K. Böröczky, L. Szabó, Arrangements of 14, 15, 16 and 17 Points on a Sphere. Studi. Sci. Math. Hung. **40**, 407–421 (2003)
14. J. Cantarella, J.H. Fu, R. Kusner, J.M. Sullivan, N.C. Wrinkle, Criticality for the Gehring link problem. Geom. Topol. **10**, 2055–2116 (2006)
15. G. Carlsson, J. Gorham, M. Kahle, J. Mason, Computational topology for configuration spaces of hard disks. Phys. Rev. E **85**, 011303 (2012)
16. F.R. Cohen, Artin's braid groups, classical homotopy theory, and sundry other curiosities, 167–206, in *Braids*, Contemporary Mathematics, vol. 78 (American Mathatical Society, 1988)

17. F.R. Cohen, *Introduction to Configuration Spaces and Their Applications*. Lecture Notes Series, Institute for Mathematical Sciences, National University of Singapore, vol. 19 (World Scientific Publishing, Hackensack, 2010)
18. H. Cohn, Y. Jian, A. Kumar, S. Torquato, Rigidity of spherical codes. Geom. Topol. **15**, 2235–2273 (2011)
19. R. Connelly, Rigidity of packings. Eur. J. Comb. **29**(8), 1862–1871 (2008)
20. J.H. Conway, N.J.A. Sloane, *Sphere Packings, Lattices and Groups*, 3rd edn. (Springer, New York, 1998). (First Edition: 1988)
21. H.S.M. Coxeter, The problem of packing a number of equal non-overlapping circles on a sphere. Trans. New York Acad. Sci. Ser. **II**(24), 220–231 (1962)
22. L. Danzer, *Endliche Punktmengen auf der 2-Sphäre mit möglichst grossen Minimalabstand, Habilitationsschrift* (University of Göttingen, Göttingen, 1963)
23. L. Danzer, Finite point-sets on $\mathbf{S}^2$ with minimum distance as large as possible. Discret. Math. **60**, 3–66 (1986). [English translation of Danzer Habilitationsschrift, with extra references added.]
24. D.M. Dennison, The crystal structure of ice. Phys. Rev. **17**, 20–22 (1921). (Science 24 Sept. 1920 **52**(1343), 296–297)
25. A. Donev, S. Torquato, F.H. Stillinger, R. Connelly, Jamming in hard sphere and hard disk packings. J. Appl. Phys. **95**(3), 989–999 (2004)
26. M.D. Ediger, C.A. Angell, S.R. Nagel, Supercooled liquids and glasses. J. Phys. Chem. **100**, 13200–13212 (1996)
27. A.C. Edmondson, *A Fuller Explanation: The Synergetic Geometry of R* (Buckminster Fuller, Birkhäuser, Boston, 1987)
28. E.R. Fadell, Homotopy groups of configuration spaces and the string problem of Dirac. Duke Math. J. **29**, 231–242 (1962)
29. E.R. Fadell, S.Y. Husseini, *Geometry and Topology of Configuration Spaces, Springer Monographs in Mathematics* (Springer, Berlin, 2001)
30. E.R. Fadell, L. Newirth, Configuration spaces. Math. Scand. **10**, 111–118 (1962)
31. M. Farber, *Invitation to Topological Robotics, Zürich Lectures in Advanced Mathematics* (European Mathematical Society, Switzerland, 2008)
32. E.M. Feichtner, G.M. Ziegler, The integral cohomology algebras of ordered configuration spaces of spheres. Doc. Math. **5**, 115–139 (2000)
33. L. Fejes Tóth, *Über die Abschätzung des kürzesten Abstandes zweier Punkte eines auf einer Kugelfäches liegenden Punktsystems*. Jber. Deutsch. Math. Verein. **53**, 66–68 (1943)
34. L. Fejes, Tóth, Über die dichteste Kugellagerung. Math. Z. **48**, 676–684 (1943)
35. L. Fejes, Tóth, On the densest packing of spherical caps. Am. Math. Mon. **56**, 330–331 (1949)
36. L. Fejes Tóth, *Lagerungen in der Ebene, auf der Kugel und in Raum* (Springer, Berlin, 1953). (2nd edn. 1972)
37. L. Fejes Tóth, Kugelunterdeckungen und Kugelüberdeckungen in Räumen konstanter Krümmung. Archiv der Math. **10**, 307–313 (1959)
38. L. Fejes Tóth, Eräitä "kauniita" extremaalikuvioita, (Finnish) [On some "nice" extremal figures] Arkhimedes **1959**(2), 1–10 (1959)
39. L. Fejes, Tóth, Remarks on a theorem of R. M. Robinson. Studia Scientiarum Mathematicarum Hungarica **4**, 441–445 (1969)
40. F.C. Frank, Supercooling of liquids. Proc. R. Soc. Lond. A Math. Phys. Sci. **215**, 43–46 (1952)
41. D.B. Fuks, Cohomologies of the braid group mod 2. Funct. Anal. Appl. **4**, 143–151 (1970)
42. R.B. Fuller, *Synergetics: The Geometry of Thinking* (Macmillan, New York, 1976)
43. V. Gershkovich, H. Rubinstein, Morse theory for Min-type functions. Asian J. Math. **1**(4), 696–715 (1997)
44. M. Goresky, R. MacPherson, *Stratified Morse Theory, Ergebnisse der Mathematik und ihrer Grenzgeiete 14* (Springer, Berlin, 1988)
45. R.L. Graham, D. Knuth, O. Patashnik, *Concrete Mathematics: A Foundation For Computer Science* (Addison-Wesley, Reading, 1994)
46. D. Gregory, *The Elements of Astronomy, Physical and Geometrical. Done into English, with Additions and Corrections. To which is annex'd Dr. Halley's Synopsis of the Astronomy of Comets. In Two Volumes*, Printed for John Morphew near Stationers Hall: London MDCCXV

47. S. Günther, Ein sterometrisches problem. Archiv Math. Physik (Grunert) **57**, 209–215 (1875)
48. W. Habicht, B.L. van der Waerden, Lagerungen von Punkten auf der Kugel. Math. Ann. **123**, 223–234 (1951)
49. T. Hales, The status of the Kepler conjecture. Math. Intell. **16**, 47–58 (1994)
50. T. Hales, *The strong dodecahedral conjecture and Fejes Tóth's conjecture on sphere packings with kissing number twelve*, pp. 121–132 in: *Discrete Geometry and Optimization*, (K. Bezdek, A. Deza, Y. Ye, eds.) Fields Inst. Commun. **69**: Fields Institute, Toronto (2013)
51. T. Hales et. al., M. Adams, G. Bauer, Dat Tat Dang, T. Harrison, Truong Le Hoang, C. Kaliszk, V. Magron, S. McLaughlin, Thang Tat Nguyen, Truong Quang Nguyen, T. Nipkow, S. Obua, J. Pleso, J. Rute, A. Solovyev, Hoai Thi Ta, Trung Nam Tran, Diep Thi Trieu, J. Urban, Ky Khac Vu, R. Zumkeller, *A Formal Proof of the Kepler Conjecture*, arXiv:1501.02155
52. T. Hales, S. McLaughlin, The dodecahedral conjecture. J. Am. Math. Soc. **23**(2), 299–344 (2010)
53. T. Hariot, *A Briefe and True Report of the New Found Land of Virginia* (Frankfort, Johannis Wecheli, 1590)
54. L. Hárs, The Tammes problem for $n = 10$. Studia Sci. Math. Hungar. **21**(3–4), 439–451 (1986)
55. N.J. Hicks, *Notes on Differential Geometry* (Van Nostrand Co Inc, Princeton, 1965)
56. W.G. Hiscock (ed), *David Gregory, Isaac Newton and the Circle. Extracts from David Gregory's Memoranda 1677–1708* (Oxford, Printed for the Editor 1937)
57. M. Holmes-Cerfon, Enumerating rigid sphere packings. SIAM Rev. **58**(2), 229–244 (2016)
58. R. Hoppe, Bemerkung der Redaktion. Archiv der Mathematik und Physik (Grunert) **56**, 307–312 (1874)
59. M.A. Hoskin, Newton, providence and the universe of stars. J. Hist. Astron. (JHA) **8**, 77–101 (1977)
60. R.H. Kargon, *Atomism in England from Hariot to Newton* (Clarendon Press, Oxford, 1966)
61. J. Kepler, *Strena seu de nive Sexangula*, Frankfurt, Jos. Tampach 1611. Translation as: *The Six-Cornered Snowflake: A New Year's Gift* (Colin Hardie, Translator) (Clarendon Press, Oxford, 1966)
62. J. Kepler, *Epitome Astronomiae Copernicae, usitatâ formâ Quaestionum & Responsionum conscripta, inq; VII. Libros digesta, quorum TRES hi priores sunt de Doctrina Sphaericâ* , Lentijs ad Danubium, excudebat Johannes Plancus, MDCXVIII
63. A. Koyré, *From the Closed World to the Infinite Universe* (The Johns Hopkins Press, Baltimore, 1957)
64. R. Kusner, W. Kusner, J.C. Lagarias, S. Shlosman, *Max-min Morse Theory for Configurations on the 2-Sphere*, Paper in Preparation
65. J.C. Lagarias (ed) *The Kepler Conjecture: The Hales-Ferguson Proof, by Thomas C. Hales, Samuel P. Ferguson* (Springer, New York, 2011)
66. J. Leech, The problem of the thirteen spheres. Math. Gaz. **40**, 22–23 (1956)
67. A.J. Liu, S.R. Nagel, The jamming transition and the marginally jammed solid. Ann. Rev. Condens. Matter Phys. **1**, 347–369 (2010)
68. H. Löwen, Fun with hard spheres, in *Statistical Physics and Spatial Statistics (Wuppertal, 1999)*. Lecture Notes in Physics, vol. 554 (Springer, Berlin, 2000), pp. 295–331
69. B. Lubachevsky, R.L. Graham, Dense packings of $3k(3k+1)+1$ equal disks, in *a circle for* $k = 1, 2, 3, 4$, *and 5, in Computing and Combinatorics, First Annual Conference, COCOON '95*, Lecture Notes in Computer Science, ed. by Du Ding-Zhu, Ming Li, vol. 959, (Springer, New York, 1995), pp. 302–311
70. B. Lubachevsky, F.H. Stillinger, Geometric properties of hard disk packings. J. Stat. Phys. **60**(5–6), 561–583 (1990)
71. H. Maehara, Isoperimetric problem for spherical polygons and the problem of 13 spheres. Ryukyu Math. J. **14**, 41–57 (2001)
72. H. Maehara, The problem of thirteen spheres-a proof for undergraduates. Eur. J. Combin. **28**, 1770–1778 (2007)
73. T.W. Melnyk, O. Knop, W.R. Smith, Extremal arrangements of points and unit charges on a sphere: equilibrium configurations revisited. Canad. J. Chem. **55**, 1745–1761 (1977)

74. J. Milnor, *Morse Theory*. Based on Lecture Notes by M. Spivak, R. Wells. Annals of Mathematics Studies vol. 51 (Princeton University Press, Princeton, 1963)
75. O. Musin, The kissing problem in three dimensions. Discret. Comput. Geom. **35**, 375–384 (2006)
76. O. Musin, A.S. Tarasov, The strong thirteen spheres problem. Discret. Comput. Geom. **48**(1), 128–141 (2012)
77. O. Musin, A.S. Tarasov, Enumerations of irreducible contact graphs on the sphere. Fundam. Prikl. Mat. **18**(2), 125–145 (2013)
78. O. Musin, A.S. Tarasov, The Tammes problem for $N = 14$. Exper. Math. **24**, 460–468 (2015)
79. C.S. O'Hern, L.E. Silbert, A.J. Liu, S.R. Nagel, Jamming at zero temperature and zero applied stress: the epitome of disorder. Phys. Rev. E **68**, 011306 (2003)
80. I. Newton, *The Correspondence of Isaac Newton*, ed. by H.W. Turnbull, J.F. Scott, vol. 9 (Cambridge University Press, Cambridge, 1961)
81. L. Pauling, The structure and entropy of ice and other crystals with some randomness of atomic arrangement. J. Am. Chem. Soc. **57**, 2680–2684 (1935)
82. C.L. Phillips, E. Jankowski, M. Marval, S.C. Glotzer, Self-assembled clusters of spheres related to spherical codes. Phys. Rev. E **86**, 041124 (2012)
83. C.L. Phillips, E. Jankowski, B.J. Krishnatreya, K.V. Edmond, S. Sacanna, D.G. Grier, D.J. Pine, S.C. Glotzer, Digital colloids: reconfigurable clusters as high information density elements. Soft Matter **10**, 7468–7479 (2014)
84. A. Postnikov, R. Stanley, *Deformations of Coxeter hyperplane arrangements. In memory of Gian-Carlo Rota.* J. Comb Theory Ser. A **91**(1–2), 544–597 (2000)
85. R.M. Robinson, Arrangements of 24 points on a sphere. Math. Ann. **144**, 17–48 (1961)
86. R.M. Robinson, Finite sets of points on a sphere with each nearest to five others. Math. Ann. **179**, 296–318 (1969)
87. K. Schütte, B.L. van der Waerden, Auf welcher Kugel haben 5, 6, 7, 8 oder 9 Punkte mit Mindestabstand 1 Platz? Math. Ann. **123**, 96–124 (1951)
88. K. Schütte, B.L. van der Waerden, Das problem der dreizehn Kugeln. Math. Ann. **125**, 325–334 (1953)
89. C. Schwabe, Eureka and Serendipity: The Rudolf van Laban Icosahedron and Buckminster Fuller's Jitterbug, Bridges, Mathematics. Music, Art, Architecture, Culture **2010**, 271–278 (2010)
90. G.D. Scott, D.M. Kilgour, *The density of random close packing of spheres*. Brit. J. Appl. Phys. (J. Phys. D) **2**, 863–866 (1969)
91. J.W. Shirley, *Thomas Hariot: A Biography* (Clarendon Press, Oxford, 1983)
92. P.M.L. Tammes, On the origin of number and arrangement of the places of exit on the surface of pollen-grains. Recueil des travaux botaniques néerlandais **27**, 1–84 (1930)
93. T. Tarnai, Zs. Gáspár, Improved packing of equal circles on a sphere and rigidity of its graph. Math. Proc. Camb. Phil. Soc. **93**, 191–218 (1983)
94. T. Tarnai, Zs. Gáspár, Arrangements of 23 points on a sphere (on a conjecture of R.M. Robinson). Proc. R. Soc. Lond. Ser. A **433**, 257–267 (1991)
95. S. Torquato, F. Stillinger, Jammed hard-particle packings: from Kepler to Bernal and beyond. Rev. Mod. Phys. **82**, 2633–2672 (2010)
96. B. Totaro, Configuration spaces of algebraic varieties. Topology **35**(4), 1057–1067 (1996)
97. H. Verheyen, The complete set of jitterbug transformers and the analysis of their motion, symmetry 2: unifying human understanding. Comput. Math. Appl. **17**(1–3), 203–250 (1989)
98. H. Whitney, Tangents to an analytic variety. Ann. Math. **81**, 496–549 (1964)

Spaces of Convex n-Partitions

Emerson León and Günter M. Ziegler

Abstract We construct and study the space $\mathcal{C}(\mathbb{R}^d, n)$ of all partitions of $\mathbb{R}^d$ into n non-empty open convex regions (n-partitions). A representation on the upper hemisphere of an n-sphere is used to obtain a metric and thus a topology on this space. We show that the space of partitions into possibly empty regions $\mathcal{C}(\mathbb{R}^d, \leq n)$ yields a compactification with respect to this metric. We also describe faces and face lattices, combinatorial types, and adjacency graphs for n-partitions, and use these concepts to show that $\mathcal{C}(\mathbb{R}^d, n)$ is a union of elementary semialgebraic sets.

1 Introduction

In 2006, R. Nandakumar and N. Ramana Rao [13] asked whether any convex polygon for any integer $n \geq 2$ can be cut into n convex pieces of equal area that also have the same perimeter. More generally, is there for any probability measure on $\mathbb{R}^d$ with a continuous density function a partition of $\mathbb{R}^d$ into n convex regions that capture equal parts of the measure and equalize some $d - 1$ additional functions? This problem got a lot of attention (see Nandakumar and Ramana Rao [14], Bárány et al. [1], Karasev et al. [17], and Blagojević and Ziegler [4]), but even the original version of the problem, for polygons in the plane, is still open in the case when n is not a power of a prime. A similar and related problem asks for partitions of space into convex pieces that equipart several measures in a d-dimensional space, generalizing

The first author was funded by DFG through the *Berlin Mathematical School*. Research by the second author was supported by the DFG Collaborative Research Center TRR 109 "Discretization in Geometry and Dynamics".

E. León (✉)
Depto. de Matemáticas, Universidad de los Andes, Bogotá, Colombia
e-mail: emersonleon@gmail.com

G. M. Ziegler
Inst. Mathematics, FU Berlin, Arnimallee 2, 14195 Berlin, Germany
e-mail: ziegler@math.fu-berlin.de

G. Ambrus et al. (eds.), *New Trends in Intuitive Geometry*, Bolyai Society Mathematical Studies 27, https://doi.org/10.1007/978-3-662-57413-3_11

the Ham Sandwich Theorem; see Karasev, Aronov and Hubard [17] and Soberón [18].

All approaches to such problems start with constructing suitable configuration spaces, that is, spaces of partitions of $\mathbb{R}^d$ into n convex regions. In particular, Karasev, Aronov and Hubard [17, Sect. 2] observed that the classical configuration spaces $F(\mathbb{R}^d, n)$ of n distinct labelled points in $\mathbb{R}^d$ can—via optimal transport—be used to parameterize *regular* n-partitions (that is, weighted Voronoi partitions), while Gromov [11] and Nandakumar and Ramana Rao [14] used products of spheres to parameterize the partitions that arise from nested hyperplane 2-partitions.

Motivated by these problems we here consider the set $\mathcal{C}(\mathbb{R}^d, n)$ of *all* partitions of $\mathbb{R}^d$ into n convex regions, for positive integers d and n. We introduce the *spherical representations* of convex n-partitions (Definition 2.5). Using this, we describe a natural metric on this set, and thus can treat $\mathcal{C}(\mathbb{R}^d, n)$ as the *space of all convex n-partitions of* $\mathbb{R}^d$, which in particular is a topological space. These spaces $\mathcal{C}(\mathbb{R}^d, n)$, for $n \geq 1$ and $d \geq 1$, are our main object of study.
The main results presented in this paper are as follows:

- A natural compactification of the space $\mathcal{C}(\mathbb{R}^d, n)$ is given by the space $\mathcal{C}(\mathbb{R}^d, \leq n)$ of possibly non-proper convex n-partitions (Theorem 3.3).
- Each space $\mathcal{C}(\mathbb{R}^d, n)$ can be described as a finite union of semialgebraic sets, so in particular it has a well-defined dimension (Theorem 4.9).
- We define the face structure for each partition and use this to define and distinguish *combinatorial types*, see Definition 4.28. (As the polyhedra in a convex n-partition do not need to be pointed, these definitions are not straightforward.)
- *Realization spaces* arise as the spaces of all partitions that have the same combinatorial type (Definition 4.35). These also yield semialgebraic pieces (Theorem 4.36), from which the whole space $\mathcal{C}(\mathbb{R}^d, n)$ arises as a finite union.

This is made concrete in Sect. 5, where we describe the spaces $\mathcal{C}(\mathbb{R}^d, \leq 2)$ and $\mathcal{C}(\mathbb{R}^1, \leq n)$ as finite cell complexes.

Finally, in Sect. 6 we summarize the relationship between the space $\mathcal{C}(\mathbb{R}^d, n)$ and its subspace $\mathcal{C}_{\text{reg}}(\mathbb{R}^d, n)$ of *regular* subdivisions. The spaces $\mathcal{C}(\mathbb{R}^d, n)$ and $\mathcal{C}_{\text{reg}}(\mathbb{R}^d, n)$ are equivalent in terms of equivariant cohomology; we do not tackle the question whether the spaces $\mathcal{C}(\mathbb{R}^d, n)$ and $\mathcal{C}_{\text{reg}}(\mathbb{R}^d, n)$ are (equivariantly) homotopy equivalent. However, we point out that there is a remarkable difference between the cases $d = 2$ and $d > 2$: In the plane case ($d = 2$) the simple subdivisions are dense in the space of all convex n-partitions, but the space of regular subdivisions has a smaller dimension for $n \geq 4$. In contrast, for $d \geq 3$ all simple n-partitions are regular, and the regular n-partitions are not dense in the space of all n-partitions. Nevertheless, we close our discussion with the curious conjecture that $\dim \mathcal{C}_{\text{reg}}(\mathbb{R}^d, n) = \dim \mathcal{C}(\mathbb{R}^d, n)$ for $d \geq 3$.

2 Convex n-Partitions

We begin here with the definition of convex n-partitions.

Definition 2.1 (*Convex partitions of* $\mathbb{R}^d$*, regions, n-partitions*) Let n and d be positive integers. A *convex partition of* $\mathbb{R}^d$ is an ordered list $\mathcal{P} = (P_1, P_2, \ldots, P_n)$ of non-empty open convex subsets $P_i \subseteq \mathbb{R}^d$ that are pairwise disjoint, so that the union $\bigcup_{i=1}^n \overline{P_i}$ equals $\mathbb{R}^d$, where $\overline{P_i}$ denotes the closure of P_i. Each of the sets P_i is called a *region* of $\mathcal{P}$. The partitions of $\mathbb{R}^d$ into n convex regions are called *n-partitions*.

Since all partitions we are dealing with here are convex, we will often omit this word. The regions of an n-partition are labeled from 1 to n, where the order is important.

Definition 2.2 (*Space of convex n-partitions*) The set of all convex n-partitions of $\mathbb{R}^d$ is denoted by $\mathcal{C}(\mathbb{R}^d, n)$.

As any two regions can be separated by a hyperplane, we get that each region in an n-partition can be described as the set of all points that satisfy a finite set of strict linear inequalities.

Proposition 2.3 *Let* $\mathcal{P} = (P_1, P_2, \ldots, P_n)$ *be an n-partition of* $\mathbb{R}^d$*. Then each region* P_i *is the solution set of a system of* $n-1$ *strict linear inequalities, so it is the interior of a (possibly unbounded) n-dimensional polyhedron with at most* $n-1$ *facets.*

2.1 Spherical Representation and Partitions of S^d

We now introduce convex partitions of the unit d-sphere $S^d \subset \mathbb{R}^{d+1}$. Even if we are primarily interested in partitions of $\mathbb{R}^d$ (as in Definition 2.1), partitions of the sphere appear naturally, generalize partitions of the Euclidean space $\mathbb{R}^d$, and provide a natural setting for the definition and discussion of faces (including "faces at infinity") as well as for the construction of the metric structure and compactification of the space of n-partitions.

A *convex subset* of the unit sphere $S^d \subset \mathbb{R}^{d+1}$ is the intersection of S^d with a convex cone in $\mathbb{R}^{d+1}$. This set is *strictly convex* if in addition it is contained in an open hemisphere of S^d. This notion of spherical convexity could be weaker than what one might expect. For example, two diametrically opposite points on the sphere form a non-connected set that is convex but not strictly convex.

Definition 2.4 (*Convex partitions of* S^d) Let n and d be two positive integers. A *convex n-partition of* S^d is a list $\mathcal{Q} = (Q_1, Q_2, \ldots, Q_n)$ of non-empty open convex subsets $Q_i \subseteq S^d$ that are pairwise disjoint, so that the union $\bigcup_{i=1}^n \overline{Q_i}$ equals S^d.

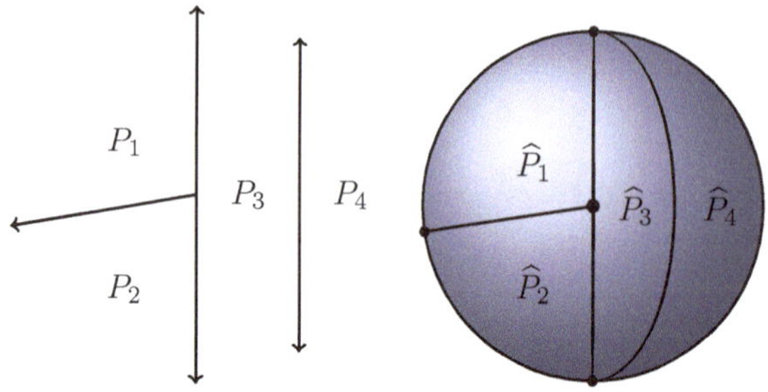

Fig. 1 A 4-partition $\mathcal{P} \in \mathcal{C}(\mathbb{R}^2, 4)$ and the upper hemisphere of its spherical representation

A unit vector $\boldsymbol{v} = (v_0, \ldots, v_d) \in S^d$ lies in the *upper hemisphere* S^d_+ if its first coordinate is positive, $v_0 > 0$. Respectively, $\boldsymbol{v}$ is in the *lower hemisphere* S^d_- if $\boldsymbol{v} \in S^d$ and $v_0 < 0$. The *equator* S^d_0 of S^d is formed by all $\boldsymbol{v} \in S^d$ with $v_0 = 0$. For any $\boldsymbol{x} \in \mathbb{R}^d$ we construct the point

$$\widehat{\boldsymbol{x}} = \frac{1}{\sqrt{1 + \|\boldsymbol{x}\|^2}} \binom{1}{\boldsymbol{x}} \in \mathbb{R}^{d+1},$$

that is, the intersection of the ray $r(\boldsymbol{x}) = \{\lambda \binom{1}{\boldsymbol{x}} \in \mathbb{R}^{d+1} : 0 \leq \lambda \in \mathbb{R}\}$ with S^d. The map $\boldsymbol{x} \mapsto \widehat{\boldsymbol{x}}$ gives a bijection between $\mathbb{R}^d$ and S^d_+.

Definition 2.5 (*Spherical representation*) The *spherical representation* of an n-partition $\mathcal{P}$ of $\mathbb{R}^d$ is the convex $(n+1)$-partition $\widehat{\mathcal{P}} = (\widehat{P}_1, \ldots, \widehat{P}_n, \widehat{P}_\infty)$ of S^d, with regions $\widehat{P}_i = \{\widehat{\boldsymbol{x}} : \boldsymbol{x} \in P_i\}$ for $i = 1, \ldots, n$ and an extra region $\widehat{P}_\infty := S^d_-$. Thus for the spherical representation $\widehat{\mathcal{P}}$ of an n-partition $\mathcal{P} \in \mathcal{C}(\mathbb{R}^d, n)$, we denote the lower index $n+1$ by ∞.

Proposition 2.6 *The spherical representation $\widehat{\mathcal{P}}$ of an n-partition $\mathcal{P}$ of $\mathbb{R}^d$ is a convex partition of S^d with $n+1$ regions.*

Example 2.7 Figure 1 shows a 4-partition $\mathcal{P}$ of $\mathbb{R}^2$ together with the upper hemisphere S^2_+ of its spherical representation $\widehat{\mathcal{P}}$, so the face $\widehat{P}_\infty$ is hidden from view. This partition includes two parallel lines as the boundary of P_3 that in the spherical representation meet at two points "at infinity," that is, on the boundary of S^2_+.

2.2 Faces and the Face Poset

Since we want to study the behavior of n-partitions at infinity as part of the face structure, it will be convenient to use for this the spherical representation. First we introduce the faces of spherical partitions. Faces will be in correspondence with index sets. The faces of an n-partition $\mathcal{P}$ ordered by inclusion will form the face poset of a regular CW ball.

Definition 2.8 (*Index sets and faces of spherical partitions*) Let $\mathcal{Q} = (Q_1, \dots, Q_n)$ be a convex partition of S^d. Let $\overline{Q}_i$ be the closure of Q_i in S^d and $C_i = \text{cone}(\overline{Q}_i)$ for $1 \leq i \leq n$. For any point $\boldsymbol{x}$ in $\mathbb{R}^{d+1}$, we define the *index set* $I(\boldsymbol{x})$ to be the set of values $i \in \{1, 2, \dots, n\}$ such that $\boldsymbol{x} \in C_i$. We define $\mathcal{I}(\mathcal{Q})$ to be the set of all index sets $I(\boldsymbol{x})$ for $\boldsymbol{x} \in \mathbb{R}^{d+1}$.

The *faces of a spherical partition* $\mathcal{Q}$ are all sets $F_I \subseteq S^d$ that can be obtained as an intersection of the form $F_I = \bigcap_{i \in I} \overline{Q_i}$ for some $I \in \mathcal{I}(\mathcal{Q})$. That is, for each $\boldsymbol{x} \in \mathbb{R}^{d+1}$ we obtain the spherical face

$$F_{I(\boldsymbol{x})} = \bigcap_{i \in I(\boldsymbol{x})} \overline{Q_i} \subseteq S^d.$$

Lemma 2.9 *If* $I(\boldsymbol{x}) \subsetneq I(\boldsymbol{x}')$ *then* $F_{I(\boldsymbol{x}')} \subsetneq F_{I(\boldsymbol{x})}$.

Proof The inclusion $F_{I(\boldsymbol{x}')} \subseteq F_{I(\boldsymbol{x})}$ is clear since the intersection $F_{I(\boldsymbol{x}')} = \bigcap_{i \in I(\boldsymbol{x}')} \overline{Q_i}$ includes all terms involved in computing $F_{I(\boldsymbol{x}')}$. Also if $I(\boldsymbol{x}) \neq I(\boldsymbol{x}')$ then $\boldsymbol{x} \notin F_{I(\boldsymbol{x}')}$, since there is at least one $i \in I(\boldsymbol{x}') - I(\boldsymbol{x})$ such that $\boldsymbol{x} \notin \overline{Q_i}$. As $\boldsymbol{x} \in F_{I(\boldsymbol{x})}$ we get a strict inclusion. □

Definition 2.10 (*Faces of partitions of* $\mathbb{R}^d$, *faces at infinity, interior faces, bounded faces*) The *faces of an n-partition* $\mathcal{P}$ of $\mathbb{R}^d$ are all the faces of the spherical representation $\widehat{\mathcal{P}}$, with the exception of $F_{\{\infty\}} = \overline{S^d_-}$. Faces $F_{I(\boldsymbol{x})}$ of $\mathcal{P}$ with $\infty \in I(\boldsymbol{x})$ are called *faces at infinity* of $\mathcal{P}$. All other faces are called *interior faces*. A face is *bounded* if it does not contain any face at infinity.

With this definition, faces of an n-partition $\mathcal{P}$ of $\mathbb{R}^d$ are *not* subsets of $\mathbb{R}^d$, but they are subsets of the closure of S^d_+. Faces at infinity are precisely the faces of $\mathcal{P}$ contained on the boundary of S^d_+, which is the equator S^d_0. For a convex n-partition $\mathcal{P}$ we set $\mathcal{I}(\mathcal{P}) = \mathcal{I}(\widehat{\mathcal{P}}) \setminus \big\{\{\infty\}\big\}$ to be the set of indices of faces of $\mathcal{P}$.

Each n-partition has only finitely many faces $I(\boldsymbol{x})$, as they are subsets of the finite set $I(\mathbf{0}) = \{1, \dots, n, \infty\}$ (where $\mathbf{0}$ represents the origin in $\mathbb{R}^{d+1}$). The union of all faces of $\mathcal{P}$ will be precisely S^d_+, since any point $\boldsymbol{x} \in S^d_+$ is contained in a face, namely in $F_{I(\boldsymbol{x})}$.

Definition 2.11 (*Face poset*) The *face poset* of an n-partition $\mathcal{P}$ is the set of all faces of $\mathcal{P}$, partially ordered by inclusion. It is denoted as $\mathcal{F}(\mathcal{P})$.

Example 2.12 In Fig. 2 we show the face poset of the partition $\mathcal{P}$ on Example 2.7. Here we denote by F_{123} the face $F_{\{1,2,3\}}$, and similarly for other sets of indices, where $F_{I(\mathbf{0})} = F_{1234\infty} = \emptyset$. To obtain the face poset of $\widehat{\mathcal{P}}$ we have to add the face F_∞ as another maximal face above all faces at infinity (appearing with dotted lines in the figure).

Here is a summary of the notation that we have introduced so far related to an n-partition:

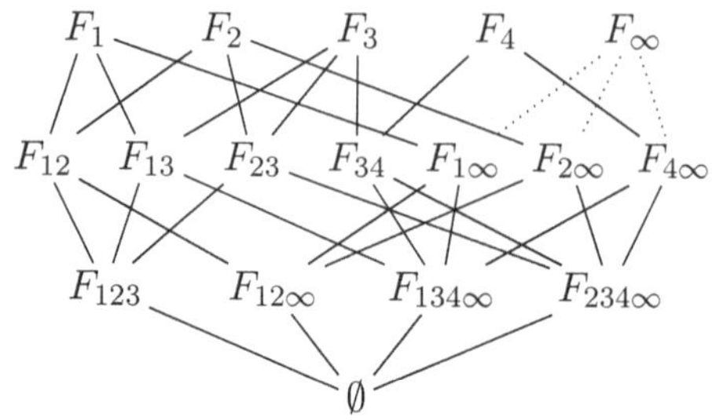

Fig. 2 Face poset of the partition $\mathcal{P}$ on Example 2.7

- $\mathcal{P} = (P_1, P_2, \ldots, P_n)$ denotes an n-partition of $\mathbb{R}^d$, $\mathcal{P} \in \mathcal{C}(\mathbb{R}^d, n)$.
- $\widehat{\mathcal{P}} = (\widehat{P}_1, \ldots, \widehat{P}_n, \widehat{P}_\infty)$ is the spherical representation of $\mathcal{P}$, a partition of S^d into $n+1$ regions.
- $\mathcal{I}(\mathcal{P}) = \mathcal{I}(\widehat{\mathcal{P}}) \setminus \{\{\infty\}\}$ is the set of indices of faces of $\mathcal{P}$.
- $F_I \subset S^d$ are the faces of $\mathcal{P}$, for $I = I(\boldsymbol{x}) \in \mathcal{I}(\mathcal{P})$ and $\boldsymbol{x} \in \mathbb{R}^{d+1}$.
- $C_I = \text{cone}(F_I)$ are the corresponding cones in $\mathbb{R}^{d+1}$.

Faces of $\mathcal{P}$ of dimension k are also known as *k-faces*. The 0-faces of $\mathcal{P}$ are called the *vertices* and the 1-faces are called *edges*, but only in case they are contractible. As an example, a partition of $\mathbb{R}^2$ given by parallel lines has a 0-face that is not a vertex, but rather consists of two separate points at infinity. We introduce now a set of partitions where such strange effects do not occur.

Definition 2.13 (*Essential partitions*) An n-partition $\mathcal{P}$ is *essential* if $F_{I(\mathbf{0})}(\mathcal{P}) = \emptyset$.

Since all $I \in \mathcal{I}(\mathcal{P})$ are contained in $I(\mathbf{0}) = \{1, \ldots, n, \infty\}$, the face $F_{I(\mathbf{0})}$ is the minimal face of the partition. It is easy to check that an n-partition $\mathcal{P}$ is essential if and only if it has a bounded face, and moreover, it is essential if and only if it has an interior vertex.

Definition 2.14 (*Subfaces*) The *subfaces* of a face F_I of an n-partition $\mathcal{P}$ are the faces of F_I considered as a convex spherical polyhedron, i. e. the faces of the cone C_I intersected with S^d for $I \in \mathcal{I}(\mathcal{P})$. Subfaces of a face of $\mathcal{P}$ are also called subfaces of $\mathcal{P}$.

Subfaces of dimension k are denoted as k-subfaces. It is easy to verify that each subface is a union of faces; see [10, Lemma 3.23].

Example 2.15 For the partition in Fig. 1, the region $\widehat{P}_3$ is bounded by two subfaces of dimension 1. One of these subfaces is the face F_{34} of the partition, while the other one is the union of the faces F_{13} and F_{23}.

A *regular CW complex* (a.k.a. *regular cell complex*) is a topological space constructed as the union of a collection of cells homeomorphic to closed balls such that the relative interiors of the cells are disjoint and the boundary of each k-cell is the union of finitely many cells of dimension smaller than k. See for example Munkres [12] or Björner [3]. All the CW complexes we consider are finite and thus compact.

Theorem 2.16 *If $\mathcal{P}$ is an essential n-partition of $\mathbb{R}^d$, then the faces of $\mathcal{P}$ form a regular CW complex homeomorphic to $\overline{S^d_+}$.*

If the partition is not essential, one can see it as a partition of a lower-dimensional subspace.

Proposition 2.17 *The order complex of the face poset of an n-partition $\mathcal{P}$ is homeomorphic to a ball of dimension $d - k$, where $k = \dim F_{I(\mathbf{0})}$.*

Proofs for these results are given in [10]. To obtain a CW-complex homeomorphic to a d-ball, we could instead make a refinement of the faces of any non-essential n-partition. One way to construct such refinement will be presented in Definition 4.20, where node systems are introduced.

3 Metric Structure, Topology and Compactification

In this section we investigate the basic structure of the space $\mathcal{C}(\mathbb{R}^d, n)$ of all convex n-partitions of $\mathbb{R}^d$. First we define a metric on $\mathcal{C}(\mathbb{R}^d, n)$, which induces a topology. Then we introduce a natural compactification, the space $\mathcal{C}(\mathbb{R}^d, \leq n)$.

For non-empty compact convex sets there are two standard ways to measure the distance between them. The *Hausdorff distance* between two compact convex sets $A, B \subset \mathbb{R}^d$ is defined as

$$\delta(A, B) = \max\left(\max_{a\in A}\min_{b\in B}\|a-b\|, \max_{b\in B}\min_{a\in A}\|a-b\|\right) \tag{1}$$

and the *symmetric difference distance*

$$\theta(A, B) = \mathrm{vol}_d(A \triangle B), \tag{2}$$

where $A \triangle B$ denotes the symmetric difference of sets A and B. These metrics induce the same topology, see [7]. They cannot be used directly for unbounded regions, since then the distances would be typically infinite. To remedy this, instead of the usual volume vol_d on $\mathbb{R}^d$ we use a continuous measure μ that is finite, i.e. such that $\mu(\mathbb{R}^d) < \infty$. A measure is *positive* if it is supported on the whole space $\mathbb{R}^d$. Throughout our discussion the measures we consider are positive, continuous, and finite.

The standard d-volume $\mu(P) = \mathrm{vol}_d(\widehat{P})$ of the projection to the sphere is a natural choice for the measure that can be used for any measurable set $P \subseteq \mathbb{R}^d$. This volume is bounded by $\mathrm{vol}_d(S^d_+) = \frac{1}{2}\mathrm{vol}_d(S^d)$. Using this measure μ, we fix a metric on $\mathcal{C}(\mathbb{R}^d, n)$ as follows.

Definition 3.1 Given two n-partitions $\mathcal{P} = (P_1, \ldots, P_n)$ and $\mathcal{P}' = (P_1', \ldots, P_n')$ of $\mathbb{R}^d$, their *distance* $d_\mu(\mathcal{P}, \mathcal{P}')$ is the sum of the measures of the symmetric differences of the corresponding regions,

$$d_\mu(\mathcal{P}, \mathcal{P}') = \sum_{i=1}^{n} \mu(P_i \vartriangle P_i').$$

This distance d_μ is a metric and endows $\mathcal{C}(\mathbb{R}^d, n)$ with the topology that we use for our study. There is a natural compactification for $\mathcal{C}(\mathbb{R}^d, n)$ that is obtained by considering generalized n-partitions that are allowed to have empty regions.

Definition 3.2 (*Non-proper and proper n-partitions*) Let n and d be two positive integers. A *non-proper n-partition of* $\mathbb{R}^d$ is a list $\mathcal{P} = (P_1, P_2, \ldots, P_n)$ of n open convex subsets $P_i \subseteq \mathbb{R}^d$ that are pairwise disjoint, so that the union $\bigcup_{i=1}^n \overline{P_i}$ equals $\mathbb{R}^d$, where now the P_i are allowed to be empty and at least one of the P_i is empty. The convex n-partitions as introduced in Definition 2.1 are called *proper* in this context. We denote by $\mathcal{C}(\mathbb{R}^d, {\leq}n)$ the set of all proper or non-proper n-partitions.

Thus $\mathcal{C}(\mathbb{R}^d, n)$ is the subset of proper partitions in $\mathcal{C}(\mathbb{R}^d, {\leq}n)$. A non-proper partition can also be seen as a k-partition with $k < n$, whose regions have distinct labels in the range from 1 to n, while labels that are not used correspond to empty regions.

Most of the results and definitions we have introduced up to now can be extended to non-proper partitions. The distance d_μ can be extended to $\mathcal{C}(\mathbb{R}^d, {\leq}n)$, so that it is also a metric and topological space. Non-proper partitions also have polyhedral regions as claimed by Theorem 2.3 (now possibly empty). We can also talk about non-proper partitions of a d-sphere, spherical representation of non-proper partitions and face structure, where now the labels of the faces $I(\boldsymbol{x})$ are contained in $I(\boldsymbol{0}) = \{i : C_i \neq \emptyset\}$, the set of labels of all non-empty regions. For a region $P_i = \emptyset$, we define C_i to be empty as well, so that we don't get new faces by adding extra empty regions. As before, a non-proper n-partition is *essential* if $F_{I(\boldsymbol{0})} = \emptyset$.

Theorem 3.3 *The space $\mathcal{C}(\mathbb{R}^d, {\leq}n)$ is compact.*

Proof For the proof we introduce additional spaces that will also be important for the discussion of the semialgebraic structure in the next section. The first one is $(S^d)^{\binom{n}{2}}$, a compact subset of $\mathbb{R}^{(d+1)\times\binom{n}{2}}$. Each of the points $\boldsymbol{c} \in (S^d)^{\binom{n}{2}}$ is represented by $\binom{n}{2}$ unit vectors $\boldsymbol{c}_{ij} \in S^d$ for $1 \leq i < j \leq n$. Each point $\boldsymbol{c}$ can be identified with a central oriented hyperplane arrangement $\mathcal{A}_c$ in $\mathbb{R}^{d+1}$, with $\binom{n}{2}$ hyperplanes H_{ij}. Each hyperplane $H_{ij} \in \mathcal{A}_c$ is given by the linear equation $\boldsymbol{c}_{ij} \cdot \boldsymbol{x} = 0$ and comes with an orientation given by the vector $\boldsymbol{c}_{ij} \in \mathbb{R}^{d+1}$. To keep the symmetry of the notation, H_{ji} denotes the same hyperplane H_{ij} with the opposite orientation, whose normal vector is $\boldsymbol{c}_{ji} = -\boldsymbol{c}_{ij}$.

Now let $\mathcal{D}(\mathbb{R}^d, \leq n)$ be the set of n labeled, disjoint, possibly empty, open polyhedral subsets $(Q_1, \ldots, Q_n)$ of $\mathbb{R}^d$. We fix the topological structure of $\mathcal{D}(\mathbb{R}^d, \leq n)$

in the same way as we did for $\mathcal{C}(\mathbb{R}^d, \leq n)$, using the metric from Definition 3.1. For this, we take the metric on $\mathcal{D}(\mathbb{R}^d, \leq n)$ where the distance of two lists $(Q_1, \ldots, Q_n)$ and $(Q'_1, \ldots, Q'_n)$ in $\mathcal{D}(\mathbb{R}^d, \leq n)$ is given by

$$\sum_{i=1}^{n} \mathrm{vol}_d(\widehat{Q}_i \vartriangle \widehat{Q}'_i),$$

that is, the sum of the measures of the symmetric differences of the projections to S^d of the pairs of corresponding polyhedra in both lists. In this way $\mathcal{C}(\mathbb{R}^d, \leq n)$ is a subspace of $\mathcal{D}(\mathbb{R}^d, \leq n)$, with the corresponding subspace topology.

Equivalently, $\mathcal{D}(\mathbb{R}^d, \leq n)$ can be considered the space of n labeled, disjoint, possibly empty, open spherical polyhedral subsets of S^d_+, the upper hemisphere. The space $\mathcal{C}(\mathbb{R}^d, \leq n)$ is thus the subspace of all lists $(Q_1, \ldots, Q_n) \in \mathcal{D}(\mathbb{R}^d, \leq n)$ for which the union of the closures of the Q_i is the whole $\mathbb{R}^d$.

We define a map $\pi : (S^d)^{\binom{n}{2}} \to \mathcal{D}(\mathbb{R}^d, \leq n)$ by taking for each $\boldsymbol{c} \in (S^d)^{\binom{n}{2}}$ the polyhedra

$$Q_i = \{\boldsymbol{x} \in \mathbb{R}^d : \boldsymbol{c}_{ij} \cdot \tbinom{1}{\boldsymbol{x}} < 0 \text{ for } 1 \leq j \leq n, j \neq i\} \qquad \text{for } 1 \leq i \leq n$$

where $\binom{1}{\boldsymbol{x}} \in \mathbb{R}^{d+1}$ is the vector obtained by adding to $\boldsymbol{x}$ a first coordinate equal to 1. In other words, each Q_i is determined by intersecting the halfspaces $\boldsymbol{c}_{ij} \cdot \binom{1}{\boldsymbol{x}} < 0$ determined by the affine hyperplanes $H_{ij} \in \mathcal{A}$ in $\mathbb{R}^d$, where the orientation of the $\boldsymbol{c}_{ij}$ indicates the side of H_{ij} that must be taken. We recall the convention that $\boldsymbol{c}_{ji} = -\boldsymbol{c}_{ij}$, which implies that all Q_i are disjoint. The polyhedral sets Q_i might be empty.

The map $\pi : (S^d)^{\binom{n}{2}} \to \mathcal{D}(\mathbb{R}^d, \leq n)$ is continuous: If we move the hyperplanes a small amount, the polyhedra projected to the sphere also change slightly and the sum of the d-volume of the symmetric differences must be small.

With this we can now complete the proof of Theorem 3.3. Since the space $(S^d)^{\binom{n}{2}}$ is compact, the image of the continuous map π is also compact. On this image we have a continuous function f to $\mathbb{R}$, given by $f(Q_1, \ldots, Q_n) = \sum_{i=1}^{n} \mathrm{vol}_d(\widehat{Q}_i)$.

This is a continuous function, so the preimage of the maximal value, namely the d-volume of S^d_+, is a closed subset of a compact space, so it is compact as well. This preimage is denoted by $\mathcal{H}(R^d, \leq n)$ (as explained later in Definition 4.7). We conclude that $\mathcal{C}(R^d, \leq n)$ is compact, as it is the image under π of the compact space $\mathcal{H}(R^d, \leq n)$. □

The space $\mathcal{C}(\mathbb{R}^d, n)$ is not compact for $n > 1$, since the limit of a sequence of proper partitions might have empty regions. On the other hand, any non-proper partition can be obtained as a limit of proper partitions. To see this, take a non-proper partition and subdivide one of its regions into one big and some small convex pieces, to get a proper n-partition out of it. If the measure of the small pieces goes to zero, in the limit we end up at the non-proper partition we started with. Therefore we can think of $\mathcal{C}(\mathbb{R}^d, \leq n)$ as a compactification of $\mathcal{C}(\mathbb{R}^d, n)$.

4 Semialgebraic Structure

A subset of $\mathbb{R}^m$ is semialgebraic if it can be described as a finite union of solution sets of systems given by finitely many polynomial equations and strict inequalities on the coordinates of $\mathbb{R}^m$. In this section we prove that each of the spaces $\mathcal{C}(\mathbb{R}^d, n)$ and $\mathcal{C}(\mathbb{R}^d, \leq n)$ is a union of finitely many pieces that can be parameterized by semialgebraic sets.

We refer to Bochnak, Coste and Roy [5] and Basu, Pollack and Roy [2] as general references on semialgebraic sets. We will use here some basic results about semialgebraic sets, such as the fact that finite unions and intersections of semialgebraic sets are semialgebraic, and the fact that the complements of semialgebraic sets are again semialgebraic. Most notably, we will use the Tarski–Seidenberg Theorem, which says that semialgebraic sets are closed under projections.

Theorem 4.1 (Tarski–Seidenberg [5, Theorem 2.2.1]) *If $X \subset \mathbb{R}^n \times \mathbb{R}^m$ is a semi-algebraic set, and if p is the projection onto the first n coordinates, then $p(X) \subseteq \mathbb{R}^n$ is also semialgebraic.*

We will also use some of the notation introduced in the proof of Theorem 3.3, such as the map $\pi : (S^d)^{\binom{n}{2}} \to \mathcal{D}(\mathbb{R}^d, \leq n)$. Note that the space $(S^d)^{\binom{n}{2}} \subset \mathbb{R}^{(d+1)\binom{n}{2}}$ is semialgebraic.

4.1 Hyperplane Description

Definition 4.2 (*Hyperplane arrangement carrying a partition*) Let $\mathcal{P}$ be an n-partition of $\mathbb{R}^d$. An oriented hyperplane arrangement $\mathcal{A}_c$ for $\boldsymbol{c} \in (S^d)^{\binom{n}{2}}$ *carries the partition* $\mathcal{P}$ if $\pi(\boldsymbol{c}) = \mathcal{P}$.

In other words, an oriented hyperplane arrangement $\mathcal{A}_c$ for $\boldsymbol{c} \in (S^d)^{\binom{n}{2}}$ carries the partition $\mathcal{P} = (P_1, \ldots, P_n)$ if the regions $\widehat{P_i}$ and $\widehat{P_j}$ are separated by the hyperplane H_{ij}, so that $\boldsymbol{c}_{ij} \cdot \boldsymbol{x} < 0$ for $\boldsymbol{x} \in \widehat{P_i}$ and $\boldsymbol{c}_{ij} \cdot \boldsymbol{x} > 0$ for $\boldsymbol{x} \in \widehat{P_j}$.

Following the proof of Proposition 2.3, we can see that for each n-partition $\mathcal{P} \in \mathcal{C}(\mathbb{R}^d, n)$ there is at least one hyperplane arrangement $\mathcal{A}$ that carries it. This arrangement is usually not unique. Similarly, each non-proper partition $\mathcal{P} \in \mathcal{C}(\mathbb{R}^d, \leq n)$ is carried by a hyperplane arrangement: If a region P_i is empty, any hyperplane H_{ij} that doesn't intersect P_j is good enough to separate these two regions. In the case $P_i = \mathbb{R}^d$ we can still take $\boldsymbol{c}_{ij} = (1, 0, \ldots, 0)$.

Example 4.3 Figure 3 (left) shows a partition of $\mathbb{R}^2$ into four regions, but to make the example more interesting we consider it as a non-proper partition in $\mathcal{C}(\mathbb{R}^2, \leq 5)$, with an extra empty region $P_5 = \emptyset$.

In Fig. 3 (right) we show an affine picture of a hyperplane arrangement carrying $\mathcal{P}$. For adjacent regions P_i and P_j, with $\{i, j\} \in A(\mathcal{P})$, there is only one possible hyperplane H_{ij}^{aff} that separates them, namely the affine span of the points on the

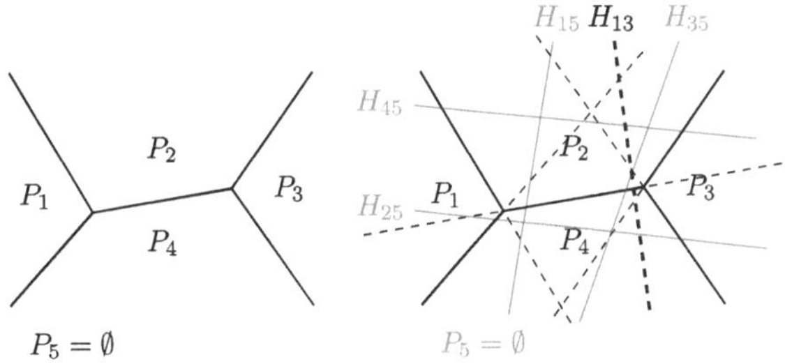

Fig. 3 Non-proper partition $\mathcal{P}$ in $\mathcal{C}(\mathbb{R}^2, \leq 5)$ together with a possible hyperplane arrangement carrying it

intersection of the boundaries. The extension of these hyperplanes appears on the figure as dashed lines. For all other hyperplanes there is some freedom to choose them. In the figure, a label appears next to each of them. For the hyperplanes involving the region P_5, it is only necessary that the other region lies entirely on one side of the hyperplane.

The unit vector $\boldsymbol{c}_{ij}$ is uniquely determined by the hyperplane H_{ij} and the requirement that P_i and P_j lie on the correct sides of H_{ij}, unless $P_i = P_j = \emptyset$. We remind the reader that an affine hyperplane H^{aff} given by the points $\boldsymbol{x} \in \mathbb{R}^d$ that satisfy an equation of the form $\boldsymbol{a} \cdot \boldsymbol{x} = b$ for $\boldsymbol{a} \in \mathbb{R}^d$ and $b \in \mathbb{R}$ is represented projectively by its corresponding vector $\boldsymbol{c} = (-b, a_1, \ldots, a_d) \in \mathbb{R}^{d+1}$ or by the vector $(b, -a_1, \ldots, -a_d)$ in case that the opposite orientation is required. This vector might later be normalized.

Definition 4.4 (*Regions of a hyperplane arrangement*) Let $\mathcal{A}_{\boldsymbol{c}}$ for $\boldsymbol{c} \in (S^d)^{\binom{n}{2}}$ be a hyperplane arrangement with hyperplanes $H_{ij} = \{\boldsymbol{x} \in \mathbb{R}^{d+1} : \boldsymbol{c}_{ij} \cdot \boldsymbol{x} = 0\}$. Let $\boldsymbol{s} \in \{+1, -1\}^{\binom{n}{2}}$ be a sign vector with coordinates $s_{ij} \in \{+1, -1\}$ for $1 \leq i < j \leq n$. A *region* $R_{\boldsymbol{s}}$ of the affine hyperplane arrangement $\mathcal{A}_{\boldsymbol{c}}^{aff}$ is a subset of the form

$$R_{\boldsymbol{s}} = \{\boldsymbol{x} \in \mathbb{R}^d : s_{ij}\boldsymbol{c}_{ij} \cdot \tbinom{1}{\boldsymbol{x}} < 0 \text{ for all } i < j\}.$$

Thus $R_{\boldsymbol{s}}$ is determined by intersecting the halfspaces $s_{ij}\boldsymbol{c}_{ij} \cdot \binom{1}{\boldsymbol{x}} < 0$ determined by the affine hyperplanes H_{ij} of $\mathcal{A}_{\boldsymbol{c}}^{aff}$ in $\mathbb{R}^d$, where each coordinate s_{ij} indicates the side of H_{ij} that contains $R_{\boldsymbol{s}}$, for $1 \leq i < j \leq n$. Regions $R_{\boldsymbol{s}}$ of a hyperplane arrangement may be empty.

We symmetrize the notation by setting $s_{ji} = -s_{ij}$ for $i > j$. To avoid confusion with the regions of partitions, we always talk about regions $R_{\boldsymbol{s}}$ when we refer to a region of hyperplane arrangements. Also, regions $R_{\boldsymbol{s}}$ of $\mathcal{A}_{\boldsymbol{c}}$ simply denote regions of $\mathcal{A}_{\boldsymbol{c}}^{aff}$.

The *complete graph* K_n is the graph with vertex set $\{1, \ldots, n\}$ and with an edge between each pair of vertices. It has $\binom{n}{2}$ edges. An *orientation* of K_n is obtained by taking the graph K_n and fixing a direction to each edge e of K_n, by choosing which of the vertices of e is the *tail* and which is the *head*. A graph with all its edges oriented is also known as a *directed graph*. Each sign vector $\boldsymbol{s} \in \{+1, -1\}^{\binom{n}{2}}$ generates an orientation $G_{\boldsymbol{s}}$ of the complete graph K_n, where the edge ij is directed from i to j if $s_{ij} = +1$, and from j to i otherwise, for $1 \leq i < j \leq n$. A *source* of $G_{\boldsymbol{s}}$ is a vertex

v of K_n that is not the head of any of the edges involving v in G_s. Since the graph K_n is complete, there can be at most one source in the directed graph G_s.

Lemma 4.5 *An oriented hyperplane arrangement $\mathcal{A}_c$ for $c \in (S^d)^{\binom{n}{2}}$ carries a (possibly non-proper) n-partition $\mathcal{P}$ if and only if for each non-empty region R_s of $\mathcal{A}_c$ the corresponding oriented complete graph G_s has a source. The partition is proper if and only if for each $i \in \{1, \dots, n\}$, there is at least one non-empty region R_s such that the source of G_s is the vertex i.*

Proof If $\mathcal{P}$ is an n-partition carried by $\mathcal{A}_c$, all non-empty regions R_s of $\mathcal{A}_c$ must be contained in some fixed region P_i of $\mathcal{P}$. If $R_s \subseteq P_i$, then we have that $s_{ij} = +1$ for all $j \neq i$, so i is a source in G_s.

On the other hand, if the directed graphs of all non-empty regions R_s have a source, then we obtain an n-partition by taking

$$P_i = \bigcap_{j \neq i} \{x \in \mathbb{R}^d : c_{ij} \cdot \begin{pmatrix} 1 \\ x \end{pmatrix} \leq 0\}.$$

The regions P_i are clearly disjoint, and their union covers all regions R_s of the hyperplane arrangement, since $R_s \subseteq P_i$ whenever i is the unique source of G_s. Therefore the union of the closures of the regions must be $\mathbb{R}^d$. The regions P_i as defined might still be empty, but if there is a non-empty region R_s in $\mathcal{A}_c$ with G_s having as source the vertex i for each i, then $R_s \subseteq P_i$ is non-empty and the partition is proper. □

Lemma 4.6 *For any $s \in \{+1, -1\}^{\binom{n}{2}}$, the set of points $c \in (S^d)^{\binom{n}{2}}$ for which the region R_s in the hyperplane arrangement $\mathcal{A}_c$ is empty is semialgebraic. The set of points c such that the region R_s in the hyperplane arrangement $\mathcal{A}_c$ is non-empty is also semialgebraic.*

Proof The region R_s is non-empty if and only if there is some $x \in \mathbb{R}^{d+1}$ such that $s_{ij} c_{ij} \cdot x < 0$ for each pair $i < j$. We can add the coordinates of x as slack variables and construct a semialgebraic set X on the coordinates of c_{ij} for $1 \leq i < j \leq n$ and of x, so that all inequalities $s_{ij} c_{ij} \cdot x < 0$ are satisfied. The parameterization of the set of all hyperplane arrangements $\mathcal{A}_c$ with $R_s \neq \emptyset$ can be obtained as a projection of X to the coordinates $c \in (S^d)^{\binom{n}{2}}$ and by Theorem 4.1, we conclude that the set of arrangements with $R_s \neq \emptyset$ is semialgebraic.

Since the complement of a semialgebraic set is semialgebraic, the set of arrangements such that $R_s = \emptyset$ is semialgebraic. Alternatively, one can use a version of the Farkas Lemma [20, Sect. 1.4] to get a semialgebraic description of this set of arrangements. □

Definition 4.7 (*The spaces $\mathcal{H}(\mathbb{R}^d, \leq n)$ and $\mathcal{H}(\mathbb{R}^d, n)$*) Let $\mathcal{H}(\mathbb{R}^d, \leq n)$ denote the space of all $c \in (S^d)^{\binom{n}{2}}$ such that the hyperplane arrangement $\mathcal{A}_c$ carries a possibly non-proper n-partition of $\mathbb{R}^d$. The subset of $\mathcal{H}(\mathbb{R}^d, \leq n)$ corresponding to hyperplane arrangements carrying a proper n-partition is denoted as $\mathcal{H}(\mathbb{R}^d, n)$.

We have the following chain of inclusions.

$$\mathcal{H}(\mathbb{R}^d, n) \subseteq \mathcal{H}(\mathbb{R}^d, \leq n) \subseteq (S^d)^{\binom{n}{2}} \subseteq \mathbb{R}^{(d+1)\times\binom{n}{2}}.$$

Theorem 4.8 *The spaces $\mathcal{H}(\mathbb{R}^d, \leq n)$ and $\mathcal{H}(\mathbb{R}^d, n)$ are semialgebraic sets.*

Proof By Lemma 4.5, a hyperplane arrangement $\mathcal{A}_c$ for $\boldsymbol{c} \in (S^d)^{\binom{n}{2}}$ carries an n-partition $\mathcal{P}$ if and only if for all regions R_s in $\mathcal{A}_c$ the oriented graph G_s have a source. Therefore, we need to characterize all hyperplane arrangements $\mathcal{A}_c$ such that all regions R_s of $\mathcal{A}_c$ are empty for all sign vector $\boldsymbol{s}$ in $S = \{\boldsymbol{s} \in \{+1, -1\}^{\binom{d}{2}} : G_s \text{ has no source}\}$. By Lemma 4.6 we find that $\mathcal{H}(\mathbb{R}^d, \leq n)$ is a finite intersection of semialgebraic sets over the coordinates of $\boldsymbol{c}_{ij}$ as variables, and therefore semialgebraic.

Also the set $\mathcal{H}(\mathbb{R}^d, n)$ of hyperplane arrangements carrying a proper n-partition, where at least one region R_s has source i for each $i \leq n$, is semialgebraic, again by Lemma 4.6. □

From Theorem 4.8 we can see that $\mathcal{H}(\mathbb{R}^d, \leq n)$ is the union of all sets of arrangements with an adjacency graph that satisfies the source conditions specified in Lemma 4.5.

Theorem 4.9 *The space $\mathcal{C}(\mathbb{R}^d, \leq n)$ is the union of finitely many subspaces indexed by adjacency graphs $A(\mathcal{P})$, which can be parameterized as semialgebraic sets. The same statement is true for the space $\mathcal{C}(\mathbb{R}^d, n)$.*

Proof The map $\pi : \mathcal{H}(\mathbb{R}^d, \leq n) \to \mathcal{C}(\mathbb{R}^d, \leq n)$ is a surjective continuous map taking each oriented hyperplane arrangement $\mathcal{A}$ in $\mathcal{H}(\mathbb{R}^d, \leq n)$ to its corresponding partition. The pieces of $\mathcal{C}(\mathbb{R}^d, \leq n)$ are given by the partitions in $\mathcal{C}(\mathbb{R}^d, \leq n)$ that share the same adjacency graph $A(\mathcal{P})$, for any given n-partition $\mathcal{P}$. Each of these pieces is denoted as $\mathcal{C}_{A(\mathcal{P})}(\mathbb{R}^d, n)$ for each partition $\mathcal{P}$, and the inverse image $\pi^{-1}(\mathcal{C}_{A(\mathcal{P})}(\mathbb{R}^d, n))$ is denoted as $\mathcal{H}_{A(\mathcal{P})}(\mathbb{R}^d, n)$.

To see that $\mathcal{H}_{A(\mathcal{P})}(\mathbb{R}^d, n)$ is a semialgebraic set, we take the description of $\mathcal{H}(\mathbb{R}^d, n)$ and add extra restrictions to express the fact that certain hyperplanes do not determine any $(d-1)$-face of the partition. These extra restrictions are described in what follows.

A pair $\{i, j\}$ is in $A(\mathcal{P})$ for $\mathcal{P} = \pi(\mathcal{A})$ if and only if there are $\boldsymbol{s}, \boldsymbol{s}' \in \{+1, -1\}^{\binom{n}{2}}$ with exactly the same entries, except only by the entry $s_{ij} = -s'_{ij}$, with oriented graphs G_s, $G_{s'}$ having sources i and j respectively, so that the regions R_s, $R_{s'}$ are non-empty.

Using Lemma 4.6 we find that the subset of arrangements $\mathcal{A}' \in \mathcal{H}(\mathbb{R}^d, n)$ with $\{i, j\} \in A(\pi(\mathcal{A}))$ for a given $\mathcal{A} \in \mathcal{H}(\mathbb{R}^d, n)$ is semialgebraic, since it is the union over all pairs $\boldsymbol{s}, \boldsymbol{s}'$ that differ only in the ij-coordinate and with respective graphs sources i and j of the subsets of $\mathcal{H}(\mathbb{R}^d, n)$ where R_s and $R_{s'}$ are non-empty. The complements of those subsets, which represent hyperplane arrangements with $\{i, j\} \notin A(\pi(\mathcal{A}))$, are also semialgebraic.

Finally $\mathcal{H}_{A(\mathcal{P})}(\mathbb{R}^d, n)$ is the intersection of subsets of $\mathcal{H}(\mathbb{R}^d, n)$ where $\{i, j\} \in A(\pi(\mathcal{A}))$ for $\{i, j\} \in A(\mathcal{P})$ and $\{k, \ell\} \notin A(\pi(\mathcal{A}))$ for $\{k, \ell\} \notin A(\mathcal{P})$ and thus it is also a semialgebraic set. Since the map $\pi : \mathcal{H}(\mathbb{R}^d, \leq n) \to \mathcal{C}(\mathbb{R}^d, \leq n)$ is a projection obtained by deleting the coordinates $\boldsymbol{c}_{ij}$ of the hyperplanes H_{ij} for $\{i, j\} \notin A(\mathcal{P})$, by Theorem 4.1 we conclude that $\mathcal{C}_{A(\mathcal{P})}(\mathbb{R}^d, n)$ is a semialgebraic set on the coordinates of the vectors $\boldsymbol{c}_{ij}$ for $\{i, j\} \in A(\mathcal{P})$ and $\mathcal{C}(\mathbb{R}^d, \leq n)$ is a union of semialgebraic pieces.

If there are two or more non-empty regions in $\mathcal{P}$, the vertices of $A(\mathcal{P})$ contained in at least one edge correspond to the non-empty regions of $\mathcal{P}$. Therefore, we can obtain $\mathcal{C}(\mathbb{R}^d, n)$ as the union of the semialgebraic pieces of the form $\mathcal{C}_{A(\mathcal{P})}(\mathbb{R}^d, n)$ where $A(\mathcal{P})$ is a connected graph on the vertices from 1 to n. □

It is not enough to know these semialgebraic pieces in order to reconstruct the spaces $\mathcal{C}(\mathbb{R}^d, n)$ and $\mathcal{C}(\mathbb{R}^d, \leq n)$. We also need the topological structure induced by the metric given in Sect. 3 to know how to glue the different semialgebraic pieces of the form $\mathcal{C}_{A(\mathcal{P})}(\mathbb{R}^d, n)$ in order to obtain the spaces of n-partitions $\mathcal{C}(\mathbb{R}^d, n)$ and $\mathcal{C}(\mathbb{R}^d, \leq n)$.

On each semialgebraic piece $\mathcal{C}_{A(\mathcal{P})}(\mathbb{R}^d, n)$ we have a topological structure by viewing it as a subset of $\mathbb{R}^{(d+1)\times E}$ given by the parameterization through the $\boldsymbol{c}_{ij}$, where E is the number of edges in $A(\mathcal{P})$. This topological structure is equivalent to the one that is obtained by considering $\mathcal{C}_{A(\mathcal{P})}(\mathbb{R}^d, n)$ as a subset of $\mathcal{C}(\mathbb{R}^d, \leq n)$: To see this, note that a sequence of partitions $(\mathcal{P}^k)_{k\in\mathbb{N}}$ in $\mathcal{C}_{A(\mathcal{P})}(\mathbb{R}^d, n)$ converges to a partition $\mathcal{P}$ in the δ_μ-topology if and only if each sequence of coordinates c_{ij}^k of the parameterizations of $\mathcal{P}^k$ for $\{i, j\} \in A(\mathcal{P})$ converges to the corresponding coordinate of $\boldsymbol{c}_{ij}$.

4.2 Node Systems and Combinatorial Types

Pointed partitions are an important class of partitions, where every face is completely determined by its set of vertices. For general partitions the same doesn't hold. We introduce "node systems" to get similar properties for any n-partition (Definition 4.16).

Definition 4.10 (*Pointed partitions*) A cone C is *pointed* if it doesn't contain a linear subspace of positive dimension. An n-partition $\mathcal{P} = (P_1, \ldots, P_n)$ of $\mathbb{R}^d$ is *pointed* if for each region P_i the cone C_i is pointed.

Here we state some simple results about pointed partitions; proofs are on record in [10].

Proposition 4.11 *If $\mathcal{P}$ is a pointed n-partition, then every face F_I of $\mathcal{P}$ can be obtained as the spherical convex hull of all vertices in F_I.*

Proposition 4.12 *Pointed n-partitions are essential.*

The converse of Proposition 4.12 is not true: Example 2.7 shows a 4-partition that is essential but not pointed.

Now we define the node systems of an n-partition, in order to get that every face is the spherical convex hull of its corresponding nodes. For a pointed partition $\mathcal{P}$, the nodes will coincide with the vertices of $\mathcal{P}$.

Definition 4.13 (*Half-linear faces*) A face F of a partition $\mathcal{P}$ is *half-linear* if it is the intersection of S^d with a linear subspace of $\mathbb{R}^{d+1}$ and a unique closed halfspace given by a linear inequality. The set of half-linear faces of a partition is denoted as $\mathcal{F}^H(\mathcal{P})$.

The *relative boundary* of a polyhedral set $S \subseteq \mathbb{R}^d$ is its boundary as a subset of its affine span aff S. Define similarly the *relative interior* of S. These concepts differ from the usual boundary and interior in case that S has codimension greater than one. If a face F is half-linear, then it has a unique linear subface F' in its relative boundary. The subface F' is the union of some faces of $\mathcal{P}$, and is the intersection of a linear subspace with S^d. F' cannot have any boundary since it is topologically a sphere (of dimension $\dim F' = \dim F - 1$) and is the union of some faces at infinity of $\mathcal{P}$. Since $\hat{P}_\infty$ is not a face of $\mathcal{P}$, in particular it is not a half-linear face of $\mathcal{P}$ (but it is a half-linear face of the spherical partition $\hat{\mathcal{P}}$).

The only face F_I of $\mathcal{P}$ such that its corresponding cone C_I is a linear subspace is the minimal face $F_{I(\mathbf{0})}$ [10, Lemma 3.22]. This face has no boundary and no subfaces. All faces covering $F_{I(\mathbf{0})}$ in the face poset are half-linear. If a partition is essential, all vertices are half-linear faces.

Example 4.14 For the 4-partition $\mathcal{P}$ of Example 2.7/Fig. 1, every vertex is half-linear (there are four of them). Besides, there are two more half-linear faces in the figure, namely the faces F_{34} and $F_{4\infty}$. For these two 1-faces, there is a unique linear subface that covers the relative boundary and is the union of two vertices of $\mathcal{P}$.

Example 4.15 A 4-partition $\mathcal{P}'$ of the plane given by four regions separated by three parallel lines is non-essential: Its minimal face $F_{I(\mathbf{0})}(\mathcal{P}')$ consists of two antipodal points. Here all 1-faces are half-linear, since they cover $F_{I(\mathbf{0})}(\mathcal{P}')$ and there are no other half-linear faces on this partition.

Definition 4.16 (*Node systems, nodes*) Let $\mathcal{P}$ be a partition in $\mathcal{C}(\mathbb{R}^d, \leq n)$. If $\mathcal{P}$ is essential, a *node system* N of $\mathcal{P}$ is a set of points $\boldsymbol{v}_F$, one in the relative interior of each half-linear face F of $\mathcal{P}$. If the partition $\mathcal{P}$ is non-essential, with $\dim F_{I(\mathbf{0})} = k \geq 0$, then a node system again contains one point $\boldsymbol{v}_F$ in the relative interior of each half-linear face F of $\mathcal{P}$, and additionally an ordered sequence of $k+2$ extra points $\boldsymbol{v}_1, \ldots, \boldsymbol{v}_{k+2}$ on the face $F_{I(\mathbf{0})}$ such that they positively span the linear subspace $C_{I(\mathbf{0})}$.

The points in a node system are referred as *nodes*. We denote by $N(\mathcal{P})$ the set of all node systems of $\mathcal{P}$. Note that all vertices of $\mathcal{P}$ are also nodes in any node system of $\mathcal{P}$.

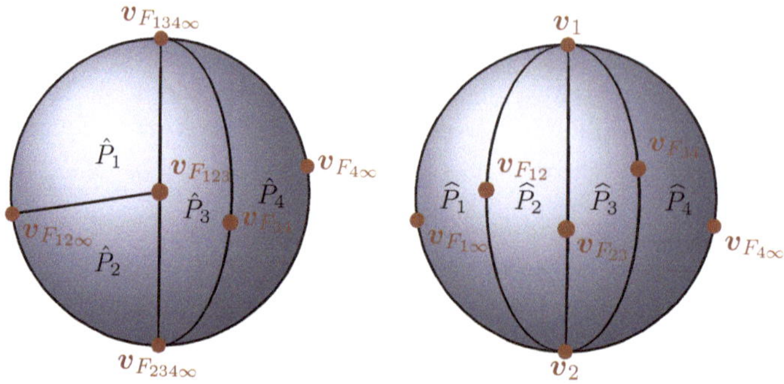

Fig. 4 Node systems for two different 4-partitions. They have a node in the relative interior of each half-linear face

We sometimes write $\boldsymbol{v}_F(N) = \boldsymbol{v}_F \in N$, in case it might not be clear which node system we are using. If $\mathcal{P}$ is non-essential, the same applies to the nodes $\boldsymbol{v}_i$ in the minimal face.

Example 4.17 Here we construct node systems for both partitions of the Examples 4.14 and 4.15. In the first case, every vertex of $\mathcal{P}$ must be a node. We need to include two more nodes $\boldsymbol{v}_{F_{34}}$ and $\boldsymbol{v}_{F_{4\infty}}$ in the relative interior of the faces F_{34} and $F_{4\infty}$ respectively. We have one degree of freedom to choose each of these two nodes. In Fig. 4 (left) we depict one possible choice for a node system N of $\mathcal{P}$.

For the second partition, in Fig. 4 (right), we need to have two nodes $\boldsymbol{v}_1$ and $\boldsymbol{v}_2$ on the linear face $F_{I(\mathbf{0})}(\mathcal{P}')$, so that they positively span $C_{I(\mathbf{0})}$. There are two possibilities to choose $\boldsymbol{v}_1$, and $\boldsymbol{v}_2$ must be the antipodal point $-\boldsymbol{v}_1$. Besides these two nodes, we need five more nodes, one in the relative interior of each half-linear face, to get a node system N' of $\mathcal{P}'$.

Proposition 4.18 *If $\mathcal{P}$ is an essential n-partition, then the set $N(\mathcal{P})$ of all node systems is a semialgebraic set of dimension* $\dim N(\mathcal{P}) = \sum_{F\in\mathcal{F}^H} \dim(F)$.

If $\mathcal{P}$ is non-essential and $k = \dim(F_{I(\mathbf{0})})$, the set $N(\mathcal{P})$ has dimension

$$\dim N(\mathcal{P}) = k(k+2) + \sum_{F\in\mathcal{F}^H} \dim(F).$$

The proof of this proposition and a more precise description of the set of all node systems for a given partition can be found in [10, Proposition 4.29].

Lemma 4.19 *If N is a node system of a partition $\mathcal{P}$, then any face F of $\mathcal{P}$ can be obtained as the spherical convex hull of the set of nodes in N contained in F.*

Proof By induction on the dimension of F, first take $F = F_{I(\mathbf{0})}$ to be the minimal face of $\mathcal{P}$, with $\dim(F_{I(\mathbf{0})}) = k$. We have $k+2$ nodes in $F_{I(\mathbf{0})}$ that positively span the

$(k+1)$-dimensional linear subspace $C_{I(\mathbf{0})}$, and therefore its spherical convex hull is equal to $F_{I(\mathbf{0})}$. If the partition is essential, $F_{I(\mathbf{0})} = \emptyset$ doesn't contain any node, and its convex hull is also empty.

Now suppose that $\dim(F) = m$ and every face F' of $\mathcal{P}$ with $\dim F' < m$ is equal to the convex hull of the nodes contained in F'. If F is half-linear, we have an extra node $\boldsymbol{v}_F$ in the interior of F, and any other point $\boldsymbol{x}$ in F is in an interval between $\boldsymbol{v}_F$ and a point $\boldsymbol{x}'$ on the boundary of F. Since $\boldsymbol{v}_F$ cannot be antipodal to $\boldsymbol{x}'$, then $\boldsymbol{x}$ can be written as a positive combination of $\boldsymbol{v}_F$ and $\boldsymbol{x}'$. By the induction hypothesis, $\boldsymbol{x}'$ is a positive combination of the nodes in the face where it belongs that are also contained in F. We use here that subfaces are union of faces. Therefore $\boldsymbol{p}$ is in the spherical convex hull of the nodes in F.

If F is not half-linear, we can find a node $\boldsymbol{v}$ on the boundary such that its antipodal point is not in F. Now we can repeat the argument given before, and the result follows. □

Definition 4.20 (*Cell complex from a node system*) For any node system N of a partition $\mathcal{P}$, there is a CW complex $\mathcal{P}_N$ such that the vertices of this complex are precisely the nodes in N, and such that each face of $\mathcal{P}$ is union of faces of $\mathcal{P}_N$. The *complex* $\mathcal{P}_N$ is obtained recursively as follows:

- Include a face F_S in $\mathcal{P}_N$ for every subset S of nodes contained in the minimal face $F_{I(\mathbf{0})}$, with $|S| \leq k+1$, where F_S is the spherical convex hull of S and $k = \dim F_{I(\mathbf{0})}$. For essential partitions, only the empty set is included in this step.
- For every half-linear face F of $\mathcal{P}$ such that the boundary is already covered by faces of $\mathcal{P}_N$, the spherical convex hull of every face G of $\mathcal{P}_N$ contained on the boundary of F together with the node $\boldsymbol{v}_F$ is also a face of $\mathcal{P}_N$. (These faces of $\mathcal{P}_N$ are pyramids over the faces on the boundary of F.)
- All other faces of $\mathcal{P}$ that are not linear or half-linear are also faces of $\mathcal{P}_N$.

Example 4.21 For the two partitions given in Fig. 4, the cell complex obtained from this construction coincides precisely with what is shown in the picture, where every half-linear 1-face is subdivided in two segments and every non-pointed region forms a 2-cell with four nodes and four 1-faces on the boundary. For a more illustrative example, consider the 1-partition of $\mathbb{R}^2$ into one region. This "partition" is non-essential, with minimal face $F_{I(\mathbf{0})} = F_{1\infty}$ of dimension one, equals to the boundary of $\overline{S}^d_+$ (this face is homeomorphic to S^1 and cannot be a cell). There is also one half-linear face F_1, that coincides with $\overline{S}^d_+$. Therefore a node system here would have four nodes, three on the boundary face $F_{I(\mathbf{0})}$ that positively span the plane containing that face, and one more node n in the interior of $\overline{S}^d_+$. The cell complex in this case is obtained by first taking the spherical convex hull of every subset of nodes on the boundary with two or less elements, that form a subdivision of $F_{I(\mathbf{0})}$ in three edges and three vertices, and then taking the pyramid over all those faces, with apex on the interior node n, to obtain a cell decomposition as shown in Fig. 5.

Lemma 4.22 *The complex $\mathcal{P}_N$ is a regular CW complex homeomorphic to a d-ball.*

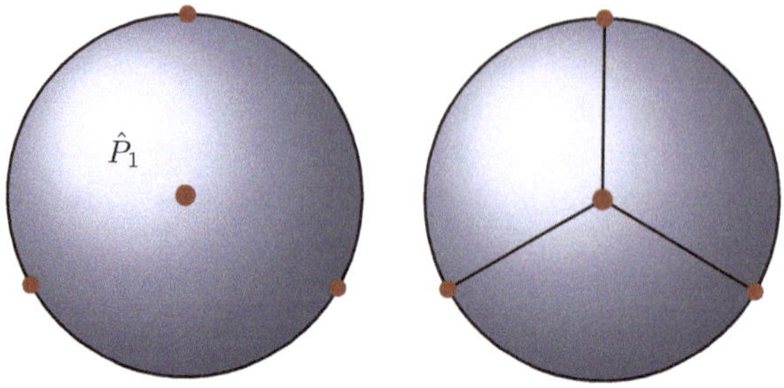

Fig. 5 Node system and cell complex $\mathcal{P}_N$ corresponding to the partition of $\mathbb{R}^2$ with only one region

Proposition 4.23 *For a pointed partition $\mathcal{P}$, the complex $\mathcal{P}_N$ coincides with the cell complex $\mathcal{P}$ described in Theorem 2.16. The vertices of $\mathcal{P}_N$ are precisely the vertices of $\mathcal{P}$.*

Proof For essential partitions all vertices are half-linear faces, and the corresponding node must be precisely at the vertex. If the partition $\mathcal{P}$ is pointed, there are no other half-linear faces, since the cone of a half-linear face F of dimension $\dim F \geq 1$ contains antipodal points on its boundary and therefore is not pointed. Then no other nodes are included, and all faces of $\mathcal{P}_N$ are precisely the faces of $\mathcal{P}$, so that we end up with the same complex. □

Lemma 4.24 *Let $\mathcal{P}$ be a fixed n-partition. Then the combinatorial structure of the complex $\mathcal{P}_N$ does not depend on the choice of the nodes in N, i. e. for any two node systems $N, N' \in N(\mathcal{P})$ the face posets of the complexes $\mathcal{P}_N$ and $\mathcal{P}_{N'}$ are isomorphic and the complexes are cellularly homeomorphic.*

Proof The face poset of $\mathcal{P}_N$ can be obtained from the face poset of $\mathcal{P}$, once we know which are the linear and half-linear faces, independently of the choice of the node system N. Following the construction in Definition 4.20, we can obtain the face poset of $\mathcal{P}_N$ from the face poset of $\mathcal{P}$. □

Definition 4.25 (*Flags, node frames, node bases, flats*) Let $\mathcal{P}$ be an n-partition, together with a node system N. A *flag* of faces of $\mathcal{P}_N$ is a list of faces $G_0 \subset \cdots \subset G_d$ completely ordered by containment. A *node frame* of N is a list $(\boldsymbol{v}_0, \ldots, \boldsymbol{v}_d)$ of $d+1$ different nodes in N such that the nodes $\boldsymbol{v}_0, \ldots, \boldsymbol{v}_k$ are contained on a k-face G_k of $\mathcal{P}_N$ for all $k \leq d$ and the faces $G_0 \subset \cdots \subset G_d$ form a flag. A *node basis* is a node frame whose vectors are linearly independent and a *flat* is a node frame whose vectors are linearly dependent.

As the vertices of $\mathcal{P}_N$ are precisely the nodes in N and the face poset of $\mathcal{P}_N$ is the same for any node system N, for any node frame $(\boldsymbol{v}_0, \ldots, \boldsymbol{v}_d)$ and any other node system N' of $\mathcal{P}$, the corresponding list of nodes $(v_0(N'), \ldots, v_d(N'))$ is a node frame of N'. Also any two partitions $\mathcal{P}$ and $\mathcal{P}'$ with the same face poset and the same corresponding half-linear faces have a bijection between node frames, as node frames can be read completely from the combinatorial structure of $\mathcal{P}_N$.

Lemma 4.26 *Let $G_0 \subset \cdots \subset G_d$ be a complete flag of faces in the complex $\mathcal{P}_N$. Then for any list $\boldsymbol{x}_0, \ldots, \boldsymbol{x}_k$ of linearly independent vectors in S^d such that $\boldsymbol{x}_i \in G_i$ for all $0 \leq i \leq d$, the sign of the determinant* $\det(\boldsymbol{x}_0, \ldots, \boldsymbol{x}_d)$ *is given uniquely by the flag $G_0 \subset \cdots \subset G_d$.*

Proof Let $\boldsymbol{b}_0, \ldots, \boldsymbol{b}_d$ be the basis of $\mathbb{R}^{d+1}$ where $\boldsymbol{b}_0 \in G_0$ and every $\boldsymbol{b}_i$ for $0 < i \leq d$ is the vector in the linear space spanned by the face G_i orthogonal to the subspace spanned by G_{i-1}, such that any point $\boldsymbol{x} \in G_i$ satisfy the inequality $\boldsymbol{b}_i \cdot \boldsymbol{x} \geq 0$. This basis is uniquely defined by the flag $G_0 \subset \cdots \subset G_d$.

Then the vectors $(\boldsymbol{b}_0, \ldots, \boldsymbol{b}_i)$ span the same linear subspace as the face G_i. In terms of this basis, the list of vectors $(\boldsymbol{x}_0, \ldots, \boldsymbol{x}_d)$ is represented by an upper triangular matrix

$$\begin{pmatrix} 1 & a_{01} & \cdots & a_{0d} \\ 0 & a_{11} & \cdots & a_{1d} \\ \vdots & \vdots & \ddots & \vdots \\ 0 & 0 & \cdots & a_{dd} \end{pmatrix}$$

where all diagonal entries a_{ii} are greater than zero. Then we conclude that

$$\det(\boldsymbol{x}_0, \ldots, \boldsymbol{x}_d) = \Big(\prod_{i=1}^{d} a_{ii}\Big) \det(\boldsymbol{b}_0, \ldots, \boldsymbol{b}_d).$$

will always have the same sign, independently of the choice of the points $\boldsymbol{x}_i$. (This determinant cannot be 0 as we require that the vectors $\boldsymbol{x}_0, \ldots, \boldsymbol{x}_d$ are linearly independent.) □

Now we will explain a second approach to prove that $\mathcal{C}(\mathbb{R}^d, n)$ is a union of semialgebraic pieces. With the different concepts we have now, we can define when two partitions are combinatorially equivalent, and use this to construct the realization space of any partition $\mathcal{P}$ (made by all partitions that are combinatorially equivalent to $\mathcal{P}$). This will be useful in the discussion about the dimension of the spaces of convex n-partitions.

Given an n-partition $\mathcal{P}$, we want to describe all n-partitions that are combinatorially equivalent to $\mathcal{P}$. Two partitions $\mathcal{P}$ and $\mathcal{P}'$ have the same face poset if $\mathcal{I}(\mathcal{P}) = \mathcal{I}(\mathcal{P}')$. They have the same corresponding half-linear faces if the indices $I \in \mathcal{I}(\mathcal{P})$ such that $F_I(\mathcal{P})$ is half-linear are the same indices for which $F_I(\mathcal{P}')$ is half-linear.

Definition 4.27 (*Orientation of a partition*) The *orientation* of a partition $\mathcal{P}$ of $\mathbb{R}^d$ is given by the signs of the determinants $\det(\boldsymbol{v}_0, \ldots, \boldsymbol{v}_d)$ of all node frames of a node system N of $\mathcal{P}$.

Orientations of partitions are closely related with orientations of cell complexes. If we consider the barycentric subdivision $S_N = \operatorname{sd} \mathcal{P}_N$ obtained by taking a point $\boldsymbol{y}_G$ in the relative interior of each face G of $\mathcal{P}_N$, and with maximal simplices that are the spherical convex hull of sets $y_{G_0}, \dots, y_{G_d}$ for each complete flag $G_0 \subset \cdots \subset G_d$ in $\mathcal{P}_N$, then by Lemma 4.26, we can read an orientation of the simplicial complex S_N from the orientation of $\mathcal{P}$.

Since orientations of oriented simplicial complexes are determined after fixing the orientation of one simplex, then it is enough to know the sign of one node basis to determine the sign of all other node bases of $\mathcal{P}_N$. In particular, if $\mathcal{P}$ is an essential partition, then any node system on $\mathcal{P}$ will give rise to the same orientation. If $\mathcal{P}$ is non-essential, there are two possible orientations, depending on the choice of the nodes on the minimal face $F_{I(\mathbf{0})}$.

Orientations also keep track of which node frames are node basis and which are flats. Two partitions $\mathcal{P}$ and $\mathcal{P}'$ with the same face poset and corresponding half-linear faces have the same orientation if there are node systems N and N' on each of them, so that the sign of the determinants of corresponding node basis are always the same and they have the same corresponding flats.

Definition 4.28 (*Combinatorial type of a partition*) The *combinatorial type* of an n-partition $\mathcal{P}$ is given by the following information: the set $\mathcal{I}(\mathcal{P})$ of labels of the face poset, the set of half-linear faces of $\mathcal{P}$, and the orientation given by a node system of $\mathcal{P}$.

Orientations allow us to distinguish the combinatorial type of an essential partition and its reflection on a hyperplane. If a partition has some reflection symmetry, it implies that it is non-essential. Orientations also make sure that combinatorially equivalent partitions have the same π-angles, as defined next.

Definition 4.29 (*π-angles*) Two $(d-1)$-faces F_{ij} and F_{ik} form a *π-angle* if they belong to the same $(d-1)$-subface of a d-face F_i of $\mathcal{P}$ and their intersection is $(d-2)$-dimensional. This means that the dihedral angle between these two $(d-1)$-faces is equal to π.

For the proof of Theorem 4.31 we need a characterization for cone partitions: A *cone partition* of a cone C is a collection of cones $C_1, \dots, C_r$ contained in C such that every point of C is contained in one of the subcones C_i, where also the intersection of any two subcones $C_i \cap C_k$ is a face of both cones.

Lemma 4.30 (De Loera et al. [8, Sect. 3], Firla–Ziegler [16, Theorem 4]) *A set of cones $C_1, \dots, C_r$ of dimension $d+1$ contained in a bigger cone $C \subset \mathbb{R}^{d+1}$ form a cone partition of C if and only if the following two conditions are satisfied:*

- *there is a generic vector $\mathbf{g}$ (i.e., in the interior of exactly one of the cones and not contained in the boundary of any other one of the cones C_k), and*
- *for any d-face F of a $(d+1)$-cone C_i that is not contained on the boundary of C there is a second cone C_j with $C_i \cap C_j = F$ such that F is a face of C_j.*

Theorem 4.31 *Let $\mathcal{P}$ be a partition of $\mathbb{R}^d$ together with a node system N. Consider a list of vectors $\boldsymbol{x}_v \in \mathbb{R}^{d+1}$ for every node $\boldsymbol{v} \in N$ that satisfy the following algebraic relationships and inequalities:*

(i) $\|\boldsymbol{x}_{\boldsymbol{v}}\| = 1$ *for every node* $\boldsymbol{v}$ *in* N.
(ii) $\det(\boldsymbol{x}_{\boldsymbol{v}_0}, \ldots, \boldsymbol{x}_{\boldsymbol{v}_d}) > 0$, *for every node basis* $(\boldsymbol{v}_0, \ldots, \boldsymbol{v}_d)$ *with* $\det(\boldsymbol{v}_0, \ldots, \boldsymbol{v}_d) > 0$.
(iii) $\det(\boldsymbol{x}_{\boldsymbol{v}_0}, \ldots, \boldsymbol{x}_{\boldsymbol{v}_d}) = 0$, *for every node flat* $(\boldsymbol{v}_0, \ldots, \boldsymbol{v}_d)$.
(iv) $\boldsymbol{e}_0 \cdot \boldsymbol{x}_{\boldsymbol{v}} = 0$, *for any node* $\boldsymbol{v} \in N$ *at infinity (i.e. on the boundary of* S^d_+*).*
(v) $\boldsymbol{e}_0 \cdot \boldsymbol{x}_{\boldsymbol{v}} > 0$, *for any other node* $\boldsymbol{v} \in N$, *not at infinity.*

Assume also that there is a generic vector $\boldsymbol{g} \in \mathbb{R}^{d+1}$, *not contained on any of the hyperplanes spanned by d vectors* $\boldsymbol{x}_{\boldsymbol{v}_i}$, *so that* $\boldsymbol{g}$ *belongs to the interior of exactly one of the cones spanned by all vectors* $\boldsymbol{x}_{\boldsymbol{v}}$ *corresponding to the nodes* $\boldsymbol{v}$ *that belong to a d-face of* $\mathcal{P}_N$.

Then there is a partition $\mathcal{P}'$ *that is combinatorially equivalent to* $\mathcal{P}$ *with a node system given by the points* $x_{\boldsymbol{v}}$ *for* $\boldsymbol{v} \in N$.

Proof We want to see first that we can construct a regular CW complex $\mathcal{P}_X$ by taking a face G' for each face G in $\mathcal{P}_N$, where G' is the spherical convex hull of the points $\boldsymbol{x}_{\boldsymbol{v}}$ for all nodes $\boldsymbol{v} \in G$. Then we will obtain the partition $\mathcal{P}'$ out of the complex $\mathcal{P}_X$.

Consider the barycentric subdivision sd $\mathcal{P}_N$ of the complex $\mathcal{P}_N$ obtained by taking points $\boldsymbol{y}_G$ in the relative interior of each face G of $\mathcal{P}_N$. The maximal simplices of sd $\mathcal{P}_N$ correspond to complete flags $G_0 \subset \cdots \subset G_d$ in $\mathcal{P}_N$ and have $\boldsymbol{y}_{G_0}, \ldots, \boldsymbol{y}_{G_d}$ as vertices. Then take a point $\boldsymbol{y}'_G$ in the relative interior of each spherical polyhedral set G' in $\mathcal{P}_X$. We want to see that if we construct the family S_X of simplicial cones over the sets $\boldsymbol{y}'_{G_0}, \ldots, \boldsymbol{y}'_{G_d}$ for each complete flag $G_0 \subset \cdots \subset G_d$ in $\mathcal{P}_N$, then we obtain a cone partition of the upper halfspace of $\mathbb{R}^{d+1}$ (with first coordinate $x_0 \geq 0$), by making use of Lemma 4.30.

Lemma 4.32 *Let G be a d-face of* $\mathcal{P}_N$. *Then the algebraic relationships and inequalities for node frames (of type* (ii) *and* (iii)*) imply that* G' *is combinatorially equivalent to G as a polyhedral cone.*

Proof The relationships of type (iii) coming from flats tell us that the points $\boldsymbol{x}_{\boldsymbol{v}}$ corresponding to nodes $\boldsymbol{v}$ on the same d-subface of G are all on the same hyperplane and the inequalities of type (ii) for node bases tell us that this hyperplane defines a facet of G'. Moreover, for each node $\boldsymbol{v}$, the set of facets on G where it belongs must be similar to the set of facets of G' where the point $\boldsymbol{x}_{\boldsymbol{v}}$ is contained.

We can tell which nodes are vertices of G from the set of facets where each node belong. Vertices are in the maximal sets under inclusion, because if a node $\boldsymbol{v}$ is not a vertex, the set $A_{\boldsymbol{v}}$ of facets of G containing $\boldsymbol{v}$ is determined by the subface of G that contains it, and this is a subset of the set $A_{\boldsymbol{v}'}$ of facets of G containing a vertex $\boldsymbol{v}'$ of that subface. Therefore G and G' have the same vertex-facet incidences, and this imply that they are combinatorially equivalent (this is a direct consequence of the analogous result for convex polytopes, see [20, Lect. 2]). □

The cone over G' is subdivided by all cones of the form $\text{cone}(\boldsymbol{y}'_{G_0}, \ldots, \boldsymbol{y}'_{G_d})$ associated to complete flags $G_0 \subset \cdots \subset G_d$ on with $G = G_d$. By assumption, there is a generic vector $\boldsymbol{g}$ contained in exactly one of the cones spanned by the vectors

$\boldsymbol{x}_{\boldsymbol{v}}$ for all nodes $\boldsymbol{v}$ that belong to a d-face G of the complex $\mathcal{P}_N$. This is precisely the cone over the spherical polyhedron G'.

Since the vector $\boldsymbol{g}$ is generic, it will belong to the interior of exactly one of the subcones $\operatorname{cone}(\boldsymbol{y}'_{G_0}, \ldots, \boldsymbol{y}'_{G_d})$ corresponding to a complete flag with $G = G_d$. By a similar argument, if $G_d \neq G$, it is not possible that $\boldsymbol{g}$ belong to any other cone corresponding to a flag ending in G_d and $\boldsymbol{g}$ is in the interior of a unique cone from S_X. We conclude that the vector $\boldsymbol{g}$ belong to the interior of exactly one of the subcones $\operatorname{cone}(\boldsymbol{y}'_{G_0}, \ldots, \boldsymbol{y}'_{G_d})$ corresponding to a complete flag with $G = G_d$, and therefore $\boldsymbol{g}$ is in the interior of a unique cone in S_X.

Now we want to see that for any d-face F of a $(d+1)$-cone C_i in S_X that is not contained on the boundary of the upper halfplane in $\mathbb{R}^{d+1}$ there is a second cone C_j with $C_i \cap C_j = F$ such that F is a face of C_j. Notice that the cones spanned by $\boldsymbol{y}_{G_0}, \ldots, \boldsymbol{y}_{G_d}$ form a simplicial cone partition S_N of the upper halfspace of $\mathbb{R}^{d+1}$, since they arise from a barycentric subdivision of $\mathcal{P}_N$.

Lemma 4.33 *For any complete flag $G_0 \subset \cdots \subset G_d$, the determinant* $\det(\boldsymbol{y}'_{G_0}, \ldots, \boldsymbol{y}'_{G_d})$ *has the same sign as the determinant* $\det(\boldsymbol{y}_{G_0}, \ldots, \boldsymbol{y}_{G_d})$.

Proof By Lemma 4.26 we know that the sign of the determinant $\det(\boldsymbol{y}_{G_0}, \ldots, \boldsymbol{y}_{G_d})$ is the same than the sign of $\det(\boldsymbol{v}_0, \ldots, \boldsymbol{v}_d)$ for any node basis $(\boldsymbol{v}_0, \ldots, \boldsymbol{v}_d)$ in N with $\boldsymbol{v}_i \in G_i$.

By the algebraic conditions on the $\boldsymbol{x}_{\boldsymbol{v}}$, this sign is also the same as that of the determinant $\det(\boldsymbol{x}_{v_0}, \ldots, \boldsymbol{x}_{v_d})$ for any node basis $(\boldsymbol{v}_0, \ldots, \boldsymbol{v}_d)$ in N with $\boldsymbol{v}_i \in G_i$. We want to see that the determinant $\det(\boldsymbol{y}'_{G_0}, \ldots, \boldsymbol{y}'_{G_d})$ also has the same sign.

The fact that $\boldsymbol{y}'_G \in \operatorname{relint} G'$ can be expressed by a linear combination

$$\boldsymbol{y}'_G = \sum_{\boldsymbol{v} \in N \cap G} \alpha_{\boldsymbol{v}} \boldsymbol{x}_{\boldsymbol{v}},$$

where all $\alpha_{\boldsymbol{v}} > 0$. Since determinants are multilinear, we can expand as follows.

$$\det(\boldsymbol{y}'_{G_0}, \ldots, \boldsymbol{y}'_{G_d}) = \sum_{(\boldsymbol{v}_0, \ldots, \boldsymbol{v}_d)} \Big(\prod_{i=0}^{d} \alpha_{\boldsymbol{v}_i}\Big) \det(\boldsymbol{x}_{v_0}, \ldots, \boldsymbol{x}_{v_d}),$$

where the sum goes over all lists $(\boldsymbol{v}_0, \ldots, \boldsymbol{v}_d)$ such that $\boldsymbol{v}_i \in G_i$, namely the node systems on the flag $G_0 \subset \cdots \subset G_d$. We can see that all summands on the right have the same sign as $\det(\boldsymbol{y}_{G_0}, \ldots, \boldsymbol{y}_{G_d})$ or are zero. □

Lemma 4.33 implies that two adjacent cones in S_X don't overlap on their interiors, since the corresponding cones in S_N don't overlap. All d-faces of S_X corresponding to faces on the boundary of S_N are also on the boundary of the upper halfspace (due to relationships of type (iv)) while a d-face of a cone $C_i \in S_X$ corresponding to an interior d-face of S_N are always interior (due to the inequalities of type (v)), and by the lemma we can find that there is a second cone in S_X such that its intersection with C_i is the corresponding d-face, by looking at the cone with analogous property in S_N.

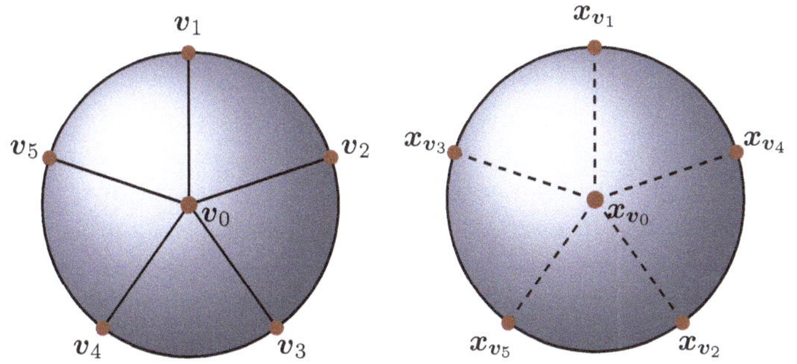

Fig. 6 Nodes of a 5-partition together with points $\boldsymbol{x}_{\boldsymbol{v}}$ that satisfy all algebraic relationships and inequalities in Theorem 4.31 but don't define a new 5-partition

Now we are in position to use Lemma 4.30 to conclude that the cones in S_X don't overlap and make a cone partition of the upper hemisphere.

Each of the faces G' of the partition $\mathcal{P}_X$ can be obtained as the intersection of S^d with the union of the cones over sets $\boldsymbol{y}'_{G_0}, \ldots, \boldsymbol{y}'_{G_k}$ where $G_0 \subset \cdots \subset G_k = G$ are partial flags on $\mathcal{P}_N$. We can see that $\mathcal{P}_X$ is a CW complex since the relative interiors of its faces are pairwise disjoint and that the boundary of each face G' is covered by the faces of $\mathcal{P}_X$ contained in G', since by construction we have inclusion between faces $G'_1 \subset G'_2$ if and only if the corresponding faces in $\mathcal{P}_N$ satisfy that $G_1 \subset G_2$. Also the resulting complex $\mathcal{P}_X$ will have the same face poset as $\mathcal{P}_N$. Half-linear faces F of $\mathcal{P}$ can be obtained as unions of faces of $\mathcal{P}_N$. The union F' of the corresponding faces of $\mathcal{P}_X$ has to be in a linear subspace of the right dimension, due to equations of type (iii), which say that points $\boldsymbol{x}_{\boldsymbol{v}}$ for $\boldsymbol{v} \in F$ have to be coplanar for all facets of all regions of $\mathcal{P}$ containing F. In addition, the F' have on the boundary the same faces at infinity as F (due to equations of type (iv)), so F' will be a half-linear face for a new partition $\mathcal{P}'$ that has as faces in its spherical representation the same faces as $\mathcal{P}_X$, but where those faces corresponding to the same half-linear face of $\mathcal{P}$ are glued together.

The fact that $\mathcal{P}'$ is a partition of $\mathbb{R}^d$ is a consequence that S_X is a cone partition of the upper halfspace. By Lemma 4.33 we can find that the $\mathcal{P}$ and $\mathcal{P}'$ have the same orientations, and we conclude that they are combinatorially equivalent as we wanted. This completes the proof of Theorem 4.31. □

The condition of the existence of a vector $\boldsymbol{g}$ in the interior of only one of the d-faces of $\mathcal{P}_X$ is important and cannot be omitted. To see this, consider a 5-partition of $\mathbb{R}^2$ as in the left of Fig. 6, and the choice of points $x_{\boldsymbol{v}_i}$ depicted on the right. For simplicity we called the vertices $\boldsymbol{v}_i$ and all nodes are vertices since the partition is pointed. In that example, all conditions from Theorem 4.31 are satisfied, except the existence of the point $\boldsymbol{g}$. In this case we get that the expected spherical regions form a double covering of the upper hemisphere.

Proposition 4.34 *Let $\mathcal{P}$ be an n-partition of $\mathbb{R}^d$. The space of pairs $(\mathcal{P}', N')$ of partitions $\mathcal{P}'$ combinatorially equivalent to together with a node system N' on $\mathcal{P}'$ is a semialgebraic set.*

Proof Theorem 4.31 gives an algebraic description by equations and inequalities of a set that parameterizes all these pairs, under the condition of the existence of the point $\boldsymbol{g}$. Notice that if a partition $\mathcal{P}'$ is combinatorially equivalent to $\mathcal{P}$, then any node system give rise to an equivalent system of equations and inequalities, and therefore it satisfies the system given by $\mathcal{P}$. The condition about the point $\boldsymbol{g}$ can be also given as a system of algebraic conditions after introducing new slack variables for $\boldsymbol{g}$. Then by Theorem 4.1 (and some unions and intersections) we find that the set we are interested in is semialgebraic. □

Definition 4.35 (*Realization spaces*) The *realization space* of an n-partition $\mathcal{P}$ is the subspace of $\mathcal{C}(\mathbb{R}^d, n)$ of all partitions $\mathcal{P}'$ with the same combinatorial type as $\mathcal{P}$. It is denoted as $\mathcal{C}_{\mathcal{P}}(\mathbb{R}^d, n)$.

Theorem 4.36 *Let $\mathcal{P}$ be an n-partition of $\mathbb{R}^d$. Then the realization space $\mathcal{C}_{\mathcal{P}}(\mathbb{R}^d, n)$ is a semialgebraic set.*

Proof Proposition 4.34 shows that for pointed partitions $\mathcal{P}$ the space $\mathcal{C}_{\mathcal{P}}(\mathbb{R}^d, n)$ is semialgebraic, since all vertices are nodes, and there is a unique node system on each partition in the realization space. In general, the realization space of $\mathcal{P}$ can be obtained as the image of the space of pairs $(\mathcal{P}', N')$ described in Proposition 4.34 to the space $\mathbb{R}^{h(d+1)}$ describing by the equations of the h hyperplanes that define $(d-1)$-faces of the partition, where each partition corresponds a unique point. We make use of an equivalent formulation of Theorem 4.1 that claims that the image under a polynomial mapping $f : \mathbb{R}^m \to \mathbb{R}^{m'}$ of a semialgebraic set is semialgebraic (see [2, Prop. 2.83]). □

This result gives us an alternative proof of the fact (Theorem 4.9) that space of n-partitions $\mathcal{C}(\mathbb{R}^d, n)$ is a union of finitely-many semialgebraic pieces, namely the union of all realization spaces of n-partitions of $\mathbb{R}^d$.

5 Examples

We will analyze here the spaces of n-partitions for small values of n and d. For $n = 1$ the space of partitions $\mathcal{C}(\mathbb{R}^d, 1)$ simply consists of one point. A more interesting but still easy case is $n = 2$.

Proposition 5.1 *The space $\mathcal{C}(\mathbb{R}^d, \leq 2)$ is homeomorphic to the sphere S^d. The space of partitions $\mathcal{C}(\mathbb{R}^d, 2)$ is homotopy equivalent to S^{d-1} and is obtained from $\mathcal{C}(\mathbb{R}^d, \leq 2)$ by removing two points.*

Proof To parameterize our space of 2-partitions for fixed d we only need to choose the coordinates $\boldsymbol{c}_{1,2}$, that describe the normal to the hyperplane H_{ij} by a point in S^n. Two special cases have to be taken into account that characterize the cases when the combinatorial type of the 2-partition is not the generic one. These are precisely when $\boldsymbol{c}_{ij} = \pm(1, 0, \ldots, 0)$. In those cases, there is no hyperplane in $\mathbb{R}^d$, representing the partitions with only one (labeled) non-empty region. These extreme partitions can be obtained as a limit of proper 2-partitions, and S^d will handle the topological structure of $\mathcal{C}(\mathbb{R}^d, \leq 2)$ in the right way. □

For $n \geq 3$, things begin to be more complicated, even in the case of $d = 1$.

Proposition 5.2 *The space $\mathcal{C}(\mathbb{R}^1, \leq n)$ is homeomorphic to a CW complex with n vertices and $k!\binom{n}{k}$ simplicial $(k-1)$-cells for $0 \leq k \leq n$. It is made out of $n!$ simplices of dimension $(n-1)$ glued appropriately on the boundaries. The space $\mathcal{C}(\mathbb{R}^1, n)$ is homeomorphic to $n!$ open $(n-1)$-balls.*

Proof For a combinatorial type with k non-empty regions, its realization space is contractible and can be realized as a $(k-1)$-simplex. To do this, take an order preserving homeomorphism from $\mathbb{R}$ to the open interval $(0, 1)$. Then the coordinates of the $k-1$ interior vertices (hyperplanes!) $\boldsymbol{v}_{i,j} = F_{i,j} \in \mathbb{R}$ need to be in a prescribed order, and via the homeomorphism we can map any partition to a point inside a $(k-1)$-simplex contained in the unit cube $(0, 1)^{k-1}$.

For example, if the n-partition has the region i at the left of region $i+1$ for all $i < n$ (and no empty region) then we only need to specify the coordinates of the vertices $\boldsymbol{v}_{i,i+1}$ such that $\boldsymbol{v}_{1,2} \leq \cdots \leq \boldsymbol{v}_{n-1,n}$. Mapping these $n-1$ values to the unit cube $(0, 1)^{n-1}$ via the homeomorphism, we identify the realization space of this particular n-partition with the interior of an $(n-1)$-simplex.

The boundary of each of those simplices will represent the case when some of the points coincide, and can be naturally identified with the realization spaces of other combinatorial types with some extra empty regions. In this way we give to $\mathcal{C}(\mathbb{R}^1, \leq n)$ the structure of a regular cell complex (start with n vertices corresponding to the realization spaces of partitions with only one non-empty region, and then for higher dimensions, identify the boundary with a subspace of the union of the cells of smaller dimension).

There will be $n!$ combinatorial types without empty regions. The space $\mathcal{C}(\mathbb{R}^1, n)$ of proper partitions of $\mathbb{R}$ is the union of the interior of all those simplices. All other combinatorial types can be obtained in the limit (on the boundary) of those proper combinatorial types and therefore $\mathcal{C}(\mathbb{R}^1, \leq n)$ will have $n!$ top-dimensional simplicial $(n-1)$-cells, and $\binom{n}{k}k!$ cells of dimension $k-1$. □

Example 5.3 The space $\mathcal{C}(\mathbb{R}^1, \leq 3)$ is homeomorphic to a 2-dimensional space made out topologically by gluing six simplices along the boundaries in a special way, since there are two different edges joining each pair of vertices. The vertices represent the partitions with one non-empty region, and the edges represent the partitions with two non-empty regions.

In Fig. 7 we can see the six simplices of this CW complex with the corresponding labels on the different cells. These simplices have to be glued along the edges

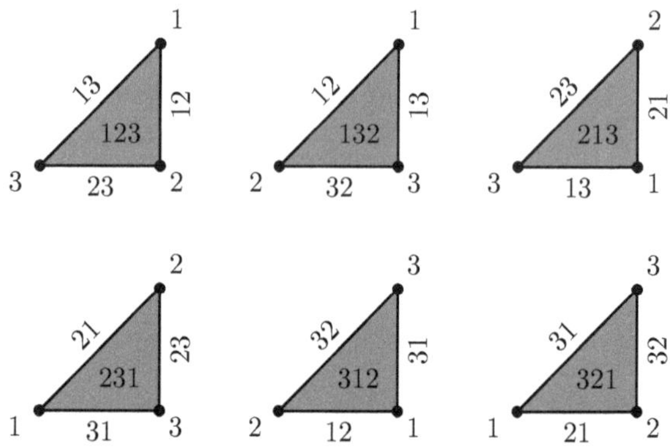

Fig. 7 Simplices to build a cell complex homeomorphic to $\mathcal{C}(\mathbb{R}^1, \leq 3)$

corresponding to the same partitions, in such way that the corresponding vertices coincide. Each edge appears in three of the simplices.

As a further example, in [10, Sect. 7.3] we give a complete description of the space $\mathcal{C}(\mathbb{R}^2, 3)$ as well as a cell complex model for its closure $\mathcal{C}(\mathbb{R}^2, \leq 3)$.

6 The Subspaces of Simple and of Regular n-Partitions

Inside the space $\mathcal{C}(\mathbb{R}^d, n)$, there are other spaces that catch our attention.

In particular, there is the open subspace of *simple* n-partitions, where the closures of the parts are polyhedra such that the intersection of any two of them is a face of both of them, the intersection of any k of them has co-dimension at least $k-1$ for $k \leq d+2$, so in particular the intersection of any $d+2$ of them is empty.

Moreover, there is the important subspace $\mathcal{C}_{\text{reg}}(\mathbb{R}^d, n)$ of *regular* partitions in the sense of Gelfand et al. (see [20, Sect. 5.1]; in Computational Geometry these are also known as *generalized Voronoi diagrams* or as *power diagrams*, cf. [6]): These are given by

$$P_i = \{x \in \mathbb{R}^d : |x - s_i|^2 - w_i \leq |x - s_j|^2 - w_j \text{ for } 1 \leq i \leq n\}$$

for distinct points ("sites") $s_1, \ldots, s_n$ and real numbers ("weights") $w_1, \ldots, w_n$ that may be assumed to sum to zero. Such partitions may be visualized as arising by projection of the facets of a convex polyhedron of dimension $d+1$. Regular partitions appear in many different contexts and are much better understood than general partitions, as they can be parameterized by the classical configuration spaces $F(\mathbb{R}^d, n)$ of n distinct labelled points in $\mathbb{R}^d$, see e.g. [17, Sect. 2]. As there are S_n-equivariant maps

$$\mathcal{C}_{\text{reg}}(\mathbb{R}^d, n) \longrightarrow \mathcal{C}(\mathbb{R}^d, n) \longrightarrow F(\mathbb{R}^d, n) \longrightarrow \mathcal{C}_{\text{reg}}(\mathbb{R}^d, n)$$

given by inclusion, by the map to the barycenter (in the spherical representation), and then by the traditional Voronoi diagram, these spaces are equivalent in terms of equivariant cohomology; in particular they have the same Fadell–Husseini index. We have not tackled the question whether these spaces are (equivalently) homotopy equivalent, although the above maps certainly suggest that. Instead, we have studied how the space of regular partitions is embedded in the space of all convex n-partitions.

It turns out that, somewhat surprisingly, there is a big difference between the case $d = 2$ and the case when $d \geq 3$. For $d = 2$ and large n, the subspace $\mathcal{C}_{\text{reg}}(\mathbb{R}^2, n)$ of regular n-partitions has much smaller dimension than $\mathcal{C}(\mathbb{R}^2, n)$, as it can be seen from the following results (see [10]).

Theorem 6.1 *For $n \geq 3$ the space $\mathcal{C}(\mathbb{R}^2, n)$ of partitions of $\mathbb{R}^2$ into n convex pieces has dimension* $\dim \mathcal{C}(\mathbb{R}^2, n) = 4n - 7$. *The partitions whose realization spaces attain the top dimension are simple with exactly three unbounded regions.*

Theorem 6.2 *For $d \geq 2$ and $n \geq 2$, the space $\mathcal{C}_{\text{reg}}(\mathbb{R}^d, n)$ of regular partitions is a semialgebraic set of dimension*

$$\dim \mathcal{C}_{\text{reg}}(\mathbb{R}^d, n) = (d+1)(n-1) - 1.$$

In particular, the space of regular n-partitions of the plane has dimension $\dim \mathcal{C}_{\text{reg}}(\mathbb{R}^2, n) = 3n - 4$.

For $d \geq 3$, however, a theorem by Whiteley [19] and Rybnikov [9] asserts that all simple n-partitions are regular.

Conjecture 6.3 $\dim \mathcal{C}(\mathbb{R}^d, n) = \dim \mathcal{C}_{\text{reg}}(\mathbb{R}^3, n)$ *for $n \geq 2$ and $d \geq 3$.*

However, $\mathcal{C}_{\text{reg}}(\mathbb{R}^3, n)$ is *not* a dense subset in $\mathcal{C}(\mathbb{R}^3, n)$ for $n > 3$, and there are non-simple combinatorial types whose realization spaces have the same dimension as $\mathcal{C}_{\text{reg}}(\mathbb{R}^3, n)$, where partitions are generically non-regular.

In general, realization spaces of partitions of a given combinatorial type are expected to be complicated objects. There is a close relation to the work by Richter-Gebert [15] on realization spaces of polytopes, where the main result is the *Universality Theorem*, according to which the realization spaces of d-dimensional polytopes, for any fixed $d \geq 4$, are as complicated as arbitrary semialgebraic sets defined over $\mathbb{Z}$. A similar result holds for realization spaces of regular partitions [10, Theorem 5.17]:

Theorem 6.4 *For any primary basic semialgebraic set X and $d \geq 3$, there is an n-partition $\mathcal{P}$ of $\mathbb{R}^d$ such that the set of regular partitions combinatorially equivalent to $\mathcal{P}$, up to affine equivalence, form a semialgebraic set stably equivalent to X.*

Acknowledgements This paper presents main results of the doctoral thesis of the first author [10]. We are very grateful to both referees for very valuable and thoughtful comments.

References

1. I. Bárány, P.V.M. Blagojević, A. Szűcs, Equipartitioning by a convex 3-fan. Adv. Math. **223**, 579–593 (2010)
2. S. Basu, R. Pollack, M.-F. Roy, *Algorithms in Real Algebraic Geometry*, 2nd edn., Algorithms and Computation in Mathematics, vol 10 (Springer, Berlin, 2006)
3. A. Björner, in *Topological Methods*, vol II, ed. by R. Graham, M. Grötschel, L. Lovász (North-Holland/Elsevier, Amsterdam, 1995), Handbook of Combinatorics, pp. 1819–1872
4. P.V.M. Blagojević, G.M. Ziegler, Convex equipartitions via equivariant obstruction theory. Isr. J. Math. **200**, 49–77 (2014)
5. J. Bochnak, M. Coste, M.F. Roy, *Géométrie Algébrique Réelle*, Ergebnisse Math. Grenzgebiete (3), vol 12 (Springer, Berlin, 1987)
6. F. Aurenhammer, A criterion for the affine equivalence of cell complexes in $\mathbb{R}^d$ and convex polyhedra in $\mathbb{R}^{d+1}$. Discret. Comput. Geom. **2**, 49–64 (1987)
7. P.M. Gruber, P. Kenderov, Approximation of convex bodies by polytopes. Rend. Circ. Mat. Palermo (2) **31**(2), 195–225 (1982)
8. J.A. De Loera, S. Hoşten, F. Santos, B. Sturmfels, The polytope of all triangulations of a point configuration. Doc. Math. **1**, 103–119 (1996)
9. K. Rybnikov, Stresses and liftings of cell-complexes. Discret. Comput. Geom. **21**, 481–517 (1999)
10. E. León, Spaces of convex n-partitions. Ph.D. thesis, vol vi (Freie Universität Berlin, 2015), p. 101 published at www.diss.fu-berlin.de
11. M. Gromov, Isoperimetry of waists and concentration of maps. Geom. Funct. Anal. (GAFA) **13**, 178–215 (2013)
12. J.R. Munkres, *Elements of Algebraic Topology* (Addison-Wesley, Menlo Park, 1984)
13. R. Nandakumar, Fair partitions. Blog entry, http://nandacumar.blogspot.de/2006/09/cutting-shapes.html, 28 September 2006
14. R. Nandakumar, N.R. Rao, Fair partitions of polygons: an elementary introduction. Proc. Indian Acad. Sci.–Math. Sci. **122**, 459–467 (2012)
15. J. Richter-Gebert, *Realization Spaces of Polytopes*, Lecture Notes in Mathematics, vol 1643 (Springer, Heidelberg, 1996)
16. R.T. Firla, G.M. Ziegler, Hilbert bases, unimodular triangulations, and binary covers of rational polyhedral cones. Discret. Comput. Geom. **21**, 205–216 (1999)
17. R.N. Karasev, A. Hubard, B. Aronov, Convex equipartitions: the spicy chicken theorem. Geometriae Dedicata **170**, 263–279 (2014)
18. P. Soberón, Balanced convex partitions of measures in $\mathbb{R}^d$. Mathematika **58**, 71–76 (2012)
19. W. Whiteley, 3-diagrams and Schlegel diagrams of simple 4-polytopes. Preprint 1994
20. G.M. Ziegler, *Lectures on Polytopes*, Graduate Texts in Math., vol 152 (Springer, New York, 1995). Revised edition, 1998; seventh updated printing 2007

New Regular Compounds of 4-Polytopes

Peter McMullen

Abstract This note describes six apparently new regular compounds of polytopes in $\mathbb{E}^4$, that were missed by Coxeter in his systematic faceting procedure. In fact, three have the same symbols as ones in Coxeter's list (and two are nearly the same); however, their symmetry groups are much smaller. The treatment relies heavily on the use of quaternions.

MSC (2010) Primary 51M20 · Secondary 52B15

1 Introduction

In his classical monograph [1, Sect. 14.3 and Table VI(iv)], in a procedure that he calls *systematic faceting*, Coxeter goes through various sections of the 120-cell $\{5, 3, 3\}$ from a vertex, thereby listing the regular polytopes that can be inscribed in its vertices. The relevant sections themselves will consist of one or more regular polyhedra and, if they are favourably situated, are then the vertex-figures of such inscribed polytopes. While regular star-polytopes can make contributions, only convex polytopes concern us in this note. In any event, it is well known that a regular star-polytope has the same vertices as a convex regular polytope; this was treated in [1, Chap. XIV], and given a more direct proof in [4] or [5, Theorem 7D6].

What we point out here is that one of the sections was incompletely analysed. On closer inspection, it turns out that, rather than containing a single regular tetrahedron, it actually contains six others as well. As a consequence, there exists a new regular compound, namely,

$$6\{5, 3, 3\}[720\{3, 3, 3\}]6\{3, 3, 5\}.$$

P. McMullen (✉)
University College London, Gower street, WC1E6BT London, England
e-mail: p.mcmullen@ucl.ac.uk

G. Ambrus et al. (eds.), *New Trends in Intuitive Geometry*, Bolyai Society Mathematical Studies 27, https://doi.org/10.1007/978-3-662-57413-3_12

We shall explain the notation in the next section but, roughly speaking, the compound consists of 720 regular simplices $\{3, 3, 3\}$ in $\mathbb{E}^4$ inscribed in a regular 120-cell $\{5, 3, 3\}$ and escribed to a regular 600-cell $\{3, 3, 5\}$. Moreover, from this one can extract a sub-compound

$$\{5, 3, 3\}[\mathit{120}\{3, 3, 3\}]\{3, 3, 5\}^{(\mathrm{var})},$$

which is distinct from the well-known one with a similar symbol in Coxeter's list.

In addition, while the central section was described completely, Coxeter failed to notice that sub-compounds can be extracted from the compound $\mathit{4}\{5, 3, 3\}[\mathit{300}\{3, 3, 4\}]\mathit{8}\{3, 3, 5\}$ and its relatives. These are

$$\{5, 3, 3\}[\mathit{75}\{3, 3, 4\}]\mathit{2}\{3, 3, 5\}^{(\mathrm{var})},$$
$$\mathit{2}\{5, 3, 3\}[\mathit{75}\{4, 3, 3\}]\{3, 3, 5\}^{(\mathrm{var})},$$
$$\{5, 3, 3\}[\mathit{25}\{3, 4, 3\}]^{(\mathrm{var})}.$$

The last is only vertex-regular, in contrast to Coxeter's vertex- and facet-regular (because self-dual) compound $\{5, 3, 3\}[\mathit{25}\{3, 4, 3\}]\{3, 3, 5\}$; we have not listed its facet-regular dual, which is the sixth of the compounds.

The first of the six new compounds in the lists has full symmetry group [3,3,5], while the symmetry groups of the other five have orders 600 or 1200; the former is only just large enough to be transitive on the vertices of the 120-cell or facets of the 600-cell.

In what follows, we shall actually describe *all* the regular compounds in $\mathbb{E}^4$, for two reasons. First, and most important, we want to establish completeness of the enumeration. Second, the techniques illustrate how quickly we can arrive at the list of compounds. Indeed, our analysis will be based entirely on an extensive use of quaternions. Du Val's monograph [2] is our main source for these; in [2, Sects. 26 and 27] he describes some compounds, but does not attempt to list them all.

2 Regular Compounds

We should make clear what we regard as a regular compound. We describe the vertex-regular case; facet-regularity is the dual concept, and regularity (without qualification) combines the two notions. (However, we often use 'regular' to mean 'vertex-regular', since we concentrate on this case.) Our condition is stronger than that stated by Coxeter [1, 3.6 and 14.3], but in practice seems to be what is actually demanded.

Formally, we say that one polytope Q is *inscribed* in another P if $\operatorname{vert} \mathsf{Q} \subset \operatorname{vert} \mathsf{P}$, where vert denotes the vertex-set; we denote this by $\mathsf{Q} \prec \mathsf{P}$. We shall always assume that $\dim \mathsf{Q} = \dim \mathsf{P}$. Then a *regular compound* $\mathcal{C}$ consists of copies of a regular polytope Q inscribed in a fixed regular polytope P, in such a way that some subgroup

$\mathbf{G} = \mathbf{G}(\mathcal{C})$ of the symmetry group $\mathbf{G}(\mathsf{P})$ of P is transitive both on vert P and on the copies of Q.

Remark 2.1 In less formal terms, our definition of a regular compound ensures that it looks the same at all vertices of P, and that all copies of Q are inscribed in P in the same way.

It follows that there are numbers m and n such that $\mathcal{C}$ consists of n copies of Q, together covering vert P m times; we then write $\mathcal{C} = m\mathsf{P}[n\mathsf{Q}]$. For facet-regularity, the notation is $[n\mathsf{Q}]k\mathsf{R}$; what this means is that there is a vertex-regular dual compound $k\mathsf{R}^\delta[n\mathsf{Q}^\delta]$, where P^δ denotes the geometric dual (or polar) polytope to P. Thus $m\mathsf{P}[n\mathsf{Q}]k\mathsf{R}$ denotes a compound that is both vertex- and facet-regular. In what follows, we shall take for granted the duals of compounds that are only vertex-regular.

When P, Q are regular polytopes such that $\mathsf{Q} \prec \mathsf{P}$, we write $\mathbf{K} = \mathbf{K}(\mathsf{P}, \mathsf{Q}) := \mathbf{G}(\mathsf{P}) \cap \mathbf{G}(\mathsf{Q})$ for their *common subgroup*. We emphasize here that Q may be inscribed in P in essentially different ways, so that $\mathbf{K}$ depends on the manner of the inscription; the larger $\mathbf{K}$ is, the more symmetrically is Q inscribed in P. If $n := [\mathbf{G}(\mathsf{P}) : \mathbf{K}]$ is the index of $\mathbf{K}$ in $\mathbf{G}(\mathsf{P})$, then there is a compound $m\mathsf{P}[n\mathsf{Q}]$ that is *fully regular*, in that its symmetry group is $\mathbf{G}(\mathsf{P})$. However, as we in effect pointed out in Sect. 1, we specifically allow the group $\mathbf{G}(\mathcal{C})$ of a compound $\mathcal{C} = m\mathsf{P}[n\mathsf{Q}]$ *not* be the whole symmetry group $\mathbf{G}(\mathsf{P})$. In the same way, it is not reasonable to ask that every symmetry of a copy of Q be one of P, so that $\mathbf{K}$ may be a proper subgroup of $\mathbf{G}(\mathsf{Q})$ (see the next section).

3 Compounds of Polyhedra

Although this note is directed towards compounds of regular 4-polytopes, it is worth illustrating the concepts that we have introduced in the more familiar arena of regular polyhedra. These compounds should be well known; they are discussed in (for example) [1, Sect. 3.6], [2, Sect. 11] and [3, Sect. 10].

The compound

$$\{4, 3\}[2\{3, 3\}]\{3, 4\}$$

of two tetrahedra inscribed in a cube and escribed to an octahedron is fully regular; the common subgroup is just the symmetry group [3, 3] of the tetrahedron itself. This is the well-known *stella octangula* of Kepler.

There are three compounds with the vertices of the dodecahedron. These are

$$\{5, 3\}[5\{3, 3\}]\{3, 5\},$$
$$2\{5, 3\}[10\{3, 3\}]2\{3, 5\},$$
$$2\{5, 3\}[5\{4, 3\}].$$

The first is a vertex- and facet-regular compound on which only the rotation groups of the tetrahedron $\{3, 3\}$ and dodecahedron $\{5, 3\}$ (or icosahedron $\{3, 5\}$) act; the common subgroup of this and the next is the former, namely, $[3, 3]^+$. Since its symmetry group consists of rotations only, it occurs in two enantiomorphic varieties. The second and third are fully regular. The common subgroup of the latter is $[3, 3]^+ \times \mathbf{C}_2$, the rotation group of the tetrahedron with the central inversion adjoined. The corresponding compound of five cubes is only vertex-regular; its facet-regular dual $[5\{3, 4\}]2\{3, 5\}$ consists of five octahedra escribed to an icosahedron.

Remark 3.1 There are pictures of the compounds in [1, Plate III, 6] or [2, Figs. 9, 15a,b]. More spectacular are the Anaglyphs I–III at the end of [3], which display the compounds three-dimensionally.

4 Quaternions

We begin the discussion of the 4-dimensional cases with a brief overview of the quaternions to establish notation and conventions; these basically follow [2, Chap. 3]. A general quaternion is of the form $\mathbf{g} = \gamma_1 + \gamma_2\mathbf{i} + \gamma_3\mathbf{j} + \gamma_4\mathbf{k}$, with $(\gamma_1, \gamma_2, \gamma_3, \gamma_4) \in \mathbb{E}^4$ and $\mathbf{i}, \mathbf{j}, \mathbf{k}$ satisfying

$$\mathbf{i}^2 = \mathbf{j}^2 = \mathbf{k}^2 = \mathbf{ijk} = -1.$$

Multiplication is associative, but not necessarily commutative. The *real part* of $\mathbf{g}$ is $\Re\mathbf{g} = \gamma_1$ and the *imaginary part* is $\Im\mathbf{g} = \gamma_2\mathbf{i} + \gamma_3\mathbf{j} + \gamma_4\mathbf{k}$. The *conjugate* of $\mathbf{g}$ is $\widetilde{\mathbf{g}} = \gamma_1 - \gamma_2\mathbf{i} - \gamma_3\mathbf{j} - \gamma_4\mathbf{k}$, so that the usual inner product of $\mathbf{x} = \xi_1 + \xi_2\mathbf{i} + \xi_3\mathbf{j} + \xi_4\mathbf{k}$ and $\mathbf{y} = \eta_1 + \eta_2\mathbf{i} + \eta_3\mathbf{j} + \eta_4\mathbf{k}$ in $\mathbb{E}^4$ is $\langle \mathbf{x}, \mathbf{y} \rangle = \Re(\widetilde{\mathbf{x}}\mathbf{y}) = \Re(\mathbf{x}\widetilde{\mathbf{y}})$; note that $\widetilde{\mathbf{x}\mathbf{y}} = \widetilde{\mathbf{y}}\widetilde{\mathbf{x}}$. Hence $\widetilde{\mathbf{g}}\mathbf{g} = \mathbf{g}\widetilde{\mathbf{g}} = \gamma_1^2 + \gamma_2^2 + \gamma_3^2 + \gamma_4^2 = \|(\gamma_1, \gamma_2, \gamma_3, \gamma_4)\|^2 =: \|\mathbf{g}\|^2$, the square of the usual euclidean norm.

It is clear that the set of non-zero quaternions forms a group under multiplication. A finite subgroup $\mathbf{G}$ of these must consist of *unit* quaternions, namely, those with $\|\mathbf{g}\| = 1$; the inverse of $\mathbf{g} \in \mathbf{G}$ is $\mathbf{g}^{-1} = \widetilde{\mathbf{g}}$. If $\mathbf{u}, \mathbf{v}, \mathbf{w}$ are mutually orthogonal pure imaginary unit quaternions, then $\{\pm 1, \pm\mathbf{u}, \pm\mathbf{v}, \pm\mathbf{w}\}$ is a subgroup conjugate to $\mathbf{V} := \{\pm 1, \pm\mathbf{i}, \pm\mathbf{j}, \pm\mathbf{k}\}$. A general unit quaternion can be expressed in the form $\mathbf{g} = \cos\vartheta + \sin\vartheta\mathbf{u}$ for some angle ϑ and some pure imaginary unit quaternion $\mathbf{u}$. If $\mathbf{v}$ is any pure imaginary unit quaternion orthogonal to $\mathbf{u}$, then $\widetilde{\mathbf{v}}\mathbf{g}\mathbf{v} = \cos\vartheta - \sin\vartheta\mathbf{u} = \widetilde{\mathbf{g}}$, so that unit quaternions are always conjugate to their inverses.

Quaternions are related to rotations in $\mathbb{E}^3$ in the following way. If we identify the pure imaginary quaternions with $\mathbb{E}^3$, then the mapping induced by $\mathbf{x} \mapsto \widetilde{\mathbf{g}}\mathbf{x}\mathbf{g}$, with $\mathbf{g}$ as just previously, is the rotation through -2ϑ about the axis in direction $\mathbf{u}$. The mapping identifies $\mathbf{g}$ and $-\mathbf{g}$, and so is 2-to-1 on each finite group $\mathbf{G}$, except for a cyclic group of odd order.

The *binary icosahedral group* $\mathbf{I}$ consists of the quaternions $\mathbf{g} = \gamma_1 + \gamma_2\mathbf{i} + \gamma_3\mathbf{j} + \gamma_4\mathbf{k}$, with $(\gamma_1, \gamma_2, \gamma_3, \gamma_4)$ all even permutations with all changes of sign of

$$(1, 0, 0, 0), \quad \tfrac{1}{2}(1, 1, 1, 1), \quad \tfrac{1}{2}(\tau, 1, \tau^{-1}, 0), \tag{1}$$

where $\tau = \frac{1}{2}(1 + \sqrt{5})$; thus $\mathbf{I}$ has order $|\mathbf{I}| = 120$. Its name tells us that the corresponding rotation group in $\mathbb{E}^3$ is the icosahedral group $[3, 5]^+$, the rotation group of the regular icosahedron. The binary icosahedral group plays a central part in our discussion; as a point-set in $\mathbb{E}^4$ it forms the vertex-set of a 600-cell $\{3, 3, 5\}$.

We have already seen the subgroup $\mathbf{V}$ corresponding to the first eight points of (1); the *binary tetrahedral group* $\mathbf{T}$ corresponds to the first 24. We shall also have recourse to the *cyclic group* $\mathbf{C} := \mathbf{C}_{10}$ of order 10 and *binary dihedral group* $\mathbf{D} := \mathbf{D}_5$ of order 20; we shall specify these later.

An important rôle is also played by the group $\mathbf{I}^{\dagger} \cong \mathbf{I}$, obtained from $\mathbf{I}$ by the change of sign of $\sqrt{5}$, that is, $\tau \leftrightarrow -\tau^{-1}$; on appropriate quaternions $\mathbf{q}$ (such as those in $\mathbf{I}$ or $\mathbf{I}^{\dagger}$), we denote this induced mapping by $\mathbf{q} \mapsto \mathbf{q}^{\dagger}$. Another way of obtaining $\mathbf{I}^{\dagger}$ from $\mathbf{I}$ is by an interchange of two coordinates (that is, an *odd* permutation of the vectors of (1)).

The *binary octahedral group* $\mathbf{O}$ consists of $\mathbf{T}$ together with the set $\mathbf{U}$, which corresponds to the permutations of the vectors

$$\tfrac{1}{\sqrt{2}}(1, 1, 0, 0), \tag{2}$$

with all changes of sign; thus $|\mathbf{O}| = 48$. Observe that $\mathbf{U}$ is the other (left and right) coset of $\mathbf{T}$ in $\mathbf{O}$. Further, we have

Lemma 4.1 *Conjugacy between* $\mathbf{I}$ *and* $\mathbf{I}^{\dagger}$ *is induced by the mapping* $\mathbf{x} \mapsto \widetilde{\mathbf{u}}\mathbf{x}\mathbf{u}$, *with* $\mathbf{u} \in \mathbf{U}$ *any element.*

A crucial observation is the following, which can be seen from the tables of [2].

Lemma 4.2 *Subgroups of* $\mathbf{I}$ *isomorphic to* $\mathbf{D}$ *and* $\mathbf{T}$ *are maximal.*

Remark 4.3 Note that $\mathbf{I}$ does not contain dihedral subgroups; its only involution is -1. Observe also that $\mathbf{I} \cap \mathbf{I}^{\dagger} = \mathbf{T}$.

As shown in [2], an orthogonal mapping on $\mathbb{E}^4$ can be represented in terms of quaternions by

$$\mathbf{x} \mapsto \widetilde{\mathbf{a}}\mathbf{x}\mathbf{b} \text{ or } \widetilde{\mathbf{a}}\widetilde{\mathbf{x}}\mathbf{b}, \tag{3}$$

with $\mathbf{a}, \mathbf{b}$ unit quaternions; the first kind of mapping is *direct* or a *rotation*, and the second is *opposite*. If $\mathbf{G}$ is a (finite) orthogonal group represented in this way, then the set of quaternions acting on the left forms a subgroup $\mathbf{L}$; similarly, there is a right acting subgroup $\mathbf{R}$. In each case here, $\mathbf{L}$ and $\mathbf{R}$ will contain -1. If $\mathbf{G}$ contains rotations only and $\mathbf{L} \neq \mathbf{R}$, then – for what we need here – the only identifications will be between $(\mathbf{a}, \mathbf{b})$ and $(-\mathbf{a}, -\mathbf{b})$ which give rise to the same isometry. Following [2] we write $\mathbf{G} := (\mathbf{L}/\mathbf{L}; \mathbf{R}/\mathbf{R})$, whose order is then $|\mathbf{L}| \cdot |\mathbf{R}|/2$. If $\mathbf{G}$ contains opposite mappings, then necessarily $\mathbf{L} = \mathbf{R} = \mathbf{K}$, say. Here, it will happen that there is a normal subgroup $\mathbf{N}$ (possibly $\mathbf{K}$ itself) such that $\widetilde{\mathbf{a}}\mathbf{b} \in \mathbf{N}$; we then write $\mathbf{G} := (\mathbf{K}/\mathbf{N}; \mathbf{K}/\mathbf{N})^*$, whose order is $|\mathbf{K}| \cdot |\mathbf{N}|$. The corresponding rotation subgroup (of half the order) is denoted $(\mathbf{K}/\mathbf{N}; \mathbf{K}/\mathbf{N})$.

Remark 4.4 The general finite orthogonal group $\mathbf{G}$ has a more complicated description than this, and we have tailored our approach to suit our particular circumstances.

5 The 24-Cell

From now on, all polytopes mentioned will be regular and 4-dimensional; following the conventions of [5] we shall thus write cube to mean regular 4-cube, and so on.

There is no problem about the compounds

$$\{4, 3, 3\}[2\{3, 3, 4\}], \quad [2\{4, 3, 3\}]\{3, 3, 4\}, \tag{4}$$

$$\{3, 4, 3\}[3\{3, 3, 4\}]2\{3, 4, 3\}, \quad 2\{3, 4, 3\}[3\{4, 3, 3\}]\{3, 4, 3\} \tag{5}$$

that involve only the rational polytopes, but we shall briefly discuss them anyway since they do have bearing on the others.

The usual vertex-set of the 24-cell $\mathsf{S} := \{3, 4, 3\}$ can be identified with $\mathbf{T}$, but observe that the vertices of the dual 24-cell S^δ are then identified with $\mathbf{U}$. The symmetry group $[3, 4, 3]$ of these dual polytopes is $(\mathbf{O}/\mathbf{T}; \mathbf{O}/\mathbf{T})^*$, which we recall consists of the mappings (3) with $\mathbf{a}, \mathbf{b} \in \mathbf{O}$ such that $\widetilde{\mathbf{a}}\mathbf{b} \in \mathbf{T}$. It follows from the general description in Sect. 4 that $[3, 4, 3]$ has order $|[3, 4, 3]| = 48 \cdot 24 = 1152$. The rotation subgroup is $[3, 4, 3]^+ = (\mathbf{O}/\mathbf{T}; \mathbf{O}/\mathbf{T})$, consisting of the mappings of the first type (that is, not involving $\widetilde{\mathbf{x}}$), and thus has order 576.

The compounds of (5) arise in the following way. First, $\mathbf{V}$ is a normal subgroup of $\mathbf{T}$ of index 3. Regarded as vertex-sets, this inscribes three copies of the cross-polytope $\mathsf{X} = \{3, 3, 4\}$ in S. Since the vertex-set of the dual cube $\mathsf{C} = \{4, 3, 3\}$ is $\operatorname{vert} \mathsf{C} = \operatorname{vert} \mathsf{S} \setminus \operatorname{vert} \mathsf{X} = \mathbf{T} \setminus \mathbf{V} =: \mathbf{W}$, say, we immediately obtain the dual pair of fully regular compounds of (5).

This construction has also inscribed two copies of X in C (that is, the other two cosets of $\mathbf{V}$ in $\mathbf{T}$ as subsets of $\mathbf{W}$), giving the first (only) vertex-regular compound of (4); the second is its facet-regular dual, and again both are fully regular.

The symmetry group of X and C is $[3, 3, 4] = (\mathbf{O}/\mathbf{V}; \mathbf{O}/\mathbf{V})^*$ (a subgroup of $[3, 4, 3]$ of index 3) of order $2 \cdot 48 \cdot 8/2 = 384$. The common subgroup of a copy of X inscribed in C is isomorphic to $(\mathbf{T}/\mathbf{V}; \mathbf{T}/\mathbf{V})^*$ of order 192, and is that denoted $[3^{1,1,1}]$ in [1, Sect. 11.9].

Remark 5.1 A useful observation is that a given cross-polytope X is inscribed in a unique 24-cell S. Indeed, if X has vertices $\pm v_j$ for $j = 1, \dots, 4$, where $v_1, \dots, v_4$ are mutually orthogonal vectors of the same length, then the additional vertices of S are all those of the form $\frac{1}{2}(\pm v_1 \pm v_2 \pm v_3 \pm v_4)$. Note that these extra vertices are those of a cube C dual to X. Moreover, the same result holds for a cube: given a cube C, there is a unique 24-cell S for which $\mathsf{C} \prec \mathsf{S}$.

6 The 600-Cell

Even though the compounds inscribed in the 600-cell $\mathsf{P} = \{3, 3, 5\}$ are well known, it helps to set the scene if we go over them again. The symmetry group [3, 3, 5] of P consists of all mappings of the form (3) with $\mathbf{a}, \mathbf{b} \in \mathbf{I}$, and so is $(\mathbf{I}/\mathbf{I}; \mathbf{I}/\mathbf{I})^*$. Again from Sect. 4, we see that [3, 3, 5] has order $|[3, 3, 5]| = 120 \cdot 120 = 14400$. Recall that we have identified vert P with $\mathbf{I}$.

Since $\mathbf{T} \subset \mathbf{I}$, we can obviously inscribe $\mathsf{S} = \{3, 4, 3\}$ in P. However, it is worth noting that only half its symmetry group survives into the common subgroup, namely, $[3, 4, 3]^* := (\mathbf{T}/\mathbf{T}; \mathbf{T}/\mathbf{T})^*$ consisting of those mappings with $\mathbf{a}, \mathbf{b} \in \mathbf{T}$. As groups, we have $\mathbf{T} < \mathbf{I}$; if we define

$$\mathbf{p} := -\tfrac{1}{2}(\tau - \mathbf{i} + \tau^{-1}\mathbf{j}), \tag{6}$$

which is a typical element of $\mathbf{I}$ of period 5, then the powers of $\mathbf{p}$ act as both left and right coset representatives of $\mathbf{T}$ in $\mathbf{I}$. Indeed, the 25 subsets $\mathbf{p}^{-r}\mathbf{T}\mathbf{p}^{s} = \widetilde{\mathbf{p}}^{r}\mathbf{T}\mathbf{p}^{s} \subset \mathbf{I}$ are distinct, and yield a compound

$$5\{3, 3, 5\}[25\{3, 4, 3\}]\{3, 3, 5\}; \tag{7}$$

we shall justify the facet-regularity in Sect. 9.

At this point, it is worth introducing a binary dihedral group that will appear quite often. Observe that $\widetilde{\mathbf{kpk}} = \widetilde{\mathbf{p}}$ and, since $\mathbf{k}^2 = -1$, it should be clear that $\langle \mathbf{p}, \mathbf{k} \rangle$ is binary dihedral of order 20, and so a copy of $\mathbf{D}$.

Now we can clearly consider the (say) right cosets $\mathbf{T}\mathbf{p}^{s}$ alone, which must give rise to a compound of type $\mathsf{P}[5\mathsf{S}]$. If we dualize, keeping track of the vertices as quaternions, then for the typical copy of S we have

$$\delta\colon \mathbf{T}\mathbf{p}^{s} \mapsto \mathbf{U}\mathbf{p}^{s} = \widetilde{\mathbf{u}}\mathbf{T}\mathbf{p}^{s} = \mathbf{T}\widetilde{\mathbf{u}}\mathbf{p}^{s},$$

with $\mathbf{u} \in \mathbf{U}$ as before. The last two expressions tell us two things. First, the compound is also facet-regular, with the facets touching those of $\widetilde{\mathbf{u}}\mathsf{P}^{\delta}$, a copy of the 120-cell {5, 3, 3}; hence it is of kind

$$\{3, 3, 5\}[5\{3, 4, 3\}]\{5, 3, 3\}. \tag{8}$$

Second, note that we can only apply elements of the subgroup $\mathbf{T}$ on the left; of course, elements of $\mathbf{I}$ applied on the right just permute the cosets of $\mathbf{T}$. Following Sect. 4, therefore, the symmetry group $\mathbf{G}$ of the compound is

$$\mathbf{G} = (\mathbf{T}/\mathbf{T}; \mathbf{I}/\mathbf{I}),$$

of order $24 \cdot 120/2 = 1440$. Observe that $\mathbf{G}$ only acts on the rotation subgroup $(\mathbf{T}/\mathbf{T}; \mathbf{T}/\mathbf{T})$ of $[3, 4, 3]^*$ of order $24 \cdot 24/2 = 288$, which accounts for there being $1440/288 = 5$ copies of S in the compound.

Dealing with the left cosets of **T** in **I** instead results in an enantiomorphic copy of the compound; the two can be swapped by a suitable hyperplane reflexion.

The compounds of the cross-polytope and cube in the 24-cell lead to further compounds

$$\mathit{5}\{3,3,5\}[\mathit{75}\{3,3,4\}]\mathit{10}\{5,3,3\},\quad \mathit{10}\{3,3,5\}[\mathit{75}\{4,3,3\}]\mathit{5}\{5,3,3\}, \tag{9}$$
$$\{3,3,5\}[\mathit{15}\{3,3,4\}]2\{5,3,3\},\quad 2\{3,3,5\}[\mathit{15}\{4,3,3\}]\{5,3,3\}. \tag{10}$$

The symmetry group of the dual pair of (9) remains the whole group [3, 3, 5]; the common subgroup is $[3^{1,1,1}] = (\mathbf{T}/\mathbf{V};\ \mathbf{T}/\mathbf{V})^*$ of order $2 \cdot 24 \cdot 8/2 = 192$. The group of the two compounds of (10) remains $(\mathbf{T}/\mathbf{T};\ \mathbf{I}/\mathbf{I})$, but this is now acting on the rotation subgroup $(\mathbf{T}/\mathbf{V};\ \mathbf{T}/\mathbf{V})$.

7 The 120-Cell

There are two ways of describing the vertex-set of the 120-cell $\mathsf{Q} := \{5, 3, 3\}$, of which we shall find the second more useful. If we write $\mathbf{u} := (1 - \mathbf{k})/\sqrt{2} \in \mathbf{U}$, then $\mathbf{u}$ can be taken as the initial vertex of Q so that – as unit quaternions – the vertex-set of Q is

$$\mathbf{IuI} = \mathbf{IUI}.$$

Using Lemma 4.1, we can rewrite this set as

$$\mathbf{u} \cdot \widetilde{\mathbf{u}}\mathbf{Iu} \cdot \mathbf{I} = \mathbf{u} \cdot \mathbf{I}^{\dagger}\mathbf{I}.$$

Applying left multiplication by $\widetilde{\mathbf{u}}$ then gives $\mathbf{H} := \mathbf{I}^{\dagger}\mathbf{I}$. Particularly observe that $\mathbf{H}$ contains $\mathbf{I}$ and $\mathbf{I}^{\dagger}$ themselves as subsets.

In this form, the rotational symmetries in $[3, 3, 5]^+$ are the mappings $\mathbf{x} \mapsto \mathbf{a}^*\mathbf{x}\mathbf{b}$ with $\mathbf{a}, \mathbf{b} \in \mathbf{I}$, where we write $\mathbf{g}^* := \widetilde{\mathbf{g}}^{\dagger} = (\mathbf{g}^{\dagger})^{-1} \in \mathbf{I}^{\dagger}$. However, the opposite symmetries are now $\mathbf{x} \mapsto \mathbf{a}^*(-\mathbf{u}\widetilde{\mathbf{x}}\mathbf{u})\mathbf{b}$, with $\mathbf{a}, \mathbf{b} \in \mathbf{I}$ and $\mathbf{u} \in \mathbf{U}$ as before. (In terms of coordinates, $\mathbf{x} \mapsto -\mathbf{u}\widetilde{\mathbf{x}}\mathbf{u}$ is $(\xi_1, \ldots, \xi_4) \mapsto (\xi_4, \xi_2, \xi_3, \xi_1)$.)

Remark 7.1 Multiplied by 4 and regarded as vectors, our coordinates can be obtained from those of [1, Table V(v)] by changing the sign of one (fixed) coordinate. In other words, the latter actually correspond to $\mathbf{II}^{\dagger}$ rather than $\mathbf{I}^{\dagger}\mathbf{I}$,

Remark 7.2 We have chosen $\mathbf{p}$ in (6) and $\mathbf{u}$ to be compatible, in the sense that, if we write $\mathbf{q} := \mathbf{p}^{\dagger}$, then $\widetilde{\mathbf{u}}\mathbf{p}\mathbf{u} = \mathbf{q}^2$ (and $\widetilde{\mathbf{u}}\mathbf{q}\mathbf{u} = \mathbf{p}^2$). This interaction between $\mathbf{p}$ and $\mathbf{q}$ is central in what follows.

We immediately find our next compounds. With the same $\mathbf{p}$, the five copies $\mathbf{I}^{\dagger}\mathbf{p}^k$ of $\mathbf{I}^{\dagger}$ are disjoint, and thus form a vertex-regular compound

$$\{5,3,3\}[\mathit{5}\{3,3,5\}]. \tag{11}$$

If $\mathbf{b} \in \mathbf{I}$ and $j = 0, \ldots, 4$, then we can express $\mathbf{p}^j\mathbf{b} = \mathbf{c}\mathbf{p}^k$ for some $\mathbf{c} \in \mathbf{T}$ and $k = 0, \ldots, 4$. It follows that, if $\mathbf{a}, \mathbf{b} \in \mathbf{I}$, then under a general rotation $\mathbf{x} \mapsto \mathbf{a}^*\mathbf{x}\mathbf{b}$, we have

$$\mathbf{I}^\dagger\mathbf{p}^j \mapsto \mathbf{a}^*\mathbf{I}^\dagger\mathbf{c}\mathbf{p}^k = \mathbf{I}^\dagger\mathbf{p}^k;$$

thus the symmetry group of the compound is the rotation group $[3, 3, 5]^+$.

Adjoining the analogous sets $\mathbf{q}^k\mathbf{I}$ then gives a fully regular compound

$$2\{5, 3, 3\}[\mathit{10}\{3, 3, 5\}]. \tag{12}$$

Naturally, there are the corresponding dual facet-regular compounds.

8 Compounds of Simplices

The previously known regular compound of simplices in the 120-cell is

$$\{5, 3, 3\}[\mathit{120}\{3, 3, 3\}]\{3, 3, 5\}. \tag{13}$$

But before we show how to find it, we need to make one more remark about $\mathbf{I}$ and $\mathbf{I}^\dagger$. As in (6), we define $\mathbf{p} := -\frac{1}{2}(\tau - \tau^{-1}\mathbf{i} + \mathbf{j})$, and further set $\mathbf{q} := \mathbf{p}^\dagger$ as in Remark 7.2; thus $\mathbf{p}^* = \widetilde{\mathbf{q}}$. We first have (see [2, Sect. 21])

Lemma 8.1 *If* $\mathbf{g} \in \mathbf{I}$, *then* $\mathbf{g}^*\mathbf{g} = \mathbf{q}^{-k}\mathbf{p}^k = (\mathbf{p}^*)^k\mathbf{p}^k$ *for some* $k = 0, \ldots, 4$.

Proof As we saw in Sect. 6, the powers of $\mathbf{p}$ are right coset representatives of $\mathbf{T}$ in $\mathbf{I}$; thus we can write $\mathbf{g} = \mathbf{a}\mathbf{p}^k$ for some k and some $\mathbf{a} \in \mathbf{T}$. Since $\mathbf{a}^\dagger = \mathbf{a}$, it follows that

$$\mathbf{g}^*\mathbf{g} = (\mathbf{a}\mathbf{p}^k)^*(\mathbf{a}\mathbf{p}^k) = ((\mathbf{a}\mathbf{p}^k)^{-1})^\dagger(\mathbf{a}\mathbf{p}^k) = (\mathbf{p}^{-k}\mathbf{a}^{-1})^\dagger(\mathbf{a}\mathbf{p}^k) = \mathbf{q}^{-k}\mathbf{p}^k,$$

as claimed.

Following [2], we write $\mathbf{P} := \{\mathbf{q}^{-k}\mathbf{p}^k \mid k = 0, \ldots, 4\}$. We next show

Theorem 8.2 *The set* $\mathbf{P}$ *is the vertex-set of a regular* 4*-simplex* A, *whose full symmetry group is a subgroup of the group* $[3, 3, 5]$ *of* $Q = \{5, 3, 3\}$. *In particular,* $[3, 3, 3]$ *is a subgroup of* $[3, 3, 5]$ *of index* 120.

Proof First, the mapping $\mathbf{x} \mapsto \mathbf{p}^*\mathbf{x}\mathbf{p}$ permutes the five points of $\mathbf{P}$ cyclically. Second, if $\mathbf{v} \in \mathbf{U}$ is pure imaginary, so that $\widetilde{\mathbf{v}} = -\mathbf{v}$, then the mapping $\mathbf{x} \mapsto \widetilde{\mathbf{v}}\widetilde{\mathbf{x}}\mathbf{v}$ interchanges $\mathbf{I}$ and $\mathbf{I}^\dagger$, and fixes 1. For $\mathbf{g} \in \mathbf{I}$, define $\mathbf{h} := \widetilde{\mathbf{v}}\mathbf{g}^\dagger\mathbf{v}$; we see that

$$\mathbf{h}^* = \widetilde{\mathbf{h}}^\dagger = (\mathbf{v}\widetilde{\mathbf{g}}^\dagger\widetilde{\mathbf{v}})^\dagger = \mathbf{v}\widetilde{\mathbf{g}}\widetilde{\mathbf{v}} = \widetilde{\mathbf{v}}\widetilde{\mathbf{g}}\mathbf{v}.$$

Hence, under this mapping,

$$\mathbf{g}^*\mathbf{g} = \widetilde{\mathbf{g}}^\dagger\mathbf{g} \mapsto \widetilde{\mathbf{v}}\widetilde{\mathbf{g}}\mathbf{g}^\dagger\mathbf{v} = (\widetilde{\mathbf{v}}\widetilde{\mathbf{g}}\mathbf{v})(\widetilde{\mathbf{v}}\mathbf{v})(\widetilde{\mathbf{v}}\mathbf{g}\mathbf{v}) = \mathbf{h}^*\mathbf{h},$$

which is again some element of **P**. These two kinds of symmetry generate the whole group of A, as we had to demonstrate. □

As we said at the beginning of the section, we deduce at once the existence of the compound of (13). That the compound is self-dual is easy to see, because the (suitably scaled) dual of A is −A, its point-reflexion in the origin o, which is also inscribed in the 120-cell Q.

Remark 8.3 The subgroup relationship of Theorem 8.2 is well known. As vectors, the five points of **P** are $(1, 0, 0, 0)$ and all $\frac{1}{4}\big(-1, \sqrt{5}(\varepsilon_2, \varepsilon_3, \varepsilon_4)\big)$ with $\varepsilon_j = \pm 1$ for each j and $\varepsilon_2\varepsilon_3\varepsilon_4 = 1$. We have arranged things so that $\mathbf{p}^*\mathbf{p} = \frac{1}{4}\big(-1 + \sqrt{5}(\mathbf{i} + \mathbf{j} + \mathbf{k})\big)$.

What seems to have escaped notice until now is that there are regular simplices of quite a different kind inscribed in {5, 3, 3}. With **p** as in (6) and $\mathbf{q} = \mathbf{p}^\dagger$, the mapping $\mathbf{x} \mapsto \mathbf{qxp}$ is cyclic of period 5, and takes 1 into the set of five points $\mathbf{Q} := \{\mathbf{q}^k\mathbf{p}^k \mid k = 0, \ldots, 4\}$. Moreover, since **q** is conjugate (actually within $\mathbf{I}^\dagger$) to $\widetilde{\mathbf{q}}$, we should not be surprised to learn that **Q** is also the vertex-set of a regular 4-simplex, which we denote by B.

Apart from 1, the points of **Q** are

$$\tfrac{1}{4}\big(-1 \pm (\mathbf{i} + 3\mathbf{j}) - \sqrt{5}\mathbf{k}\big), \quad \tfrac{1}{4}\big(-1 \pm (3\mathbf{i} - \mathbf{j}) + \sqrt{5}\mathbf{k}\big),$$

with $\mathbf{qp} = -\frac{1}{4}(1 + \mathbf{i} + 3\mathbf{j} + \sqrt{5}\mathbf{k})$. As well as the cyclic symmetry $\mathbf{x} \mapsto \mathbf{qxp}$, we have the opposite symmetry $(\xi_1, \ldots, \xi_4) \mapsto (\xi_1, \xi_3, -\xi_2, -\xi_4)$ of period 4; in terms of quaternions, this is

$$\mathbf{x} \mapsto \widetilde{\mathbf{u}}\widetilde{\mathbf{x}}\mathbf{u}, \tag{14}$$

where $\mathbf{u} = (1 - \mathbf{k})/\sqrt{2} \in \mathbf{U}$ as earlier. Of course, this must be compatible with the other symmetries. Remark 7.2 shows why this is so, since

$$\mathbf{q}^k\mathbf{p}^k \mapsto \widetilde{\mathbf{u}}\mathbf{p}^{-k}\mathbf{q}^{-k}\mathbf{u} = \mathbf{q}^{-2k}\mathbf{p}^{-2k} \in \mathbf{Q},$$

as required.

We thus see that the common group $\mathbf{K}(\mathsf{Q}, \mathsf{B})$ has order 20 and so is of index $14400/20 = 720$ in [3,3,5]; hence we obtain the first of our compounds involving B, namely,

$$6\{5, 3, 3\}[720\{3, 3, 3\}]6\{3, 3, 5\}. \tag{15}$$

By construction, it is fully regular. The compound is self-dual, since the dual −B of B is also inscribed in {5, 3, 3}.

However, we can extract a sub-compound from (15). If we just apply elements of **I** to the right, we find that we obtain $120 \cdot 5 = 600$ points in **H**. These points are distinct, because the angle $\arccos(-\frac{1}{4})$ subtended by two vertices of B is not among those subtended by vertices of the 600-cell {3, 3, 5}, and so **H** is covered exactly once. We have thus found a new compound

$$\{5,3,3\}[\mathit{120}\{3,3,3\}]\{3,3,5\}^{(\mathrm{var})}, \tag{16}$$

where $^{(\mathrm{var})}$ indicates that it is not the previous compound of (13).

Now B does not retain all the symmetries of $\mathbf{K}(\mathsf{Q},\mathsf{B})$ in the compound. In fact, since the square of the mapping (14) is $\mathbf{x} \mapsto \mathbf{kx}\widetilde{\mathbf{k}} = \widetilde{\mathbf{k}}\mathbf{xk}$, all the mappings of the kind $\mathbf{x} \mapsto \mathbf{a}^*\mathbf{xb}$ with $\mathbf{a} \in \mathbf{D} := \langle \mathbf{p}, \mathbf{k} \rangle$ and $\mathbf{b} \in \mathbf{I}$ are allowed, but just these; here, $\mathbf{D}$ is binary dihedral of order 20 and, as we have already noted, we have $\widetilde{\mathbf{k}}\mathbf{pk} = \mathbf{p}^{-1}$ ($\widetilde{\mathbf{k}}\mathbf{qk} = \mathbf{q}^{-1}$ follows similarly). We conclude that the symmetry group of the compound (16) is $(\mathbf{D}/\mathbf{D}; \mathbf{I}/\mathbf{I})$, of order $20 \cdot 120/2 = 1200$ (observe that the cyclic symmetries $\mathbf{x} \mapsto \mathbf{q}^k\mathbf{xp}^k$ are needed to ensure the transitivity of the compound on vert{5, 3, 3}).

It is clear that we can equally well use left multiplication by $\mathbf{I}^\dagger$ rather than right multiplication by $\mathbf{I}$; this leads to a compound enantiomorphic to the original. Apart from this, we naturally ask whether there is a larger sub-compound of (15) than (16). When we bear in mind Lemma 4.2, we see that the only possibility is to replace the left group $\mathbf{D}$ by $\mathbf{I}^\dagger$. But then we acquire all the symmetries $\mathbf{x} \mapsto \widetilde{\mathbf{w}}\mathbf{xw}$ with $\mathbf{w} \in \mathbf{T}$, and hence (with a little inspection) the other five copies of B containing 1. For instance, the case $\mathbf{w} = \mathbf{i}$ changes the signs of $\mathbf{j}$ and $\mathbf{k}$, while $\mathbf{w} = \mathbf{c} := -\frac{1}{2}(1 + \mathbf{i} + \mathbf{j} + \mathbf{k})$ permutes $\mathbf{i}, \mathbf{j}, \mathbf{k}$ cyclically (the group of the vertex-figure consists of all permutations of $\mathbf{i}, \mathbf{j}, \mathbf{k}$ with an even number of changes of their signs). An important factor here is that the common subgroup $\mathbf{K}(\mathsf{Q},\mathsf{B})$ already has opposite symmetries.

Remark 8.4 The fact that the 24 points of the second kind in [1, Table V(v)] in section 12_0 opposite to section 18_0 do not look particularly symmetric (apart from the obvious symmetries induced by the group of the vertex-figure of {5, 3, 3}) may account for Coxeter failing to notice the inscribed tetrahedra.

9 Compounds of Cross-Polytopes

Just as with simplices, there are two quite different ways of inscribing a regular cross-polytope $\mathsf{X} = \{3,3,4\}$ in a 120-cell $\mathsf{Q} = \{5,3,3\}$; similarly, we can inscribe cubes {4, 3, 3} and 24-cells {3, 4, 3} in two different ways. We shall treat compounds of all three polytopes in Q together.

The first way to inscribe X in Q illustrates the subgroup relationship $[3^{1,1,1}] < [3,3,5]$. Working with the same Q as hitherto, the copy of $[3^{1,1,1}]$ now consists of the mappings $\mathbf{x} \mapsto \widetilde{\mathbf{a}}\mathbf{xb}$ with $\mathbf{a}, \mathbf{b} \in \mathbf{T}$ and $\mathbf{a} \mapsto \widetilde{\mathbf{a}\mathbf{x}}\mathbf{b}$ with $\mathbf{a}, \mathbf{b} \in \mathbf{U}$, such that $\widetilde{\mathbf{a}}\mathbf{b} \in \mathbf{V}$ in each case; the order remains 192. This copy of X has vertex-set $\mathbf{V}$, which we identify as vectors with $\pm e_j$ for $j = 1, \ldots, 4$. (Note that the obvious necessary condition for the mappings to preserve $\mathbf{V}$ is also sufficient.)

Since the index is $\big[[3,3,5]:[3^{1,1,1}]\big] = 14400/192 = 75$, we obtain a fully regular compound

$$\{5,3,3\}[75\{3,3,4\}]2\{3,3,5\}; \tag{17}$$

we shall justify the facet-regularity shortly. Note that we cannot extract any sub-compound, since we need all $75 \cdot 8 = 600$ vertices of the copies of X.

Now we notice that $\mathsf{X} \prec \mathsf{Q}$ implies that $\mathsf{C} = \mathsf{X}^\delta \prec \mathsf{Q}$ also. Hence, in the same way, $\mathsf{S} = \{3,4,3\} \prec \mathsf{Q}$, with vert S identified with $\mathbf{T}$. The common subgroup $\mathbf{K}(\mathsf{Q},\mathsf{S})$ consists of those mappings of type (3) with $\mathbf{a},\mathbf{b} \in \mathbf{T}$ when $\mathbf{x} \mapsto \widetilde{\mathbf{a}}\mathbf{x}\mathbf{b}$ or $\mathbf{a},\mathbf{b} \in \mathbf{U}$ when $\mathbf{x} \mapsto \widetilde{\mathbf{a}}\widetilde{\mathbf{x}}\mathbf{b}$; this group, of index 2 in $[3,4,3]$, is conjugate to the subgroup $[3,4,3]^*$ of Section 6. We then obtain the fully regular compound

$$\{5,3,3\}[25\{3,4,3\}]5\{5,3,3\}; \tag{18}$$

this is the dual to the compound (7). As with (17), this does not admit any sub-compounds.

Completing this particular picture, we have the dual of (17), namely,

$$2\{5,3,3\}[75\{4,3,3\}]\{3,3,5\}; \tag{19}$$

The other way of inscribing X is given typically by $\mathbf{V}' := \widetilde{\mathbf{p}}\mathbf{V}\mathbf{p} \subset \mathbf{I} \subset \mathbf{H}$. The new vertex-set is (as a subgroup) conjugate to $\mathbf{V}$. However, it is clear that rather few of the symmetries of Q induce ones of this copy, which we shall call Y. Indeed, we cannot allow opposite symmetries of Q at all, since they take $\mathbf{V}'$ into a subset of $\mathbf{I}^\dagger$. Easy calculations show that, apart from ± 1, the elements of $\mathbf{V}'$ are the cyclic permutations (under $\mathbf{i} \mapsto \mathbf{j} \mapsto \mathbf{k} \mapsto \mathbf{i}$) of $\pm\frac{1}{2}(\tau\mathbf{i} - \tau^{-1}\mathbf{j} + \mathbf{k}) = \pm\widetilde{\mathbf{p}}\mathbf{i}\mathbf{p}$. We see from this that the common subgroup $\mathbf{K}(\mathsf{Q},\mathsf{Y})$ consists of right multiplication by $\mathbf{V}'$ itself, extended by the element $\mathbf{x} \mapsto \widetilde{\mathbf{c}}\mathbf{x}\mathbf{c}$ of order 3, where $\mathbf{c} = -\frac{1}{2}(1 + \mathbf{i} + \mathbf{j} + \mathbf{k})$ as earlier; hence $|\mathbf{K}| = 3 \cdot 8 = 24$.

Now we can allow the full group [3,3,5] or its rotation subgroup $[3,3,5]^+$ to act on Y, yielding the two compounds

$$8\{5,3,3\}[600\{3,3,4\}]16\{3,3,5\}, \tag{20}$$
$$4\{5,3,3\}[300\{3,3,4\}]8\{3,3,5\}. \tag{21}$$

In contrast to the inscription $\mathsf{B} \prec \mathsf{Q}$ of Sect. 8, $\mathbf{K}(\mathsf{Q},\mathsf{Y})$ does not contain opposite symmetries, so that we genuinely do have a sub-compound under rotations alone.

The notation above comes from the fact that we can dualize, to obtain corresponding compounds of cubes:

$$16\{5,3,3\}[600\{4,3,3\}]8\{3,3,5\}, \tag{22}$$
$$8\{5,3,3\}[300\{4,3,3\}]4\{3,3,5\}; \tag{23}$$

once again, bear in mind Remark 5.1.

Completing this pattern, we have compounds of 24-cells:

$$8\{5,3,3\}[200\{3,4,3\}], \tag{24}$$
$$4\{5,3,3\}[100\{3,4,3\}]. \tag{25}$$

These two compounds are only vertex-regular; the images of the facet normals of the inscribed copies of $\{3,4,3\}$ do not fall into the vertices of either $\{5,3,3\}$ or $\{3,3,5\}$.

Up to here, the compounds in this section are already listed in [1]. We can now see that the only compounds that could possibly give rise to sub-compounds are those of (21), (23) and (25). Let us analyse the geometry of the last compound more carefully. Exactly as for $\mathsf{Y} \prec \mathsf{Q}$, the common subgroup $\mathbf{K}$ of the inscribed copy of $\{3,4,3\}$ in $\{5,3,3\}$, whose vertex-set is identified with $\widetilde{\mathbf{p}}\mathbf{T}\mathbf{p}$, consists of right multiplication by $\widetilde{\mathbf{p}}\mathbf{T}\mathbf{p}$ itself, extended by the element $\mathbf{x} \mapsto \widetilde{\mathbf{c}}\mathbf{x}\mathbf{c}$, where $\mathbf{c}$ is as before. Hence $|\mathbf{K}| = 3 \cdot 24 = 72$, and this accounts for the numbers of copies of $\{3,4,3\}$ in the compounds, namely, $14400/72 = 200$ for the fully regular one, and $7200/72 = 100$ for that with rotational symmetry.

We now refer to the compound (11), in the alternative form as composed of the $\mathbf{q}^j\mathbf{I}$, with $\mathbf{q} = \mathbf{p}^\dagger$ as before. Since $\widetilde{\mathbf{p}}\mathbf{T}\mathbf{p} \subset \mathbf{I}$, feeding the compound (8) into $\widetilde{\mathbf{p}}\mathbf{I}\mathbf{p} = \mathbf{I}$ as a vertex-set, it follows that we have a new compound of copies of $\{3,4,3\}$, whose vertex-sets are identified with the $\mathbf{q}^j\widetilde{\mathbf{p}}\mathbf{T}\mathbf{p}^k$ for $j,k = 0,\dots,4$. For this, all $\mathbf{I}$ acts on the right, but only powers of $\mathbf{q}$ act on the left. Thus the symmetry group is $(\mathbf{C}/\mathbf{C};\mathbf{I}/\mathbf{I})$ of order $10 \cdot 120/2 = 600$, where $\mathbf{C} = \mathbf{C}_{10} := \langle -\mathbf{q} \rangle$, and we obtain a compound

$$\{5,3,3\}[25\{3,4,3\}]^{(\mathrm{var})}. \tag{26}$$

This is only vertex-regular; its dual has vertex-set $\mathbf{q}^j\widetilde{\mathbf{p}}\mathbf{U}\mathbf{p}^k$ for $j,k = 0,\dots,4$, whose 600 distinct points cannot be fitted into the vertices of a copy of $\{5,3,3\}$. (As usual, we take the facet-regular dual for granted.)

In exactly the same way, we derive the two dual compounds

$$\{5,3,3\}[75\{3,3,4\}]2\{3,3,5\}^{(\mathrm{var})}, \tag{27}$$
$$2\{5,3,3\}[75\{4,3,3\}]\{3,3,5\}^{(\mathrm{var})}; \tag{28}$$

by applying the same symmetries of $(\mathbf{C}/\mathbf{C};\mathbf{I}/\mathbf{I})$ to the initial copies of the cross-polytope and cube, respectively.

Although $\mathbf{C}$ is not maximal, we cannot extend it to $\mathbf{D}$ as for the exceptional compound of 120 simplices. Rather, if we do, then we find that we have to include (for example) the copies of $\{3,3,4\}$ obtained by changing signs of any two of the last three coordinates, which results in the compound of (21). We conclude from this that our enumeration is now complete.

Remark 9.1 Observe that the common symmetry group of the last compounds is as small as it possibly can be; it is simply transitive on vert$\{5,3,3\}$.

10 Amendments

The foregoing implies that three of the tables in [1] need to be modified:

- in Table V(v), against 12_0 replace "Tetrahedron" by "$1+6$ tetrahedra";
- in Table VI(iv), against 18_0 replace "$\{3,3\}$" by "$(1+6)\{3,3\}$";
- in Table VII(i), insert (in appropriate places)

$$\begin{aligned}
&\mathit{6}\{5,3,3\}[\mathit{720}\{3,3,3\}]\mathit{6}\{3,3,5\}\\
&\{5,3,3\}[\mathit{120}\{3,3,3\}]\{3,3,5\}^{(\mathrm{var})},\\
&\{5,3,3\}[\mathit{75}\{3,3,4\}]\mathit{2}\{3,3,5\}^{(\mathrm{var})},\\
&\mathit{2}\{5,3,3\}[\mathit{75}\{4,3,3\}]\{3,3,5\}^{(\mathrm{var})},\\
&\{5,3,3\}[\mathit{25}\{3,4,3\}]^{(\mathrm{var})},\\
&[\mathit{25}\{3,4,3\}]\{3,3,5\}^{(\mathrm{var})}
\end{aligned}$$

with the indication that 'var' marks out those that are *different* from others with a similar symbol.

It is also worth pointing out an omission and a mistake in the tables of [2] that we have referred to, although these are not immediately relevant to our discussion:

- in the list at the bottom of p. 56, insert "$\mathbf{C}_{2n}$ in $\mathbf{D}_{nr}$ $(\mathfrak{D}_{2r})$";
- in the list on p. 57, item 18 should read "$(\mathbf{D}_{3m}/\mathbf{C}_{2m};\ \mathbf{O}/\mathbf{V})\quad 48m$".

References

1. H.S.M. Coxeter, *Regular Polytopes*, 3rd edn. (Dover, New York, 1973)
2. P. Du Val, *Homographies, Quaternions and Rotations* (Oxford University Press, Oxford, 1964)
3. L. Fejes Tóth, *Reguläre Figuren.* Akadémiai Kiadó (Budapest, 1965). (English translation: Regular Figures) (Pergamon Press, Oxford, 1964)
4. P. McMullen, Regular star-polytopes, and a theorem of Hess. Proc. Lond. Math. Soc. **18**(3), 577–596 (1968)
5. P. McMullen, E. Schulte, *Abstract Regular Polytopes*, Encyclopedia of Mathematics and its Applications (Cambridge University Press, Cambridge, 2002)

Five Essays on the Geometry of László Fejes Tóth

Oleg R. Musin

Abstract In this paper we consider the following topics related to results of László Fejes Tóth: (1) The Tammes problem and Fejes Tóth's bound on circle packings; (2) Fejes Tóth's problem on maximizing the minimum distance between antipodal pairs of points on the sphere; (3) Fejes Tóth's problem on the maximum kissing number of packings on the sphere; (4) The Fejes Tóth–Sachs problem on the one-sided kissing numbers; (5) Fejes Tóth's papers on the isoperimetric problem for polyhedra.

1 Tammes' Problem and Fejes Tóth's Bound on Circle Packings

1.1 Tammes' Problem

We start with the following classical problem: How should N points be distributed on a unit sphere so that the minimum distance between two points of the set attains its maximum value d_N? This problem was first asked by the Dutch botanist Tammes [58] while examining the distribution of openings on the pollen grains of different flowers. This question is also known as the problem of the "inimical dictators" [40], namely *"where should N dictators build their palaces on a planet so as to be as far away from each other as possible?"* The problem is equivalent with the problem of densest packing of congruent circles on the sphere (see e.g. [11, Sect. 1.6: Problem 6]): How are N congruent, non-overlapping circles distributed on a sphere when the common radius of the circles has to be as large as possible? The higher dimensional analogue of the problem has applications in information theory [60]. This justifies the terminology that a finite subset X of $\mathbb{S}^n$ with

This research is partially supported by the NSF grant DMS-1400876 and the RFBR grant 15-01-99563.

O. R. Musin (✉)
School of Mathematical and Statistical Sciences, University of Texas Rio Grande Valley, One West University Boulevard, Brownsville, TX 78520, USA
e-mail: oleg.musin@utrgv.edu

G. Ambrus et al. (eds.), *New Trends in Intuitive Geometry*, Bolyai Society Mathematical Studies 27, https://doi.org/10.1007/978-3-662-57413-3_13

$$\psi(X) := \min_{x,y \in X, x \neq y} \operatorname{dist}(x, y)$$

is called a spherical $\psi(X)$-code.

Tammes' problem is presently solved only for $N \le 14$ and $N = 24$. L. Fejes Tóth [18] solved the problem for $N = 3, 4, 6, 12$. Schütte and van der Waerden [55] settled the cases $N = 5, 7, 8, 9$. The cases $N = 10$ and 11 were solved by Danzer [15][1] (see also the papers by Böröczky [7] for $N = 11$ and Hárs [36] for $N = 10$). Robinson [51] solved the problem for $N = 24$. In my recent papers with Tarasov [46, 49] we gave a computer–assisted solution for $N = 13$ and $N = 14$.

Robinson extended Fejes Tóth's method and gave a bound valid for all N that is sharp besides the cases $N = 3, 4, 6, 12$ also for $N = 24$. The solution of all other cases is based on the investigation of the so called contact graphs associated with a finite set of points. For a finite set X in $\mathbb{S}^2$ the *contact graph* $\operatorname{CG}(X)$ is the graph with vertices in X and edges (x, y), $x, y \in X$, such that $\operatorname{dist}(x, y) = \psi(X)$. The concept of contact graphs was first used by Schütte and van der Waerden [55]. They used the method also for the solution of the thirteen spheres (Newton–Gregory) problem [56].

In Chap. VI of the book [21] the concept of irreducible contact graphs is considered in details. The method of irreducible spherical contact graphs was used also [8–10, 34, 59] for obtaining bounds for the kissing number and Tammes problem.

1.2 The Fejes Tóth Bound

Now we consider a theorem on bounds of equal–circle packing and covering of the sphere proved by László Fejes Tóth in 1943 [18, 21, 25].

Theorem 1.1 (L. Fejes Tóth [18]) *If Θ is the density of a packing of the unit sphere $\mathbb{S}^2$ in $\mathbb{R}^3$ by N congruent spherical caps then*

$$\Theta \le \frac{N}{4}\left(2 - \csc \omega_N\right), \quad \textit{where } \omega_N := \frac{N\pi}{6N - 12}$$

If Ω is the density of a covering of $\mathbb{S}^2$ by N congruent spherical caps then

$$\Omega \ge \frac{N}{2}\left(1 - \frac{1}{\sqrt{3}} \cot \omega_N\right).$$

Denote by $A(n, \varphi)$ the maximum cardinality of a φ–code in $\mathbb{S}^{n-1}$. In other words, $A(n, \varphi)$ is the maximum cardinality of a packing in $\mathbb{S}^{n-1}$ by spherical caps of radius $\varphi/2$.

---

[1] Actually, Danzer's paper [15] is the English translation of his Habilitationsschrift "Endliche Punktmengen auf der 2-sphäre mit möglichst großem Minimalabstand". Universität Göthingen, 1963.

The bound in Theorem 1.1 yields

$$A(3, \varphi) \leq \frac{2\pi}{\Delta(\varphi)} + 2,$$

where

$$\Delta(\varphi) = 3 \arccos\left(\frac{\cos\varphi}{1+\cos\varphi}\right) - \pi$$

is the area of a spherical regular triangle with side length φ.

The bound is tight for $N = 3, 4, 6$ and 12. So for these N it gives a solution of the Tammes problem. It is also tight asymptotically. However, is not tight for any other cases.

1.3 Coxeter's Bound

In 1963 Coxeter [14] proposed an extension of Fejes Tóth's bound for all dimensions. His bound was based on the conjecture that in n-dimensional spherical space equal size balls cannot be packed denser than the density of $n + 1$ mutually touching balls of the same size with respect to the simplex spanned by the centers of the balls. This conjecture has been stated by Fejes Tóth for the 3-dimensional case in [22] and for all dimensions in [23, 24]. Assuming the correctness of the conjecture Coxeter calculated the upper bounds 26, 48, 85, 146, and 244 for the kissing numbers $k(n) = A(n, \pi/3)$ for $n = 4, 5, 6, 7$, and 8, respectively. The conjecture, which was finally confirmed by Böröczky [6] in 1978 also yields that

$$A(4, \pi/5) = 120.$$

Theorem 1.2 (Böröczky [6] and Coxeter [14])

$$A(n, \varphi) \leq 2F_{n-1}(\alpha)/F_n(\alpha),$$

where

$$\sec 2\alpha = \sec\varphi + n - 2,$$

and the function F is defined recursively by

$$F_{n+1}(\alpha) = \frac{2}{\pi} \int\limits_{arcsec(n)/2}^{\alpha} F_{n-1}(\beta)\, d\theta, \ \ \sec 2\beta = \sec 2\theta - 2,$$

with the initial conditions $F_0(\alpha) = F_1(\alpha) = 1$.

2 The Problem on Maximizing the Minimum Distance Between Antipodal Pairs of Points on the Sphere

L. Fejes Tóth [26] considered Tammes' problem for antipodal sets on $\mathbb{S}^2$ i.e. for sets X that are invariant under the antipodal mapping $A : \mathbb{S}^d \to \mathbb{S}^d$, where $A(x) = -x$. Let

$$a_M := \max_{X=-X\subset\mathbb{S}^2} \{\psi(X)\}, \quad \text{for } |X| = 2M.$$

For a given M, Fejes Tóth's problem for antipodal sets is to find all configurations

$$X = \{x_1, -x_1, \ldots, x_M, -x_M\}$$

on $\mathbb{S}^2$ such that $\psi(X) = a_M$.

This problem is presently solved only for $M \leq 7$. It is clear that $a_M \leq d_{2M}$. Therefore, if $\psi(X) = d_{2M}$, $|X| = 2M$, and X is antipodal then $a_M = d_{2M}$. Thus, for $M = 3$ and $M = 6$ we have this equality. The following theorem is the main result of [26].

Theorem 2.1 (L. Fejes Tóth, [26]) *Let $P_M \subset \mathbb{S}^2$ be a maximal set for the Fejes Tóth problem for antipodal configurations, i. e. $\psi(P_M) = a_M$. Then*

1. *P_2 is the set of vertices of a square on the equator, $a_2 = 90°$;*
2. *P_3 is the set of vertices of a regular octahedron, $a_3 = 90°$;*
3. *P_4 is the set of vertices of a cube, $a_4 = \arccos(1/3)$;*
4. *P_5 consists of five pairs of antipodal vertices of a regular icosahedron, $a_5 = \arccos(1/\sqrt{5})$.*
5. *P_6 is the set of vertices of a regular icosahedron, $a_6 = \arccos(1/\sqrt{5})$.*

In our paper with Tarasov [48] we gave an alternative proof of this theorem. In [47] we found the list of all irreducible contact graphs with N vertices on the sphere $\mathbb{S}^2$, where $6 \leq N \leq 11$. Since the contact graph of P_M is irreducible the theorem (for $M < 6$) follows from this list.

Fejes Tóth conjectured that the solution of the problem for seven pairs of antipodal points consists of the vertices of a rhombic dodecahedron (see the second edition of [21], page 210). This was proved by Cohn and Woo [12] as a consequence of a more general theorem.

3 Problems on the Maximum Contact Number of Packings on the Sphere

In [28] (pages 86 and 87) Fejes Tóth raised three problems about the number of touching pairs in a packing of congruent circles on the sphere.

Consider a packing P of $\mathbb{S}^2$ by N circles $c_1, \ldots, c_N$ of diameter d. In the packing P let c_i be touched by k_i circles. The first problem is to *find the maximum number of contact points:*

$$\mathrm{K}_N(d) := \max_{P:|P|=N} \frac{k_1 + \cdots + k_N}{2}.$$

In other words, $\mathrm{K}_N(d)$ is the maximum number of touching pairs in a packing of N spherical caps of diameter d.

It is clear that if $d = d_N$, then $\mathrm{K}_N(d)$ is realized by the solution of the Tammes problem. It seems that the case $d < d_N$ is not well considered. There is only one paper in this direction [30], where this problem is considered for $N = 12$ and $d = 60°$. There, it is proved that

$$\mathrm{K}_{12}(60°) = 24.$$

Fejes Tóth also proposed the problem of *finding the maximum*

$$\bar{K}(d) := \max_N \frac{\mathrm{K}_N(d)}{N}$$

of the average number of points of contacts over all packings of circles of diameter d.

The third problem is: *For a given N, find the maximum kissing number* K_N *over all packings of equal circles,* i.e., find

$$\mathrm{K}_N := \max_{d \le d_N} \mathrm{K}_N(d).$$

Let X be the set of centers of a packing of congruent circles on $\mathbb{S}^2$. Denote by $e(X)$ the number of edges of the contact graph $\mathrm{CG}(X)$. It is easy to see that

$$\mathrm{K}_N = \max_{X \in \mathbb{S}^2, |X|=N} e(X).$$

This number is currently known only for $N \le 12$ and $N = 24, 48, 60, 120$.

Denote by $\kappa(d)$ the kissing number of the spherical cap with diameter d in $\mathbb{S}^2$, i.e. it is the maximum number of non-overlapping circles of diameter d that can touch a circle of the same diameter. Note that if $d \le \arccos(1/\sqrt{5})$, then $\kappa(d) = 5$.

We say that a packing of N spherical caps with diameter d is *maximal* if

$$\mathrm{K}_N(d) = N\kappa(d)/2.$$

The following theorem has been proved by Robinson [52] and Fejes Tóth [27].

Theorem 3.1 (Robinson [52], Fejes Tóth [27]) *A maximal packing of N equal spherical caps exists only if* $N = 2, 3, 4, 6, 8, 9, 12, 24, 48, 60$ *or* 120.

This theorem implies

Corollary 3.1 $\mathrm{K}_2 = 1$, $\mathrm{K}_3 = 3$, $\mathrm{K}_4 = 6$, $\mathrm{K}_6 = 12$, $\mathrm{K}_8 = 16$, $\mathrm{K}_9 = 18$ *and for* $N = 12, 24, 48, 60$ *or* 120 *we have* $\mathrm{K}_N = 5N/2$.

In our paper [48] we considered K_N for $N < 12$. In particular, we proved that

Theorem 3.2 (Musin and Tarasov [48]) $\mathrm{K}_5 = 8$, $\mathrm{K}_7 = 12$, $\mathrm{K}_{10} = 21$, *and* $\mathrm{K}_{11} = 25$.

Note that K_5 is attained by the set of vertices of a square pyramid. For $N = 7$ and $N = 11$, $\mathrm{K}_N(d)$ achieves its maximum on optimal configurations for Tammes' problem. However, the arrangement realizing the optimal value K_{10} is obtained by removing from the set of vertices of a regular icosahedron two adjacent vertices. In this case the contact graph $\mathrm{CG}(X)$ is not irreducible.

Our proof of Theorem 3.2 in [48] is based on two lemmas.

Lemma 3.1 *Let X be a finite set on the sphere $\mathbb{S}^2$. If every face of the contact graph $\mathrm{CG}(X)$ is either a triangle or a quadrilateral, then this graph is irreducible.*

Lemma 3.2 *Let X be a finite set on the sphere $\mathbb{S}^2$, where $|X| = N$ and $N > 6$. Suppose that $e(X) \geq 3N - 8$. Then the contact graph $\mathrm{CG}(X)$ is irreducible.*

Using these lemmas, Theorem 3.2 follows by checking the list of irreducible contact graphs for $N \leq 11$ [47].

4 The Fejes Tóth–Sachs Problem on the One-Sided Kissing Numbers

Let H be a closed half-space of $\mathbb{R}^n$. Suppose S is a unit sphere in H that touches the bounding hyperplane of H. The *one-sided kissing number* $B(n)$ is the maximal number of unit non-overlapping spheres in H that can touch S.

The problem of finding $B(3)$ was raised by Fejes Tóth and Sachs in 1976 [29] in another context. K. Bezdek and Brass [5] studied the problem in a more general setting and they introduced the term "one-sided Hadwiger number", which in the case of a ball is the same as the one-sided kissing number. The term "one-sided kissing number" has been introduced by K. Bezdek [4].

Clearly, $B(2) = 4$. The Fejes Tóth–Sachs problem in three dimensions was solved by G. Fejes Tóth [17]. He proved that $B(3) = 9$ (see also Sachs [53] and A. Bezdek and K. Bezdek [3] for other proofs). Finally, Kertész [37] proved that the maximal one-sided kissing arrangement is unique up to isometry.

The first upper bound for $B(4)$ was given by Szabó [57]. He used the Odlyzko–Sloane bound $k(4) \leq 25$ for the kissing number of the four-dimensional ball to show that $B(4) \leq 20$. Next K. Bezdek [4], based on the result that $k(4) = 24$ [42, 44], lowered the bound to $B(4) \leq 19$.

In [43] I proved that $B(4) = 18$. This proof relies on the extension of Delsarte's method that was developed in [44]. However, technically the proof is more complicated than the proof of the fact that $k(4) = 24$. An alternate proof was given in [2] using semidefinite programming. The problem of uniqueness of the maximal one-sided kissing arrangement in four dimensions is still open.

In [43] I conjectured that $B(5) = 32,\ B(8) = 183$ and $B(24) = 144855$. This conjecture for $n = 8$ was proved by Bachoc anf Vallentin [2]. In [1, 45] we proposed several upper bounds on $B(n)$. However, all these bounds were improved in [2].

It is clear that there are some relations between kissing numbers and one-sided kissing numbers. Look at these nice equalities:

$$n = 2, \quad 4 = B(2) = \frac{k(1) + k(2)}{2} = \frac{2+6}{2};$$

$$n = 3, \quad 9 = B(3) = \frac{k(2) + k(3)}{2} = \frac{6+12}{2};$$

$$n = 4, \quad 18 = B(4) = \frac{k(3) + k(4)}{2} = \frac{12+24}{2}.$$

We do not know whether the equality

$$B(n) = \bar{K}(n) := \frac{k(n-1) + k(n)}{2}$$

holds for all n. However, there are reasons to believe that $B(n) = \bar{K}(n)$ for $n = 5, 8$ and 24. We propose a weaker conjecture, namely, that the equality $B(n) = \bar{K}(n)$ holds asymptotically:

Conjecture. We have

$$\lim_{n\to\infty} \frac{B(n)}{\bar{K}(n)} = 1.$$

5 The Work of Fejes Tóth on the Isoperimetric Problem for Polyhedra and Their Extensions

5.1 Isoperimetric Problem for Polyhedra

The isoperimetric problem in spacc can be formulated as follows: *Find a convex body of given surface area F which contains the largest volume V.*

The famous isoperimetric inequality states

$$F^3 \geq 36\pi V^2.$$

For any solid P consider the *Isoperimetric Quotient*

$$\mathrm{IQ}(P) = 36\pi \frac{V^2}{F^3},$$

a term introduced by Pólya in [50, Chap. 10, Problem 43]. The isoperimetric inequality implies that $\mathrm{IQ}(P) \leq 1$ and the equality holds only if P is a sphere.

The isoperimetric problem for polyhedra was first considered by Lhuilier (1782), see [39], and Steiner (1842), see [54]. Steiner stated the following conjecture.

Steiner's conjecture [54]. *Each of the five Platonic solids is optimal (i.e. with the highest* IQ*) among all isomorphic polyhedra.*

This problem is still open for the icosahedron.

Consider the isoperimetric problem for polyhedra with given number of faces f. Actually, this problem is currently solved only for $f \leq 7$ and $f = 12$. However, the first theorem on this problem was discovered in the 19th century.

Theorem 5.1 (Lindelöf [38] and Minkowski [41]) *Of all convex polyhedra with the same number of faces, a polyhedron with the highest* IQ *is circumscribed about a sphere which touches each face in its centroid.*

Note that IQ(tetrahedron) ≈ 0.302, IQ(cube) ≈ 0.524, IQ(octahedron) ≈ 0.605, IQ(dodecahedron) ≈ 0.755, and IQ(icosahedron) ≈ 0.829 [50, Chap. 10, p. 189].

In fact, there are simple polyhedra F_{12} and C_{36} with 8 and 20 faces that have greater IQ than, respectively, the regular octahedron and icosahedron. Goldberg [33] computed that IQ(octahedron) $< \mathrm{IQ}(F_{12}) \approx 0.628$ and IQ(icosahedron) $< \mathrm{IQ}(C_{36}) \approx 0.848$.

Goldberg proved that the regular dodecahedron is the best polyhedron with 12 facets and stated the following conjecture.

Goldberg's conjecture [33]. *If a polyhedron P with $f \neq 11, 13$ faces and v vertices has the greatest* IQ, *then P is simple and its faces are $\lfloor 6 - 24/(v+4) \rfloor$–gons or $\lfloor 7 - 24/(v+4) \rfloor$– gons.*

Note that according to Goldberg's conjecture if $v \geq 20$, then the faces of a best polyhedron can be only pentagons and hexagons, in other words P is a *fullerene* (see [16]).

Let P be a convex polyhedron with f faces. In [33] Goldberg proposed the following inequality:

$$\frac{F^3}{V^2} \geq 54\,(f-2)\tan\omega_f\,(4\sin^2\omega_f - 1), \quad \omega_f := \frac{\pi f}{6f-12},$$

or equivalently

$$\mathrm{IQ}(P) \leq \frac{2\pi\cot\omega_f}{3(f-2)(4\sin^2\omega_f - 1)}, \tag{5.1}$$

where the equality holds only if $f = 4$ (regular tetrahedron), $f = 6$ (cube) or $f = 12$ (regular dodecahedron).

This inequality was independently found by Fejes Tóth [19] (see Fejes Tóth's books [21, 25] and Florian's survey [32] for references and historical remarks). However, both proofs contained a gap, namely the proof of the convexity of a certain function of two variables. A first rigorous proof of (5.1) was given by Fejes Tóth in the paper [20]. Finally, Florian [31] filled the gap in the previous proof by establishing the convexity of the respective function.

Two conjectures of Fejes Tóth on isoperimetric inequalities are still open. Let P be a convex polyhedron with v vertices. The first conjecture states that

$$\frac{F^3}{V^2} \geq \frac{27\sqrt{3}}{2}(v-2)(3\tan^2\omega_v - 1).$$

Let P be a convex polyhedron with f faces, v vertices and e edges. The second conjecture of Fejes Tóth states that

$$\frac{F^3}{V^2} \geq 9\, e \sin\frac{2\pi}{p}\left(\tan^2\frac{\pi}{p}\,\tan^2\frac{\pi}{q} - 1\right),$$

where $p := 2e/f$ and $q := 2e/v$.

Note that the validity of any of these conjectures would yield a proof of the open conjecture of Steiner concerning the isoperimetric property of the icosahedron.

5.2 The Goldberg–Fejes Tóth Inequality

Here we consider a "dual" version of the Goldberg–Fejes Tóth inequality (5.1)

Let P be a convex polyhedron with f faces. Then Euler's formula implies

$$v \leq 2f - 4, \tag{5.2}$$

where v is the number of vertices of P. Equality holds only if P is a simple polyhedron.

Suppose P is a polyhedron with highest IQ and fixed f. Then by the Lindelöf–Minkowski theorem there is a sphere S that touches each face Γ_i of P in its centroid x_i. Thus, the set $X := \{x_1, \ldots, x_f\}$ is a subset of S.

Without loss of generality it can be assumed that S is of radius $r = 1$. Since all faces Γ_i touch the unit sphere S, we have

$$V = \frac{1}{3}F.$$

Therefore, for a given f, P has the highest IQ if and only if $F = \text{area}(P)$ achieves its minimum.

Let O be the center of the sphere S. Consider the central projection $g : P \to S$, that carries a point $A \in P$ in the intersection $g(A)$ of the line OA with S. It is clear that $g(x_i) = x_i$.

Let $p_1, \ldots, p_v$ be vertices of P. Denote by Q the projection of this vertex set, i.e. $Q := \{q_1, \ldots, q_v\}$, where $q_i := g(p_i) \in S$.

The set Q coincides with the set of vertices $\{v_i\}$ of the Voronoi diagram VD(X) of X in S. Equivalently, if $G_{i1}, \ldots, G_{im}$ are faces of VD(X) with a common vertex v_i, then v_i is the circumcenter of the Delaunay cell D_i with vertices $x_{i1}, \ldots, x_{im}$ in S. It immediately follows that $|v_i x_{ij}|$ does not depend on j. Indeed, we have

$$|v_i x_{ij}|^2 = |Ov_i|^2 - |Ox_{ij}|^2, \quad \text{where } |Ox_{ij}| = 1 \text{ for all } j = 1, \ldots, m.$$

Let $G_i := g(\Gamma_i)$, $i = 1, \cdots, f$. Then G_i are Voronoi cells of VD(X). Let $D_1, \ldots, D_v$ be the cells of the Delaunay tessellation DT(X) in S. Then

$$\text{area}(G_1) + \cdots + \text{area}(G_f) = \text{area}(D_1) + \cdots + \text{area}(D_v) = 4\pi. \tag{5.3}$$

We have

$$F = \text{area}(\Gamma_1) + \cdots + \text{area}(\Gamma_f) = \text{area}(g^{-1}(G_1)) + \cdots + \text{area}(g^{-1}(G_f)).$$

It is easy to see, that equality in the Goldberg–Fejes Tóth inequality (see [33] and [21, Sect. V.4]) holds only if $v = 2f - 4$ and all G_i are congruent regular polygons. Equivalently, equality holds only if all D_k are congruent regular triangles.

Denote $\tilde{D}_i = g^{-1}(D_i)$. Then

$$F = \text{area}(\tilde{D}_1) + \cdots + \text{area}(\tilde{D}_v). \tag{5.4}$$

If $v = 2f - 4$ and all D_i are congruent triangles, then (5.3) yields

$$\text{area}(D_i) = \frac{2\pi}{f-2}, \quad i = 1, \ldots, v.$$

Let T be a regular spherical triangle in S with area$(T) = t$. Denote

$$\rho(t) := \text{area}(\tilde{T}), \quad \tilde{T} := g^{-1}(T) \subset P_T.$$

Thus, we have the following inequality that is equivalent to (5.1).

$$\text{IQ}(P) \le \frac{\tau}{\rho(\tau)}, \quad \tau = \frac{2\pi}{f-2}. \tag{5.5}$$

5.3 The Goldberg–Fejes Tóth Inequality for Higher Dimensions

Here we consider a d–dimensional analog of the inequality (5.5).

Let P be a convex polyhedron in $\mathbb{R}^d$ with v vertices and n facets, i.e. $v = f_0(P)$ and $n = f_{d-1}(P)$. Then McMullen's Upper Bound Theorem [61, p. 254] yields the extension

$$v \le h_d(n) := \binom{n - \lceil d/2 \rceil}{\lfloor d/2 \rfloor} + \binom{n - \lfloor d/2 \rfloor - 1}{\lceil d/2 \rceil - 1}. \tag{5.6}$$

of the inequality (5.2) for all dimensions.

Define

$$\mathrm{IQ}(P) := d^{d-1}\, \Omega_d\, \frac{V^{d-1}}{F^d}, \quad \Omega_d := \mathrm{area}(\mathbb{S}^{d-1}).$$

The Lindelöf–Minkowski theorem holds for all dimensions (see [35, p. 274]). So we have that $\mathrm{IQ}(P)$ achieves its maximum only if P is circumscribed about a sphere S. As above we assume that S is a unit sphere.

It is easy to see that all definitions from Sect. 5.2 can be extended to all dimensions. Therefore, equality (5.4) holds also for a d-dimensional polyhedron P. An analog of (5.3) is the following equality:

$$\mathrm{area}(D_1) + \cdots + \mathrm{area}(D_v) = \Omega_d. \tag{5.7}$$

Let D be a regular spherical simplex in S with $\mathrm{area}(D) = t$. Denote

$$\rho_d(t) := \mathrm{area}(\tilde{D}).$$

Our conjecture is that the following extension of the Goldberg–Fejes Tóth inequality (5.5) holds for all dimensions:

$$\mathrm{IQ}(P) \le \frac{\Omega_d}{v\, \rho_d(\Omega_d/v)}. \tag{5.8}$$

Since $v \le h_d(n)$, in particular we have

$$\mathrm{IQ}(P) \le \frac{\tau}{\rho_d(\tau)}, \quad \tau = \frac{\Omega_d}{h_d(n)}. \tag{5.9}$$

Perhaps, (5.8) can be proved by the same way as László Fejes Tóth proved (5.1) ≡ (5.5) in [20] and [21, Sect. V.4]. Actually, for all dimensions there are extensions of (5.2)–(5.4). We think that the most complicated step here is to prove that $\mathrm{IQ}(P)$ cannot exceed IQ of a such polyhedron with n facets that all its D_i are congruent regular spherical simplices.

References

1. A. Barg, O.R. Musin, Codes in spherical caps. Adv. Math. Commun. **1**, 131–149 (2007)
2. C. Bachoc, F. Vallentin, Semidefinite programming, multivariate orthogonal polynomials, and codes in spherical caps. Eur. J. Comb. **30**, 625–637 (2009)
3. A. Bezdek, K. Bezdek, A note on the ten-neighbour packing of equal balls. Beiträge zur Alg. und Geom. **27**, 49–53 (1988)
4. K. Bezdek, Sphere packing revisited. Eur. J. Comb. **27**, 864–883 (2006)
5. K. Bezdek, P. Brass, On k^+-neighbour packings and one-sided Hadwiger configurations. Contrib. Algebra Geom. **4**, 493–498 (2003)
6. K. Böröczky, Packing of spheres in spaces of constant curvature. Acta Math. Acad. Sci. Hung. **32**, 243–261 (1978)
7. K. Böröczky, The problem of Tammes for $n = 11$. Stud. Sci. Math. Hung. **18**, 165–171 (1983)
8. K. Böröczky, The Newton-Gregory problem revisited, in *Discrete Geometry*, ed. by A. Bezdek (Dekker, New York, 2003), pp. 103–110
9. K. Böröczky, L. Szabó, Arrangements of 13 points on a sphere, in *Discrete Geometry*, ed. by A. Bezdek (Dekker, New York, 2003), pp. 111–184
10. K. Böröczky, L. Szabó, Arrangements of 14, 15, 16 and 17 points on a sphere. Stud. Sci. Math. Hung. **40**, 407–421 (2003)
11. P. Brass, W.O.J. Moser, J. Pach, *Research Problems in Discrete Geometry* (Springer, Berlin, 2005)
12. H. Cohn, J. Woo, Three-point bounds for energy minimization. J. Am. Math. Soc. **25**, 929–958 (2012)
13. J.H. Conway, N.J.A. Sloane, *Sphere Packings, Lattices, and Groups*, 3rd edn. (Springer, New York, 1999)
14. H.S.M. Coxeter, An upper bound for the number of equal nonoverlapping spheres that can touch another of the same size. Proc. Symp. Pure Math. AMS **7**, 53–71 (1963). (= Chap. 9 of H.S.M. Coxeter, Twelve Geometric Essays, Southern Illinois Press, Carbondale Il, (1968))
15. L. Danzer, Finite point-sets on S^2 with minimum distance as large as possible. Discret. Math. **60**, 3–66 (1986)
16. A. Deza, M. Deza, V. Grishukhin, Fullerenes and coordination polyhedra versus half-cube embeddings. Discret. Math. **192**, 41–80 (1998)
17. G. Fejes Tóth, Ten-neighbor packing of equal balls. Period. Math. Hung. **12**, 125–127 (1981)
18. L. Fejes Tóth, Über die Abschätzung des kürzesten Abstandes zweier Punkte eines auf einer Kugelfläche liegenden Punktsystems. Jber. Deutch. Math. Verein, **53** (1943), 66–68
19. L. Fejes Tóth, Über einige Extremaleigenschaften der regulären Polyeder und des gleichseitigen Dreiecksgitters. Ann. Scuola. Norm. Super. Pisa (2) **13** (1944), 51–58. (1948)
20. L. Fejes Tóth, The isepiphan problem for n-hedra. Am. J. Math. **70**, 174–180 (1948)
21. L. Fejes Tóth, *Lagerungen in der Ebene, auf der Kugel und in Raum*, 2nd edn. (Springer-Verlag, 1953). (1972: Russian translation, Moscow, 1958)
22. L. Fejes Tóth, On close-packings of spheres in spaces of constant curvature. Publ. Math. Debr. **3**, 158–167 (1953)
23. L. Fejes Tóth, Kugelunterdeckungen und Kugelüberdeckungen in Räumen konstanter Krümmung. Arch. Math. **10**, 307–313 (1959)
24. L. Fejes Tóth, Neuere Ergebnisse in der diskreten Geometrie. Elem. Math. **15**, 25–36 (1960)
25. L. Fejes Tóth, *Regular Figures* (Pergamon Press, Oxford, 1964)
26. L. Fejes Tóth, Distribution of points in the elliptic plane. Acta Math. Acad. Sci. Hung. **16**, 437–440 (1965)
27. L. Fejes Tóth, Remarks on a theorem of R. M. Robinson. Stud. Sci. Math. Hung. **4**, 441–445 (1969)
28. L. Fejes Tóth, Symmetry induced by economy. Comput. Math. Appl. **12**, 83–91 (1986)
29. L. Fejes Tóth, H. Sachs, Research problem 17. Period. Math. Hung. **7**, 125–127 (1976)
30. L. Flatley, A. Tarasov, M. Taylor, F. Theil, Packing twelve spherical caps to maximize tangencies. J. Comput. Appl. Math. **254**, 220–225 (2013)

31. A. Florian, Eine Ungleichung über konvexe Polyeder. Monatsh. Math. **60**, 130–156 (1956)
32. A. Florian, Extremum problems for convex discs and polyhedra, in *Handbook of Convex Geometry*, ed. by P.M. Gruber, J.M. Wills (Elsevier, Amsterdam, 1993), pp. 177–221
33. M. Goldberg, The isoperimetric problem for polyhedra. Tohoku Math. J. **40**, 226–236 (1934)
34. W. Habicht und, B.L. van der Waerden, Lagerungen von Punkten auf der Kugel. Math. Ann. **123**, 223–234 (1951)
35. H. Hadwiger, *Vorlesungen über lnhalt, Oberfläche und lsoperimetrie* (Springer, Berlin, 1957)
36. L. Hárs, The Tammes problem for $n = 10$. Stud. Sci. Math. Hung. **21**, 439–451 (1986)
37. G. Kertész, Nine points on the hemisphere. Colloq. Math. Soc. J. Bolyai (Intuitive Geometry, Szeged 1991) **63**, 189–196 (1994)
38. L. Lindelöf, Propriétés générales des polyèdres qui, sous une étendue superficielle donnée, renferment le plus grand volume. Math. Ann. **2**, 150–159 (1869)
39. S. Lhuilier, De relatione mutua capacitatis et terminorum figuranum, etc. Varsaviae (1782)
40. H. Meschkowski, *Unsolved and Unsolvable Problems in Geometry* (Frederick Ungar Publishing Company, New York, 1966)
41. H. Minkowski, Allgemeine Lehrsätze, über konvexe Polyeder. Nachr. Ges. Wiss. Göttingen, math.-physisk. Kl. 198–219 (1897). (= Ges. Abbh. II. Leipzig und Berlin 1911 1033–121)
42. O.R. Musin, The problem of the twenty-five spheres. Russ. Math. Surv. **58**, 794–795 (2003)
43. O.R. Musin, The one-sided kissing number in four dimensions. Period. Math. Hung. **53**, 209–225 (2006)
44. O.R. Musin, The kissing number in four dimensions. Ann. Math. **168**(1), 1–32 (2008)
45. O.R. Musin, Bounds for codes by semidefinite programming. Proc. Steklov Inst. Math. **263**, 134–149 (2008)
46. O.R. Musin, A.S. Tarasov, The strong thirteen spheres problem. Discret. Comput. Geom. **48**, 128–141 (2012)
47. O.R. Musin, A.S. Tarasov, Enumeration of irreducible contact graphs on the sphere. J. Math. Sci. **203**, 837–850 (2014)
48. O.R. Musin, A.S. Tarasov, Extreme problems of circle packings on a sphere and irreducible contact graphs. Proc. Steklov Inst. Math. **288**, 117–131 (2015)
49. O.R. Musin, A.S. Tarasov, The Tammes problem for $N = 14$. Exp. Math. **24**(4), 460–468 (2015)
50. G. Pólya, *Induction and Analogy in Mathematics* (Princeton University Press, Princeton, 1954)
51. R.M. Robinson, Arrangement of 24 circles on a sphere. Math. Ann. **144**, 17–48 (1961)
52. R.M. Robinson, Finite sets on a sphere with each nearest to five others. Math. Ann. **179**, 296–318 (1969)
53. H. Sachs, No more than nine unit balls can touch a closed hemisphere. Stud. Sci. Math. Hung. **21**, 203–206 (1986)
54. J. Steiner, Uber Maximum und Minimum bei Figuren in der Ebene, auf der Kugelflache und im Raume berhaupt. J. Math Pres Appl. **6**, 105–170 (1842). (= Gesammelte Werke II 254–308, Reimer, Berlin 1882)
55. K. Schütte, B.L. van der Waerden, Auf welcher Kugel haben 5, 6, 7, 8 oder 9 Punkte mit Mindestabstand 1 Platz? Math. Ann. **123**, 96–124 (1951)
56. K. Schütte, B.L. van der Waerden, Das problem der dreizehn Kugeln. Math. Ann. **125**, 325–334 (1953)
57. L. Szabó, 21-neighbour packing of equal balls in the 4-dimensional Euclidean space. Geom. Dedicata **38**, 193–197 (1991)
58. R.M.L. Tammes, On the origin number and arrangement of the places of exits on the surface of pollengrains. Rec. Trv. Bot. Neerl. **27**, 1–84 (1930)
59. B.L. van der Waerden, Punkte auf der Kugel. Drei Zusätze. Math. Ann. **125**, 213–222 (1952)
60. B.L. van der Waerden, Pollenkörner. Punktverteilungen auf der Kugel und Informationstheorie, Die Naturwissenschaften **48**, 189–192 (1961)
61. G.M. Ziegler, *Lectures on Polytopes*, vol. 152, Graduate Texts in Mathematics (Springer, Berlin, 1995)

Flavors of Translative Coverings

Márton Naszódi

Abstract We survey results on the problem of covering the space $\mathbb{R}^n$, or a convex body in it, by translates of a convex body. Our main goal is to present a diverse set of methods. A theorem of Rogers is a central result, according to which, for any convex body K, the space $\mathbb{R}^n$ can be covered by translates of K with density around $n \ln n$. We outline four approaches to proving this result. Then, we discuss the illumination conjecture, decomposability of multiple coverings, Sudakov's inequality and some problems concerning coverings by sequences of sets.

2010 Mathematics Subject Classification 52C17 · 05B40 · 52A23

1 Introduction

The problem of covering a set by few translates of another appears naturally in several contexts. In computational applications it may be used for divide and conquer algorithms, in analysis, it yields ε–nets, in functional analysis, it is used to quantify how compact an operator between Banach spaces is. In geometry, it is simply an interesting question on its own.

Our primary focus is to describe a representative family of methods, rather than giving a complete account of the state of the art. In particular, we highlight some combinatorial ideas, and sketch some instructive probabilistic computations.

The author acknowledges the support of the János Bolyai Research Scholarship of the Hungarian Academy of Sciences, and the National Research, Development, and Innovation Office, NKFIH Grants PD104744 and K119670.

M. Naszódi (✉)
Department of Geometry, Lorand Eötvös University, Pázmány Péter Sétány 1/C, Budapest 1117, Hungary
e-mail: marton.naszodi@math.elte.hu

G. Ambrus et al. (eds.), *New Trends in Intuitive Geometry*, Bolyai Society Mathematical Studies 27, https://doi.org/10.1007/978-3-662-57413-3_14

We minimize overlap with the fundamental works of L. Fejes Tóth [30] and Rogers [83]. Böröczky's book [18] is the most recent source on finite coverings. Some of the topics covered here are discussed in more detail in the books [8, 21, 61, 68]. Many of the topics omitted, or only touched upon here (most notably, planar and three–dimensional results, lattice coverings and density) are discussed in the surveys [31, 32, 34–36].

In Sect. 3, we state Rogers' result, and a few of its relatives, on the existence of an economical covering of the whole space by translates of an arbitrary convex body. In Sect. 4, we outline three probabilistic proofs of these results. In Sect. 5, we describe a fourth approach, which is based on an algorithmic (non–probabilistic) result from combinatorics. Then, in Sect. 6, we discuss the problem of illumination. There, we sketch the proof of a result of Schramm, which is currently the best general upper bound for Borsuk's problem. In Sect. 7, we state some of the most recent results on the problem of decomposability of multiple coverings. Section 8 provides a window to how the asymptotic theory of convex bodies views translative coverings. Finally, in Sect. 9, we consider coverings by sequences of convex bodies.

We use the following notations, and terminology. For two Borel measurable sets K and L in $\mathbb{R}^n$, let $N(K, L)$ denote the *translative covering number* of K by L, that is, the minimum number of translates of L that cover K.

The Euclidean ball of radius one centered at the origin is $\mathbf{B}_2^n = \{x \in \mathbb{R}^n : |x|^2 =< x, x >\leq 1\}$, where $< ., . >$ denotes the standard scalar product on $\mathbb{R}^n$. We denote the Haar probability measure on the sphere $\mathbb{S}^{n-1} = \{x \in \mathbb{R}^n : |x| = 1\}$ by σ.

A *symmetric* convex body is a *convex body* (that is, a compact convex set with non–empty interior) that is centrally symmetric about some point. A hyperplane H *supports* a convex set K, if H intersects the boundary of K, and K is contained in one of the closed half–spaces bounded by H. The *support function* h_K of a convex set K is defined as $h_K(x) = \sup\{< x, k > : k \in K\}$ for any $x \in \mathbb{R}^n$. We denote the *polar* of a convex body K by

$$K^* = \{x \in \mathbb{R}^n \ : < x, k > \leq 1 \text{ for all } k \in K\}.$$

The cardinality of a set X is denoted by $|X|$.

2 Basics

We list a number of simple properties of covering numbers, their proofs are quite straight forward, cf. [3].

Fact 2.1 *Let K, L, M be convex sets in $\mathbb{R}^n$, $T : \mathbb{R}^n \to \mathbb{R}^n$ an invertible linear transformation. Then we have*

$$N(K, L) = N(T(K), T(L)), \tag{1}$$
$$N(K, L) \le N(K, M)N(M, L), \tag{2}$$
$$N(K + M, L + M) \le N(K, L), \tag{3}$$
$$N(K, 2(K \cap L)) \le N(K, L), \text{ if } K = -K. \tag{4}$$

We note a special property of the Euclidean ball as a covering set.

Fact 2.2 *Let K be a convex set in $\mathbb{R}^n$. If K is covered by t Euclidean balls, then K is covered by t Euclidean balls with centers in K.*

This fact follows from the observation that the intersection of $\mathbf{B}_2^n$ with a half-space not containing the origin is contained in the unit ball centered at the orthogonal projection of the origin to the bounding hyperplane of the half-space.

The following obvious lower bound is often sufficient:

$$N(K, L) \ge \frac{\text{vol}(L)}{\text{vol}(K)}. \tag{5}$$

Next, we assume that L is symmetric, and find an upper bound on $N(K, L)$. Let $X + (L/2)$ be a *saturated packing* of translates of $L/2$ in $K + (L/2)$, that is a maximal family of translates of $L/2$ with pairwise disjoint interiors. Then $K \subseteq X + L$, that is, we have a covering of K by $|X|$ translates of L. Thus,

$$N(K, L) \le |X| \le 2^n \frac{\text{vol}(K + (L/2))}{\text{vol}(L)}, \text{ if } L = -L. \tag{6}$$

Section 3, and a large part of this paper discuss how this bound can be improved.

3 Covering the Whole Space

Let K be a convex body, Λ a lattice, and T a finite set in $\mathbb{R}^n$. We call the family $\mathcal{F} = K + \Lambda + T = \{K + v + t \,:\, v \in \Lambda, t \in T\}$ a *periodic arrangement* of translates of K. The *density* of $\mathcal{F}$ is defined as $\delta(\mathcal{F}) = |T| \,\text{vol}(K)/\det \Lambda$. We say that $\mathcal{F}$ is a *covering* of $\mathbb{R}^n$ if $\cup\mathcal{F} = \mathbb{R}^n$. The *translative covering density* $\theta(K)$ of K is the infimum of the densities of periodic coverings of $\mathbb{R}^n$ by translates of K. We note that one can define the density of a non-periodic arrangement, too (cf. [18, 30, 68, 83]), and – as is easy to see – we obtain the same density infimum if we allow also non-periodic coverings of $\mathbb{R}^n$.

The first milestone in the theory of translative coverings is the following theorem of Rogers.

Theorem 3.1 (Rogers, [80]) *Let K be a bounded convex set in $\mathbb{R}^n$ with non-empty interior. Then the translative covering density of K is at most*

$$\theta(K) \le n \ln n + n \ln \ln n + 5n. \tag{7}$$

Earlier, exponential upper bounds for the covering density were obtained by Rogers, Bambah and Roth, and for the special case of the Euclidean ball by Davenport and Watson (cf. [80] for references). The last summand, $5n$ may be replaced by $3n$, if n is sufficiently large. The current best bound on $\theta(K)$ is due to G. Fejes Tóth [29], who replaced $5n$ by $n + o(n)$ (see Theorem 3.4). It is an open problem whether one can improve the bound by a multiplicative factor below 1, or, very ambitiously, if Cn is an upper bound, for some universal $C > 0$.

It is natural to ask what happens if the density is replaced by the maximum multiplicity.

Theorem 3.2 (Erdös, Rogers, [28]) *For any convex body K in $\mathbb{R}^n$ there is a periodic covering of $\mathbb{R}^n$ by translates of K such that no point is covered by more than $e(n \ln n + n \ln \ln n + 4n)$ translates, and the density is below $n \ln n + n \ln \ln n + 4n$, provided n is large enough.*

A good candidate for a "bad" convex body, that is, one that cannot cover the space economically is the Euclidean ball, $\mathbf{B}_2^n$.

Theorem 3.3 (Coxeter, Few, Rogers, [22]) $\theta(\mathbf{B}_2^n) \ge Cn$ *with a universal constant* $C > 0$.

If we restrict ourselves to *lattice coverings*, that is, coverings of $\mathbb{R}^n$ by translates of a convex body K where the translation vectors form a lattice in $\mathbb{R}^n$ (and denote the infimum of the densities by θ_L), we have a much weaker bound. Rogers [82] (see also [83]) showed that for any convex body K, we have $\theta_L(K) \le n^{\log_2 \ln n + c}$. If K has an affine image symmetric about at least $\log_2 \ln n + 4$ coordinate hyperplanes then, by a result of Gritzmann [41] (see also [68]), we have $\theta_L(K) \le cn(\ln n)^{1+\log_2 e}$. For the Euclidean ball, Rogers' estimate is the best known: $\theta_L(\mathbf{B}_2^n) \le cn(\ln n)^{2.047}$.

The original proofs of Theorems 3.1 and 3.2 yield periodic coverings without any further structure. G. Fejes Tóth gave a proof of Theorem 3.1 that yields a covering with more of a lattice–like structure, and a slightly better density bound.

Theorem 3.4 (G. Fejes Tóth, [29]) *For any convex body K in $\mathbb{R}^n$ there is a lattice Λ and a set $T \subset \mathbb{R}^n$ of $O(\ln n)$ translation vectors such that $K + \Lambda + T$ covers $\mathbb{R}^n$ with density at most $n \ln n + n \ln \ln n + n + o(n)$.*

We give an outline of the proof of this result in Sect. 4.3.

The following is a simple corollary to Theorem 3.1 (or the better bound, Theorem 3.4), which was first spelled out in [85].

Corollary 3.5 (Rogers and Zong [85]) *Let K and L be convex bodies in $\mathbb{R}^n$. Then*

$$N(K, L) \le \frac{\operatorname{vol}(K - L)}{\operatorname{vol}(L)}(n \ln n + n \ln \ln n + n + o(n)). \tag{8}$$

Indeed, consider a covering $L + G$ of a large cube C by translates of L with density close to $n \ln n + n \ln \ln n + n + o(n)$. For any $t \in \mathbb{R}^n$, let $m(t) = |\{g \in G \, : \, K \cap (g + t + L) \neq \emptyset\}| = |G \cap (K - L - t)|$. By averaging $m(t)$ over t in C, we obtain that for some $t \in C$, we have $m(t) \leq \frac{\mathrm{vol}(K-L)}{\mathrm{vol}(L)}(n \ln \ln n + n \ln n + n + o(n) + \varepsilon)$.

In [24], Dumer showed that $\mathbb{R}^n$ can be covered with Euclidean unit balls of density around $\frac{1}{2}n \ln n$. A minor error in the proof was corrected in [25].

4 Proofs of Theorems 3.1, 3.2 and 3.4

4.1 A Probabilistic Proof: Cover Randomly and then Mind the Gap

We give an outline of Rogers' proof of Theorem 3.1.

We may assume that K has volume one, and that the centroid (that is, the center of mass with respect to the Lebesgue measure) of K is the origin. It follows that $K \subset -nK$. (Bonnesen and Fenchel in §34. of [17] give several references to this fact: Minkowski [63] p. 105, Radon [79], Estermann [98] and Süss [94].)

Let C be the cube $C = [0, R]^n$, where R is large. Set $\eta = \frac{1}{n \ln n}$, and choose $N = R^n n \ln \frac{1}{\eta}$ random translation vectors $x_1, \ldots, x_N$ in C uniformly and independently. Let Λ be the lattice $\Lambda = R\mathbb{Z}^n$. Thus, we obtain the family $K + \Lambda + \{x_1, \ldots, x_N\}$ of translates of K. The expected density of the union of this family is close to one, and hence, one can choose the N translation vectors in such a way that the volume of the uncovered part of C is small (at most $R^n(1 - R^{-n})^N$).

Next, we take a saturated (that is, maximal) packing $y_1 - \frac{1}{n}K, \ldots y_M - \frac{1}{n}K$ of translates of $-\frac{1}{n}K$ inside this uncovered part of C. By the previous volume computation, we have few ($M \leq \eta^{-n} R^n (1 - R^{-n})^N$) such translates. We replace each of these copies of $-\frac{1}{n}K$ by the same translate of K, make it a periodic arrangement by Λ, and we obtain $K + \Lambda + \{y_1, \ldots, y_M\}$.

Now, we have two families of translates of K. We enlarge each member of these two families by a factor $1 + \eta$, and –as it is easy to see– obtain a covering of $\mathbb{R}^n$. The omitted computations yield the density bound, finishing the proof of Theorem 3.1.

This method (first, picking random copies, and then, filling the gap, which is small, in a greedy way), developed by Rogers can be applied for obtaining upper bounds in other situations as well. The proof of Theorem 3.2 given by Erdös and Rogers is an example of the use of this random covering technique combined with a sophisticated way of keeping track of multiply covered points using an inclusion–exclusion formula. Other examples include bounds on covering the sphere $\mathbb{S}^{n-1}$ with spherical caps.

4.2 Another Probabilistic Proof: Using the Lovász Local Lemma

Füredi and Kang [39] gave a proof of Theorem 3.2 that is essentially different from the original. Their method yields a slightly worse bound (instead of the order $en \ln n$, they obtain $10n \ln n$), but it is very elegant.

First, by considering an affine image of K, we may assume that $\mathrm{vol}(K) = 1$, and $\frac{1}{e}\mathbf{B}_2^n \subset K$ (cf. [6], see also [7] for the symmetric case). Let $h = 1/(4en\sqrt{n})$, and consider the lattice $\Lambda = h\mathbb{Z}^n$. The goal is to cover $\mathbb{R}^n$ with translates of K of the form $K + z$ with $z \in \Lambda$. Let $Q = [0, h)^n$ be the half closed, half open fundamental cube of Λ. We define a hypergraph with base set Λ. The hypergraph has two types of edges. For any $z \in \Lambda$, we define a "small edge" as $A^-(z) := \{y \in \Lambda \ : \ y + Q \subset z + K\}$, and a "big edge" as $A^+(z) := \{y \in \Lambda \ : \ (y + Q) \cap (z + K) \neq \emptyset\}$. Clearly, all big edges are of the same size (say α), and so are all small edges. One can verify that the size of a small edge is at least $\alpha/2$.

Next, to make the problem finite, let $\ell \in \mathbb{Z}^+$ be an arbitrarily large integer. Our goal is to select vectors $z_1, \ldots, z_t \in \Lambda$ in such a way that every point of $[-\ell, \ell]^n \cap \Lambda$ is covered by a small edge $A^-(z_i)$, and no point of $[-\ell, \ell]^n \cap \Lambda$ is covered by more than $10n \ln n$ large edges of the form $A^+(z_i)$. Clearly, that would suffice for proving the theorem. We will pick these vectors randomly: select each vector in $\{z \in \Lambda \ : \ A^-(z) \cap [-\ell, \ell]^n \neq \emptyset\}$ with probability p, where $p = e^{-6/5} 10n \ln n/\alpha$.

For every point of $[-\ell, \ell]^n \cap \Lambda$, we have two kinds of bad events. One is if it is not covered by a small edge, and second, if it is covered by too many big edges. Now, we state the main tool of the proof, the Lovász Local Lemma (see Alon and Spencer [1] for a good introduction of it).

Lemma 4.1 (Lovász Local Lemma, [27, 91]) *Let $A_1, A_2, \ldots, A_N$ be events in an arbitrary probability space. A directed graph $D = (V, E)$ on the set of vertices $V = \{1, 2, \ldots, N\}$ is called a* dependency digraph *for the events $A_1, A_2, \ldots, A_N$, if for each $1 \leq i \leq N$, the event A_i is mutually independent of all the events $\{A_j \ : \ (i, j) \notin E\}$. Suppose that the maximum degree of D is at most d, and that the probability of each A_i is at most p. If $ep(d + 1) \leq 1$, then with positive probability no A_i holds.*

Finally, with a geometric argument, one can bound the maximum degree in a dependency digraph of the bad events, and Lemma 4.1 yields the existence of a good covering.

4.3 Covering Using Few Lattices

G. Fejes Tóth's proof of Theorem 3.4 relies on a deep result, Theorem 10* in [86] of Schmidt. A consequence of this result is

Lemma 4.2 *Let $c_0 = 0.278\ldots$ be the root of the equation $1 + x + \ln x = 0$. Then, for any $0 < c < c_0$, and $\varepsilon > 0$, and any sufficiently large n, for any Borel set $S \subset \mathbb{R}^n$,* *there is a lattice–arrangement of S with density cn covering $\mathbb{R}^n$ with the exception of a set whose density is at most $(1+\varepsilon)e^{-cn}$ for some universal constant $c > 0$.*

By this lemma, for a given K there is a lattice Λ such that $(1 + \lfloor n \ln n \rfloor^{-1})^{-1} K + \Lambda$ covers $\mathbb{R}^n$ with the exception of a set whose density is at most e^{-cn+1}.

Lemma 4.3 *If, for some finite set T, $K + \Lambda + T$ is an arrangement of K with density $1 - \delta$, then there is a vector $t \in \mathbb{R}^n$ such that the arrangement $K + \Lambda + T'$ has density at least $1 - \delta^2$, where $T' = T \cup (T + t)$.*

The proof of Lemma 4.3 relies on considering the density of $\mathbb{R}^n \setminus (K + \Lambda + T')$ as a function of t, and averaging it over the fundamental domain of Λ.

To prove Theorem 3.4, we pick an appropriate c for Lemma 4.2, and using Lemma 4.3 roughly $\log_2(c^{-1} \ln n)$ times, we obtain a finite set T of size about $c^{-1} \ln n$ such that $(1 + \lfloor n \ln n \rfloor^{-1})^{-1} K + \Lambda + T$ has density about $n \ln n$ with the uncovered part being of density at most of order $(n \ln n)^{-n}$. Finally, one can verify that $K + \Lambda + T$ is a covering of space with the desired density.

So far, we presented three probabilistic methods that yield economical coverings. In the next section, we present a fourth method, which is not random. Instead, it relies on an algorithmic combinatorial result.

5 A Fractional Approach

5.1 A Few Words of Combinatorics

We recall some notions from the theory of hypergraphs.

Definition 5.1 Let Λ be a set, $\mathcal{H}$ a family of subsets of Λ. A *covering* of Λ by $\mathcal{H}$ is a subset of $\mathcal{H}$ whose union is Λ. The *covering number* $\tau(\Lambda, \mathcal{H})$ of Λ by $\mathcal{H}$ is the minimum cardinality of its coverings by $\mathcal{H}$.

A *fractional covering* of Λ by $\mathcal{H}$ is a measure μ on $\mathcal{H}$ with

$$\mu(\{H \in \mathcal{H} \ : \ p \in H\}) \geq 1 \quad \text{for all } p \in \Lambda.$$

The *fractional covering number* of $\mathcal{H}$ is

$$\tau^*(\Lambda, \mathcal{H}) = \inf \{\mu(\mathcal{H}) \ : \ \mu \text{ is a fractional covering of } \Lambda \text{ by } \mathcal{H}\}.$$

When Λ is a finite set, finding the value of $\tau(\Lambda, \mathcal{H})$ is an integer programming problem. Indeed, we assign a variable x_H to each member H of $\mathcal{H}$, and set x_H to 1 if H is in the covering, and 0 otherwise. Each element p of Λ yields an inequality: $\sum_{p \in H \in \mathcal{H}} x_H \geq 1$.

Computing $\tau^*(\Lambda, \mathcal{H})$ is the linear relaxation of the above integer programming problem. For more on (fractional) coverings, cf. [38] in the abstract (combinatorial) setting and [61, 68] in the geometric setting.

The gap between τ and τ^* is bounded in the case of finite set families (hypergraphs) by the following result.

Lemma 5.2 (Lovász [57], Stein [92]) *For any finite* Λ *and* $\mathcal{H} \subseteq 2^\Lambda$ *we have*

$$\tau(\Lambda, \mathcal{H}) < (1 + \ln(\max_{H \in \mathcal{H}} |H|))\tau^*(\Lambda, \mathcal{H}). \tag{9}$$

Furthermore, the greedy algorithm (always picking the set that covers the largest number of uncovered points) yields a covering of cardinality less than the right hand side in (9).

We note that a probabilistic argument yields a slightly different bound on the covering number:

$$\tau(\Lambda, \mathcal{H}) \leq \left\lfloor 1 + \frac{\ln |\Lambda|}{-\ln\left(1 - \frac{1}{\tau^*}\right)} \right\rfloor, \tag{10}$$

with the notation $\tau^* = \tau^*(\Lambda, \mathcal{H})$. When we do not have an upper bound on $\max_{H \in \mathcal{H}} |H|$ better than $|\Lambda|$, then (10) is a bit better than (9).

To prove (10), let μ be a fractional covering of Λ by $\mathcal{H}$ such that $\mu(\mathcal{H}) = \tau^* + \varepsilon$, where $\varepsilon > 0$ is very small. We normalize μ to obtain the probability measure $\nu = \mu/\mu(\mathcal{H})$ on $\mathcal{H}$. Let m denote the right hand side in (10), and pick m members of $\mathcal{H}$ randomly according to ν. Then we have

$$\mathbb{P}\left(\exists u \in \Lambda \ : \ u \text{ is not covered}\right) \leq |\Lambda| \left(1 - \frac{1}{\tau^* + \varepsilon}\right)^m < 1.$$

Thus, with positive probability, we have a covering.

We will need the duals of these notions as well. Let Λ be a set and $\mathcal{H}$ be a family of subsets of Λ. The *dual* of this set family is another set family, whose base set is $\mathcal{H}$, and the set family on $\mathcal{H}$ is $\mathcal{H}^* = \left\{\{H \in \mathcal{H} \ : \ p \in H\} \ : \ p \in \Lambda\right\}$.

We call a set $T \subset \Lambda$ a *transversal* to the set family $\mathcal{H}$, if T intersects each member of $\mathcal{H}$. One may define *fractional transversals* in the obvious way, and then define the *(fractional) transversal number*.

Clearly $\mathcal{G} \subset \mathcal{H}$ is a covering of Λ if and only if, $\mathcal{G}$ is a transversal to $\mathcal{H}^*$. Fractional coverings and fractional transversals are dual notions in the same manner. We leave it as an exercise (which will be needed later) to formulate the dual of Lemma 5.2 and of (10).

5.2 The Fractional Covering Number

Motivated by the above combinatorial notions, the fractional version of $N(K, \operatorname{int} K)$ (which is the illuminaton number of K, see Sect. 6) first appeared in [65], and in general for $N(K, L)$ in [4] and [5].

Definition 5.3 Let K and L be bounded Borel measurable sets in $\mathbb{R}^n$. A *fractional covering* of K by translates of L is a Borel measure μ on $\mathbb{R}^n$ with $\mu(x - L) \geq 1$ for all $x \in K$. The *fractional covering number* of K by translates of L is

$$N^*(K, L) =$$

$$\inf \left\{\mu(\mathbb{R}^n) \ : \ \mu \text{ is a fractional covering of } K \text{ by translates of } L\right\}.$$

Clearly,

$$N^*(K, L) \leq N(K, L). \tag{11}$$

In Definition 5.3 we may assume that a fractional cover μ is supported on $\operatorname{cl}(K - L)$. According to Theorem 1.7 of [5], we have

$$\max\left\{\frac{\operatorname{vol}(K)}{\operatorname{vol}(L)}, 1\right\} \leq N^*(K, L) \leq \frac{\operatorname{vol}(K - L)}{\operatorname{vol}(L)}. \tag{12}$$

The second inequality is easy to see: the Lebesgue measure restricted to $K - L$ with the following scaling $\mu = \operatorname{vol} / \operatorname{vol}(L)$ is a fractional covering of K by translates of L. To prove the first inequality, assume that μ is a fractional covering of K by translates of L. Then

$$\operatorname{vol}(L)\mu(\mathbb{R}^n) = \int_{\mathbb{R}^n} \operatorname{vol}(L) \mathrm{d}\mu(x) = \int_{\mathbb{R}^n}\int_{\mathbb{R}^n} \chi_L(y - x)\mathrm{d}y\mathrm{d}\mu(x) =$$

$$\int_{\mathbb{R}^n}\int_{\mathbb{R}^n} \chi_L(y - x)\mathrm{d}\mu(x)\mathrm{d}y = \int_{\mathbb{R}^n} \mu(y - L)\mathrm{d}y \geq \int_{\mathbb{R}^n} \chi_K(y)\mathrm{d}y = \operatorname{vol}(K).$$

We recall from Sect. 5.1 that computing N means solving an integer programming problem (though, in this situation, with infinitely many variables), and computing N^* is its linear relaxation. The linear relaxation is usually easier to solve, so having an inequality bounding N from above by some function of N^* is desirable. It is open whether such inequality exists in general for convex sets. More precisely, we do not know if there is a function f such that for any dimension n, and any convex bodies K and L in $\mathbb{R}^n$, we have $N(K, L) \leq f(n, N^*(K, L))$.

Using a probabilistic argument, Artstein–Avidan and Slomka [5] found a bound of $N(K, L)$ in terms of $N^*(K', L')$, where K' and L' are very close (but not identical) to K and L. A somewhat stronger bound was obtained in [78] by a non-probabilistic proof. For two sets $K, T \subset \mathbb{R}^n$, we denote their *Minkowski difference* by $K \sim T = \{x \in \mathbb{R}^n \ : \ T + x \subseteq K\}$.

Theorem 5.4 (Artstein–Avidan and Slomka [5], Naszódi [78]) *Let K, L and T be bounded Borel measurable sets in $\mathbb{R}^n$ and let $\Lambda \subset \mathbb{R}^n$ be a finite set with $K \subseteq \Lambda + T$. Then*

$$N(K, L) \leq \tag{13}$$

$$(1 + \ln(\max_{x \in K-L} |(x + (L \sim T)) \cap \Lambda|)) \cdot N^*(K - T, L \sim T).$$

If $\Lambda \subset K$, then we have

$$N(K, L) \leq \tag{14}$$

$$(1 + \ln(\max_{x \in K-L} |(x + (L \sim T)) \cap \Lambda|)) \cdot N^*(K, L \sim T).$$

We sketch a proof of Theorem 5.4 in 5.3.

For a set $K \subset \mathbb{R}^n$ and $\delta > 0$, we denote the *δ-inner parallel body* of K by $K_{-\delta} := K \sim \delta \mathbf{B}_2^n = \{x \in K \ : \ x + \delta \mathbf{B}_2^n \subseteq K\}$. As an application of Theorem 5.4, one quickly obtains the following result which, in turn, may be used to give a simple proof of Rogers' result, Theorem 3.1.

Theorem 5.5 (Naszódi [78]) *Let $K \subseteq \mathbb{R}^n$ be a bounded measurable set. Then there is a covering of $\mathbb{R}^n$ by translated copies of K of density at most*

$$\inf_{\delta > 0} \left[\frac{\operatorname{vol}(K)}{\operatorname{vol}(K_{-\delta})} \left(1 + \ln \frac{\operatorname{vol}\left(K_{-\delta/2}\right)}{\operatorname{vol}\left(\frac{\delta}{2}\mathbf{B}_2^n\right)} \right) \right].$$

A similar theorem holds if, in the definition of the δ-inner parallel body, the Euclidean ball is replaced by some other convex body.

5.3 *Proof of Theorem 5.4*

The proofs outlined so far were all probabilistic in nature. In this one, the role that probability plays elsewhere is played by the following straightforward corollary to Lemma 5.2.

Observation 5.6 Let Y be a set, $\mathcal{F}$ a family of subsets of Y, and $X \subseteq Y$. Let Λ be a finite subset of Y and $\Lambda \subseteq U \subseteq Y$. Assume that for another family $\mathcal{F}'$ of subsets of Y we have $\tau(X, \mathcal{F}) \leq \tau(\Lambda, \mathcal{F}')$. Then

$$\tau(X, \mathcal{F}) \leq \tau(\Lambda, \mathcal{F}') \leq (1 + \ln(\max_{F' \in \mathcal{F}'} \left|\Lambda \cap F'\right|)) \cdot \tau^*(U, \mathcal{F}'). \tag{15}$$

The proof is simply a substitution into (15). We set $Y = \mathbb{R}^n$, $X = K$, $\mathcal{F} = \{L + x \,:\, x \in K - L\}$, $\mathcal{F}' = \{L \sim T + x \,:\, x \in K - L\}$. One can use $U = K - T$, as any member of Λ not in $K - T$ could be dropped from Λ and Λ would still have the property that $\Lambda + T \supseteq K$. That proves (13). To prove (14), we notice that in the case when $\Lambda \subset K$, one can take $U = K$.

5.4 *Detour: Covering the Sphere by Caps*

To illustrate the applicability of the method that yields Theorem 5.4, we turn to coverings on the sphere. We denote the closed spherical cap of spherical radius ϕ centered at $u \in \mathbb{S}^{n-1}$ by $C(u, \phi) = \{v \in \mathbb{S}^{n-1} \,:\, <u, v> \geq \cos\phi\}$, and its probability measure by $\Omega(\phi) = \sigma(C(u, \phi))$. For a set $K \subset \mathbb{S}^{n-1}$ and $\delta > 0$, we denote the *δ–inner parallel body* of K by $K_{-\delta} = \{u \in K \,:\, C(u, \delta) \subseteq K\}$.

A set $K \subset \mathbb{S}^{n-1}$ is called *spherically convex*, if it is contained in an open hemisphere and for any two of its points, it contains the shorter great circular arc connecting them.

The *spherical circumradius* of a subset of an open hemisphere of $\mathbb{S}^{n-1}$ is the spherical radius of the smallest spherical cap (the *circum-cap*) that contains the set. A proof mimicking the proof of Theorem 5.4 yields

Theorem 5.7 (Naszódi [78]) *Let $K \subseteq \mathbb{S}^{n-1}$ be a measurable set. Then there is a covering of $\mathbb{S}^{n-1}$ by rotated copies of K of density at most*

$$\inf_{\delta>0}\left[\frac{\sigma(K)}{\sigma(K_{-\delta})}\left(1 + \ln\frac{\sigma\left(K_{-\delta/2}\right)}{\Omega\left(\frac{\delta}{2}\right)}\right)\right].$$

Improving an earlier result of Rogers [81], Böröczky and Wintsche [19] showed that for any $0 < \varphi < \pi/2$ and dimension n there is a covering of $\mathbb{S}^n$ by spherical caps of radius ϕ with density at most $n \ln n + n \ln\ln n + 5n$. This result follows from Theorem 5.7. Other bounds on covering the sphere by caps (or, a ball by smaller equal balls) can be found in [100] by Verger–Gaugry.

6 The Illumination Conjecture

We fix a convex body K in $\mathbb{R}^n$. Once the covering number is defined, it is fairly natural to ask what Levi [56] asked: how large may $N(K, \operatorname{int} K)$ be. We will call this quantity the *illumination number* of K, and denote it by $i(K) = N(K, \operatorname{int} K)$. The naming will become obvious in the next paragraphs.

Following Hadwiger [46], we say that a point $p \in \mathbb{R}^n \setminus K$ *illuminates* a boundary point $b \in \operatorname{bd} K$, if the ray $\{p + \lambda(b - p) \,:\, \lambda > 0\}$ emanating from p and passing through b intersects the interior of K. Boltyanski [16] gave the following slightly different definition. A direction $u \in \mathbb{S}^{n-1}$ is said to *illuminate* K at a boundary point $b \in \operatorname{bd} K$, if the ray $\{b + \lambda u \ : \ \lambda > 0\}$ intersects the interior of K. It is easy to see that the minimum number of directions that illuminate each boundary point of K is equal to the minimum number of points that illuminate each boundary point of K, which in turn is equal to the illumination number of K (as defined in the paragraph above).

Gohberg and Markus [40] asked how large $\inf\{N(K, \lambda K) : 0 < \lambda < 1\}$ can be. It also follows easily that this number is equal to $i(K)$.

The following dual formulation of the definition of the illumination number was found independently by P. Soltan, V. Soltan [90] and by Bezdek [11]. First, recall that an *exposed face* of a convex body K is the intersection of K with a supporting hyperplane. Now, let K be a convex body in $\mathbb{R}^n$ containing the origin in its interior. Then $i(K)$ is the minimum size of a family of hyperplanes in $\mathbb{R}^n$ such that each exposed face of the polar K^* of K is strictly separated from the origin by at least one of the hyperplanes in the family (for the definition of K^*, see the introduction).

Any smooth convex body (ie., a convex body with a unique support hyperplane at each boundary point) in $\mathbb{R}^n$ is illuminated by $n + 1$ directions. Indeed, for a smooth convex body, the set of directions illuminating a given boundary point is an open hemisphere of $\mathbb{S}^{n-1}$, and one can find $n + 1$ points (eg., the vertices of a regular simplex) in $\mathbb{S}^{n-1}$ with the property that every open hemisphere contains at least one of the points. Thus, these $n + 1$ points in $\mathbb{S}^{n-1}$ (ie., directions) illuminate any smooth convex body in $\mathbb{R}^n$. It is easy to see that no convex body is illuminated by less than $n + 1$ directions.

On the other hand, the illumination number of the cube is 2^n, since no two vertices of the cube share an illumination direction. An important unsolved problem in Discrete Geometry is the *Gohberg–Markus–Levi–Boltyanski–Hadwiger Conjecture* (or, Illumination Conjecture), according to which *for any convex body K in $\mathbb{R}^n$, we have $i(K) = 2^n$, where equality is attained only when K is an affine image of the cube.*

In this section, we mention some results on illumination. For a more complete account of the current state of the problem, see [8, 10, 21, 59, 95]. In Chap. VI. of [15], among many other facts on illumination, one can find a proof of the equivalence of the first four definitions of $i(K)$ given at the beginning of this section. Quantitative versions of the illumination number are discussed in the article of Bezdek and Khan in this volume [13]. Connections of the illumination number to other quantities are discussed in [102, 103].

One detail of the history of the conjecture may tell a lot about it. It was asked several times in different formulations (see the different definitions of $i(K)$ above), first in 1960 (though, Levi's study of $N(K, \operatorname{int} K)$ on the plane is from 1955). Several partial results appeared solving the conjecture for special families of convex bodies. Yet, the best general bound is an immediate consequence of Rogers' Theorem 3.1 (more precisely, Corollary 3.5) dating 1957 combined with the *Rogers–Shepard inequality* [84], according to which $\operatorname{vol}(K - K) \leq \binom{2n}{n} \operatorname{vol}(K)$ for any convex body K in $\mathbb{R}^n$.

Theorem 6.1 (Rogers [80]) *Let K be a convex body in $\mathbb{R}^n$. Then*

$$i(K) \leq \begin{cases} 2^n(n\ln n + n\ln\ln n + 5n) & \text{if } K = -K, \\ \binom{2n}{n}(n\ln n + n\ln\ln n + 5n) & \text{otherwise.} \end{cases}$$

By [56], the Illumination Conjecture holds on the plane. Papadoperakis [75] proved $i(K) \leq 16$ in dimension three. The upper bound in the conjecture (that is, not the equality case) was verified in the following cases: if $K = -K \subset \mathbb{R}^3$ (Lassak [54]), if $K \subset \mathbb{R}^3$ is a convex polyhedron with at least one non-trivial affine symmetry (Bezdek [11]), if $K \subset \mathbb{R}^3$ is symmetric about a plane (Dekster [23]).

6.1 Borsuk's Problem and Illuminating Sets of Constant Width

The problem of illumination is closely related to another classical question in geometry. *Borsuk's problem* [20] (or, Borsuk's Conjecture, though, he formulated it as a question) asks whether every bounded set X in $\mathbb{R}^n$ can be partitioned into $n+1$ sets of diameter less than the diameter of X (cf. [60] for a comprehensive survey). The minimum number of such parts is the *Borsuk number* of X, and clearly, it is at most the illumination number of $\text{conv}(X)$. Since any bounded set in $\mathbb{R}^n$ is contained in a set of constant width of the same diameter, it follows that any upper bound on the illumination number of sets of constant width in a certain dimension is also a bound on the maximum Borsuk number in the same dimension.

The affirmative answer to Borsuk's problem in the plane was proved by Borsuk, then, in three–space by Perkal [76] and Eggleston [26] (in the case of finite, three-dimensional sets, see Grünbaum [45], Heppes–Révész [48] and Heppes [47]). It was first shown by Lassak [53] (see also [14, 101]) that sets of constant width in $\mathbb{R}^3$ can be illuminated by three pairs of opposite directions. It would be a nice alternative proof of the bound 4 on the Borsuk number in three–space, if one could show that three–dimensional sets of constant width have illumination number 4 (see Conjecture 3.3.5. in [8]).

In 1993 by an ingenious proof, Kahn and Kalai [52] (based on a deep combinatorial result of Frankl and Wilson [37]) showed that if n is large enough, then there is a finite set in $\mathbb{R}^n$ whose Borsuk number is greater than $(1.2)^{\sqrt{n}}$, thus answering Borsuk's question in the negative. That result made the following bound on the illumination number by Schramm [87] all the more relevant. Currently, this is also the best general bound for the Borsuk number.

Theorem 6.2 (Schramm [87]) *In any dimension n for any set W of constant width in $\mathbb{R}^n$, we have*

$$i(W) \leq 5n\sqrt{n}(4 + \ln n)\left(\frac{3}{2}\right)^{n/2}.$$

By a fine analysis of Schramm's method, Bezdek (Theorem 6.8.3. of [8]) extended Theorem 6.2 to the class of those convex bodies W that can be obtained as $W = \bigcap_{x \in X}(x + \mathbf{B}_2^n)$ for some $X \subset \mathbb{R}^n$ compact set with diam $X \leq 1$. Note that a set W is of constant width one if and only if, $W = \bigcap_{x \in W}(x + \mathbf{B}_2^n)$.

We sketch the proof. First, we give yet another way to compute the illumination number of a convex body K. Let b be a boundary point of K, and consider its *Gauss image* $\beta(b) \subset \mathbb{S}^{n-1}$ consisting of the inner unit normal vectors of all hyperplanes supporting K at b. It is a closed, spherically convex set. We denote the *open polar* of a subset of the sphere $F \subset \mathbb{S}^{n-1}$ by $F^+ = \{u \in \mathbb{S}^{n-1} \ : < u, f >> 0 \text{ for all } f \in F\}$. Consider the set family $\mathcal{F} = \{(\beta(b))^+ \ : \ b \in \operatorname{bd} K\}$. Clearly, the directions $u_1, \dots, u_m \in \mathbb{S}^{n-1}$ illuminate K if and only if, each member of $\mathcal{F}$ contains at least one u_i. In other words, we are looking for a small cardinality transversal to the set family $\mathcal{F}$ (for definitions, see Sect. 5.1). We note that the idea of considering the Gauss image and $\mathcal{F}$ to bound the illumination number also appears in [9, 11, 14].

Now, consider a set $W = \bigcap_{x \in X}(x + \mathbf{B}_2^n)$ with a compact set $X \subset \mathbb{R}^n$ of diameter at most one. To make the problem of bounding $i(W)$ finite, we take a covering of $\mathbb{S}^{n-1}$ by spherical caps of Euclidean diameter $\varepsilon := \sqrt{\frac{2n}{2n-1}} - 1$, say $C_1 \cup \ldots \cup C_N = \mathbb{S}^{n-1}$. Such covering exists with $N \leq (1 + \frac{4}{\varepsilon})^n$ by the simple bound (6). We could use a better bound, but that would not yield any visible improvement on the bound on $i(W)$. Let

$$U_i := \bigcup_{\beta(b) \cap C_i \neq \emptyset} \beta(b),$$

and consider the set family $\mathcal{G} = \{U_i^+ \ : \ i = 1, \dots, N\}$. Clearly, any transversal to the finite set family $\mathcal{G}$ is a transversal to $\mathcal{F}$, and hence, is a set that illuminates K. One can show that

$$\operatorname{diam}(U_i) \leq 1 + \varepsilon. \tag{16}$$

Let $V(t) := \inf\{\sigma(F^+) \ : \ F \subset \mathbb{S}^{n-1}, \operatorname{diam} S \leq t\}$. A key element of the proof is the highly non-trivial claim that

$$V(t) \geq \frac{1}{\sqrt{8\pi n}} \left(\frac{3}{2} + \frac{\left(2 - \frac{1}{n}\right) t^2 - 2}{4 - \left(2 - \frac{2}{n}\right) t^2} \right)^{-\frac{n-1}{2}} \tag{17}$$

for all $0 < t < \sqrt{\frac{2n}{n-1}}$ and $n \geq 3$.

We notice that by (16), $\frac{\sigma}{V(1+\varepsilon)}$ is a fractional transversal to $\mathcal{G}$. Now, the original proof is completed by applying the dual of (10) to get $i(W) \leq \left\lfloor 1 + \frac{\ln N}{-\ln(1 - V(1+\varepsilon))} \right\rfloor$. Substituting the bound on N and (17), the theorem follows. Another way to complete the proof is to use the dual of Lemma 5.2, which yields the slightly worse bound $i(W) < \frac{1 + \ln N}{V(1+\varepsilon)}$.

6.2 *Fractional Illumination*

The notion of fractional illumination was defined in [65], and then further studied in [4].

Definition 6.3 The *fractional illumination number* of a convex body K in $\mathbb{R}^n$ is

$$i^*(K) = N^*(K, \operatorname{int} K).$$

It was observed in [65] that by (12) and the Rogers–Shepard inequality ($\operatorname{vol}(K - K) \leq \binom{2n}{n} \operatorname{vol}(K)$) we have

$$i^*(K) \leq \begin{cases} 2^n & \text{if } K = -K, \\ \binom{2n}{n} & \text{otherwise.} \end{cases} \tag{18}$$

The fractional form of the Illumination Conjecture (weaker than the original) reads: *$i^*(K) \leq 2^n$, and equality is attained by parallelotopes only.* When K is symmetric, the case of equality was settled by Artstein–Avidan and Slomka [5] using a lemma by Schneider.

Interestingly, no better bound is known, so the fractional form of the Illumination Conjecture does not seem much easier than the original. On the other hand, just as in general, for $N(K, L)$ and $N^*(K, L)$, we do not have an upper bound of $i(K)$ in terms of $i^*(K)$.

The *fractional version of Borsuk's problem* can be stated in a natural way, and was investigated in [49] using the language of multiple Borsuk coverings. We note that the example of a set in $\mathbb{R}^n$ with high Borsuk number given by Kahn and Kalai (see Sect. 6.1) is a set with high fractional Borsuk number as well.

7 Decomposability of Multiple Coverings

An *m–fold covering* of $\mathbb{R}^n$ by translates of a set K is a family $\mathcal{F}$ of translates of K such that each point is contained in at least m members. It is a natural question whether, for a particular K, if m is large enough (say, at least $m(K)$), then all m–fold coverings of $\mathbb{R}^n$ by translates of K can be *decomposed* into two coverings. That is, can $\mathcal{F}$ be colored with two colors such that each color class of $\mathcal{F}$ is a covering of $\mathbb{R}^n$?

It was proved in [66] that if K is a centrally symmetric convex polygon then such $m(K)$ exists. This was generalized to all convex polygons in [74, 97].

Arguably the most natural special case was asked by Pach [67]: consider the open unit disk. The un–published manuscript [58] was cited several times as having given a positive answer in this case, though, Pach [71] warned that the result "has not been independently verified." The following result of Mani–Levitska and Pach (see [1]) also suggested that such $m(K)$ should exist for unit disks. *For every $n \geq 2$, there is a positive constant c_n with the following property. For every positive integer m, any m–fold covering of $\mathbb{R}^n$ with unit balls can be decomposed into two coverings, provided that no point of the space belongs to more than $c_n 2^{m/n}$ balls.* This result was one of the first geometric applications of the Lovász local lemma.

Pach and Pálvölgyi [69] recently showed that, very surprisingly, *there is no such $m(K)$ for the open unit disk.*

Their proof consists of a combinatorial part followed by an intricate geometric argument. First, based on [73], they construct a finite abstract hypergraph, with a non-decomposable multiple covering. Then, the hypergraph is given a *geometric realization*, that is, the vertex set is mapped to a set of points on the plane, and the edges are mapped to open unit disks in an incidence–preserving manner. Finally, this m–fold covering by disks of this finite planar set is extended to an m–fold covering of the whole plane without adding any disk that contains any of the points in the finite set.

For more on decomposability of coverings, see [70], and the more recent paper [69].

8 An Asymptotic View

In this section, we present two topics to illustrate the point of view taken in the asymptotic theory of convex bodies on the problem of translative coverings.

8.1 Sudakov's Inequality

Sudakov's Inequality relates the minimum number of Euclidean balls that cover a symmetric convex body to the *mean width* of the body, where the latter is defined as

$$w(K) = \int_{\mathbb{S}^{n-1}} h_K(u) + h_K(-u) \ \mathrm{d}\sigma(u). \tag{19}$$

(See the definition of σ and h_K in the introduction.)

Theorem 8.1 (Sudakov's inequality, [93]) *For any symmetric convex body K in $\mathbb{R}^n$ and any $t > 0$, we have*

$$\log N(K, t\mathbf{B}_2^n) \leq cn \left(\frac{w(K)}{t} \right)^2$$

with an absolute constant $c > 0$.

It was observed by Tomczak–Jaegermann [99], that this inequality can be obtained from a dual form proved by Pajor and Tomczak–Jaegermann [72].

Theorem 8.2 (Dual Sudakov inequality) *For any symmetric convex body K in $\mathbb{R}^n$ and any $t > 0$, we have*

$$\log N(\mathbf{B}_2^n, tK) \leq cn \left(\frac{w(K^*)}{t} \right)^2$$

with an absolute constant $c > 0$.

First, we sketch a proof of Theorem 8.2 due to Talagrand [96], [55], and later turn to the proof of Theorem 8.1. The main idea is to apply a volumetric argument, but, instead of using the Lebesgue measure, one uses the Gaussian measure. Recall, that the *Gaussian measure* γ_n is an absolutely continuous probability measure on $\mathbb{R}^n$, with density

$$\mathrm{d}\gamma_n(x) = \frac{e^{-|x|^2/2}}{(2\pi)^{n/2}} \, \mathrm{d}x.$$

First, by computation one obtains that for any origin–symmetric convex body K in $\mathbb{R}^n$ and any translation vector $z \in \mathbb{R}^n$, we have

$$\gamma_n(K + z) \geq e^{-|z|^2/2} \gamma_n(K). \tag{20}$$

Next, we consider a maximal set $\{x_1, \ldots, x_N\}$ in $\mathbf{B}_2^n$ with the property that $\|x_i - x_j\|_K \geq t$ for all i, j pairs. Now, for any rescaling factor $\lambda > 0$, we have that $\{\lambda x_i + \frac{\lambda t}{2} K \ : \ i = 1, \ldots, N\}$ is a packing in $\lambda \mathbf{B}_2^n$, and thus, the total γ_n–measure of these sets is at most one. Integration in polar coordinates yields that

$$\gamma_n \left(\frac{\lambda t}{2} K \right) \geq 1 - \frac{2c\sqrt{n}}{\lambda t} w(K^*)$$

for an absolute constant $c > 0$. With the choice $\lambda = 4c\sqrt{n}w(K^*)/t$, we have $\gamma_n(K) \geq \frac{1}{2}$. Finally, using (20), we obtain the bound in Theorem 8.2.

We note that this proof yields a little more than stated in the Theorem. We obtain an upper bound on the minimum size of a covering of $\mathbf{B}_2^n$ by translates of tK with the constraint that the translation vectors are in $\mathbf{B}_2^n$.

The following Lemma is the key to reducing Theorem 8.1 to Theorem 8.2.

Lemma 8.3 (Tomczak–Jaegermann [99]) *For any origin–symmetric convex body K in $\mathbb{R}^n$, and any $t > 0$, we have*

$$N(K, t\mathbf{B}_2^n) \leq N(K, 2t\mathbf{B}_2^n) N \left(\mathbf{B}_2^n, \frac{t}{8} K^* \right).$$

Proof of Lemma 8.3 Observe that $2K \cap \left(\frac{t^2}{2}K^*\right) \subseteq t\mathbf{B}_2^n$. Thus, by (4),

$$N(K, t\mathbf{B}_2^n) \leq N\left(K, 2K \cap \frac{t^2}{2}K^*\right) \leq N\left(K, \frac{t^2}{4}K^*\right) \leq$$

$$N(K, 2t\mathbf{B}_2^n)N\left(\mathbf{B}_2^n, \frac{t}{8}K^*\right).$$

Proof of Theorem 8.1 Combining Lemma 8.3 and Theorem 8.2, we have

$$t^2 \log N(K, t\mathbf{B}_2^n) \leq \frac{1}{4}(2t)^2 \log N(K, 2t\mathbf{B}_2^n) + 64(t/8)^2 \log N\left(\mathbf{B}_2^n, \frac{t}{8}K^*\right)$$

Taking supremum over all $t > 0$, we get

$$\frac{3}{4}\sup_{t>0}\left\{t^2 \log N(K, t\mathbf{B}_2^n)\right\} \leq 64 \sup_{t>0}\left\{t^2 \log N\left(\mathbf{B}_2^n, tK^*\right)\right\} \leq 64cn\,(w(K))^2.$$

8.2 Duality of Covering Numbers

We briefly mention the following open problem in geometric analysis, for a comprehensive discussion, cf. Chap. 4 of [3].

Conjecture 8.4 There are universal constants $c, C > 0$ such that for any dimension n and any two symmetric convex bodies K and L in $\mathbb{R}^n$, we have

$$N(K, L) \leq N(L^*, cK^*)^C.$$

The problem is known as the *Duality of entropy*, and was posed by Pietsch [77]. An important special case, when K or L is a Euclidean ball (or, equivalently, an ellipsoid) was confirmed by Artstein–Avidan, Milman and Szarek [2].

9 Covering by Sequences of Sets

So far, we considered problems where a set was to be covered by translates of another fixed set. Now, we turn to problems where a family $\mathcal{F}$ of sets is given, and we need to find a translation for each set in $\mathcal{F}$ to obtain a covering of a given set C. If such translations exist, we say that $\mathcal{F}$ *permits a translative covering* of C. We call $\mathcal{F}$ a *bounded family*, if the set of diameters of members of $\mathcal{F}$ is a bounded set.

For a comprehensive account of coverings by sequences of convex sets, see the surveys [32, 43].

9.1 Covering (Almost) the Whole Space.

Clearly, for $\mathcal{F}$ to permit a translative covering of $\mathbb{R}^n$, it is necessary that the total volume of the members of $\mathcal{F}$ be infinite. It is not sufficient, though. Indeed, consider rectangles of side lengths i by $1/i^2$ for $i = 1, 2, \ldots$ on the plane. Their total area is infinite, and yet, according to Bang's theorem, they do not permit a translative covering of $\mathbb{R}^2$ [32]. On the other hand, if a family of planar convex sets is bounded and has infinite total area, then it permits a translative covering of $\mathbb{R}^2$ [42, 51]. It is an open problem whether the same holds for $n > 2$.

A covering of *almost all of* some set C is a covering of a subset of C whose complement in C is of measure zero.

Theorem 9.1 (Groemer, [44]) *Let $\mathcal{F}$ be a bounded family of Lebesgue measurable sets. Then $\mathcal{F}$ permits a translative covering of almost all of $\mathbb{R}^n$ if and only if,* $\sum_{F\in\mathcal{F}} \mathrm{vol}(F) = \infty$.

Indeed, let $\mathcal{F} = \{F_1, F_2, \ldots\}$ be a bounded family with infinite total volume. Clearly, it is sufficient to cover almost all of the cube $C = [-1/2, 1/2]^n$. We may assume that $F \subset C$ for all $F \in \mathcal{F}$.

We find the translation vectors inductively. Let $x_1 = 0$. If x_k is defined, we denote the uncovered part by $E_k = C \setminus \left(\bigcup_{j=1}^{k} (F_j + x_j)\right)$. We choose x_{k+1} in such a way that

$$\frac{\mathrm{vol}\left((F_{k+1} + x_{k+1}) \cap E_k\right)}{\mathrm{vol}(F_{k+1})} \geq \frac{1}{2^n}\,\mathrm{vol}(E_k). \tag{21}$$

It is possible, since

$$\frac{1}{2^n}\int_{2C} \mathrm{vol}\left((F_{k+1} + x) \cap E_k\right)\mathrm{d}x = \frac{1}{2^n}\int_{2C}\int_{C} \chi_{F_{k+1}}(y - x)\chi_{E_k}(y)\mathrm{d}y\mathrm{d}x$$

$$= \frac{1}{2^n}\int_{C} \chi_{E_k}(y)\int_{2C} \chi_{F_{k+1}}(y - x)\mathrm{d}x\mathrm{d}y = \frac{1}{2^n}\,\mathrm{vol}(F_{k+1})\,\mathrm{vol}(E_k).$$

It is easy to see that (21) implies that $\lim_{k\to\infty} \mathrm{vol}(E_k) = 0$.

We note that the condition that $\mathcal{F}$ is bounded may be replaced by the condition that $\mathcal{F}$ contains only convex sets, see [43].

9.2 A Sufficient Condition for a Family of Homothets

For convex bodies K and L, we define $f(K, L)$ as the infimum of those $t > 0$, such that for any family $\mathcal{F}$ of homothets of L with coefficients $0 < \lambda_1, \lambda_2, \cdots < 1$, the following holds:

$$\text{If } \sum_i \lambda_i^d \geq t \text{ then } \mathcal{F} \text{ permits a translative covering of } K.$$

We set $f(n) := \sup\{f(K, K) : K \subset \mathbb{R}^n \text{ a convex body}\}$.

The question of bounding $f(2)$ was originally posed by L. Fejes Tóth [33] (cf. Sect. 3.2 in [21]). He conjectured that $f(2) \leq 3$. Januszewski [50] showed that $f(2) \leq 6.5$. In higher dimensions Meir and Moser [62], and later, A. Bezdek and K. Bezdek [12] considered the cube and proved that $f([0, 1]^d) = 2^d - 1$. Using a simple argument based on saturated packings by half-sized copies (see (6)), the author [64] showed

$$f(K, L) \leq 2^n \frac{\operatorname{vol}\left(K + \frac{L\cap(-L)}{2}\right)}{\operatorname{vol}(L \cap (-L))},$$

from which the bound

$$f(K, K) \leq \begin{cases} 3^n, & \text{if } K = -K, \\ 6^n, & \text{in general.} \end{cases}$$

follows.

On the other hand, clearly, $f(K, K) \geq n$ since we may consider n homothetic copies of K with homothety ratios slightly below one, and use the lower bound on the illumination number of K (see Sect. 6).

9.3 A Necessary Condition for a Family of Homothets

A converse to the problem discussed above was formulated by V. Soltan [89] (cf. Sect. 3.2 in [21]). Let

$$g(K) := \inf\left\{\sum_i \lambda_i : K \subseteq \bigcup_i \lambda_i K + x_i, 0 < \lambda_i < 1\right\},$$

and $g(n) := \inf\{g(K) : K \subset \mathbb{R}^n \text{ a convex body}\}$. V. Soltan conjectured $g(n) \geq n$.

Since the n-dimensional simplex Δ can be covered by $n + 1$ translates of $\frac{n}{n+1}\Delta$, we have that $g(n) \leq g(\Delta) \leq n$. V. Soltan and É. Vásárhelyi [88] showed $g(2) \geq 2$, and also proved that the conjecture holds when only $n + 1$ homothets are allowed.

Soltan's conjecture was confirmed in an asymptotic sense in [64]: $\lim\limits_{n\to\infty} \frac{g(n)}{n} = 1$.

Acknowledgements The author is grateful for the many illuminating conversations with Gábor Fejes Tóth about covering problems in general, and about this manuscript.

References

1. N. Alon, J.H. Spencer, *The Probabilistic Method. With an Appendix on the Life and Work of Paul Erdős*, 3rd ed. (Hoboken, Wiley, 2008) (English)
2. S. Artstein, V. Milman, S.J. Szarek, Duality of metric entropy. Ann. Math. (2) **159**(3), 1313–1328 (2004). MR2113023 (2005h:47037)
3. S. Artstein-Avidan, A. Giannopoulos, Mathematical surveys and monographs, *Asymptotic Geometric Analysis. Part I* (American Mathematical Society, Providence, 2015). MR3331351
4. S. Artstein-Avidan, O. Raz, Weighted covering numbers of convex sets. Adv. Math. **227**(1), 730–744 (2011)
5. S. Artstein-Avidan, B.A. Slomka, On Weighted Covering Numbers and the Levi-Hadwiger Conjecture. Israel Journal of Mathematics **209**(1), 125–155 (September 2015), arXiv:1310.7892
6. K. Ball, Volume ratios and a reverse isoperimetric inequality. J. Lond. Math. Soc. (2) **44**(2), 351–359 (1991). MR1136445 (92j:52013)
7. K. Ball, Volumes of sections of cubes and related problems, *Geometric Aspects of Functional Analysis (1987–88)* (1989), pp. 251–260. MR1008726 (90i:52019)
8. K. Bezdek, *Classical Topics in Discrete Geometry*, CMS Books in Mathematics/Ouvrages de Mathématiques de la SMC (Springer, New York, 2010)
9. K. Bezdek, Gy Kiss, On the X-ray number of almost smooth convex bodies and of convex bodies of constant width. Can. Math. Bull. **52**(3), 342–348 (2009). MR2547800 (2010m:52009)
10. K. Bezdek, The illumination conjecture and its extensions. Period. Math. Hung. **53**(1–2), 59–69 (2006)
11. K. Bezdek, The problem of illumination of the boundary of a convex body by affine subspaces. Mathematika **38**(2), 362–375 (1991, 1992). MR1147835 (92m:52020)
12. A. Bezdek, K. Bezdek, *Eine hinreichende Bedingung für die Überdeckung des Einheitswürfels durch homothetische Exemplare im n- dimensionalen euklidischen Raum*, Beiträge Algebra Geom. **17** (1984), pp. 5–21. MR755762 (85h:52017)
13. K. Bezdek, M.A. Khan, The geometry of homothetic covering and illumination, in *Discrete geometry and symmetry (to appear)*, 2018
14. K. Bezdek, Z. Lángi, M. Naszódi, P. Papez, Ball-polyhedra. Discret. Comput. Geom. **38**(2), 201–230 (2007). MR2343304 (2008i:52001)
15. V. Boltyanski, H. Martini, P.S. Soltan, *Excursions into Combinatorial Geometry* (Universitext, Springer, Berlin, 1997). MR1439963 (98b:52001)
16. V. Boltyanski, The problem of illuminating the boundary of a convex body. Izv. Mold. Fil. AN SSSR **76**, 77–84 (1960)
17. T. Bonnesen, W. Fenchel, *Theory of convex bodies. Transl. from the German and ed. by L. Boron, C. Christenson, B. Smith, with the Collaboration of W. Fenchel* (Moscow, Idaho, USA): BCS Associates. IX, 172 p. (1987) (English)
18. K. Böröczky Jr., *Finite Packing and Covering*, Cambridge Tracts in Mathematics, vol 154 (Cambridge University Press, Cambridge, 2004). MR2078625 (2005g:52045)
19. K. Böröczky Jr., K. Böröczky Jr., Covering the spere by equal spherical balls, *Discrete and Computational Geometry* (Springer, Berlin, 2003), pp. 235–251
20. K. Borsuk, *Drei sätze über die n-dimensionale euklidische sphäre*, Fundamenta Mathematicae **20** (1933), no. 1, 177–190 (ger)
21. P. Brass, W. Moser, J. Pach, *Research Problems in Discrete Geometry* (Springer, New York, 2005)
22. H.S.M. Coxeter, L. Few, C.A. Rogers, Covering space with equal spheres. Mathematika **6**, 147–157 (1959). MR0124821 (23 #A2131)
23. B.V. Dekster, Each convex body in E^3 symmetric about a plane can be illuminated by 8 directions. J. Geom. **69**(1–2), 37–50 (2000). MR1800455 (2001m:52003)
24. I. Dumer, Covering spheres with spheres, Discrete Comput. Geom. **38**(4), 665Â–679 (2007)
25. I. Dumer, *Covering spheres with spheres* (2018), arXiv:0606002v2 [math]

26. H.G. Eggleston, Covering a three-dimensional set with sets of smaller diameter. J. Lond. Math. Soc. **30**, 11–24 (1955). MR0067473 (16,734b)
27. P. Erdős and L. Lovász, *Problems and results on 3-chromatic hypergraphs and some related questions*, Infinite and finite sets (Colloquium, Keszthely, 1973; dedicated to P. Erdős on his 60th birthday), Vol. II, 1975, pp. 609–627. Colloquium Mathematical Society, János Bolyai, vol. 10. MR0382050 (52 #2938)
28. P. Erdős, C.A. Rogers, Covering space with convex bodies. Acta Arith. **7**, 281–285 (1961/1962)
29. G. Fejes Tóth, A note on covering by convex bodies. Can. Math. Bull. **52**(3), 361–365 (2009)
30. L. Fejes Tóth, *Lagerungen in der Ebene, auf der Kugel und im Raum*, Die Grundlehren der Mathematischen Wissenschaften in Einzeldarstellungen mit besonderer Berücksichtigung der Anwendungsgebiete, (Band LXV, Springer, Berlin, 1953). MR0057566 (15,248b)
31. G. Fejes Tóth, New results in the theory of packing and covering, in *Convexity and its Applications, Collect. Surv.* (1983), 318–359; 1983 (English)
32. G. Fejes Tóth, Packing and covering, *Handbook of Discrete and Computational Geometry*, 2nd ed. (2004), pp. 25–53 (English)
33. L. Fejes Tóth, Personal communication (1984)
34. G. Fejes Tóth, Recent progress on packing and covering, in *Advances in Discrete and Computational Geometry: Proceedings of the 1996 AMSIMS-SIAM Joint Summer Research Conference on Discrete and Computational Geometry: Ten Years Later, South Hadley, USA, 14–18 July 1996* (1996), pp. 145–162 (English)
35. G. Fejes Tóth, W. Kuperberg, A survey of recent results in the theory of packing and covering, *New Trends in Discrete and Computational Geometry* (1993), pp. 251–279. (English)
36. G. Fejes Tóth, W. Kuperberg, Packing and covering with convex sets, in *Handbook of Convex Geometry*, vol. B (1993), pp. 799–860 (English)
37. P. Frankl, R.M. Wilson, Intersection theorems with geometric consequences. Combinatorica **1**(4), 357–368 (1981). MR647986 (84g:05085)
38. Z. Füredi, Matchings and covers in hypergraphs. Graphs Comb. **4**(2), 115–206 (1988)
39. Z. Füredi, J.-H. Kang, Covering the n-space by convex bodies and its chromatic number. Discret. Math. **308**(19), 4495–4500 (2008)
40. I. Gohberg, A. Markus, A problem on covering of convex figures by similar figures. Izv. Mold. Fil. AN SSSR **10**, 87–90 (1960)
41. P. Gritzmann, Lattice covering of space with symmetric convex bodies. Mathematika **32**(2), 311–315 (1985); (1986). MR834499
42. H. Groemer, Covering and packing properties of bounded sequences of convex sets. Mathematika **29**, 18–31 (1982). (English)
43. H. Groemer, Coverings and packings by sequences of convex sets, *Discrete Geometry and Convexity* (New York, 1982); (1985), pp. 262–278
44. H. Groemer, Space coverings by translates of convex sets. Pac. J. Math. **82**, 379–386 (1979). (English)
45. B. Grünbaum, A simple proof of Borsuk's conjecture in three dimensions. Math. Proc. Camb. Philos. Soc. **53**, 776–778 (1957). MR0090072 (19,763d)
46. H. Hadwiger, Ungelöste probleme, nr. 38. Elem. Math. **15**, 130–131 (1960)
47. A. Heppes, On the partitioning of three-dimensional point-sets into sets of smaller diameter. Magyar Tud. Akad. Mat. Fiz. Oszt. Közl. **7**, 413–416 (1957). MR0095450 (20 #1952)
48. A. Heppes, P. Révész, A splitting problem of Borsuk, Mat. Lapok **7** (1956), 108–111. MR0098353 (20 #4814)
49. M. Hujter, Z. Lángi, On the multiple Borsuk numbers of sets. Israel J. Math. **199**(1), 219–239 (2014). MR3219534
50. J. Januszewski, Translative covering a convex body by its homothetic copies. Stud. Sci. Math. Hung. **40**(3), 341–348 (2003). MR2036964 (2005b:52044)
51. E.jr Makai, J. Pach, Controlling function classes and covering Euclidean space. Stud. Sci. Math. Hung. **18**, 435–459 (1983). (English)

52. J. Kahn, G. Kalai, A counterexample to Borsuk's conjecture. Bull. Amer. Math. Soc. (N.S.) **29**(1), 60–62 (1993). MR1193538 (94a:52007)
53. M. Lassak, Illumination of three-dimensional convex bodies of constant width, in *Proceedings of the 4th International Congress of Geometry: Thessaloniki, 1996* (1997), pp. 246–250. MR1470984 (98g:52013)
54. M. Lassak, Solution of Hadwiger's covering problem for centrally symmetric convex bodies in E^3, J. Lond. Math. Soc. (2) **30**(3), 501–511 (1984). MR810959 (87e:52024)
55. M. Ledoux, M. Talagrand, *Probability in Banach Spaces*, Ergebnisse der Mathematik und ihrer Grenzgebiete (3) [Results in Mathematics and Related Areas (3)], vol. 23 (Springer, Berlin, 1991). Isoperimetry and Processes. MR1102015 (93c:60001)
56. F.W. Levi, Überdeckung eines Eibereiches durch Parallelverschiebung seines offenen Kerns. Arch. Math. (Basel) **6**, 369–370 (1955). MR0076368 (17,888b)
57. L. Lovász, On the ratio of optimal integral and fractional covers. Discret. Math. **13**(4), 383–390 (1975)
58. P. Mani-Levitska, J. Pach, *Decomposition Problems for Multiple Coverings with Unit Balls* (1986). Manuscript: Parts of the manuscript available at http://www.math.nyu.edu/~pach/publications/unsplittable.pdf
59. H. Martini, V. Soltan, Combinatorial problems on the illumination of convex bodies. Aequ. Math. **57**(2–3), 121–152 (1999)
60. J. Math. Sci. Around Borsuk's hypothesis. **154**(4), 604–623 (2008). (English)
61. J. Matoušek, *Lectures on discrete geometry (Graduate Texts in Mathematics)*, vol. 212 (Springer, New York, 2002)
62. A. Meir, L. Moser, On packing of squares and cubes. J. Comb. Theory **5**, 126–134 (1968). MR0229142 (37 #4716)
63. H. Minkowski, Allgemeine Lehrsätze über die convexen Polyeder. Nachr. Ges. Wiss. Göttingen, Math.-Phys. Kl. **1897**, 198–219 (1897) (German)
64. M. Naszódi, Covering a set with homothets of a convex body. Positivity **14**(1), 69–74 (2010). MR2596464
65. M. Naszódi, Fractional illumination of convex bodies. Contrib. Discret. Math. **4**(2), 83–88 (2009)
66. J. Pach, Covering the plane with convex polygons. Discret. Comput. Geom. **1**, 73–81 (1986). (English)
67. J. Pach, Decomposition of multiple packing and covering. Diskrete Geometrie, 2. Kolloq., Inst. Math. Univ. Salzburg 1980, 169–178 (1980). 1980 (English)
68. J. Pach, P.K. Agarwal, *Combinatorial Geometry* (New York, Wiley, 1995) (English)
69. J. Pach, D. Pálvölgyi, Unsplittable coverings in the plane. Adv. Math. **302**, 433–457 (2016)
70. J. Pach, D. Pálvölgyi, G. Tóth, Survey on decomposition of multiple coverings, in *Geometry—Intuitive, Discrete, and Convex* (2013), pp. 219–257. MR3204561
71. J. Pach, G. Tóth, Decomposition of multiple coverings into many parts. Comput. Geom. **42**(2), 127–133 (2009). (English)
72. A. Pajor, N. Tomczak-Jaegermann, *Remarques sur les nombres d'entropie d'un opérateur et de son transposé*, C. R. Acad. Sci. Paris Sér. I Math. **301**(15), 743–746 (1985). MR817602 (87f:47027)
73. D. Pálvölgyi, Indecomposable coverings with concave polygons. Discret. Comput. Geom. **44**(3), 577–588 (2010). (English)
74. D. Pálvölgyi, G. Tóth, Convex polygons are cover-decomposable. Discret. Comput. Geom. **43**(3), 483–496 (2010). (English)
75. I. Papadoperakis, An estimate for the problem of illumination of the boundary of a convex body in E^3, Geom. Dedicata **75**(3), 275–285 (1999). MR1689273 (2000g:52014)
76. J. Perkal, Sur la subdivision des ensembles en parties de diamétre intérieure. Colloq. Math. **1**, 45 (1947)
77. A. Pietsch, *Theorie der Operatorenideale (Zusammenfassung)*, Friedrich-Schiller-Universität, Jena, 1972. Wissenschaftliche Beiträge der Friedrich-Schiller-Universität Jena. MR0361822 (50 #14267)

78. Proc. Amer. Math. Soc. On some covering problems in geometry. **144**(8), 3555–3562 (2016). MR3503722
79. J. Radon, Über eine Erweiterung des Begriffes der konvexen Funktionen mit einer Anwendung auf die Theorie der konvexen Körper. Wien. Ber. **125**, 241–258 (1916). (German)
80. C.A. Rogers, A note on coverings. Mathematika **4**, 1–6 (1957)
81. C.A. Rogers, Covering a sphere with spheres. Mathematika **10**, 157–164 (1963)
82. C.A. Rogers, Lattice coverings of space. Mathematika **6**, 33–39 (1959)
83. C.A. Rogers, *Packing and Covering*, Cambridge Tracts in Mathematics and Mathematical Physics, vol. 54 (Cambridge University Press, New York, 1964)
84. C.A. Rogers, G.C. Shephard, The difference body of a convex body. Arch. Math. (Basel) **8**, 220–233 (1957)
85. C.A. Rogers, C. Zong, Covering convex bodies by translates of convex bodies. Mathematika **44**(1), 215–218 (1997)
86. W. Schmidt, Maßtheorie in der Geometrie der Zahlen. Acta Math. **102**, 159–224 (1959). (German)
87. O. Schramm, Illuminating sets of constant width. Mathematika **35**(2), 180–189 (1988)
88. V. Soltan, É. Vásárhelyi, Covering a convex body by smaller homothetic copies. Geom. Dedicata **45**(1), 101–113 (1993). MR1199732 (94a:52040)
89. V. Soltan, Personal Communication (1990)
90. P.S. Soltan, V.P. Soltan, Illumination through convex bodies. Dokl. Akad. Nauk SSSR **286**(1), 50–53 (1986). MR822098 (87f:52008)
91. J. Spencer, Asymptotic lower bounds for Ramsey functions. Discret. Math. **20**(1), 69–76 (1977/78). MR0491337 (58 #10600)
92. S.K. Stein, Two combinatorial covering theorems. J. Comb. Theory Ser. A **16**(3), 391–397 (1974)
93. V.N. Sudakov, Gaussian random processes, and measures of solid angles in Hilbert space, Dokl. Akad. Nauk SSSR **197**, 43–45 (1971). MR0288832 (44 #6027)
94. W. Süss, Über den Vektorenbereich eines Eikörpers. Jahresber. Dtsch. Math.-Ver. **37**, 87–90 (1928). (German)
95. L. Szabó, Recent results on illumination problems, in *Intuitive Geometry (Budapest, 1995)* (1997), pp. 207–221. MR1470759 (98h:52015)
96. M. Talagrand, A new isoperimetric inequality for product measure and the tails of sums of independent random variables. Geom. Funct. Anal. **1**(2), 211–223 (1991). MR1097260 (92j:60004)
97. G. Tardos, G. Toth, Multiple coverings of the plane with triangles. Discret. Comput. Geom. **38**(2), 443–450 (2007). (English)
98. Th. Estermann, Über den Vektorenbereich eines konvexen Körpers. Math. Z. **28**, 471–475 (1928). (German)
99. N. Tomczak-Jaegermann, Dualité des nombres d'entropie pour des opérateurs à valeurs dans un espace de Hilbert. C. R. Acad. Sci. Paris Sér. I Math. **305**(7), 299–301 (1987). MR910364 (89c:47027)
100. J.-L. Verger-Gaugry, Covering a ball with smaller equal balls in $\mathbb{R}^n$. Discret. Comput. Geom. **33**(1), 143–155 (2005). (English)
101. B. Weissbach, Invariante Beleuchtung konvexer Körper, Beiträge Algebr. Geom. **37**(1), 9–15 (1996). MR1407801 (97j:52011)
102. C.M. Zong, Some remarks concerning kissing numbers, blocking numbers and covering numbers. Period. Math. Hung. **30**(3), 233–238 (1995). MR1334968
103. C. Zong, The kissing number, blocking number and covering number of a convex body. Surv. Discret. Comput. Geom. **453**, 529–548 (2008). MR2405694

Incidences Between Points and Lines in Three Dimensions

Micha Sharir and Noam Solomon

Abstract We give a fairly elementary and simple proof that shows that the number of incidences between m points and n lines in $\mathbb{R}^3$, so that no plane contains more than s lines, is

$$O\left(m^{1/2}n^{3/4} + m^{2/3}n^{1/3}s^{1/3} + m + n\right)$$

(in the precise statement, the constant of proportionality of the first and third terms depends, in a rather weak manner, on the relation between m and n). This bound, originally obtained by Guth and Katz (Ann Math 181:155–190, 2015, [10]) as a major step in their solution of Erdős's distinct distances problem, is also a major new result in incidence geometry, an area that has picked up considerable momentum in the past decade. Its original proof uses fairly involved machinery from algebraic and differential geometry, so it is highly desirable to simplify the proof, in the interest of better understanding the geometric structure of the problem, and providing new tools for tackling similar problems. This has recently been undertaken by Guth (Discrete Comput Geom 53(2):428–444, 2015, [8]). The present paper presents a different and simpler derivation, with better bounds than those in Guth, and without the restrictive assumptions made there. Our result has a potential for applications to other incidence problems in higher dimensions.

Work on this paper by Noam Solomon and Micha Sharir was supported by Grant 892/13 from the Israel Science Foundation. Work by Micha Sharir was also supported by Grant 2012/229 from the U.S.–Israel Binational Science Foundation, by the Israeli Centers of Research Excellence (I-CORE) program (Center No. 4/11), and by the Hermann Minkowski-MINERVA Center for Geometry at Tel Aviv University. A preliminary version of this paper has appeared in *Proc. 31st Sympos. Comput. Geom.* (2015), 553–568.

M. Sharir (✉) · N. Solomon
School of Computer Science, Tel Aviv University, 69978 Tel Aviv, Israel
e-mail: michas@post.tau.ac.il

N. Solomon
e-mail: noam.solom@gmail.com

G. Ambrus et al. (eds.), *New Trends in Intuitive Geometry*, Bolyai Society Mathematical Studies 27, https://doi.org/10.1007/978-3-662-57413-3_15

1 Introduction

Let P be a set of m distinct points in $\mathbb{R}^3$ and let L be a set of n distinct lines in $\mathbb{R}^3$. Let $I(P, L)$ denote the number of incidences between the points of P and the lines of L; that is, the number of pairs (p, ℓ) with $p \in P$, $\ell \in L$, and $p \in \ell$. If all the points of P and all the lines of L lie in a common plane, then the classical Szemerédi–Trotter theorem [29] yields the worst-case tight bound

$$I(P, L) = O\left(m^{2/3}n^{2/3} + m + n\right). \tag{1}$$

This bound clearly also holds in three dimensions, by projecting the given lines and points onto some generic plane. Moreover, the bound will continue to be worst-case tight by placing all the points and lines in a common plane, in a configuration that yields the planar lower bound.

In the 2010 groundbreaking paper of Guth and Katz [10], an improved bound has been derived for $I(P, L)$, for a set P of m points and a set L of n lines in $\mathbb{R}^3$, provided that not too many lines of L lie in a common plane. Specifically, they showed[1]:

Theorem 1 (Guth and Katz [10]) *Let P be a set of m distinct points and L a set of n distinct lines in $\mathbb{R}^3$, and let $s \le n$ be a parameter, such that no plane contains more than s lines of L. Then*

$$I(P, L) = O\left(m^{1/2}n^{3/4} + m^{2/3}n^{1/3}s^{1/3} + m + n\right).$$

This bound was a major step in the derivation of the main result of [10], which was to prove an almost-linear lower bound on the number of distinct distances determined by any finite set of points in the plane, a classical problem posed by Erdős in 1946 [6]. Their proof uses several nontrivial tools from algebraic and differential geometry, most notably the Cayley–Salmon theorem on osculating lines to algebraic surfaces in $\mathbb{R}^3$, and additional properties of ruled surfaces. All this machinery comes on top of the main innovation of Guth and Katz, the introduction of the *polynomial partitioning technique*; see below.

In this paper, we provide a simple derivation of this bound, which bypasses most of the techniques from algebraic geometry that are used in the original proof. A recent related study by Guth [8] provides another simpler derivation of a similar bound, but (a) the bound obtained in [8] is slightly worse, involving extra factors of the form m^{ε}, for any $\varepsilon > 0$, and (b) the assumptions there are stronger, namely that no algebraic surface of degree at most c_{ε}, a (potentially large) constant that depends on ε, contains more than s lines of L (in fact, Guth considers in [8] only the case $s = \sqrt{n}$). It should be noted, though, that Guth also manages to derive a (slightly weaker but still) near-linear lower bound on the number of distinct distances.

---

[1]We skip over certain subtleties in their bound: They also assume that no *regulus* contains more than s input lines, but then they are able also to bound the number of intersection points of the lines. Moreover, if one also assumes that each point is incident to at least three lines then the term m in the bound can be dropped.

As in the classical work of Guth and Katz [10], and in the follow-up study of Guth [8], here too we use the polynomial partitioning method, as pioneered in [10]. The main difference between our approach and those of [8, 10] is the choice of the degree of the partitioning polynomial. Whereas Guth and Katz [10] choose a large degree, and Guth [8] chooses a constant degree, we choose an intermediate degree. This reaps many benefits from both the high-degree and the constant-degree approaches, and pays a small price in the bound (albeit much smaller than in [8]). Specifically, our main result is a relatively simple and fairly elementary derivation of the following result.

Theorem 2 *Let P be a set of m distinct points and L a set of n distinct lines in $\mathbb{R}^3$, and let $s \le n$ be a parameter, such that no plane contains more than s lines of L. Then*

$$I(P, L) \le A_{m,n}\left(m^{1/2}n^{3/4} + m\right) + B\left(m^{2/3}n^{1/3}s^{1/3} + n\right), \tag{2}$$

where B is an absolute constant, and, for another suitable absolute constant $b > 1$,

$$A_{m,n} = O\left(b^{\frac{\log(m^2 n)}{\log(n^3/m^2)}}\right), \quad \textit{for } m \le n^{3/2}, \textit{ and } O\left(b^{\frac{\log(m^3/n^4)}{\log(m^2/n^3)}}\right), \textit{for } m \ge n^{3/2}. \tag{3}$$

When m approaches $n^{3/2}$, the coefficient $A_{m,n}$ tends to infinity. We resolve this issue by establishing (2) in a suitable range, away from $n^{3/2}$, and interpolate the bounds for the middle range of m; see the proof for details.

Remarks. (a) Only the range $\sqrt{n} \le m \le n^2$ is of interest; outside this range, regardless of the dimension of the ambient space, we have the well known and trivial upper bound $O(m + n)$.

(b) The term $m^{2/3}n^{1/3}s^{1/3}$ comes from the planar Szemerédi–Trotter bound (1), and is unavoidable, as it can be attained if we densely "pack" points and lines into planes, in patterns that realize the bound in (1).

(c) Ignoring this term, the two terms $m^{1/2}n^{3/4}$ and m "compete" for dominance; the former dominates when $m \le n^{3/2}$ and the latter when $m \ge n^{3/2}$. Thus the bound in (2) is qualitatively different within these two ranges.

(d) The threshold $m = n^{3/2}$ also arises in the related problem of *joints* (points incident to at least three non-coplanar lines) in a set of n lines in 3-space; see [9].

A concise rephrasing of the bound in (2) and (3) is as follows. We partition each of the ranges $m \le n^{3/2}$, $m > n^{3/2}$ into a sequence of subranges $n^{\alpha_{j-1}} < m \le n^{\alpha_j}$, $j = 0, 1, \ldots$ (for $m \le n^{3/2}$), or $n^{\alpha_{j-1}} > m \ge n^{\alpha_j}$, $j = 0, 1, \ldots$ (for $m \ge n^{3/2}$). Within each range the bound asserted in the theorem holds for some fixed constant of proportionality (denoted as $A_{m,n}$ in the bound), where these constants vary with j, and grow, exponentially in j, as prescribed in (3), as m approaches $n^{3/2}$ (from either side). Informally, as already noted, if we keep m "sufficiently away" from $n^{3/2}$, the bound in (2) holds with a fixed constant of proportionality. Handling the "border range" $m \approx n^{3/2}$ is also fairly straightforward, although, to bypass the exponential growth of the constant of proportionality, it results in a slightly different bound; see below for details.

Our proof is elementary to the extent that, among other things, it avoids any explicit handling of *singular* and *flat* points on the zero set of the partitioning polynomial. While these notions are relatively easy to handle in three dimensions (see, e.g., [5, 9]), they become more complex notions in higher dimensions (as witnessed, for example, in our work on the four-dimensional setting [25]), making proofs based on them harder to extend.

Additional merits and features of our analysis are discussed in detail in the concluding section. In a nutshell, the main merits are:

(i) We use two separate partitioning polynomials. The first one is of "high" degree, and is used to prune away some points and lines, and to establish useful properties of the surviving points and lines. The second partitioning step, using a polynomial of "low" degree, is then applied, from scratch, to the surviving input, exploiting the properties established in the first step. This idea seems to have a potential for further applications.

(ii) Because of the way we use the polynomial partitioning technique, we need induction to handle incidences within the cells of the second partition. One of the nontrivial achievements of our technique is the ability to retain the "planar" term $O(m^{2/3}n^{1/3}s^{1/3})$ in the bound in (2) through the inductive process. Without such care, this term does not "pass well" through the induction, which has been a sore issue in several recent works on related problems (see [8, 22–24]). This is one of the main reasons for using two separate partitioning steps.

Background. Incidence problems have been a major topic in combinatorial and computational geometry for the past 35 years, starting with the aforementioned Szemerédi-Trotter bound [29] back in 1983. Several techniques, interesting in their own right, have been developed, or adapted, for the analysis of incidences, including the crossing-lemma technique of Székely [28], and the use of cuttings as a divide-and-conquer mechanism (e.g., see [3]). Connections with range searching and related algorithmic problems in computational geometry have also been noted, and studies of the Kakeya problem (see, e.g., [30]) indicate the connection between this problem and incidence problems. See Pach and Sharir [17] for a comprehensive (albeit a bit outdated) survey of the topic.

The landscape of incidence geometry has dramatically changed in the past six years, due to the infusion, in two groundbreaking papers by Guth and Katz [9, 10], of new tools and techniques drawn from algebraic geometry. Although their two direct goals have been to obtain a tight upper bound on the number of joints in a set of lines in three dimensions [9], and a near-linear lower bound for the classical distinct distances problem of Erdős [10], the new tools have quickly been recognized as useful for incidence bounds. See [5, 13, 14, 23, 27, 33, 34] for a sample of recent works on incidence problems that use the new algebraic machinery.

The simplest instances of incidence problems involve points and lines, tackled by Szemerédi and Trotter in the plane [29], and by Guth and Katz in three dimensions [10]. Other recent studies on incidence problems include incidences between points and lines in four dimensions (Sharir and Solomon [24, 25]), and incidences between points and circles in three dimensions (Sharir, Sheffer and Zahl [23]), not to mention incidences with higher-dimensional surfaces, such as in [1, 13, 27, 33, 34]. In a paper (with Sheffer) [22], we study the general case of incidences between points and curves in any dimension, and derive reasonably sharp bounds (albeit weaker in several respects than the one derived here).

That tools from algebraic geometry form the major key for successful solution of difficult problems in combinatorial geometry, came as a big surprise to the community. It has lead to intensive research of the new tools, aiming to extend them and to find new applications. A major purpose of this study, as well as of Guth [8], is to show that one can still tackle successfully the problems using simpler algebraic machinery. This offers a new, simplified, and more elementary approach, which we expect to prove potent for other applications too, such as those just mentioned. Looking for simpler, yet effective techniques that would be easier to extend to more involved contexts (such as incidences in higher dimensions) has been our main motivation for this study.

A more detailed supplementary discussion (which would be premature at this point) of the merits and other issues related to our technique is given in a concluding section.

2 Proof of Theorem 2

The proof proceeds by induction on m. As already mentioned, the bound in (2) is qualitatively different in the two ranges $m \le n^{3/2}$ and $m \ge n^{3/2}$. The analysis bifurcates accordingly. While the general flow is fairly similar in both cases, there are many differences too.

The case $m < n^{3/2}$. We partition this range into a sequence of ranges $m \le n^{\alpha_0}$, $n^{\alpha_0} < m \le n^{\alpha_1}, \ldots$, where $\alpha_0 = 1/2$ and the sequence $\{\alpha_j\}_{j \ge 0}$ is increasing and converges to 3/2. More precisely, as our analysis will show, we can take $\alpha_j = \frac{3}{2} - \frac{2}{j+2}$, for $j \ge 0$. The induction is actually on the index j of the range $n^{\alpha_{j-1}} < m \le n^{\alpha_j}$, and establishes (2) for m in this range, with a coefficient A_j (written in (2, 3) as $A_{m,n}$) that increases with j (concretely, $A_j = O(b^j)$, for a suitable constant b). This paradigm has already been used in Sharir et al. [23] and in Zahl [34], for related incidence problems, albeit in a somewhat less effective manner; see the discussion at the end of the paper.

The base range of the induction is $m \le \sqrt{n}$, where the trivial general upper bound on point-line incidences, in any dimension,[2] is $I = O(m^2 + n) = O(n)$, so (2) holds for a sufficiently large choice of the initial constant A_0.

[2]The number of lines that are incident to a fixed point and contain other points too is at most $m - 1$, for a total of at most $m(m - 1)$ incidences. The number of incidences with lines containing just one point is at most n.

Assume then that (2) holds for all $m \le n^{\alpha_{j-1}}$ for some $j \ge 1$, and consider an instance of the problem with $n^{\alpha_{j-1}} < m \le n^{3/2}$ (the analysis will force us to constrain this upper bound in order to complete the induction step, thereby obtaining the next exponent α_j).

Fix a parameter r, whose precise value will be chosen later (in fact, and this is a major novelty of our approach, there will be two different choices for r—see below), and apply the polynomial partitioning theorem of Guth and Katz (see [10] and [14, Theorem 2.6]), to obtain an r-partitioning trivariate (real) polynomial f of degree $D = O(r^{1/3})$. That is, every connected component of $\mathbb{R}^3 \setminus Z(f)$ contains at most m/r points of P, where $Z(f)$ denotes the zero set of f. By Warren's theorem [32] (see also [14]), the number of components of $\mathbb{R}^3 \setminus Z(f)$ is $O(D^3) = O(r)$.

Set $P_1 := P \cap Z(f)$ and $P_1' := P \setminus P_1$. A major recurring theme in this approach is that, although the points of P_1' are more or less evenly partitioned among the cells of the partition, no nontrivial bound can be provided for the size of P_1; in the worst case, all the points of P could lie in $Z(f)$. Each line $\ell \in L$ is either fully contained in $Z(f)$ or intersects it in at most D points (since the restriction of f to ℓ is a univariate polynomial of degree at most D). Let L_1 denote the subset of lines of L that are fully contained in $Z(f)$ and put $L_1' = L \setminus L_1$. We then have

$$I(P, L) = I(P_1, L_1) + I(P_1, L_1') + I(P_1', L_1').$$

We first bound $I(P_1, L_1')$ and $I(P_1', L_1')$. As already observed, we have

$$I(P_1, L_1') \le |L_1'| \cdot D \le nD.$$

We estimate $I(P_1', L_1')$ as follows. For each (open) cell τ of $\mathbb{R}^3 \setminus Z(f)$, put $P_\tau = P \cap \tau$ (that is, $P_1' \cap \tau$), and let L_τ denote the set of the lines of L_1' that cross τ; put $m_\tau = |P_\tau| \le m/r$, and $n_\tau = |L_\tau|$. Since every line $\ell \in L_1'$ crosses at most $1 + D$ components of $\mathbb{R}^3 \setminus Z(f)$, we have

$$\sum_\tau n_\tau \le n(1 + D), \qquad \text{and} \qquad I(P_1', L_1') = \sum_\tau I(P_\tau, L_\tau).$$

For each τ we use the trivial bound $I(P_\tau, L_\tau) = O(m_\tau^2 + n_\tau)$. Summing over the cells, we get

$$\begin{aligned} I(P_1', L_1') = \sum_\tau I(P_\tau, L_\tau) &= O\left(r \cdot (m/r)^2 + \sum_\tau n_\tau\right) \\ &= O\left(m^2/r + nD\right) = O(m^2/D^3 + nD). \end{aligned}$$

For the initial value of D, we take $r = m^{3/2}/n^{3/4}$, making D proportional to $m^{1/2}/n^{1/4}$, note that $1 \le r \le m$ because $n^{1/2} \le m \le n^{3/2}$, and get the bound

$$I(P_1', L_1') + I(P_1, L_1') = O(m^{1/2}n^{3/4}).$$

This choice of D is the one made in [10]. It is sufficiently large to control the situation in the cells, by the bound just obtained, but requires heavy-duty machinery from algebraic geometry to handle the situation on $Z(f)$.

We now turn to $Z(f)$, where we need to estimate $I(P_1, L_1)$. Since all the incidences involving any point in P_1' and/or any line in L_1' have already been accounted for, we discard these sets, and remain with P_1 and L_1 only. We "forget" the preceding polynomial partitioning step, and start afresh, applying a new polynomial partitioning to P_1 with a polynomial g of degree E, which will typically be much smaller than D, but still non-constant.

Before doing this, we note that the set of lines L_1 has a special structure, because all its lines lie on the algebraic surface $Z(f)$, which has degree D. We exploit this to derive the following lemmas. We emphasize, since this will be important later on in the analysis, that Lemmas 3–7 hold for any choice of (r and) D.

We note that in general the partitioning polynomial f may be reducible, and apply some of the following arguments to each irreducible factor separately, and, with a slight abuse of notation, denote the relevant irreducible factor by f. Clearly, there are at most D such factors.

Lemma 3 *Let π be a plane which is not a component of $Z(f)$. Then π contains at most D lines of L_1.*

Proof Suppose to the contrary that π contains at least $D+1$ lines of L. Every generic line λ in π intersects these lines in at least $D+1$ distinct points, all belonging to $Z(f)$. Hence f must vanish identically on λ, and it follows that $f \equiv 0$ on π, so π is a component of $Z(f)$, contrary to assumption. □

Lemma 4 *The number of incidences between the points of P_1 that lie in the planar components of $Z(f)$ and the lines of L_1, is $O(m^{2/3}n^{1/3}s^{1/3} + nD + m)$.*

Proof Clearly, f can have at most D linear factors, and thus $Z(f)$ can contain at most D planar components. Enumerate them as $\pi_1, \ldots, \pi_k$, where $k \le D$. Let $\tilde{P}_1$ denote the subset of the points of P_1 that lie in these planar components. Assign each point of $\tilde{P}_1$ to the first plane π_i, in this order, that contains it, and assign each line of L_1 to the first plane that fully contains it; some lines might not be assigned at all in this manner. For $i = 1, \ldots, k$, let $\tilde{P}_i$ denote the set of points assigned to π_i, and let $\tilde{L}_i$ denote the set of lines assigned to π_i. Put $m_i = |\tilde{P}_i|$ and $n_i = |\tilde{L}_i|$. Then $\sum_i m_i \le m$ and $\sum_i n_i \le n$; by assumption, we also have $n_i \le s$ for each i. Then

$$I(\tilde{P}_i, \tilde{L}_i) = O(m_i^{2/3}n_i^{2/3} + m_i + n_i) = O(m_i^{2/3}n_i^{1/3}s^{1/3} + m_i + n_i).$$

Summing over the k planes, we get, using Hölder's inequality,

$$\sum_i I(\tilde{P}_i, \tilde{L}_i) = \sum_i O(m_i^{2/3} n_i^{1/3} s^{1/3} + m_i + n_i)$$
$$= O\left(\left(\sum_i m_i\right)^{2/3} \left(\sum_i n_i\right)^{1/3} s^{1/3} + m + n \right) = O\left(m^{2/3} n^{1/3} s^{1/3} + m + n\right).$$

We also need to include incidences between points $p \in \tilde{P}_1$ and lines $\ell \in L_1$ not assigned to the same plane as p (or not assigned to any plane at all). Any such incidence (p, ℓ) can be charged (uniquely) to the intersection point of ℓ with the plane π_i to which p has been assigned. The number of such intersections is $O(nD)$, and the lemma follows. □

Lemma 5 *Each point $p \in Z(f)$ is incident to at most D^2 lines of L_1, unless $Z(f)$ has an irreducible component that is either a plane containing p or a cone[3] with apex p.*

Proof Fix any line ℓ that passes through p, and write its parametric equation as $\{p + tv \mid t \in \mathbb{R}\}$, where v is the direction of ℓ. Consider the Taylor expansion of f at p along ℓ

$$f(p + tv) = \sum_{i=1}^{D} \frac{1}{i!} F_i(p; v) t^i,$$

where $F_i(p; v)$ is the i-th order derivative of f at p in direction v; it is a homogeneous polynomial in v (p is considered fixed) of degree i, for $i = 1, \ldots, D$ (for example, $F_1(p; v) = \nabla f(p) \cdot v$). For each line $\ell \in L_1$ that passes through p, f vanishes identically on ℓ, so we have $F_i(p; v) = 0$ for each i. Assuming that p is incident to more than D^2 lines of L_1, we conclude that the homogeneous system

$$F_1(p; v) = F_2(p; v) = \cdots = F_D(p; v) = 0 \tag{4}$$

has more than D^2 (projectively distinct) roots in the variables v (regarding p as fixed). The classical Bézout's theorem, applied in the projective plane where the directions v are represented (e.g., see [4]), asserts that, since all these polynomials are of degree at most D, each pair of polynomials $F_i(p; v)$, $F_j(p; v)$ must have a common factor. The following slightly more involved inductive argument shows that in fact all these polynomials must have a common factor.[4]

Lemma 6 *Let $f_1, \ldots, f_n \in \mathbb{C}[x, y, z]$ be n homogeneous polynomials of degree at most D. If $|Z(f_1, \ldots, f_n)| > D^2$, then all the f_i's have a nontrivial common factor.*

---

[3]A cone with apex p is an algebraic surface, such that all the lines connecting p to other points on the surface are fully contained in the surface.

[4]See also [19] for a similar observation.

Proof The proof is via induction on n. The case $n = 2$ follows by the classical Bézout's theorem in the projective plane. Assume that the inductive claim holds for $n - 1$ polynomials. By assumption, $|Z(f_1, \ldots, f_{n-1})| \geq |Z(f_1, \ldots, f_n)| > D^2$, so the induction hypothesis implies that there is a polynomial g that divides f_i, for $i = 1, \ldots, n-1$; assume, as we may, that $g = GCD(f_1, \ldots, f_{n-1})$. If there are more than $\deg(g)\deg(f_n)$ points in $Z(g, f_n)$, then again, by the classical Bézout's theorem in the projective plane, g and f_n have a nontrivial common factor, which is then also a common factor of f_i, for $i = 1, \ldots, n$, completing the proof. Otherwise, put $\tilde{f}_i = f_i/g$, for $i = 1, \ldots, n-1$. Notice that $Z(f_1, \ldots, f_{n-1}) = Z(\tilde{f}_1, \ldots, \tilde{f}_{n-1}) \cup Z(g)$, implying that each point of $Z(f_1, \ldots, f_n)$ belongs either to $Z(g) \cap Z(f_n)$ or to $Z(\tilde{f}_1, \ldots, \tilde{f}_{n-1}) \cap Z(f_n)$. As $|Z(f_1, \ldots, f_n)| > D^2$ and $|Z(g, f_n)| \leq \deg(g)\deg(f_n) \leq \deg(g)D$, it follows that

$$|Z(\tilde{f}_1, \ldots, \tilde{f}_{n-1})| \geq |Z(\tilde{f}_1, \ldots, \tilde{f}_{n-1}, f_n)| \geq (D - \deg(g))D > (D - \deg(g))^2.$$

Hence, applying the induction hypothesis to the polynomials $\tilde{f}_1, \ldots, \tilde{f}_{n-1}$ (all of degree at most $D - \deg(g)$), we conclude that they have a nontrivial common factor, contradicting the fact that g is the greatest common divisor of $f_1, \ldots, f_{n-1}$. □

Continuing with the proof of Lemma 5, there is an infinity of directions v that satisfy (4), so there is an infinity of lines passing through p and contained in $Z(f)$. The union of these lines can be shown to be a two-dimensional algebraic variety,[5] contained in $Z(f)$, so $Z(f)$ has an irreducible component that is either a plane through p or a cone with apex p, as claimed. □

Lemma 7 *The number of incidences between the points of P_1 that lie in the (non-planar) conical components of $Z(f)$, and the lines of L_1, is $O(m + nD)$.*

Proof Let σ be such an (irreducible) conical component of $Z(f)$ and let p be its apex. We observe that σ cannot contain any line that is not incident to p, because such a line would span with p a plane contained in σ, contradicting the assumption that σ is irreducible and non-planar. It follows that the number of incidences between $P_\sigma := P_1 \cap \sigma$ and L_σ, consisting of the lines of L_1 contained in σ, is thus $O(|P_\sigma| + |L_\sigma|)$ (p contributes $|L_\sigma|$ incidences, and every other point at most one incidence). Applying a similar "first-come-first-serve" assignment of points and lines to the conical components of $Z(f)$, as we did for the planar components in the proof of Lemma 4, and adding the bound $O(nD)$ on the number of incidences between points and lines not assigned to the same component, we obtain the bound asserted in the lemma. □

Remark Note that in both Lemmas 4 and 7, we bound the number of incidences between points on planar or conical components of $Z(f)$ and *all* the lines of L_1.

---

[5]It is simply the variety given by the Eq. (4), rewritten as $F_1(p; x-p) = F_2(p; x-p) = \cdots = F_D(p; x-p) = 0$. It is two-dimensional because it is contained in $Z(f)$, hence at most two-dimensional, and it cannot be one-dimensional since it would then consist of only finitely many lines (see, e.g., [25, Lemma 2.3]).

Pruning. To continue, we remove all the points of P_1 that lie in some planar or conical component of $Z(f)$, and all the lines of L_1 that are fully contained in such components. With the choice of $D = m^{1/2}/n^{1/4}$, we lose in the process

$$O(m^{2/3}n^{1/3}s^{1/3} + m + nD) = O(m^{1/2}n^{3/4} + m^{2/3}n^{1/3}s^{1/3})$$

incidences (recall that the term m is subsumed by the term $m^{1/2}n^{3/4}$ for $m < n^{3/2}$). Continue, for simplicity of notation, to denote the sets of remaining points and lines as P_1 and L_1, respectively, and their sizes as m and n. Now each point is incident to at most D^2 lines (a fact that we will not use for this value of D), and no plane contains more than D lines of L_1, a crucial property for the next steps of the analysis. That is, this allows us to replace the input parameter s, bounding the maximum number of coplanar lines, by D; this is a key step that makes the induction work.

A new polynomial partitioning. We now return to the promised step of constructing a new polynomial partitioning. We adapt the preceding notation, with a few modifications. We choose a degree E, typically much smaller than D, and construct a partitioning polynomial g of degree E for P_1. With an appropriate value of $r = \Theta(E^3)$, we obtain $O(r)$ open cells, each containing at most m/r points of P_1, and each line of L_1 either crosses at most $E + 1$ cells, or is fully contained in $Z(g)$.

Set $P_2 := P_1 \cap Z(g)$ and $P_2' := P_1 \setminus P_2$. Similarly, denote by L_2 the set of lines of L_1 that are fully contained in $Z(g)$, and put $L_2' := L_1 \setminus L_2$. We first dispose of incidences involving the lines of L_2. (That is, now we first focus on incidences within $Z(g)$, and only then turn to look at the cells.) By Lemmas 4 and 7, the number of incidences involving points of P_2 that lie in some planar or conical component of $Z(g)$, and all the lines of L_2, is

$$O(m^{2/3}n^{1/3}s^{1/3} + m + nE) = O(m^{1/2}n^{3/4} + m^{2/3}n^{1/3}s^{1/3});$$

this is because both terms m and $nE < nD$ are at most $m^{1/2}n^{3/4}$. (For $E \ll D$, this might be a gross overestimation, but we do not care.) We remove these points from P_2, and remove all the lines of L_2 that are contained in such components; continue to denote the sets of remaining points and lines as P_2 and L_2. Now each point is incident to at most E^2 lines of L_2 (Lemma 5), so the number of remaining incidences involving points of P_2 is $O(mE^2)$; for E suitably small, this bound will be subsumed by $O(m^{1/2}n^{3/4})$. Adding the bound nE for $I(P_2, L_2')$, obtained as above, we get

$$I(P_2, L_2) + I(P_2, L_2') = O(m^{1/2}n^{3/4} + m^{2/3}n^{1/3}s^{1/3}). \tag{5}$$

Unlike the case of a "large" D, namely, $D = m^{1/2}/n^{1/4}$, here the difficult part is to treat incidences within the cells of the partition. Since $E \ll D$, we cannot use the naive bound $O(m^2 + n)$ within each cell, because that would make the overall bound too large. Therefore, to control the incidence bound within the cells, we proceed in the following inductive manner.

For each cell τ of $\mathbb{R}^3 \setminus Z(g)$, put $P_\tau := P_2' \cap \tau$, and let L_τ denote the set of the lines of L_2' that cross τ; put $m_\tau = |P_\tau| \le m/r$, and $n_\tau = |L_\tau|$. Since every line $\ell \in L_1$ (that is, of L_2') crosses at most $1 + E$ components of $\mathbb{R}^3 \setminus Z(g)$, we have $\sum_\tau n_\tau \le n(1 + E)$.

It is important to note that at this point of the analysis the sizes of P_1 and of L_1 might be smaller than the original respective values m and n. In particular, we may no longer assume that $|P_1| > |L_1|^{\alpha_{j-1}}$, as we did assume for m and n. Nevertheless, in what follows m and n will denote the original values, which serve as upper bounds for the respective actual sizes of P_1 and L_1, and the induction will work correctly with these values; see below for details.

In order to apply the induction hypothesis within the cells of the partition, we want to assume that $m_\tau \le {n_\tau}^{\alpha_{j-1}}$ for each τ. To ensure that, we require that the number of lines of L_2' that cross a cell be at most n/E^2. Cells τ that are crossed by $\kappa n/E^2$ lines, for $\kappa > 1$, are treated as if they occur $\lceil \kappa \rceil$ times, where each incarnation involves all the points of P_τ, and at most n/E^2 different lines of L_τ. The number of subproblems is now

$$\sum_\tau \left\lceil \frac{n_\tau}{n/E^2} \right\rceil \le \frac{\sum_\tau n_\tau}{n/E^2} + \sum_\tau 1 \le \frac{n(1+E)}{n/E^2} + O(E^3) = O(E^3).$$

Arguing similarly, we may also assume that $m_\tau \le m/E^3$ for each cell τ (by "duplicating" each cell into a constant number of subproblems, if needed).

We therefore require that $\frac{m}{E^3} \le \left(\frac{n}{E^2}\right)^{\alpha_{j-1}}$. (Note that, as already commented above, these are only upper bounds on the actual sizes of the corresponding subsets, but this will have no real effect on the induction process; for example, we can add dummy points and/or lines, so as to have exactly $\frac{m}{E^3}$ points and $\frac{n}{E^2}$ lines, use induction on the padded sets, and get an upper bound that also holds for the original smaller sets.) That is, we require

$$E \ge \left(\frac{m}{n^{\alpha_{j-1}}}\right)^{1/(3-2\alpha_{j-1})}. \tag{6}$$

With these preparations, we apply the induction hypothesis within each cell τ, recalling that no plane contains more than D lines[6] of $L_2' \subseteq L_1$, and get

$$\begin{aligned} I(P_\tau, L_\tau) &\le A_{j-1}\left(m_\tau^{1/2} n_\tau^{3/4} + m_\tau\right) + B\left(m_\tau^{2/3} n_\tau^{1/3} D^{1/3} + n_\tau\right) \\ &\le A_{j-1}\left((m/E^3)^{1/2}(n/E^2)^{3/4} + m/E^3\right) \\ &+ B\left((m/E^3)^{2/3}(n/E^2)^{1/3} D^{1/3} + n/E^2\right). \end{aligned}$$

Summing these bounds over the cells τ, that is, multiplying them by $O(E^3)$, we get, for a suitable absolute constant b,

[6] This was the main reason for carrying out the first partitioning step, as already noted.

$$I(P_2', L_2') = \sum_{\mathcal{T}} I(P_{\mathcal{T}}, L_{\mathcal{T}}) \le bA_{j-1}\left(m^{1/2}n^{3/4} + m\right) + B\left(m^{2/3}n^{1/3}E^{1/3}D^{1/3} + nE\right).$$

We now require that $E = O(D)$ (we already used this requirement in the derivation of (5)). Then, as already remarked, the last term satisfies $nE = O(nD) = O(m^{1/2}n^{3/4})$, and the term m is also subsumed by the first term. The third term, after substituting $D = O(m^{1/2}/n^{1/4})$, becomes $O(m^{5/6}n^{1/4}E^{1/3})$. Hence, with a slightly larger b, we have

$$I(P_2', L_2') \le bA_{j-1}m^{1/2}n^{3/4} + bBm^{5/6}n^{1/4}E^{1/3}.$$

Adding up all the bounds, including the one in (5), and those for the portions of P and L that were discarded during the first partitioning step, we obtain, for a suitable constant c,

$$I(P, L) \le c\left(m^{1/2}n^{3/4} + m^{2/3}n^{1/3}s^{1/3} + mE^2\right) + bA_{j-1}m^{1/2}n^{3/4} + bBm^{5/6}n^{1/4}E^{1/3}.$$

We choose E to ensure that the two E-dependent terms are dominated by the term $m^{1/2}n^{3/4}$. That is,

$$m^{5/6}n^{1/4}E^{1/3} \le m^{1/2}n^{3/4}, \quad \text{or} \quad E \le n^{3/2}/m,$$
$$\text{and} \quad mE^2 \le m^{1/2}n^{3/4}, \quad \text{or} \quad E \le n^{3/8}/m^{1/4}.$$

Since $n^{3/2}/m = \left(n^{3/8}/m^{1/4}\right)^4$, and both expressions are ≥ 1, the latter condition is stricter, and we ignore the former. As already noted, we also require that $E = O(D)$; specifically, we require that $E \le m^{1/2}/n^{1/4}$.

In conclusion, recalling (6), the two constraints on the choice of E are

$$\left(\frac{m}{n^{\alpha_{j-1}}}\right)^{1/(3-2\alpha_{j-1})} \le E \le \min\left\{\frac{n^{3/8}}{m^{1/4}}, \frac{m^{1/2}}{n^{1/4}}\right\}, \tag{7}$$

and, for these constraints to be compatible, we require that

$$\left(\frac{m}{n^{\alpha_{j-1}}}\right)^{1/(3-2\alpha_{j-1})} \le \frac{n^{3/8}}{m^{1/4}}, \quad \text{or} \quad m \le n^{\frac{9+2\alpha_{j-1}}{2(7-2\alpha_{j-1})}},$$

and that

$$\left(\frac{m}{n^{\alpha_{j-1}}}\right)^{1/(3-2\alpha_{j-1})} \le \frac{m^{1/2}}{n^{1/4}},$$

which fortunately always holds, as is easily checked, since $m \le n^{3/2}$ and $\alpha_{j-1} \ge 1/2$. Note that we have not explicitly stated any concrete choice of E; any value satisfying (7) will do. We put

$$\alpha_j := \frac{9 + 2\alpha_{j-1}}{2(7 - 2\alpha_{j-1})},$$

and conclude that if $m \le n^{\alpha_j}$ then the bound asserted in the theorem holds, with $A_j = bA_{j-1} + c$ and $B = c$. This completes the induction step. Note that the recurrence $A_j = bA_{j-1} + c$ solves to $A_j = O(b^j)$.

It remains to argue that the induction covers the entire range $m = O(n^{3/2})$. Using the above recurrence for the α_j's, with $\alpha_0 = 1/2$, it easily follows that

$$\alpha_j = \frac{3}{2} - \frac{2}{j+2},$$

for each $j \ge 0$, showing that α_j converges to $3/2$, implying that the entire range $m = O(n^{3/2})$ is covered by the induction.

To calibrate the dependence of the constant of proportionality on m and n, we note that, for $n^{\alpha_{j-1}} \le m < n^{\alpha_j}$, the constant is $O(b^j)$. We have

$$\frac{3}{2} - \frac{2}{j+1} = \alpha_{j-1} \le \frac{\log m}{\log n}, \qquad \text{or} \qquad j \le \frac{\frac{1}{2} + \frac{\log m}{\log n}}{\frac{3}{2} - \frac{\log m}{\log n}} = \frac{\log(m^2 n)}{\log(n^3/m^2)}.$$

This establishes the expression for $A_{m,n}$ given in the statement of the theorem.

Handling the middle ground $m \approx n^{3/2}$. Some care is needed when m approaches $n^{3/2}$, because of the potentially unbounded growth of the constant A_j. To handle this situation, we simply fix a value j, in the manner detailed below, write $m = kn^{\alpha_j}$, solve k separate problems, each involving $m/k = n^{\alpha_j}$ points of P and all the n lines of L, and sum up the resulting incidence bounds. We then get

$$\begin{aligned} I(P, L) &\le akb^j \left((m/k)^{1/2} n^{3/4} + (m/k)\right) + kB\left((m/k)^{2/3} n^{1/3} s^{1/3} + n\right) \\ &= ak^{1/2} b^j m^{1/2} n^{3/4} + ab^j m + k^{1/3} B m^{2/3} n^{1/3} s^{1/3} + kBn, \end{aligned}$$

for a suitable absolute constant a. Recalling that $\alpha_j = \frac{3}{2} - \frac{2}{j+2}$, we have

$$k \le m/n^{\alpha_j} \le n^{3/2}/n^{\alpha_j} = n^{2/(j+2)}.$$

Hence the coefficient of the leading term in the above bound is bounded by $an^{1/(j+2)} b^j$, and we (asymptotically) minimize this expression by choosing

$$j = j_0 := \sqrt{\log n}/\sqrt{\log b}.$$

With this choice all the other coefficients are also dominated by the leading coefficient, and we obtain

$$I(P, L) = O\left(2^{2\sqrt{\log b}\sqrt{\log n}} \left(m^{1/2} n^{3/4} + m^{2/3} n^{1/3} s^{1/3} + m + n\right)\right). \qquad (8)$$

In other words, the bound in (2) and (3) holds for any $m \le n^{3/2}$, but, for $m \ge n^{\alpha_{j_0}}$ one should use instead the bound in (8), which controls the exponential growth of the constants of proportionality within this range (which is very close to $n^{3/2}$, as is easily checked).

The case $m > n^{3/2}$. The analysis of this case is, in a sense, a mirror image of the preceding analysis, except for a new key lemma (Lemma 8). For the sake of completeness, we repeat a sizeable portion of the analysis, providing many of the relevant (often differing) details.

We partition this range into a sequence of ranges $m \ge n^{\alpha_0}$, $n^{\alpha_1} \le m < n^{\alpha_0}, \ldots,$ where $\alpha_0 = 2$ and the sequence $\{\alpha_j\}_{j\ge 0}$ is decreasing and converges to 3/2. The induction is on the index j of the range $n^{\alpha_j} \le m < n^{\alpha_{j-1}}$, and establishes (2) for m in this range, with a coefficient A_j (written in (2), (3) as $A_{m,n}$) that increases with j.

The base range of the induction is $m \ge n^2$, where the trivial general upper bound on point-line incidences in any dimension, dual to the one used in the previous case, yields $I = O(n^2 + m) = O(m)$, so (2) holds for a sufficiently large choice of the initial constant A_0.

Assume then that (2) holds for all $m \ge n^{\alpha_{j-1}}$ for some $j \ge 1$, and consider an instance of the problem with $n^{3/2} \le m < n^{\alpha_{j-1}}$ (again, the lower bound will increase, to n^{α_j}, to facilitate the induction step).

For a parameter r, to be specified later, apply the polynomial partition theorem to obtain an r-partitioning trivariate (real) polynomial f of degree $D = O(r^{1/3})$. That is, every connected component of $\mathbb{R}^3 \setminus Z(f)$ contains at most m/r points of P, and the number of components of $\mathbb{R}^3 \setminus Z(f)$ is $O(D^3) = O(r)$.

Set $P_1 := P \cap Z(f)$ and $P_1' := P \setminus P_1$. Each line $\ell \in L$ is either fully contained in $Z(f)$ or intersects it in at most D points. Let L_1 denote the subset of lines of L that are fully contained in $Z(f)$ and put $L_1' = L \setminus L_1$. As before, we have

$$I(P, L) = I(P_1, L_1) + I(P_1, L_1') + I(P_1', L_1').$$

We have

$$I(P_1, L_1') \le |L_1'| \cdot D \le nD,$$

and we estimate $I(P_1', L_1')$ as follows. For each cell τ of $\mathbb{R}^3 \setminus Z(f)$, put $P_\tau = P \cap \tau$ (that is, $P_1' \cap \tau$), and let L_τ denote the set of the lines of L_1' that cross τ; put $m_\tau = |P_\tau| \le m/r$, and $n_\tau = |L_\tau|$. As before, we have $\sum_\tau n_\tau \le n(1 + D)$, so the average number of lines that cross a cell is $O(n/D^2)$. Arguing as above, we may assume, by possibly increasing the number of cells by a constant factor, that each n_τ is at most n/D^2. Clearly, we have

$$I(P_1', L_1') = \sum_\tau I(P_\tau, L_\tau).$$

For each τ we use the trivial dual bound, mentioned above, $I(P_\tau, L_\tau) = O(n_\tau^2 + m_\tau)$. Summing over the cells, we get

$$I(P_1', L_1') = \sum_{\tau} I(P_\tau, L_\tau) = O\left(D^3 \cdot (n/D^2)^2 + m\right) = O\left(n^2/D + m\right).$$

For the initial value of D, we take $D = n^2/m$, noting that $1 \le D^3 \le m$ because $n^{3/2} \le m \le n^2$, and get the bound

$$I(P_1', L_1') + I(P_1, L_1') = O(n^2/D + m + nD) = O(m + n^3/m) = O(m),$$

where the latter bound follows since $m \ge n^{3/2}$.

It remains to estimate $I(P_1, L_1)$. Since all the incidences involving any point in P_1' and/or any line in L_1' have been accounted for, we discard these sets, and remain with P_1 and L_1 only. As before, we forget the preceding polynomial partitioning step, and start afresh, applying a new polynomial partitioning to P_1 with a polynomial g of degree E, which will typically be much smaller than D, but still non-constant.

For this case we need the following lemma, which can be regarded, in some sense, as a dual (albeit somewhat more involved) version of Lemma 5. Unlike the rest of the analysis, the best way to prove this lemma is by switching to the complex projective setting. This is needed for one key step in the proof, where we need the property that the projection of a complex projective variety is a variety. Once this is done, we can switch back to the real affine case, and complete the proof.

Here is a very quick review of the transition to the complex projective setup. A real affine algebraic variety X, defined by a collection of real polynomials, can also be regarded as a complex projective variety, in the sense that one needs to take the *projective closure* of the *complexification* of X; details about these standard operations can be found, e.g., in Bochnak et al. [2, Proposition 8.7.17] and in Cox et al. [4, Definition 8.4.6].) If f is an irreducible polynomial over $\mathbb{R}$, it might still be reducible over $\mathbb{C}$, but then it must have the form $f = g\bar{g}$, where g is an irreducible complex polynomial and $\bar{g}$ is its complex conjugate. (Indeed, if h is any irreducible factor of f, then $\bar{h}$ is also an irreducible factor of f, and therefore $h\bar{h}$ is a real polynomial dividing f. As f is irreducible over $\mathbb{R}$, the claim follows.)

In the following lemma, adapting a notation used in earlier works, we say that a point $p \in P_1$ is 1-*poor* (resp., 2-*rich*) if it is incident to at most one line (resp., to at least two lines) of L_1.

Recall also that a (real) *regulus* is a doubly-ruled surface in $\mathbb{R}^3$. It is the union of all lines that pass through three fixed pairwise skew lines; it is a quadric, which is either a hyperbolic paraboloid or a one-sheeted hyperboloid. A complex regulus is the complexification of a real regulus.

Lemma 8 *Let f be an irreducible polynomial in $\mathbb{C}[x, y, z]$ of degree D, such that $Z(f)$ is not a complex plane nor a complex regulus, and let L_1 be a finite set of real lines fully contained in $Z(f)$. Then, with the possible exception of at most two lines, each line $\ell \in L_1$ is incident to at most $O(D^3)$ 2-rich points.*

Proof The strategy of the proof is to charge each incidence of ℓ with some 2-rich point p to an intersection of ℓ with another line of L_1 that passes through p, and to argue that, in general, there can be only $O(D^3)$ such other lines. This in turn will be shown by arguing that the union of all the lines that are fully contained in $Z(f)$ and pass through ℓ is a one-dimensional variety, of degree $O(D^3)$, from which the claim will follow. As we will show, this will indeed be the case except when ℓ is one of at most two "exceptional" lines on $Z(f)$.

Fix a line ℓ as in the lemma, assume for simplicity that it passes through the origin, and write it as $\{tv_0 \mid t \in \mathbb{C}\}$; since ℓ is a real line, v_0 can be assumed to be real. Consider the union $V(\ell)$ of all the lines that are fully contained in $Z(f)$ and are incident to ℓ; that is, $V(\ell)$ is the union of ℓ with the set of all points $p \in Z(f) \setminus \ell$ for which there exists $t \in \mathbb{C}$ such that the line connecting p to $tv_0 \in \ell$ is fully contained in $Z(f)$. In other words, for such a t and for each $s \in \mathbb{C}$, we have $f((1-s)p + stv_0) = 0$. Regarding the left-hand side as a polynomial in s, we can write it as $\sum_{i=0}^{D} G_i(p;t)s^i \equiv 0$, for suitable (complex) polynomials $G_i(p;t)$ in p and t, each of total degree at most D. In other words, p and t have to satisfy the system

$$G_0(p;t) = G_1(p;t) = \cdots = G_D(p;t) = 0, \qquad (9)$$

which defines an algebraic variety $\sigma(\ell)$ in $\mathbb{P}^4(\mathbb{C})$. Note that, substituting $s = 0$, we have $G_0(p;t) \equiv f(p)$, and that the limit points (tv_0, t) (corresponding to points on ℓ) also satisfy this system, since in this case $f((1-s)tv_0 + stv_0) = f(tv_0) = 0$ for all s.

In other words, $V(\ell)$ is the projection of $\sigma(\ell)$ into $\mathbb{P}^3(\mathbb{C})$, given (in affine coordinates) by $(p,t) \mapsto p$. For each $p \in Z(f) \setminus \ell$ this system has only finitely many solutions in t, for otherwise the plane spanned by p and ℓ_0 would be fully contained in $Z(f)$, contrary to our assumption.

By the projective extension theorem (see, e.g., [4, Theorem 8.5.6]), the projection of $\sigma(\ell)$ into $\mathbb{P}^3(\mathbb{C})$, in which t is discarded, is an algebraic variety $\tau(\ell)$. We observe that $\tau(\ell)$ is contained in $Z(f)$, and is therefore of dimension at most two.

Assume first that $\tau(\ell)$ is two-dimensional. As f is irreducible over $\mathbb{C}$, we must have $\tau(\ell) = Z(f)$. This implies that each point $p \in Z(f) \setminus \ell$ is incident to a (complex) line that is fully contained in $Z(f)$ and is incident to ℓ. In particular, $Z(f)$ is ruled by complex lines.

By assumption, $Z(f)$ is neither a complex plane nor a complex regulus. We may also assume that $Z(f)$ is not a complex cone, for then each line in L_1 is incident to at most one 2-rich point (namely, the apex of $Z(f)$), making the assertion of the lemma trivial. It then follows that $Z(f)$ is an irreducible singly-ruled (complex) surface. As argued in Guth and Katz [10] (see also our companion paper [26] for an independent analysis of this situation, which caters more explicitly to the complex setting too), $Z(f)$ can contain at most two lines ℓ with this property. Informally, the proof proceeds by showing that if ℓ is a line with this property, then almost all points on $Z(f)$ are incident to a line that connects them to ℓ and is fully contained in $Z(f)$. With some extra care, one shows that if $Z(f)$ contains three lines with this property then almost all points on $Z(f)$ have an incident line that connects them to all three exceptional lines and is fully contained in $Z(f)$, and thus $Z(f)$ is a regulus. A simple, elementary proof that a *real* surface that contains three lines with this property is a regulus is given in Fuchs and Tabachnikov [7]. The arguments for the complex case are more involved; they are only sketched in [10] and are spelled out in more detail in [26].

Excluding these (at most) two exceptional lines ℓ, we may thus assume that $\tau(\ell)$ is (at most) a one-dimensional curve.

Clearly, by definition, each point $(p, t) \in \sigma(\ell)$, except for $p \in \ell$, defines a line λ, in the original 3-space, that connects p to tv_0, and each point $q \in \lambda$ satisfies $(q, t) \in \sigma(\ell)$. Hence, the line $\{(q, t) \mid q \in \lambda\}$ is fully contained in $\sigma(\ell)$, and therefore the line λ is fully contained in $\tau(\ell)$. Since $\tau(\ell)$ is one-dimensional, this in turn implies that $\tau(\ell)$ is a *finite* union of (complex) lines, whose number is at most $\deg(\tau(\ell))$ (see, e.g., [25, Lemma 2.3]). This also implies that $\sigma(\ell)$ is the union of the same number of lines, and in particular $\sigma(\ell)$ is also one-dimensional, and the number of lines that it contains is at most $\deg(\sigma(\ell))$.

We claim that this latter degree is at most $O(D^3)$. This follows from a well-known result in algebra (see, e.g., Schmid [21, Lemma 2.2]), that asserts that, since $\sigma(\ell)$ is a one-dimensional curve in $\mathbb{P}^4(\mathbb{C})$, and is the common zero set of polynomials, each of degree $O(D)$, its degree is $O(D^3)$.

This completes the proof of the lemma. (The passage from the complex projective setting back to the real affine one is trivial for this property.) □

Remark Using some more sophisticated machinery from differential and algebraic geometry, one can improve the bound in the lemma to $O(D^2)$. Specifically, the Cayley–Salmon theorem, as used in Guth and Katz [10], implies that a surface in $\mathbb{R}^3$ of degree D cannot contain more than $11D^2 - 24D$ lines, unless it is *ruled by lines*. In particular, if $Z(f)$ is not a ruled surface, a line of L can intersect at most $11D^2 - 24D = O(D^2)$ other lines. When $Z(f)$ is a ruled surface, that is not a complex plane or a complex regulus, it is a singly ruled surface. As shown by Salmon [20, Art. 485] (see also the proof of [10, Lemma 3.4]), in a singly ruled surface, except for the two exceptional lines, a line can intersect at most $O(D)$ other lines. These observations yield the improved bound. We have presented the coarser bound in the lemma because its proof is considerably simpler, and the improvement does not have any real effect on the asymptotic bound that we derive. See an additional discussion of this issue in the concluding section.

Corollary 9 *Let f be a real or complex trivariate polynomial of degree D, such that (the complexification of) $Z(f)$ does not contain any complex plane nor any complex regulus. Let L_1 be a set of n lines fully contained in $Z(f)$, and let P_1 be a set of m points contained in $Z(f)$. Then $I(P_1, L_1) = O(m + nD^3)$.*

Proof Write $f = \prod_{i=1}^{s} f_i$ for its decomposition into irreducible factors, for $s \le D$. We apply Lemma 8 to each complex factor f_i of f. By the observation preceding Lemma 8, some of these factors might be complex (non-real) polynomials, even when f is real. That is, regardless of whether the original f is real or not, we carry out the analysis in the complex projective space $\mathbb{P}^3(\mathbb{C})$, and regard $Z(f_i)$ as a variety in that space.

Note also that, by focussing on the single irreducible component $Z(f_i)$ of $Z(f)$, we consider only points and lines that are fully contained in $Z(f_i)$. We thus shrink P_1 and L_1 accordingly, and note that the notions of being 2-rich or 1-poor are now redefined with respect to the reduced sets. All of this is rectified as follows.

Assign each line $\ell \in L_1$ to the first component $Z(f_i)$, in the above order, that fully contains it, and assign each point $p \in P_1$ to the first component that contains it. If a point p and a line ℓ are incident, then either they are both assigned to the same component $Z(f_i)$, or p is assigned to some component $Z(f_i)$ and ℓ, which is assigned to a later component, is not contained in $Z(f_i)$. Each incidence of the latter kind can be charged to a crossing between ℓ and $Z(f_i)$, and the total number of these crossings is $O(nD)$. It therefore suffices to consider incidences between points and lines assigned to the same component. Moreover, if a point p is 2-rich with respect to the entire collection L_1 but is 1-poor with respect to the lines assigned to its component, then all of its incidences except one are accounted by the preceding term $O(nD)$, which thus takes care also of the single incidence within $Z(f_i)$.

By Lemma 8, for each f_i, excluding at most two exceptional lines, the number of incidences between a line assigned to (and contained in) $Z(f_i)$ and the points assigned to $Z(f_i)$ that are still 2-rich within $Z(f_i)$, is $O(\deg(f_i)^3) = O(D^3)$. Summing over all relevant lines, we get the bound $O(nD^3)$.

Finally, each irreducible component $Z(f_i)$ can contain at most two exceptional lines, for a total of at most $2D$ such lines. The number of 2-rich points on each such line ℓ is at most n, since each such point is incident to another line, so the total number of corresponding incidences is at most $O(nD)$, which is subsumed by the preceding bound $O(nD^3)$. The number of incidences with 1-poor points is, trivially, at most m. This completes the proof of the corollary. □

Pruning. In the preceding lemma and corollary, we have excluded plane, regulus and conical components of $Z(f)$. Arguing as in the case of small m, the number of incidences involving points that lie on planar components of $Z(f)$ is $O(m^{2/3}n^{1/3}s^{1/3} + m)$ (see Lemma 4), and the number of incidences involving points that lie on conical components of $Z(f)$ is $O(m + nD) = O(m)$ (see Lemma 7). A similar bound holds for points on the reguli components. Specifically, we assign each point and line to a regulus component that contains them, if one exists, in the same first-come first-serve manner used above. Any point p can be incident to at most two lines that are fully contained in the regulus to which it is assigned, and any other incidence of p with a line ℓ can be uniquely charged to the intersection of ℓ with that regulus, for a total (over all lines and reguli) of $O(nD)$ incidences; in total we get $O(m + nD) = O(m)$ incidences, as claimed.

We remove all points that lie in any component of these kinds and all lines that are fully contained in any such component. With the choice of $D = n^2/m$, we lose in the process

$$O(m^{2/3}n^{1/3}s^{1/3} + m + nD) = O(m + m^{2/3}n^{1/3}s^{1/3})$$

incidences (recall that $nD \le m$ for $m \ge n^{3/2}$). For the remainder sets, which we continue to denote as P_1 and L_1, respectively, no plane contains more than $O(D)$ lines of L_1, as argued in Lemma 3.

A new polynomial partitioning. We adapt the notation used in the preceding case of $m < n^{3/2}$, with a few modifications. We choose a degree E, typically much smaller than D, and construct a partitioning polynomial g of degree E for P_1. With an appropriate value of $r = \Theta(E^3)$, we obtain $O(r)$ cells, each containing at most m/r points of P_1, and each line of L_1 either crosses at most $E + 1$ cells, or is fully contained in $Z(g)$.

Set $P_2 := P_1 \cap Z(g)$ and $P_2' := P_1 \setminus P_2$. Similarly, denote by L_2 the set of lines of L_1 that are fully contained in $Z(g)$, and put $L_2' := L_1 \setminus L_2$. We first dispose of incidences involving the lines of L_2. By Lemma 4 and the preceding arguments, the number of incidences involving points of P_2 that lie in some planar, conical, or regulus component of $Z(g)$, and all the lines of L_2, is

$$O(m^{2/3}n^{1/3}s^{1/3} + m + nE) = O(m^{2/3}n^{1/3}s^{1/3} + m),$$

since we will choose $E \le D$, and since $nD \le m$. Arguing as above, this also bounds $I(P_2, L_2')$.

We remove these points from P_2, and remove all the lines of L_2 that are contained in such components. Continue to denote the sets of remaining points and lines as P_2 and L_2. By Corollary 9, the number of incidences between (the new) P_2 and L_2 is $O(m + nE^3)$.

To complete the estimation, we need to bound the number of incidences in the cells of the partition, which we do inductively, as before. Specifically, for each cell τ of $\mathbb{R}^3 \setminus Z(g)$, put $P_\tau := P_2' \cap \tau$, and let L_τ denote the set of the lines of L_2' that cross τ; put $m_\tau = |P_\tau| \le m/r$, and $n_\tau = |L_\tau|$. Since every line $\ell \in L_0$ crosses at most $1 + E$ components of $\mathbb{R}^3 \setminus Z(g)$, we have $\sum_\tau n_\tau \le n(1 + E)$, and, arguing as above, we may assume that each n_τ is at most n/E^2, and each m_τ is at most m/E^3. To apply the induction hypothesis in each cell, we therefore require that $\frac{m}{E^3} \ge \left(\frac{n}{E^2}\right)^{\alpha_{j-1}}$. (As before, the actual sizes of P_1 and L_1 might be smaller than the respective original values m and n. We use here the original values, and note, similar to the preceding case, that the fact that these are only upper bounds on the actual sizes is harmless for the induction process.) That is, we require

$$E \ge \left(\frac{n^{\alpha_{j-1}}}{m}\right)^{1/(2\alpha_{j-1}-3)}. \tag{10}$$

With these preparations, we apply the induction hypothesis within each cell τ, recalling that no plane contains more than D lines of $L_2' \subseteq L_1$, and get

$$\begin{aligned} I(P_\tau, L_\tau) &\le A_{j-1}\left(m_\tau^{1/2} n_\tau^{3/4} + m_\tau\right) + B\left(m_\tau^{2/3} n_\tau^{1/3} D^{1/3} + n_\tau\right) \\ &\le A_{j-1}\left((m/E^3)^{1/2}(n/E^2)^{3/4} + m/E^3\right) \\ &+ B\left((m/E^3)^{2/3}(n/E^2)^{1/3} D^{1/3} + n/E^2\right). \end{aligned}$$

Summing these bounds over the cells τ, that is, multiplying them by $O(E^3)$, we get, for a suitable absolute constant b,

$$I(P_2', L_2') = \sum_\tau I(P_\tau, L_\tau) \le bA_{j-1}\left(m^{1/2}n^{3/4} + m\right) + bB\left(m^{2/3}n^{1/3}E^{1/3}D^{1/3} + nE\right).$$

Requiring that $E \le m/n$, the last term satisfies $nE \le m$, and the first term is also at most $O(m)$ (because $m \ge n^{3/2}$). The third term, after substituting $D = O(n^2/m)$, becomes $O(m^{1/3}nE^{1/3})$. Hence, with a slightly larger b, we have

$$I(P_2', L_2') \le bA_{j-1}m + bBm^{1/3}nE^{1/3}.$$

Collecting all partial bounds obtained so far, we obtain

$$I(P, L) \le c\left(m^{2/3}n^{1/3}s^{1/3} + m + nE^3\right) + bA_{j-1}m + bBm^{1/3}nE^{1/3},$$

for a suitable constant c. We choose E to ensure that the two E-dependent terms are dominated by m. That is,

$$m^{1/3}nE^{1/3} \le m, \quad \text{or} \quad E \le m^2/n^3, \qquad \text{and} \qquad nE^3 \le m, \quad \text{or} \quad E \le m^{1/3}/n^{1/3}.$$

In addition, we also require that $E \le m/n$, but, as is easily seen, both of the above constraints imply that $E \le m/n$, so we get this latter constraint for free, and ignore it in what follows.

As is easily checked, the second constraint $E \le m^{1/3}/n^{1/3}$ is stricter than the first constraint $E \le m^2/n^3$ for $m \ge n^{8/5}$, and the situation is reversed when $m \le n^{8/5}$. So in our inductive descent of m, we first consider the second constraint, and then switch to the first constraint.

Hence, in the first part of this analysis, the two constraints on the choice of E are

$$\left(\frac{n^{\alpha_{j-1}}}{m}\right)^{1/(2\alpha_{j-1}-3)} \le E \le \frac{m^{1/3}}{n^{1/3}},$$

and, for these constraints to be compatible, we require that

$$\left(\frac{n^{\alpha_{j-1}}}{m}\right)^{1/(2\alpha_{j-1}-3)} \le \frac{m^{1/3}}{n^{1/3}}, \qquad \text{or} \qquad m \ge n^{\frac{5\alpha_{j-1}-3}{2\alpha_{j-1}}}.$$

We start the process with $\alpha_0 = 2$, and take $\alpha_1 := \frac{5\alpha_0 - 3}{2\alpha_0} = 7/4$. As this is still larger than $8/5$, we perform two additional rounds of the induction, using the same constraints, leading to the exponents

$$\alpha_2 = \frac{5\alpha_1 - 3}{2\alpha_1} = \frac{23}{14}, \quad \text{and} \quad \alpha_3 = \frac{5\alpha_2 - 3}{2\alpha_2} = \frac{73}{46} < \frac{8}{5}.$$

To play it safe, we reset $\alpha_3 := 8/5$, and establish the induction step for $m \geq n^{8/5}$ in three rounds of induction. We can then proceed to the second part, where the two constraints on the choice of E are

$$\left(\frac{n^{\alpha_{j-1}}}{m}\right)^{1/(2\alpha_{j-1}-3)} \leq E \leq \frac{m^2}{n^3},$$

and, for these constraints to be compatible, we require that

$$\left(\frac{n^{\alpha_{j-1}}}{m}\right)^{1/(2\alpha_{j-1}-3)} \leq \frac{m^2}{n^3}, \quad \text{or} \quad m \geq n^{\frac{7\alpha_{j-1}-9}{4\alpha_{j-1}-5}}.$$

We define, for $j \geq 4$, $\alpha_j = \frac{7\alpha_{j-1} - 9}{4\alpha_{j-1} - 5}$. Substituting $\alpha_3 = 8/5$ we get $\alpha_4 = 11/7$, and in general a simple calculation shows that

$$\alpha_j = \frac{3}{2} + \frac{1}{4j-2},$$

for $j \geq 3$. This sequence does indeed converge to $3/2$ as $j \to \infty$, implying that the entire range $m = \Omega(n^{3/2})$ is covered by the induction.

In both parts, we conclude that if $m \geq n^{\alpha_j}$ then the bound asserted in the theorem holds with $A_j = bA_{j-1} + c$,, and $B = c$. This completes the induction step.

Finally, we calibrate the dependence of the constant of proportionality on m and n, by noting that, for $n^{\alpha_j} \leq m < n^{\alpha_{j-1}}$, the constant is $O(b^j)$. We have

$$\frac{3}{2} + \frac{1}{4j-6} = \alpha_{j-1} \geq \frac{\log m}{\log n}, \quad \text{or} \quad j \leq \frac{3\frac{\log m}{\log n} - 4}{2\frac{\log m}{\log n} - 3} = \frac{\log\left(m^3/n^4\right)}{\log\left(m^2/n^3\right)}.$$

(Technically, this only handles the range $j \geq 4$, but, for an asymptotic bound, we can extend it to $j = 1, 2, 3$ too.) This establishes the explicit expression for $A_{m,n}$ for the range $m \geq n^{3/2}$, as stated in the theorem, and completes its proof. □

Again, as in the case of a small m, we need to be careful when m approaches $n^{3/2}$. Here we can fix a j, assume that $n^{3/2} \leq m < n^{\alpha_j}$, and set $k := m/n^{\alpha'_j}$, where $\alpha'_j = 3/2 - 2/(j+2)$ is the j-th index in the hierarchy for $m \leq n^{3/2}$. That is,

$$k \le n^{\alpha_j - \alpha'_j} = \frac{1}{4j-2} + \frac{2}{j+2}.$$

As before, we now solve k separate subproblems, each with m/k points of P and all the lines of L, and sum up the resulting incidence bounds. The analysis is similar to the one used above, and we omit its details. It yields almost the same bound as in (8), where the slightly larger upper bound on k leads to the slightly larger bound

$$I(P, L) = O\left(2^{\sqrt{4.5}\sqrt{\log b}\sqrt{\log n}}\left(m^{1/2}n^{3/4} + m^{2/3}n^{1/3}s^{1/3} + m + n\right)\right),$$

with a slightly different absolute constant b.

3 Discussion

In this paper we derived an asymptotically tight bound for the number of incidences between a set P of points and a set L of lines in $\mathbb{R}^3$. This bound has already been established by Guth and Katz [10], where the main tool was the use of partitioning polynomials. As already mentioned, the main novelty here is to use two separate partitioning polynomials of different degrees; the one with the higher degree is used as a pruning mechanism, after which the maximum number of coplanar lines of L can be better controlled (by the degree D of the polynomial), which is a key ingredient in making the inductive argument work.

The second main tool of Guth and Katz was the Cayley–Salmon theorem. This theorem says that a surface in $\mathbb{R}^3$ of degree D cannot contain more than $11D^2 - 24D$ lines, unless it is *ruled by lines*. This is an "ancient" theorem, from the 19th century, combining algebraic and differential geometry, and its re-emergence in recent years has kindled the interest of the combinatorial geometry community in classical (and modern) algebraic geometry. New proofs of the theorem were obtained (see, e.g., Terry Tao's blog [31]), and generalizations to higher dimensions have also been developed (see Landsberg [16]). However, the theorem only holds over the complex field, and adapting it to apply over the reals requires some care.

There is also an alternative way to bound the number of point-line incidences using flat and singular points. However, as already remarked, these two, as well as the Cayley–Salmon machinery, are non-trivial constructs, especially in higher dimensions, and their generalization to other problems in combinatorial geometry (even incidence problems with curves other than lines or incidences with lines in higher dimensions) seems quite difficult (and are mostly open). It is therefore of considerable interest to develop alternative, more elementary interfaces between algebraic and combinatorial geometry, which is a primary goal of the present paper (as well as of Guth's recent work [8]).

In this regard, one could perhaps view Lemma 5 and Corollary 9 as certain weaker analogs of the Cayley–Salmon theorem, which are nevertheless easier to derive, without having to use differential geometry. Some of the tools in Guth's paper [8] might also be interpreted as such weaker variants of the Cayley–Salmon theorem. It would be interesting to see suitable extensions of these tools to higher dimensions.

Besides the intrinsic interest in simplifying the Guth–Katz analysis, the present work has been motivated by our study of incidences between points and lines in four dimensions. This has begun in a companion paper [24], where we have used the polynomial partitioning method, with a polynomial of constant degree. This, similarly to Guth's work in three dimensions [8], has resulted in a slightly weaker bound and considerably stricter assumptions concerning the input set of lines. In a more involved follow-up study [25], we have managed to improve the bound, and to get rid of the restrictive assumptions, using two partitioning steps, with polynomials of non-constant degrees, as in the present paper. However, the analysis in [25] is not as simple as in the present paper, because, even though there are generalizations of the Cayley–Salmon theorem to higher dimensions (due to Landsberg, as mentioned above), it turns out that a thorough investigation of the variety of lines fully contained in a given hypersurface of non-constant degree is a fairly intricate and challenging problem, raising many deep questions in algebraic geometry, some of which are still unresolved.

One potential application of the techniques used in this paper, mainly the interplay between partitioning polynomials of different degrees, is to the problem, recently studied by Sharir, Sheffer and Zahl [23], of bounding the number of incidences between points and circles in $\mathbb{R}^3$. That paper uses a partitioning polynomial of constant degree, and, as a result, the term that caters to incidences within lower-dimensional spaces (such as our term $m^{2/3}n^{1/3}s^{1/3}$) does not go well through the induction mechanism, and consequently the bound derived in [23] was weaker in that aspect. We believe that our technique can improve the bound of [23] in terms of this "lower-dimensional" term.

A substantial part of the present paper (half of the proof of the theorem) was devoted to the treatment of the case $m > n^{3/2}$. However, under the appropriate assumptions, the number of points incident to at least two lines was shown by Guth and Katz [10] to be bounded by $O(n^{3/2})$. A recent paper by Kollár [15] gives a simplified proof, including an explicit multiplicative constant. In his work, Kollár employs more advanced algebraic-geometric tools, like the *arithmetic genus* of a curve, which serves as an upper bound for the number of singular points. If we accept (pedagogically) the upper bound $O(n^{3/2})$ for the number of 2-rich points as a "black box", the regime in which $m > n^{3/2}$ becomes irrelevant, and can be discarded from the analysis, thus greatly simplifying and shortening the paper.

A challenging problem is thus to find an elementary proof that the number of points incident to at least two lines is $O(n^{3/2})$ (e.g., without the use of the Cayley–Salmon theorem or the tools used by Kollár). Another challenging (and probably harder) problem is to improve the bound of Guth and Katz when the bound s on the maximum number of mutually coplanar lines is $\ll n^{1/2}$: In their original derivation, Guth and Katz [10] consider mainly the case $s = n^{1/2}$, and the lower bound construction in [10] also has $s = n^{1/2}$. Another natural further research direction is to find further applications of partitioning polynomials of intermediate degrees.

Acknowledgements We would like to express our gratitude to an anonymous referee for providing very careful and helpful comments that helped in improving the presentation in the paper.

References

1. S. Basu, M. Sombra, Polynomial partitioning on varieties of codimension two and point-hypersurface incidences in four dimensions. Discrete Comput. Geom. **55**(1), 158–184 (2016)
2. J. Bochnak, M. Coste, M.F. Roy, *Real Algebraic Geometry* (Springer, Heidelberg, 1998)
3. K. Clarkson, H. Edelsbrunner, L. Guibas, M. Sharir, E. Welzl, Combinatorial complexity bounds for arrangements of curves and spheres. Discrete Comput. Geom. **5**, 99–160 (1990)
4. D. Cox, J. Little, D. O'Shea, *Ideals, Varieties, and Algorithms: An Introduction to Computational Algebraic Geometry and Commutative Algebra* (Springer, Heidelberg, 2007)
5. G. Elekes, H. Kaplan, M. Sharir, On lines, joints, and incidences in three dimensions. J. Comb. Theory, Ser. A **118**, 962–977 (2011)
6. P. Erdős, On sets of distances of n points. Am. Math. Mon. **53**, 248–250 (1946)
7. D. Fuchs, S. Tabachnikov, *Mathematical Omnibus: Thirty Lectures on Classic Mathematics* (American Mathematical Society Press, Providence, RI, 2007)
8. L. Guth, Distinct distance estimates and low-degree polynomial partitioning. Discrete Comput. Geom. **53**(2), 428–444 (2015)
9. L. Guth, N.H. Katz, Algebraic methods in discrete analogs of the Kakeya problem. Adv. Math. **225**, 2828–2839 (2010)
10. L. Guth, N.H. Katz, On the Erdős distinct distances problem in the plane. Ann. Math. **181**, 155–190 (2015)
11. J. Harris, *Algebraic Geometry: A First Course*, vol. 133 (Springer, New York, 1992)
12. R. Hartshorne, *Algebraic Geometry* (Springer, New York, 1983)
13. H. Kaplan, J. Matoušek, Z. Safernová, M. Sharir, Unit distances in three dimensions. Comb. Probab. Comput. **21**, 597–610 (2012)
14. H. Kaplan, J. Matoušek, M. Sharir, Simple proofs of classical theorems in discrete geometry via the Guth-Katz polynomial partitioning technique. Discrete Comput. Geom. **48**, 499–517 (2012)
15. J. Kollár, Szemerédi-Trotter-type theorems in dimension 3. Adv. Math. **271**, 30–61 (2015)
16. J.M. Landsberg, Is a linear space contained in a submanifold? On the number of derivatives needed to tell. J. Reine Angew. Math. **508**, 53–60 (1999)
17. J. Pach, M. Sharir, Geometric incidences, in J. Pach (ed.) *Towards a Theory of Geometric Graphs*, *Contemporary Mathematics*, vol. 342 (American Mathematical Society, Providence, RI, 2004), pp. 185–223
18. A. Pressley, *Elementary Differential Geometry*, Springer Undergraduate Mathematics Series (Springer, London, 2001)
19. O. Raz, M. Sharir, F. De Zeeuw, Polynomials vanishing on Cartesian products: the Elekes–Szabó Theorem revisited, Duke Math. J. 165(18), 3517–3566 (2016)
20. G. Salmon, *A Treatise on the Analytic Geometry of Three Dimensions*, vol. 2, 5th edn. (Hodges, Figgis and co. Ltd, Dublin, 1915)
21. J. Schmid, On the affine Bézout inequality. Manuscripta Mathematica **88**(1), 225–232 (1995)
22. M. Sharir, A. Sheffer, N. Solomon, Incidences with curves in $\mathbb{R}^d$, Electron. J. Combin. **23**(4), P4.16. Also in Proc. Eur. Sympos. Algorithms, 977–988. Also in **1501**, 02544 (2015)
23. M. Sharir, A. Sheffer, J. Zahl, Improved bounds for incidences between points and circles. Comb. Probab. Comput. **24**, 490–520 (2015)
24. M. Sharir, N. Solomon, Incidences between points and lines in $\mathbb{R}^4$, in *Proceedings of 30th Annual ACM Symposium Computational Geometry* (2014), pp. 189–197
25. M. Sharir, N. Solomon, Incidences between points and lines in four dimensions, in *Proceedings of 56th IEEE Symposium on Foundations of Computer Science* Discrete Comput. Geom. **57**, 702–756 (2017). Also in arXiv:1411.0777

26. M. Sharir, N. Solomon, Incidences between points and lines on a two- and three-dimensional varieties, Discrete Comput. Geom. **59**, 88–130 (2018). Also in arXiv:1609.09026
27. J. Solymosi, T. Tao, An incidence theorem in higher dimensions. Discrete Comput. Geom. **48**, 255–280 (2012)
28. L. Székely, Crossing numbers and hard Erdős problems in discrete geometry. Comb. Probab. Comput. **6**, 353–358 (1997)
29. E. Szemerédi, W.T. Trotter, Extremal problems in discrete geometry. Combinatorica **3**, 381–392 (1983)
30. T. Tao, From rotating needles to stability of waves: Emerging connections between combinatorics, analysis, and PDE. Notices AMS **48**(3), 294–303 (2001)
31. T. Tao, The Cayley–Salmon theorem via classical differential geometry (2014), http://terrytao.wordpress.com
32. H.E. Warren, Lower bound for approximation by nonlinear manifolds. Trans. Am. Math. Soc. **133**, 167–178 (1968)
33. J. Zahl, An improved bound on the number of point-surface incidences in three dimensions. Contrib. Discrete Math. **8**(1), 100–121 (2013)
34. J. Zahl, A Szemerédi-Trotter type theorem in $\mathbb{R}^4$. Discrete Comput. Geom. **54**(3), 513–572 (2015)

Incidence Bounds for Complex Algebraic Curves on Cartesian Products

József Solymosi and Frank de Zeeuw

Abstract We prove bounds on the number of incidences between a set of algebraic curves in $\mathbb{C}^2$ and a Cartesian product $A \times B$ with finite sets $A, B \subset \mathbb{C}$. Similar bounds are known under various restrictive conditions, but we show that the Cartesian product assumption leads to a simpler proof and lets us remove these conditions. This assumption holds in a number of interesting applications, and with our bound these applications can be extended from $\mathbb{R}$ to $\mathbb{C}$. We also obtain more precise information in the bound, which is used in several recent papers (Raz et al., 31st international symposium on computational geometry (SoCG 2015), pp 522–536, 2015, [17], Valculescu, de Zeeuw, Distinct values of bilinear forms on algebraic curves, 2014, [25]). Our proof works via an incidence bound for surfaces in $\mathbb{R}^4$, which has its own applications (Raz, Sharir, 31st international symposium on computational geometry (SoCG 2015), pp 569–583, 2015, [15]). The proof is a new application of the polynomial partitioning technique introduced by Guth and Katz (Ann Math 181: 155–190, 2015, [11]).

1 Introduction

Not many incidence bounds have been proved over the complex numbers. The quintessential incidence bound of Szemerédi and Trotter for points and lines in $\mathbb{R}^2$ was generalized to $\mathbb{C}^2$ by Tóth [24] and Zahl [26]. It states that for a finite set P

The first author was supported by ERC Advanced Research Grant no. 321104, by Hungarian National Research Grant NK 104183, and by NSERC. The second author was partially supported by Swiss National Science Foundation Grants 200020-144531 and 200021-137574. Part of this research was performed while the authors visited the Institute for Pure and Applied Mathematics (IPAM) in Los Angeles, which is supported by the National Science Foundation.

J. Solymosi (✉)
Vancouver, BC, Canada
e-mail: solymosi@math.ubc.ca

F. de Zeeuw
EPFL, Lausanne, Switzerland
e-mail: fdezeeuw@gmail.com

G. Ambrus et al. (eds.), *New Trends in Intuitive Geometry*, Bolyai Society Mathematical Studies 27, https://doi.org/10.1007/978-3-662-57413-3_16

of points in $\mathbb{C}^2$ and a finite set L of lines in $\mathbb{C}^2$, the set of *incidences*, denoted by $I(P, L) := \{(p, \ell) \in P \times L : p \in \ell\}$, satisfies

$$|I(P, L)| = O\left(|P|^{2/3}|L|^{2/3} + |P| + |L|\right).$$

The Szemerédi–Trotter bound was generalized to algebraic (and even continuous) curves in $\mathbb{R}^2$ by Pach and Sharir [14], but their result has not yet been fully extended to $\mathbb{C}^2$. Solymosi and Tao [22] and Zahl [26] did prove complex versions, but only for algebraic curves satisfying certain restrictions. These restrictions include the requirement that the curves are smooth and that the intersections of the curves are transversal (i.e., the curves have distinct tangent lines at their intersection points); both restrictions do not hold in many potential applications. Concurrently with this paper, Sheffer and Zahl [19] proved a complex version of the Pach-Sharir bound without such restrictions, but with a slightly weaker bound.

Incidence bounds are often easier to prove when the point set has the structure of a Cartesian product. This observation was used by Solymosi and Vu [23] to obtain incidence bounds in $\mathbb{R}^D$. It was noted by Solymosi [20] that the Szemerédi–Trotter bound in $\mathbb{C}^2$ can be proved more easily (compared to [24]) when the point set is a Cartesian product; this statement was then used to obtain a sum-product bound over $\mathbb{C}$. Solymosi and Tardos [22] used the same observation to obtain bounds on rich Möbius transformation from $\mathbb{C}$ to $\mathbb{C}$.

We prove a Pach-Sharir-like incidence bound for algebraic curves in $\mathbb{C}^2$, under the assumption that the point set is a Cartesian product $A \times B$ with $A, B \subset \mathbb{C}$. We do not require the curves to satisfy the restrictions that were needed in [22, 26], and the bound is slightly stronger than in [19]. Like in [14, 19, 22, 26], the curves must satisfy a degrees-of-freedom condition, which can come in different forms. Theorem 1 states our main result with what is probably the most convenient condition; several other versions can be found in Sect. 6.

Theorem 1 *Let A and B be finite subsets of $\mathbb{C}$ with $|A| = |B|$, and set $\mathcal{P} := A \times B$. Let $\mathcal{C}$ be a finite set of algebraic curves in $\mathbb{C}^2$ of degree at most d, such that any two points of $\mathcal{P}$ are contained in at most M curves of $\mathcal{C}$. Then*

$$I(\mathcal{P}, \mathcal{C}) = O(d^{4/3}M^{1/3}|\mathcal{P}|^{2/3}|\mathcal{C}|^{2/3} + M(\log M + \log d)|\mathcal{P}| + d^4|\mathcal{C}|).$$

We have worked out in detail the dependence of the bound on the parameters, which is of interest in certain applications, in particular [17, 25]. Our proof works via an incidence bound for well-behaved surfaces in $\mathbb{R}^4$, which is interesting in its own right; it was used by Raz and Sharir [15] to improve the best known bound on the number of unit area triangles determined by a point set in $\mathbb{R}^2$.

Although the assumption that the point set is a Cartesian product is very restrictive, it is satisfied in a number of interesting problems. We give several examples of such applications in Sect. 7. These include an answer to a question of Elekes [7] related to sum-product estimates, and a generalization to $\mathbb{C}$ of a recent result of Sharir, Sheffer, and Solymosi [18] on distinct distances between lines. More sophisticated applications can be found in the already mentioned works [15, 17, 25], which were in fact the original motivation for this paper.

We begin in Sect. 2 with the elementary proof of the real analogue of our main theorem, which is not a new result, but serves as an introduction to our main proof. In Sect. 3 we collect the technical tools that we use, and in Sect. 4 we prove our main bound, which concerns point-surface incidences in $\mathbb{R}^4$. In Sect. 5 we deduce some corollaries for surfaces in $\mathbb{R}^4$, and in Sect. 6 we prove corollaries for curves in $\mathbb{C}^2$, including Theorem 1 above. Finally, in Sect. 7, we give three applications.

2 Warmup: Points and Curves in $\mathbb{R}^2$

As a warmup for the complex case, we first give a proof of the corresponding statement for incidences between real algebraic curves and a Cartesian product in $\mathbb{R}^2$. This is not a new result, as it follows from the work of Pach and Sharir [14]. The proof given here is, however, much simpler, because the product structure allows for a trivial partitioning of the plane; the proof is self-contained up to a few basic facts about algebraic curves. Moreover, it provides a blueprint for our main proof in Sect. 4.

Throughout, given a set $\mathcal{P}$ of points and a set $\mathcal{C}$ of geometric objects, we define the set of *incidences* by $I(\mathcal{P}, \mathcal{C}) := \{(p, c) \in \mathcal{P} \times \mathcal{C} : p \in c\}$.

Theorem 2 *Let A and B be finite subsets of $\mathbb{R}$ and $\mathcal{P} := A \times B$. Let $\mathcal{C}$ be a finite set of algebraic curves in $\mathbb{R}^2$ of degree at most d such that any two points of $\mathcal{P}$ are contained in at most M curves of $\mathcal{C}$. We assume that no curve in $\mathcal{C}$ contains a horizontal or vertical line, that $d^4|\mathcal{C}| \leq M|\mathcal{P}|^2$, and that $|A| \leq |B|$ and $d|\mathcal{C}| \geq M|B|^2/|A|$. Then*

$$|I(\mathcal{P}, \mathcal{C})| = O(d^{2/3}M^{1/3}|\mathcal{P}|^{2/3}|\mathcal{C}|^{2/3}).$$

Proof Let r be a real number, to be chosen at the end of the proof, satisfying $d \leq r \leq |A|$. We "cut" $\mathbb{R}$ in $O(r)$ points that are not in A, splitting $\mathbb{R}$ into $O(r)$ intervals so that each interval contains $O(|A|/r)$ elements of A (this is possible because $r \leq |A|$). Similarly, we choose $O(r)$ cutting points not in B that split $\mathbb{R}$ into at most $O(r)$ intervals, each containing $O(|B|/r)$ elements of B (using $r \leq |A| \leq |B|$). This gives a partition of $\mathbb{R}^2$ into $O(r^2)$ open cells (which are rectangles) and a closed boundary (consisting of $O(r)$ lines). Each cell contains $O(|A||B|/r^2) = O(|\mathcal{P}|/r^2)$ points of $\mathcal{P}$, while the boundary is disjoint from $\mathcal{P}$.

We need to bound the number of cells that a curve $C \in \mathcal{C}$ can intersect. The curve has $O(d^2)$ connected components by Harnack's Theorem (see Lemma 6 below), and it has at most d intersection points with each of the $O(r)$ boundary lines by Bézout's Inequality (see Lemma 4 below), using the fact that the curve contains no horizontal or vertical line. Thus the $O(d^2)$ connected components are cut in $O(dr)$ points. By wiggling the cutting lines slightly, we can ensure that they do not hit a curve of $\mathcal{C}$ in a singularity, since algebraic curves have finitely many singularities (see Sect. 3).

Therefore, each cut increases the number of connected components by at most one.[1] Thus any $C \in \mathcal{C}$ intersects $O(d^2 + dr) = O(dr)$ (using $d \leq r$) of the $O(r^2)$ cells.[2]

Let I_1 be the subset of incidences $(p, C) \in I(\mathcal{P}, \mathcal{C})$ such that (p, C) is the only incidence of C in the cell containing p, and let I_2 be the subset of incidences $(p, C) \in I(\mathcal{P}, \mathcal{C})$ such that C has at least one other incidence in the cell that contains p. Then, since a curve intersects $O(dr)$ cells, we have

$$|I_1| = O(dr|\mathcal{C}|).$$

On the other hand, given two points in one cell, there are by assumption at most M curves in $\mathcal{C}$ that contain both points. Thus, in a cell with k points there are at most $2M\binom{k}{2} = O(Mk^2)$ incidences from I_2. Therefore, summing over all cells, we have

$$|I_2| = O\left(r^2 \cdot M\left(\frac{|\mathcal{P}|}{r^2}\right)^2\right) = O\left(\frac{M|\mathcal{P}|^2}{r^2}\right).$$

Choosing $r^3 := \frac{M}{d}\frac{|\mathcal{P}|^2}{|\mathcal{C}|}$ gives

$$|I(\mathcal{P}, \mathcal{C})| = I_1 + I_2 = O\left(d^{2/3}M^{1/3}|\mathcal{P}|^{2/3}|\mathcal{C}|^{2/3}\right).$$

We have to verify that our choice of r satisfies $d \leq r$ and $r \leq |A|$. The first follows from the assumption $d^4|\mathcal{C}| \leq M|\mathcal{P}|^2$ and

$$r = \left(\frac{M|\mathcal{P}|^2}{d|\mathcal{C}|}\right)^{1/3} \geq \left(\frac{d^4|\mathcal{C}|}{d|\mathcal{C}|}\right)^{1/3} = d.$$

The second follows from the assumption $d|\mathcal{C}| \geq M|B|^2/|A|$ and

$$r^3 = \frac{M|\mathcal{P}|^2}{d|\mathcal{C}|} \leq \frac{M|\mathcal{P}|^2}{M|B|^2/|A|} = |A|^3.$$

This completes the proof. □

Theorem 2 has several conditions which can be simplified in various ways to make the statement more suitable for application. We state one version here as an example, but we refer to Sect. 6 for the proof (which is identical to that of the complex version presented there).

---

[1] This could fail if a cutting point were a singularity (although even then the number of branches could be controlled with some more effort).

[2] This fact can be obtained more directly using Lemma 6 below, by noting that the union of the lines is a curve defined by a polynomial f of degree $O(r)$, so $\mathcal{C}\backslash Z(f)$ has $O(dr)$ connected components. However, we have used the argument above because it will play a crucial role in the proof of our main theorem.

Corollary 3 *Let A and B be finite subsets of $\mathbb{R}$ with $|A| = |B|$, and $\mathcal{P} := A \times B$. Let $\mathcal{C}$ be a finite set of algebraic curves in $\mathbb{R}^2$ of degree at most d, such that any two points of $\mathcal{P}$ are contained in at most M curves of $\mathcal{C}$. Then*

$$|I(\mathcal{P}, \mathcal{C})| = O\left(d^{2/3}M^{1/3}|\mathcal{P}|^{2/3}|\mathcal{C}|^{2/3} + M(\log M + \log d)|\mathcal{P}| + d^2|\mathcal{C}|\right).$$

3 Definitions and Tools

3.1 Definitions

We introduce a few definitions and basic facts from algebraic geometry in some detail, because the subtle differences between real and complex varieties play a role in our proof.

A *variety* in $\mathbb{C}^D$ is a set of the form

$$Z_{\mathbb{C}^D}(f_1, \ldots, f_m) := \left\{(z_1, \ldots, z_D) \in \mathbb{C}^D : f_i(z_1, \ldots, z_D) = 0 \text{ for } i = 1, \ldots, m\right\},$$

for polynomials $f_i \in \mathbb{C}[z_1, \ldots, z_D]$. Such sets are normally called *affine* varieties (or just *zero sets*), but since this is the only type of variety that we consider, we refer to them simply as varieties. Similarly, we define a *real variety* to be a zero set of the form

$$Z_{\mathbb{R}^D}(f_1, \ldots, f_m) := \left\{(x_1, \ldots, x_D) \in \mathbb{R}^D : f_i(x_1, \ldots, x_D) = 0 \text{ for } i = 1, \ldots, m\right\}$$

with polynomials $f_i \in \mathbb{R}[x_1, \ldots, x_D]$. We refer to [12, 13] for definitions of the dimension $\dim_{\mathbb{C}}(V)$ and the degree $\deg(V)$ of a complex variety V, and we refer to [2, Sect. 5.3] or [4, Sect. 2.8] for a careful definition of the real dimension of a real variety W, denoted by $\dim_{\mathbb{R}}(W)$. One can locally view a real variety as a real manifold (around any nonsingular point, see below), and the real dimension equals the dimension in the manifold sense (more precisely, it is the *maximum* dimension at any nonsingular point).

A *complex algebraic curve* in $\mathbb{C}^2$ is a variety V with $\dim_{\mathbb{C}}(V) = 1$. In our definition,[3] a curve $C \subset \mathbb{C}^2$ of degree d has the form $Z_{\mathbb{C}^2}(f) \cup P$ for a polynomial $f \in \mathbb{C}[x, y]$ of degree $d - k$ and a finite set P of size k. A variety $W \subset \mathbb{R}^D$ is a *real algebraic curve* in $\mathbb{R}^D$ if $\dim_{\mathbb{R}}(W) = 1$, and it is a *real algebraic surface* if $\dim_{\mathbb{R}}(W) = 2$. A real algebraic curve $C \subset \mathbb{R}^2$ can be written as $Z_{\mathbb{R}^2}(f)$ for a poly-

[3]Note that the dimension of a reducible variety is the maximum of the dimensions of its components, so a curve can have zero-dimensional components. The degree of a reducible variety is the sum of the degrees of its components, so a curve of degree d with k zero-dimensional components has a purely one-dimensional component of degree $d - k$.

nomial $f \in \mathbb{R}[x, y]$[4]; we define the *degree* of C to be minimum degree of such an f. For convenience, we will occasionally use the notion of a *semialgebraic curve* in $\mathbb{R}^2$, which is a subset of a real curve defined by polynomial inequalities; in particular, if we remove a finite point set from a real curve, the connected components of the remainder are semialgebraic curves.

A curve $C \subset \mathbb{C}^2$ is *irreducible* if there is an irreducible f such that $C = Z_{\mathbb{C}^2}(f)$. An *irreducible component* of an algebraic curve $C \subset \mathbb{C}^2$ is an irreducible algebraic curve C' such that $C' \subset C$. A curve in $\mathbb{C}^2$ of degree d has a decomposition as a union of at most d irreducible components (some of which may be points).

We also need to consider singularities of curves, but only for real curves in $\mathbb{R}^2$ or $\mathbb{R}^4$. For a curve in $\mathbb{R}^D$, we define a point on the curve to be a *singularity* if it does not have a neighborhood in which the curve is a real manifold of dimension one (for details see [12, Lecture 14] or [4, Sect. 3.3]). A key fact that we need is that the number of singularities of a curve in $\mathbb{R}^2$ of degree d is less than d^2. More precisely, define the *branches* of a singularity in a small neighborhood to be the connected components of the curve in that neighborhood after removing the singularity. Then the total number of branches over all singularities, for any choice of sufficiently small neighborhoods, is at most d^2 (see [10, Chap. 3]). For a curve in $\mathbb{R}^4$, we only need the fact that the number of singularities is finite (see [4, Proposition 3.3.14]).

3.2 Intersection Bounds

In the proof of our main theorem we will frequently have to bound the size of the intersection of two varieties, both over $\mathbb{C}$ and over $\mathbb{R}$. The prototype for such intersection bounds is the following lemma. Here we consider the degree of a finite point set to be its size, so the lemma says that the intersection of two curves is either a finite set of bounded size, or a curve of bounded degree.

Lemma 4 (Bézout's Inequality) *If C_1 and C_2 are algebraic curves in $\mathbb{C}^2$ or $\mathbb{R}^2$, then*

$$\deg(C_1 \cap C_2) \leq \deg(C_1) \cdot \deg(C_2).$$

With the right definition of degree, this inequality can be extended to varieties in $\mathbb{C}^D$, but in $\mathbb{R}^D$, the inequality may fail in this form. Nevertheless, various cautious bounds on the number of connected components of intersections of real varieties have been proved, which can often serve the same purpose; see for instance [2, Chap. 7]. We will use the following recent result of Barone and Basu [1]. It gives a refined bound when the defining polynomials of the variety have different degrees, which is crucial in our proofs. We state it in a similar way to Basu and Sombra [3, Theorem 2.5], with some modifications based on the more general form in [1]. We simplify the statement of the bound somewhat using the following (non-standard) definition.

[4]Here too a curve may have zero-dimensional components (isolated points), but in $\mathbb{R}^2$ a point (a, b) is defined by a single polynomial $(x - a)^2 + (y - b)^2$.

Definition 5 Let $V := Z_{\mathbb{R}^D}(g_1, \ldots, g_m)$ have dimension k_m, with $\deg(g_1) \le \cdots \le \deg(g_m)$. Write $k_i := \dim_{\mathbb{R}}(Z_{\mathbb{R}^D}(g_1, \ldots, g_i))$ and $k_0 := D$. We define the *Barone–Basu degree* of V by

$$\deg_{\mathrm{BB}}(V) := \prod_{i=1}^{m} \deg(g_i)^{k_{i-1}-k_i}.$$

Note that, for example, a two-dimensional variety $Z_{\mathbb{R}^4}(g_1, \ldots, g_m)$ in $\mathbb{R}^4$ that is defined by any number of polynomials of degree at most d has Barone–Basu degree d^2. Indeed, we have $k_0 = 4$ and $k_m = 2$, and either there are two g_i such that $k_{i-1} - k_i = 1$, or there is one g_i such that $k_{i-1} - k_i = 2$; in both cases the Barone–Basu degree comes out to d^2.

Lemma 6 (Barone–Basu) *Let $V := Z_{\mathbb{R}^D}(g_1, \ldots, g_m)$ with $\deg(g_1) \le \cdots \le \deg(g_m)$. Let $h \in \mathbb{R}[x_1, \ldots, x_D]$ with $\deg(h) \ge \deg(g_m)$. Then the number of connected components of both $V \cap Z_{\mathbb{R}^D}(h)$ and $V \backslash Z_{\mathbb{R}^D}(h)$ is*

$$O\left(\deg_{BB}(V) \cdot \deg(h)^{\dim_{\mathbb{R}}(V)}\right).$$

In the ideal case where each $k_i = D - i$, this would be a natural generalization of Lemma 4. On the other hand, if $\deg(g_i) \le d$ for each i, we get the bound $O(d^{D-k_m} \deg(h)^{k_m})$, without any individual conditions on the g_i. The fact that h is arbitrary allows for the following trick to deal with more polynomials in the role of h: To bound the number of connected components of, say, $Z_{\mathbb{R}^D}(g_1, \ldots, g_m) \backslash Z_{\mathbb{R}^D}(h_1, h_2)$, one can simply set $h := h_1^2 + h_2^2$ and use the lemma.

Finally, we record a simple fact about the surface in $\mathbb{R}^4$ associated to a curve in $\mathbb{C}^2$.

Lemma 7 *Let $C \subset \mathbb{C}^2$ be an algebraic curve of degree d. Then the associated real surface S in $\mathbb{R}^4$ is defined by two polynomials of degree at most $2d$, and $\deg_{BB}(S) \le 4d^2$.*

Proof There is a finite set P and a polynomial $f(x, y)$ of degree $d - |P|$ such that we can write $C = Z_{\mathbb{C}^2}(f) \cup P$. The real polynomials

$$h_1(x_1, x_2, x_3, x_4) := \mathrm{Re} f(x_1 + ix_2, x_3 + ix_4), \quad h_2(x_1, x_2, x_3, x_4) := \mathrm{Im} f(x_1 + ix_2, x_3 + ix_4)$$

define the surface in $\mathbb{R}^4$ associated to $Z_{\mathbb{C}^2}(f)$; both have degree at most $d - |P|$. The set P, viewed as a subset of $\mathbb{R}^4$, is defined by a polynomial h_3 of degree $2|P|$. If we set $g_1 = h_1 h_3$ and $g_2 = h_2 h_3$, then we have $S = Z_{\mathbb{R}^4}(g_1, g_2)$, and g_1, g_2 have degree at most $d - |P| + 2|P| \le 2d$.

Write $k_0 := 4$, $k_1 := \dim_{\mathbb{R}}(Z_{\mathbb{R}^4}(g_1))$, and $k_2 := \dim_{\mathbb{R}}(Z_{\mathbb{R}^4}(g_1, g_2))$ as in Definition 5. We clearly have $k_2 = 2$. Then $k_1 \in \{2, 3, 4\}$, and, whichever it is, we get

$$\deg_{\mathrm{BB}}(S) \le (2d)^{k_0-k_1} \cdot (2d)^{k_1-k_2} \le 4d^2.$$

This completes the proof. □

3.3 Polynomial Partitioning

Our proof relies on the following technique introduced by Guth and Katz [11].

Lemma 8 (Polynomial partitioning) *Let A be a finite subset of $\mathbb{R}^2$. For any $r \in \mathbb{R}$ with $1 \le r \le |A|^{1/2}$ there exists a polynomial $f \in \mathbb{R}[x, y]$ of degree $O(r)$ such that $\mathbb{R}^2 \backslash Z_{\mathbb{R}^2}(f)$ has $O(r^2)$ connected components, each containing $O(|A|/r^2)$ points of A.*

In the proof of Theorem 2, we in fact used a trivial partitioning on $\mathbb{R}$: For $A \subset \mathbb{R}$ and any $1 \le r \le |A|$, there is a subset $X \subset \mathbb{R}\backslash A$ of size $O(r)$ such that $\mathbb{R}\backslash X$ has $O(r)$ connected components, each containing $O(|A|/r)$ points of A. Moreover, we used the fact that the points of X have some "wiggle room", in the sense that they can be varied in some small neighborhood without affecting the partitioning property. We now show that a point set on a real algebraic curve can be partitioned in a similar way.

Such a partitioning would not be possible for arbitrary continuous curves with self-intersections, or for algebraic curves of arbitrary degree. If we take an arbitrary point set in general position and connect any two points by a line, the union of the lines is an algebraic curve of high degree that cannot be partitioned with a small number of cutting points on the curve. However, on an algebraic curve of bounded degree δ, one can control the number of self-intersections (singularities) of the curve in terms of δ, and this allows us to partition it into $O(\delta^2)$ pieces.

Lemma 9 (Partitioning a real algebraic curve) *Let $C \subset \mathbb{R}^2$ be an algebraic curve of degree δ, containing a finite set A. Then there is a subset $X \subset C\backslash A$ of $O(\delta^2)$ points, such that $C\backslash X$ consists of $O(\delta^2)$ connected semialgebraic curves, each containing $O(|A|/\delta^2)$ points of A. Moreover, each point $p \in X$ has an open neighborhood on C such that any point of that neighborhood could replace p without affecting the partitioning property.*

Proof Around every singularity p of C, choose a sufficiently small closed ball B_p with boundary circle R_p, so that B_p contains no other singularities of C, and no point of A other than possibly p itself. We put the points of $C \cap R_p$ into X for each p. For each singularity p, $|C \cap R_p|$ is at most the number of branches of C at p in the neighborhood B_p (as defined at the end of Sect. 3.1), and the total sum of these numbers is at most δ^2. Hence we have put at most δ^2 points into X. The points of X are themselves not singularities, so removing a point of X increases the number of connected components by at most one, since around such a point C is a one-dimensional manifold.

By Lemma 6, C has at most $O(\delta^2)$ connected components, so removing the points of X cuts C into $O(\delta^2)$ connected semialgebraic curves. Each of these semialgebraic curves either contains at most one point of A (a singularity), or it is simple (i.e., it has no self-intersections). We can cut these simple curves at a total of $O(\delta^2)$ points, so that every resulting curve contains $O(|A|/\delta^2)$ points of A, and no cutting point is in A. Adding these cutting points to X completes the proof. It should be clear that shifting the cutting points within a sufficiently small open neighborhood will not affect the proof. □

4 Main Bound for Surfaces in $\mathbb{R}^4$

In this section we prove our main incidence bound for points and surfaces in $\mathbb{R}^4$, from which we will deduce our incidence bounds for complex algebraic curves in Sect. 6. It only applies to surfaces that are well-behaved in the following way.

Definition 10 A surface S in $\mathbb{R}^4$ *has good fibers* if for every $p \in \mathbb{R}^2$, the fibers $(p \times \mathbb{R}^2) \cap S$ and $(\mathbb{R}^2 \times p) \cap S$ are finite.

Note that if a curve in $\mathbb{C}^2$ contains no horizontal or vertical line, then its associated surface in $\mathbb{R}^4$ has good fibers. Since it is easy to remove a line from a curve, this property is easily ensured. On the other hand, for a surface S in $\mathbb{R}^4$, the fiber $(p \times \mathbb{R}^2) \cap S$ may be a one-dimensional curve, which is not so easily removed. Nevertheless, see [15] for an example of a situation where the surfaces have this property. We also note that for surfaces with a limited set of bad fibers, the proof below might still be made to work.

In the statement that we prove here, we make the degrees-of-freedom condition a bit more flexible. We view the set $I(\mathcal{P}, \mathcal{S})$ as an *incidence graph*, i.e., the bipartite graph with vertex sets $\mathcal{P}$ and $\mathcal{S}$, where $p \in \mathcal{P}$ is connected to $S \in \mathcal{S}$ if $p \in S$. The condition that any two points are in at most M surfaces can then be rephrased as $I(\mathcal{P}, \mathcal{S})$ containing no complete bipartite subgraph $K_{2,M}$. Here we weaken that condition by considering a *subgraph* of $I(\mathcal{P}, \mathcal{S})$; we show that if that subgraph contains no $K_{2,M}$, then its number of edges (denoted by $|I|$) is bounded. This formulation is often convenient in applications; see [22, 26] for incidence bounds that are also stated in this way.

Theorem 11 *Let A_1 and A_2 be finite subsets of $\mathbb{R}^2$ and $\mathcal{P} := A_1 \times A_2$. Let $\mathcal{S}$ be a finite set of algebraic surfaces in $\mathbb{R}^4$ that have good fibers and are defined by polynomials of degree at most d. Let $I \subset I(\mathcal{P}, \mathcal{S})$ be an incidence subgraph containing no $K_{2,M}$. Assume that $d^8|\mathcal{S}| \le M|\mathcal{P}|^2$, and that $|A_1| \le |A_2|$ and $d^2|\mathcal{S}| \ge M|A_2|^2/|A_1|$. Then*

$$|I| = O(d^{4/3}M^{1/3}|\mathcal{P}|^{2/3}|\mathcal{S}|^{2/3}).$$

Our proof uses the Guth-Katz polynomial partitioning technique from Lemma 8 in a special way that is adjusted to the Cartesian product structure. Specifically, $\mathbb{R}^4$ is viewed as a product $\mathbb{R}^2 \times \mathbb{R}^2$, and we partition each copy of $\mathbb{R}^2$ separately. We first partition $\mathbb{R}^2$ using a curve provided by Lemma 8, and then we partition that curve using Lemma 9. The partitions of the two copies of $\mathbb{R}^2$ are then combined into a cell decomposition of $\mathbb{R}^4$.

To make the bookkeeping of these partitions a bit easier to follow, we use the following terminology for our cell decomposition of $\mathbb{R}^4$: a *k-cell* is a connected set of dimension k that will be used in the final cell decomposition; a *k-wall* is a k-dimensional variety that cuts out the $(k+1)$-cells, but that is itself to be decomposed into lower-dimensional cells; a *k-gap* is a k-dimensional variety that also helps to cut out the $(k+1)$-cells, but does not contain any incidences, so does not need to be partitioned further. To summarize: $\mathbb{R}^4$ is partitioned into 4-cells by 3-walls and 3-gaps; each 3-wall is then partitioned into 3-cells by 2-walls and 2-gaps; the 2-walls are then partitioned into 2-cells using only 1-gaps.

Proof Every surface $S \in \mathcal{S}$ has $\deg_{BB}(S) \le d^2$ by the remark just after Definition 5. As in the proof of Theorem 2, we partition the space, see how the varieties intersect the cells, and then use a simple counting argument. We note that the counting is exactly as in Theorem 2, except that the parameters d and r from that proof are replaced by d^2 and r^2 in this proof.

Partitioning. We partition $\mathbb{R}^4$ into $O(r^4)$ cells and some gaps, so that each cell contains $O(|\mathcal{P}|/r^4)$ points of $\mathcal{P}$, and the gaps contain no points of $\mathcal{P}$. We assume that $d^2 \le r^2 \le |A_1|$.

We use Lemma 8 to get polynomials f_1, f_2 of degree $r \le |A_1|^{1/2}$ so that $C_i := Z_{\mathbb{R}^2}(f_i)$ partitions $\mathbb{R}^2$ into r^2 cells, each containing $O(|A_i|/r^2)$ points of A_i. Then we use Lemma 9 to partition C_1 and C_2, obtaining sets $X_i \subset C_i \backslash A_i$ with $|X_i| = O(r^2)$, so that $C_i \backslash X_i$ consists of $O(r^2)$ connected components, each containing $O(|A_i|/r^2)$ points of A_i.

The 3-walls $C_1 \times \mathbb{R}^2$ and $\mathbb{R}^2 \times C_2$ partition $\mathbb{R}^4$ into $O(r^4)$ 4-cells, each containing $O(|\mathcal{P}|/r^4)$ points of $\mathcal{P}$. The 3-wall $C_1 \times \mathbb{R}^2$ is partitioned by the 2-wall $C_1 \times C_2$, combined with the 2-gap $X_1 \times \mathbb{R}^2$; similarly, $\mathbb{R}^2 \times C_2$ is partitioned by the 2-wall $C_1 \times C_2$ and the 2-gap $\mathbb{R}^2 \times X_2$. Thus the 3-walls are partitioned into $O(r^4)$ 3-cells, each containing $O(|\mathcal{P}|/r^4)$ points of $\mathcal{P}$. The gaps are not partitioned further. The 2-wall $C_1 \times C_2$ is partitioned by the 1-gaps $X_1 \times C_2$ and $C_1 \times X_2$, again resulting in $O(r^4)$ cells, each containing $O(|\mathcal{P}|/r^4)$ points of $\mathcal{P}$. This completes the partitioning. Altogether there are $O(r^4)$ cells, each containing $O(|\mathcal{P}|/r^4)$ points of $\mathcal{P}$.

Intersections. We now show that any surface $S \in \mathcal{S}$ intersects $O(d^2r^2)$ of the $O(r^4)$ cells, and we do this separately for the 4-cells, 3-cells, and 2-cells.

4-cells: The 4-cells are cut out by the 3-wall $(C_1 \times \mathbb{R}^2) \cup (\mathbb{R}^2 \times C_2) = Z_{\mathbb{R}^4}(f_1 f_2)$. To get an upper bound on the number of 4-cells intersected by S, we want an upper bound on the number of connected components of $S \backslash Z_{\mathbb{R}^4}(f_1 f_2)$. We apply Lemma 6 to deduce that $S \backslash Z_{\mathbb{R}^4}(f_1 f_2)$ has

$$O(\deg_{BB}(S) \cdot \deg(f_1 f_2)^{\dim_{\mathbb{R}}(S)}) = O(d^2 \cdot (2r)^2)$$

connected components; here $\deg_{BB}(S) \le d^2$ by assumption, and the condition of Lemma 6 (that the degree of $f_1 f_2$ is at least the degree of the polynomials defining S) follows from the assumption $d^2 \le r^2$. This means that S intersects $O(d^2r^2)$ of the 4-cells.

3-cells: Set $S_1 := S \cap (C_1 \times \mathbb{R}^2)$. Note that $\dim_{\mathbb{R}}(S_1) \leq 1$, because for any $p \in \mathbb{R}^2$ the fiber $(p \times \mathbb{R}^2) \cap S$ is finite, since S has good fibers. If $\dim_{\mathbb{R}}(S_1) = 0$, then by Lemma 6 $S_1 = S \cap Z_{\mathbb{R}^4}(f_1)$ consists of $O(d^2 \cdot r^2)$ points, so it intersects at most that many cells. Hence we can assume $\dim_{\mathbb{R}}(S_1) = 1$. To see how many 3-cells inside $C_1 \times \mathbb{R}^2$ are intersected by S_1, we separately consider its intersection with the 2-wall $C_1 \times C_2$, and with the 2-gap $X_1 \times \mathbb{R}^2$.

The fact that $\dim_{\mathbb{R}}(S_1) = 1$ implies that $\deg_{\mathrm{BB}}(S_1) = O(d^2 r)$. Therefore, by Lemma 6, $S_1 \backslash (C_1 \times C_2) = S_1 \backslash Z_{\mathbb{R}^4}(f_2)$ has

$$O(\deg_{\mathrm{BB}}(S_1) \cdot \deg(f_2)^{\dim_{\mathbb{R}}(S_1)}) = O(d^2 r \cdot r)$$

connected components. Hence the wall $C_1 \times C_2$ cuts S_1 into $O(d^2 r^2)$ connected semialgebraic curves.

Now consider the gap $X_1 \times \mathbb{R}^2$.[5] For $p \in X_1$, we have $S_1 \cap (p \times \mathbb{R}^2) \subset S \cap (p \times \mathbb{R}^2)$, and $S \cap (p \times \mathbb{R}^2)$ is finite, again because S has good fibers. Since we can write $p \times \mathbb{R}^2 = Z_{\mathbb{R}^4}((x_1 - p_x)^2 + (x_2 - p_y)^2)$, it follows from Lemma 6 that $|S \cap (p \times \mathbb{R}^2)| = O(d^2 \cdot 2^2)$. Thus the curve S_1 has

$$|S_1 \cap (X_1 \times \mathbb{R}^2)| = O(d^2 \cdot |X_1|) = O(d^2 r^2)$$

points of intersection with this gap. Moreover, using the "wiggle room" for the points in X_1 mentioned in Lemma 9, and the fact that S_1 has finitely many singularities, we can assume that none of the points in $S_1 \cap (X_1 \times \mathbb{R}^2)$ is a singularity of S_1. Hence, removing such a point increases the number of connected components by at most one (which would not quite be true at a singularity). Since the wall $C_1 \times C_2$ cuts S_1 into $O(d^2 r^2)$ connected components, and we remove $O(d^2 r^2)$ further points, it finally follows that S_1 intersects $O(d^2 r^2)$ of the 3-cells inside $C_1 \times \mathbb{R}^2$. Note that X_1 should be chosen so that $X_1 \times \mathbb{R}^2$ avoids the singularities of S_1 for all $S \in \mathcal{S}$ simultaneously, but this is possible since there are finitely many points to avoid, while there is infinite wiggle room.

A symmetric argument gives the same bounds for $S_2 := S \cap (\mathbb{R}^2 \times C_2)$, so altogether we get that S intersects $O(d^2 r^2)$ of the 3-cells inside $\mathbb{R}^2 \times C_2$.

2-cells: Set $S_3 := S \cap (C_1 \times C_2)$. The 2-wall $C_1 \times C_2$ is partitioned by the 1-gaps $X_1 \times C_2$ and $C_1 \times X_2$. As above we have $|S_3 \cap (p \times C_2)| = O(d^2)$ for $p \in X_1$, so we get $|S_3 \cap (X_1 \times C_2)| = O(d^2 r^2)$ and similarly $|S_3 \cap (C_1 \times X_2)| = O(d^2 r^2)$. Finally, we can write $S_3 = S \cap Z_{\mathbb{R}^4}(f_1^2 + f_2^2)$, so S_3 has $O(d^2 \cdot (2r)^2)$ connected components. Again each cut increases the number of connected components by at most one, so altogether S_3 intersects $O(d^2 r^2)$ of the 2-cells.

Counting. Let I_1 be the subset of incidences $(p, S) \in I$ such that (p, S) is the only incidence of S from I in the cell containing p, and let I_2 be the subset of incidences $(p, S) \in I$ such that S has at least one other incidence from I in the cell that contains p. The fact that a surface from $\mathcal{S}$ intersects $O(d^2 r^2)$ cells implies

---

[5] Note that $X_1 \times \mathbb{R}^2$ is defined by a polynomial g of degree $2|X_1| = O(r^2)$. Thus, applying Lemma 6 to $S_1 \backslash Z_{\mathbb{R}^4}(g)$ gives $O(d^2 r \cdot r^2)$, which is too large. This is why we need a more refined argument, using the specific nature of $X_1 \times \mathbb{R}^2$.

$$|I_1| = O\left(d^2r^2|\mathcal{S}|\right).$$

On the other hand, given two points p_1, p_2 in one cell, there are by assumption fewer than M surfaces $S \in \mathcal{S}$ such that $(p_1, S), (p_2, S) \in I$. Thus we have

$$|I_2| = O\left(r^4 \cdot M \cdot \left(\frac{|\mathcal{P}|}{r^4}\right)^2\right) = O\left(M \cdot \frac{|\mathcal{P}|^2}{r^4}\right).$$

Choosing $r^6 := \frac{M}{d^2}\frac{|\mathcal{P}|^2}{|\mathcal{S}|}$ gives $|I(\mathcal{P}, \mathcal{S})| = O\left(d^{4/3}M^{1/3}|\mathcal{P}|^{2/3}|\mathcal{S}|^{2/3}\right)$. We need to ensure that $d^2 \le r^2 \le |A_1|$; this follows from the two assumptions of the theorem, by the same calculation as in the proof of Theorem 2 (with d and r replaced by d^2 and r^2). □

5 Corollaries for Surfaces in $\mathbb{R}^4$

We now deduce some more practical corollaries of Theorem 11 for surfaces in $\mathbb{R}^4$, without the awkward conditions on the sizes of the sets of points and surfaces. To remove these conditions we use the Kővári-Sós-Turán theorem, a commonly used tool in incidence geometry. It gives a bound on the number of edges in a graph not containing a complete bipartite graph $K_{s,t}$; see Bollobás [5, Theorem IV.10] for the version stated here.

Lemma 12 *Let $G \subset X \times Y$ be a bipartite graph. Suppose that G contains no $K_{s,t}$, i.e., for any s vertices in X, there are fewer than t vertices in Y connected to both. Then the number of edges of G is bounded by*

$$O(t^{1/s}|X||Y|^{1-1/s} + s|Y|).$$

We now use this lemma to obtain a more convenient version of Theorem 11. See [15] for an application of this corollary.

Corollary 13 *Let A and B be finite subsets of $\mathbb{R}^2$ with $|A| = |B|$, and $\mathcal{P} = A \times B \subset \mathbb{R}^4$. Let $\mathcal{S}$ be a finite set of surfaces in $\mathbb{R}^4$ that have good fibers and are defined by polynomials of degree at most d. Let $I \subset I(\mathcal{P}, \mathcal{S})$ be an incidence subgraph containing no $K_{2,M}$ or $K_{M,2}$. Then*

$$|I| = O_{d,M}\left(|\mathcal{P}|^{2/3}|\mathcal{S}|^{2/3} + |\mathcal{P}| + |\mathcal{S}|\right).$$

Proof If $d^{-2}M|\mathcal{P}|^{1/2} \le |\mathcal{S}| \le d^{-8}M|\mathcal{P}|^2$, we can apply Theorem 11 directly, which results in the first term of the bound. If $|\mathcal{S}| > d^{-8}M|\mathcal{P}|^2$, then we can apply Lemma 12 with $X := \mathcal{P}$ and $Y := \mathcal{S}$ to get

$$|I| = O_M(|\mathcal{P}||\mathcal{S}|^{1/2} + |\mathcal{S}|) = O_M(|\mathcal{S}|).$$

On the other hand, if $|\mathcal{S}| < d^{-2}M|\mathcal{P}|^{1/2}$, then Lemma 12 with $X := \mathcal{S}$ and $Y := \mathcal{P}$ gives

$$|I| = O_M(|\mathcal{S}||\mathcal{P}|^{1/2} + |\mathcal{P}|) = O_M(|\mathcal{P}|).$$

Combining these bounds proves the corollary.

Next we prove a version of Theorem 11 where the condition on the excluded complete bipartite subgraph is weakened in a different way: Instead of requiring every two points to lie in a bounded number of surfaces, we only require this for any s points. Such a bound was given for curves in [14]; see [26] for a similar statement for surfaces in $\mathbb{R}^4$. To prove it we only have to modify the counting step in the proof of Theorem 11.

Theorem 14 *Let A_1 and A_2 be finite subsets of $\mathbb{R}^2$ and $\mathcal{P} := A_1 \times A_2$. Let $\mathcal{S}$ be a finite set of algebraic surfaces in $\mathbb{R}^4$ that have good fibers and are defined by polynomials of degree at most d. Let $I \subset I(\mathcal{P}, \mathcal{S})$ be an incidence subgraph containing no $K_{s,t}$. Assume that $|\mathcal{S}| \le d^{-(4s-2)}|\mathcal{P}|^s$, and that $|A_1| \le |A_2|$ and $|\mathcal{S}| \ge |A_1|^{1-s}|A_2|^s$. Then*

$$|I| = O_{d,s,t}(|\mathcal{P}|^{\frac{s}{2s-1}}|\mathcal{C}|^{\frac{2s-2}{2s-1}}).$$

Proof As said, we reuse the partitioning and intersection steps from the proof of Theorem 11, and we jump in at the counting step.

Let I_1 be the subset of incidences $(p, S) \in I$ such that S has at most $s - 1$ incidences from I in the cell that p lies in. Let I_2 be the subset of incidences $(p, S) \in I$ such that C has at least s incidences from I in the cell that p lies in. The fact that a surface from $\mathcal{S}$ intersects $O(d^2r^2)$ cells implies

$$|I_1| = O_{d,s}\left(r^2|\mathcal{S}|\right).$$

On the other hand, given s points in one cell, there are by assumption fewer than t surfaces $S \in \mathcal{S}$ containing all s points. Thus we have

$$|I_2| = O\left(r^4 \cdot t \cdot \left(\frac{|\mathcal{P}|}{r^4}\right)^s\right) = O\left(t \cdot \frac{|\mathcal{P}|^s}{r^{4s-4}}\right).$$

Setting $r^{4s-2} = \frac{|\mathcal{P}|^s}{|\mathcal{S}|}$ gives

$$|I| = O_{d,s,t}(|\mathcal{P}|^{\frac{s}{2s-1}}|\mathcal{C}|^{\frac{2s-2}{2s-1}}).$$

We need to ensure that $d^2 \le r^2 \le |A_1|$. The assumption that $|\mathcal{S}| \le d^{-(4s-2)}|\mathcal{P}|^s$ gives $r^{4s-2} \ge d^{4s-2}$, and the assumption that $|\mathcal{S}| \ge |A_1|^{1-s}|A_2|^s$ gives $r^{4s-2} \le |A_1|^{2s-1}$. □

Again, we can prove a version with more practical conditions.

Corollary 15 *Let A and B be finite subsets of $\mathbb{R}^2$ with $|A| = |B|$, and $\mathcal{P} := A \times B \subset \mathbb{R}^4$. Let $\mathcal{S}$ be a finite set of surfaces in $\mathbb{R}^4$ that have good fibers and are defined by polynomials of degree at most d. Let $I \subset I(\mathcal{P}, \mathcal{S})$ be an incidence subgraph containing no $K_{s,t}$ or $K_{t,2}$. Then*

$$|I| = O_{d,s,t}\left(|\mathcal{P}|^{\frac{s}{2s-1}}|\mathcal{S}|^{\frac{2s-2}{2s-1}} + |\mathcal{P}| + |\mathcal{S}|\right).$$

Proof If $|\mathcal{P}|^{1/2} \le |\mathcal{S}| \le d^{-(4s-2)}|\mathcal{P}|^s$, we can apply Theorem 11 directly, which results in the first term of the bound. If $|\mathcal{S}| > d^{-(4s-2)}|\mathcal{P}|^s$, then Lemma 12 gives $|I| = O_{d,s,t}(|\mathcal{S}|)$, while if $|\mathcal{S}| < |\mathcal{P}|^{1/2}$, then Lemma 12 gives $|I| = O_M(|\mathcal{P}|)$. Combining these bounds proves the corollary. □

6 Corollaries for Curves in $\mathbb{C}^2$

In this section we deduce several incidence bounds for complex algebraic curves from Theorem 11, including Theorem 1. There are many different ways to vary these statements, and we certainly do not cover all combinations, but we focus on those that have turned out useful in applications (see Sect. 7 and [17, 25]).

We make some effort to determine the dependence of the bounds on the parameters d and M, because this is of interest in the applications [17, 25]. In the incidence bounds for curves in $\mathbb{C}^2$ that were proved in [19, 22, 26], determining this dependence seems challenging.

The following corollary is our first practical incidence bound for curves in $\mathbb{C}^2$. Note that the second term in the bound is a bit awkward, but this seems unavoidable when $|A| \neq |B|$.

Corollary 16 *Let A and B be finite subsets of $\mathbb{C}$ with $|A| \le |B|$, and let $\mathcal{P} := A \times B$. Let $\mathcal{C}$ be a finite set of algebraic curves in $\mathbb{C}^2$ of degree d such that no two have a common component. Let $I \subset I(\mathcal{P}, \mathcal{C})$ be an incidence subgraph containing no $K_{2,M}$. Then*

$$|I| = O(d^{4/3}M^{1/3}|\mathcal{P}|^{2/3}|\mathcal{C}|^{2/3} + d^{-1}M|A|^{-1/2}|B|^{5/2} + d^4|\mathcal{C}|).$$

Proof Let I_1 be the subset of incidences $(p, C) \in I$ such that p lies on a horizontal or vertical line contained in C. Since the curves have no common components, each horizontal or vertical line occurs at most once, and any point is contained in at most two such lines, so

$$|I_1| \le 2|\mathcal{P}|.$$

Let I_2 be the subset of incidences $(p, C) \in I$ such that p does not lies on a horizontal or vertical line contained in C. Let $\mathcal{C}^*$ be the set of curves obtained by removing all the horizontal and vertical lines from the curves in $\mathcal{C}$; we have $|\mathcal{C}^*| \le |\mathcal{C}|$. We can view I_2 as a subgraph of the incidence graph $I(\mathcal{P}, \mathcal{C}^*)$. The fact that the curves in $\mathcal{C}^*$ contain no horizontal or vertical lines implies that the associated surfaces in $\mathbb{R}^4$ have good fibers, and by Lemma 7 the surfaces are defined by polynomials of degree at most $2d$. Hence we can apply Theorem 11 to obtain

$$|I_2| = O(d^{4/3}M^{1/3}|\mathcal{P}|^{2/3}|\mathcal{C}|^{2/3}),$$

unless we have $d^8|\mathcal{C}^*| > M|\mathcal{P}|^2$ or $d^2|\mathcal{C}^*| < M|B|^2/|A|$.

Suppose that $d^8|\mathcal{C}^*| > M|\mathcal{P}|^2$. Since I contains no $K_{2,M}$, we can use Lemma 12 to get

$$|I| = O(M^{1/2}|\mathcal{P}||\mathcal{C}^*|^{1/2} + |\mathcal{C}^*|) = O(d^4|\mathcal{C}|).$$

Suppose that $d^2|\mathcal{C}^*| < M|B|^2/|A|$. Because the curves do not have common components, any two curves intersect in at most d^2 points by Bézout's Inequality (Lemma 4). Thus I_2 contains no $K_{d^2+1,2}$, so by Lemma 12 we have

$$|I| \le |I(\mathcal{P}, \mathcal{C}^*)| = O((d^2)^{1/2}|\mathcal{C}^*||\mathcal{P}|^{1/2} + |\mathcal{P}|) = O(d^{-1}M|A|^{-1/2}|B|^{5/2}).$$

Combining these bounds finishes the proof.

With a little more work, we can remove the condition that no two curves have a common component, with almost no effect on the bound; this is Theorem 1. Note that we lose the flexibility of allowing a subgraph of the incidence graph. Indeed, the curves could all share a common component, and on that component the incidence subgraph could be any bipartite graph $K_{2,M}$, which need not satisfy the desired bound.

Theorem 1 *Let A and B be finite subsets of $\mathbb{C}$ with $|A| = |B|$, and $\mathcal{P} := A \times B$. Let $\mathcal{C}$ be a finite set of algebraic curves in $\mathbb{C}^2$ of degree at most d, such that any two points of $\mathcal{P}$ are contained in at most M curves of $\mathcal{C}$. Then*

$$|I(\mathcal{P}, \mathcal{C})| = O\left(d^{4/3}M^{1/3}|\mathcal{P}|^{2/3}|\mathcal{C}|^{2/3} + M(\log M + \log d)|\mathcal{P}| + d^4|\mathcal{C}|\right).$$

Proof We have to deal with horizontal or vertical lines in the curves, and with the case $d^2|\mathcal{C}| < M|\mathcal{P}|^{1/2}$; the other cases can be treated as in Corollary 16.

Let $\mathcal{C}_1$ be the *multiset* of horizontal and vertical lines contained in curves of $\mathcal{C}$, and let $\mathcal{C}_2$ be the curves that remain after these lines have been removed. In total there are at most $d|\mathcal{C}|$ lines in $\mathcal{C}_1$ (counted with multiplicity). The lines that contain at most one point of $\mathcal{P}$ together give at most $d|\mathcal{C}|$ incidences. The lines that contain at least two points of $\mathcal{P}$ have multiplicity at most M by assumption, so a point on such a line is contained in at most $2M$ such lines (counted with multiplicity), resulting in at most $2M|\mathcal{P}|$ incidences. Hence

$$|I(\mathcal{P}, \mathcal{C}_1)| = O(M|\mathcal{P}| + d|\mathcal{C}|).$$

Now suppose $d^2|\mathcal{C}| < M|\mathcal{P}|^{1/2}$. We split each curve in $\mathcal{C}_2$ into its at most d irreducible components. The components that contain at most one point of $\mathcal{P}$ give altogether at most $d|\mathcal{C}|$ incidences. Let $\mathcal{C}^*$ be the *multiset* of components that contain at least two points of $\mathcal{P}$. A curve in $\mathcal{C}^*$ has multiplicity at most M by assumption.

Let $\mathcal{C}_{ij}$ be the *set* of curves in $\mathcal{C}^*$ that have multiplicity between 2^i and 2^{i+1} and degree between 2^j and 2^{j+1}. The sum of all the degrees of all the components of the curves in $\mathcal{C}$ is at most $d|\mathcal{C}|$, so the number of curves of degree at least 2^j that occur with multiplicity at least 2^i is bounded by $d|\mathcal{C}|/2^{i+j}$. Thus

$$|\mathcal{C}_{ij}| \le d|\mathcal{C}|/2^{i+j} \le d^{-1}M|\mathcal{P}|^{1/2}/2^{i+j},$$

and two distinct curves in $\mathcal{C}_{ij}$ intersect in at most $2^{j+1}\cdot 2^{j+1} = 4\cdot 2^{2j}$ points by Lemma 4. Hence the incidence graph $I(\mathcal{P},\mathcal{C}_{ij})$ contains no $K_{(4\cdot 2^{2j}+1),2}$, so Lemma 12 gives

$$|I(\mathcal{P},\mathcal{C}_{ij})| = O\left((2^{2j})^{1/2}(d^{-1}M|\mathcal{P}|^{1/2}/2^{i+j})|\mathcal{P}|^{1/2} + |\mathcal{P}|\right) = O\left(2^{-i}d^{-1}M|\mathcal{P}| + |\mathcal{P}|\right).$$

Therefore,

$$\begin{aligned} |I(\mathcal{P},\mathcal{C}^*)| &\le \sum_{i=1}^{\log M}\sum_{j=1}^{\log d} 2^{i+1} I(\mathcal{P},\mathcal{C}_{ij}) = O\left(\sum_{i=1}^{\log M}\sum_{j=1}^{\log d} d^{-1}M|\mathcal{P}| + 2^i|\mathcal{P}|\right) \\ &= O\left(d^{-1}M\log M\log d|\mathcal{P}| + M\log d|\mathcal{P}|\right) = O\left(M(\log M + \log d)|\mathcal{P}|\right). \end{aligned}$$

Together with $|I(\mathcal{P},\mathcal{C}_2)| = |I(\mathcal{P},\mathcal{C}^*)| + O(d|\mathcal{C}|)$ this completes the proof. □

Finally, we state a curve version of Corollary 15.

Corollary 17 *Let A and B be finite subsets of $\mathbb{C}$ with $|A| = |B|$, and $\mathcal{P} := A\times B$. Let $\mathcal{C}$ be a finite set of algebraic curves in $\mathbb{C}^2$ of degree at most d, such that any s points of $\mathcal{P}$ are contained in at most t curves of $\mathcal{C}$. Then*

$$|I(\mathcal{P},\mathcal{C})| = O_{d,s,t}\left(|\mathcal{P}|^{\frac{s}{2s-1}}|\mathcal{C}|^{\frac{2s-2}{2s-1}} + |\mathcal{P}| + |\mathcal{C}|\right).$$

Proof If $|\mathcal{P}|^{1/2} \le |\mathcal{C}| \le d^{-(4s-2)}|\mathcal{P}|^s$, we can apply Theorem 11 to the surfaces associated to the curves, which results in the first term of the bound. If $|\mathcal{C}| > d^{-(4s-2)}|\mathcal{P}|^s$, then Lemma 12 gives

$$|I(\mathcal{P},\mathcal{C})| = O_{s,t}(|\mathcal{P}||\mathcal{C}|^{1-1/s} + |\mathcal{C}|) = O_{d,s,t}(|\mathcal{C}|).$$

If $|\mathcal{C}| < |\mathcal{P}|^{1/2}$, then arguing as in the proof of Theorem 1 gives $|I(\mathcal{P},\mathcal{C})| = O_{d,t}(|\mathcal{P}|)$. □

7 Applications

We now show several examples of applications in which the assumption that the point set is a Cartesian product is satisfied. We do not work out these applications in the greatest generality here, but merely give some samples that should illustrate the usefulness of our bounds.

Rich transformations. Elekes [6, 7] introduced various questions of the following form: *Given a group G of transformations on some set X and an integer k, what is the maximum size of*

$$R_k(S) := \{\varphi \in G : |\varphi(S) \cap S| \geq k\}$$

for a finite subset $S \subset X$? The work of Guth and Katz [11] involved this question for $X := \mathbb{R}^2$ and G the group of Euclidean isometries of $\mathbb{R}^2$. Solymosi and Tardos [22] gave the bound $|R_k(A)| = O(|A|^4/k^3)$ when $X := \mathbb{C}$ and G is the group of linear transformations from $\mathbb{C}$ to $\mathbb{C}$, and the bound $|R_k(A)| = O(|A|^6/k^5)$ when $X := \mathbb{C}$ and G is the group of Möbius transformations from $\mathbb{C}$ to $\mathbb{C}$.

We give one example to illustrate how our incidence bound can be used for this type of problem. We consider the group of Möbius transformations $(cz + a)/(dz + b)$ for which $c = 0, d = 1$; i.e., the *inversion transformations*.

Theorem 18 *Let $A \subset \mathbb{C}$ be finite and let $R_k(A)$ be the set of inversion transformations $\varphi_{ab}(z) = a/(z + b)$, with $a, b \in \mathbb{C}$ and $a \neq 0$, for which $|\varphi_{ab}(A) \cap A| \geq k$. Then*

$$|R_k(A)| = O\left(\frac{|A|^4}{k^3}\right).$$

Proof Let $\mathcal{P} := A \times A$. Define $C_{ab} := Z_{\mathbb{C}^2}(y(x + b) - a)$ and set $\mathcal{C} := \{C_{ab} : \varphi_{ab} \in R_k(A)\}$. Then for every $C_{ab} \in \mathcal{C}$ we have $|C_{ab} \cap \mathcal{P}| \geq k$.

The curves C_{ab} are clearly distinct. Suppose that two points $(x, y), (x', y') \in \mathbb{C}^2$ lie on the curve C_{ab}. Then we have $y(x + b) = a = y'(x' + b)$, so

$$(y - y')b = y'x' - yx.$$

If $y \neq y'$, then b is determined by this equation, and a is determined by $a = y(x + b)$. If $y' = y \neq 0$, then we have $x' = x$, a contradiction. If $y = 0$, we would have $a = 0$, also a contradiction. Thus at most two curves C_{ab} pass through any two points $(x, y), (x', y')$.

Theorem 1 then gives

$$k \cdot |\mathcal{C}| \leq |I(\mathcal{P}, \mathcal{C})| = O((|A|^2)^{2/3}|\mathcal{C}|^{2/3} + |A|^2 + |\mathcal{C}|).$$

This implies $|R_k(A)| = |\mathcal{C}| = O\left(|A|^4/k^3\right)$. □

Elekes–Nathanson–Ruzsa-type problems. In [9], Elekes, Nathanson, and Ruzsa considered generalizations of sum-product inequalities over $\mathbb{R}$. Their proofs converted these problems into incidence problems for points and curves over $\mathbb{R}$, with the point set being a Cartesian product. So these problems are well-suited to our incidence bounds over $\mathbb{C}$.

For instance, one of the main results of [9] stated that if $f : \mathbb{R} \to \mathbb{R}$ is a convex function and $A \subset \mathbb{R}$ a finite set, then

$$\max\{|A+A|, |f(A)+f(A)|\} = \Omega(n^{5/4}) \quad \text{and} \quad |A+f(A)| = \Omega(n^{5/4}),$$

where $f(A) := \{f(a) : a \in A\}$. For a rational function $f \in \mathbb{R}(x)$, the same bounds can be deduced by splitting up the graph of f into convex and concave pieces (the number of which is bounded in terms of the degree of f). Over $\mathbb{C}$, it is not clear what the analogue of a convex function would be, but for polynomials or rational functions, these bounds could be generalized to $\mathbb{C}$. We do this for the specific function $f(x) = 1/x$, thereby solving Problem 2.10 in Elekes's survey [7].

Theorem 19 *Let* $A \subset \mathbb{C}$ *be finite and write* $1/A := \{1/a : a \in A\}$. *Then*

$$\max\{|A+A|, |1/A+1/A|\} = \Omega(n^{5/4}) \quad \textit{and} \quad |A+1/A| = \Omega(n^{5/4}).$$

Proof Set $\mathcal{P} := (A+A) \times (1/A+1/A)$,

$$C_{ab} := Z_{\mathbb{C}^2}((x-a)(y-1/b)-1),$$

and $\mathcal{C} := \{C_{ab} : a, b \in A\}$. The curve C_{ab} equals the graph $y = 1/(x-a) + 1/b$. Then each of the $|A|^2$ curves C_{ab} has $|C_{ab} \cap \mathcal{P}| \geq |A|$, since for every $a' \in A$ we have $x = a + a' \in A + A$ and

$$\frac{1}{x-a} + \frac{1}{b} = \frac{1}{a'} + \frac{1}{b} = y \in 1/A + 1/A,$$

so $(x, y) \in C_{ab}$.

We now check that the curves in $\mathcal{C}$ meet the conditions of Theorem 1. Suppose that two points $(x, y), (x', y') \in \mathbb{C}^2$ lie on the curve C_{ab}. Then $x \neq a, x' \neq a$. We have

$$y - y' = \frac{1}{x-a} - \frac{1}{x'-a},$$

so $(y-y')(x-a)(x'-a) = x' - x$. This implies $y \neq y'$, since otherwise we would also get $x' = x$. Then we get

$$a^2 - (x+x')a + \left(xx' + \frac{x-x'}{y-y'}\right) = 0.$$

At most two a satisfy this equation, and a determines b by $y = 1/(x-a) - 1/b$. Thus at most two curves C_{ab} pass through the points $(x, y), (x', y')$.

By Theorem 1, we get (the second and third term have no effect)

$$|A| \cdot |A|^2 \leq |I(\mathcal{P}, \mathcal{C})| = O\left((|A|^2)^{2/3}(|A+A| \cdot |1/A + 1/A|)^{2/3}\right).$$

This gives $|A+A| \cdot |1/A + 1/A| = \Omega(|A|^{5/2})$.

For the second statement, we define $C^*_{ab} := Z((x - 1/a)(y-b) - 1)$, which is $|A|$-rich on $(A + 1/A) \times (A + 1/A)$. The conditions of Theorem 1 can be checked in a similar way, so

$$|A|^3 = O\left((|A|^2)^{2/3}(|A + 1/A|^2)^{2/3}\right),$$

which gives $|A + 1/A| = \Omega(|A|^{5/4})$. □

Elekes–Rónyai-type problems. Elekes and Rónyai [8] introduced another class of questions that lead to incidence problems on Cartesian products in a natural way. The strongest result in this direction was recently obtained by Raz, Sharir, and Solymosi in [16], and it states the following. Let $f(x, y) \in \mathbb{R}[x, y]$ be a polynomial of constant degree, let $A, B \subset \mathbb{R}$ with $|A| = |B| = n$, and write $f(A, B) := \{f(a, b) : a \in A, b \in B\}$. Then we have $|f(A, B)| = \Omega(n^{4/3})$, unless f is of the form $g(h(x) + k(y))$ or $g(h(x) \cdot k(y))$, with $g, h, k \in \mathbb{R}[z]$. In other words, f is an "expander" unless it has a special form. A generalization of this statement is proved in [17], using our Theorem 1.

Again, the typical approach to these problems is by converting them into incidence problems between points and curves, with the points forming a Cartesian product. The general analysis is considerably more difficult; in particular, the curves can actually have many common components, and one needs to show that they do not have too many common components, unless f has a special form.

To illustrate how our incidence bounds can extend such results to $\mathbb{C}$, we establish a simple case, where the polynomial does not have the special form. Moreover, this case is a nice geometric question. It was first considered in [8], and the real equivalent of the bound below was obtained by Sharir, Sheffer, and Solymosi [18], whose proof we follow here.

Consider the "Euclidean distance" defined by $D(p, q) := (p_x - q_x)^2 + (q_x - q_y)^2$ for $p = (p_x, p_y), q = (q_x, q_y) \in \mathbb{C}^2$, and write $D(A, B) := \{D(a, b) : a \in A, b \in B\}$.

Theorem 20 *Let L_1, L_2 be two lines in $\mathbb{C}^2$, and $A \subset L_1$, $B \subset L_2$ with $|A| = |B| = n$. Then*

$$|D(A, B)| = \Omega(n^{4/3}),$$

unless L_1 and L_2 are parallel or orthogonal.

Proof If the lines are not parallel or orthogonal, we can assume that L_1 is the x-axis, and that L_2 contains the origin and is not vertical. Then the lines can be parametrized by $p(x) = (x, 0)$ and $q(y) = (y, my)$, for some $m \in \mathbb{C}\backslash\{0\}$, so that the distance is given by

$$f(x, y) := D(p(x), q(y)) = (x - y)^2 + m^2y^2.$$

We will show that the polynomial f is an expander in the sense of Elekes and Rónyai.

Set $\mathcal{P} := A \times A$,

$$C_{bb'} := Z_{\mathbb{C}^2}(f(x, b) - f(y, b')),$$

and $\mathcal{C} := \{C_{bb'} : b, b' \in B\}$. The equation of $C_{bb'}$ is

$$(x - b)^2 - (y - b')^2 = m^2(b'^2 - b^2),$$

which defines a hyperbola, unless $b = b'$. The curves of the form C_{bb} have altogether at most n^2 incidences, so we can safely ignore them. A quick calculation shows that any two points are contained in at most two hyperbolas $C_{bb'}$ with $b \neq b'$. Thus, by Theorem 1, we have

$$|I(\mathcal{P}, \mathcal{C})| = O\left((|A|^2)^{2/3}(|B|^2)^{2/3} + |A|^2 + |B|^2\right) = O(n^{8/3}).$$

Writing $f^{-1}(c) := \{(a, b) \in A \times B : f(a, b) = c\}$ and using Cauchy–Schwarz gives

$$|I(\mathcal{P}, \mathcal{C})| + n^2 \geq |\{(a, b, a', b') \in (A \times B)^2 : f(a, b) = f(a', b')\}|$$

$$= \sum_{c \in f(A,B)} |f^{-1}(c)|^2 \geq \frac{1}{|f(A, B)|}\left(\sum_{c \in f(A,B)} |E_c|\right)^2 = \frac{n^4}{|f(A, B)|}.$$

Therefore $|f(A, B)| = \Omega(n^4/I(\mathcal{P}, \mathcal{C})) = \Omega(n^{4/3})$. □

References

1. S. Barone, S. Basu, On a real analogue of Bezout inequality and the number of connected components of sign conditions, in Proceedings of the London Mathematical Society, vol. 112, n. 1 (1 January 2016), pp. 115–145, arXiv:1303.1577
2. S. Basu, R. Pollack, M.-F. Roy, *Algorithms in Real Algebraic Geometry* (Springer, Berlin, 2003)
3. S. Basu, M. Sombra, Polynomial partitioning on varieties and point-hypersurface incidences in four dimensions, Discrete Comput. Geom. **55**(1), 158–184 (January 2016), arXiv:1406.2144
4. J. Bochnak, M. Coste, M.-F. Roy, *Real Algebraic Geometry* (Springer, Berlin, 1998)
5. B. Bollobás, *Modern Graph Theory* (Springer, Berlin, 1998)
6. G. Elekes, *On the Dimension of Finite Point Sets II. "Das Budapester Programm"* (2011), arXiv:1109.0636
7. G. Elekes, SUMS versus PRODUCTS in Number Theory. Algebra and Erdős Geometry, Paul Erdős and his Mathematics II, Bolyai Society Mathematical Studies **11**, 241–290 (2002)

8. G. Elekes, L. Rónyai, A combinatorial problem on polynomials and rational functions. J. Comb. Theory, Ser. A **89**, 1–20 (2000)
9. G. Elekes, M. Nathanson, I.Z. Ruzsa, Convexity and sumsets. J. Number Theory **83**, 194–201 (2000)
10. G. Fischer, *Plane Algebraic Curves* (American Mathematical Society, Providence, 2001)
11. L. Guth, N.H. Katz, On the Erdős distinct distances problem in the plane. Ann. Math. **181**, 155–190 (2015)
12. J. Harris, *Algebraic Geometry: A First Course* (Springer, Berlin, 1992)
13. J. Heintz, Definability and fast quantifier elimination in algebraically closed fields. Theor. Comput. Sci. **24**, 239–277 (1983)
14. J. Pach, M. Sharir, On the number of incidences between points and curves. Comb. Probab. Comput. **7**, 121–127 (1998)
15. O.E. Raz, M. Sharir, The number of unit-area triangles in the plane: Theme and variations. Combinatorica **37**(6), 1221–1240 (December 2017). Also in arXiv:1501.00379
16. O.E. Raz, M. Sharir, J. Solymosi, Polynomials vanishing on grids: The Elekes-Rónyai problem revisited, in *Proceedings of the Thirtieth Annual Symposium on Computational Geometry* (2014), pp. 251–260, arXiv:1401.7419
17. O.E. Raz, M. Sharir, F. de Zeeuw, Polynomials vanishing on Cartesian products: The Elekes-Szabó Theorem revisited, in *31st International Symposium on Computational Geometry (SoCG 2015)* (2015), pp. 522–536. Also in arXiv:1504.05012
18. M. Sharir, A. Sheffer, J. Solymosi, Distinct distances on two lines. J. Comb. Theory, Ser. A **120**, 1732–1736 (2013)
19. A. Sheffer, E. Szabó, J. Zahl, Point-curve incidences in the complex plane. Combinatorica **38**(2), 487–499 (April 2018), arXiv:1502.07003
20. J. Solymosi, On the number of sums and products. Bull. Lond. Math. Soc. **37**, 491–494 (2005)
21. J. Solymosi, T. Tao, An incidence theorem in higher dimensions. Discrete Comput. Geom. **48**, 255–280 (2012)
22. J. Solymosi, G. Tardos, On the number of k-rich transformations, in *Proceedings of the Twenty-Third Annual Symposium on Computational Geometry* (2007), pp. 227–231
23. J. Solymosi, V. Vu, Distinct distances in high dimensional homogeneous sets, Towards a theory of geometric graphs. Contemp. Math. **342**, 259–268 (American Mathematical Society, 2004)
24. C.D. Tóth, The Szemerédi-Trotter theorem in the complex plane. Combinatorica **35**, 95–126 (2015)
25. C. Valculescu, F. de Zeeuw, *Distinct values of bilinear forms on algebraic curves*. Contributions to Discrete Mathematics, **11**(1), (July 2016). Also in arXiv:1403.3867
26. J. Zahl, A Szemerédi-Trotter type theorem in $\mathbb{R}^4$. Discrete Comput. Geom. **54**, 513–572 (2015)

Combinatorial Distance Geometry in Normed Spaces

Konrad J. Swanepoel

Abstract We survey problems and results from combinatorial geometry in normed spaces, concentrating on problems that involve distances. These include various properties of unit-distance graphs, minimum-distance graphs, diameter graphs, as well as minimum spanning trees and Steiner minimum trees. In particular, we discuss translative kissing (or Hadwiger) numbers, equilateral sets, and the Borsuk problem in normed spaces. We show how to use the angular measure of Peter Brass to prove various statements about Hadwiger and blocking numbers of convex bodies in the plane, including some new results. We also include some new results on thin cones and their application to distinct distances and other combinatorial problems for normed spaces.

1 Introduction

Paul Erdős [58] introduced many combinatorial questions into geometry. Progress in solving these and many subsequent problems went hand-in-hand with corresponding advances in combinatorics and combinatorial number theory. Recently, some spectacular results were obtained using the polynomial method, which introduced strong connections to algebra and algebraic geometry. In this survey, we would like to explore a different direction, and consider combinatorial questions for other norms. There have been sporadic attempts at generalising geometric questions of Erdős to other normed spaces, an early example being a paper of Fullerton [73]. According to Erdős [61], Ulam was also interested in generalizing certain distance problems to other metrics. This survey is an attempt at presenting the literature in a systematic way.

K. J. Swanepoel (✉)
Department of Mathematics, London School of Economics and Political Science, Houghton Street, London WC2A 2AE, United Kingdom
e-mail: k.swanepoel@lse.ac.uk

G. Ambrus et al. (eds.), *New Trends in Intuitive Geometry*, Bolyai Society Mathematical Studies 27, https://doi.org/10.1007/978-3-662-57413-3_17

We will also present new proofs of known results and give results that have not appeared in the literature before. Since we will confine ourselves to normed spaces, it is natural that problems involving distances will play a special role. However, many of these problems have alternative formulations in terms of packings and coverings of balls, or involve packings and coverings in their solutions, so there is some overlap with the general theory of packing and covering, as conceived by László Fejes Tóth [67] and others. Nevertheless, we make no attempt here to give a systematic treatment of packing and covering, apart from reviewing what is known about Hadwiger numbers (or translative kissing numbers) of convex bodies and some close relatives, as these numbers show up when we consider minimum-distance graphs and minimal spanning trees.

We have left out many topics with a combinatorial flavour, due to limitations on space and time. These include results on vector sums in normed spaces (such as in the papers [6, 14, 106, 121, 186]), embeddings of metric spaces into normed spaces, a topic with applications in computer science (see [130, Chap. 15] and [149]), Menger-type results [11, 12, 125], and isometries and variants such as unit-distance preserving maps and random geometric graphs (for instance [9, 79]). For recent surveys on covering and illumination, see Bezdek and Khan [23] and Naszódi [145]. For a recent survey on discrete geometry in normed spaces, see Alonso, Martini, and Spirova [5].

1.1 Outline

After setting out some terminology in the next subsection, we will survey the Hadwiger number of a convex body, as well as some variants of this notion in Sect. 2. In Sect. 3 we survey recent results on equilateral sets. Although these two sections may at first not seem central to this paper, Hadwiger and equilateral numbers are often the best known general estimates for various combinatorial quantities. Then we consider three graphs that can be defined on a finite point set in a normed space: the minimum-distance graph, the unit-distance graph and the diameter graph. Section 4 covers minimum-distance graphs. Since many results on unit-distance and diameter graphs have a similar flavour, we cover them together in Sect. 5. We briefly consider some other graphs such as geometric minimum spanning trees, Steiner minimum trees and sphere-of-influence graphs in Sect. 6. Then in Sect. 7, we present some applications of an angular measure introduced by Brass [34], in order to give simple proofs of various two-dimensional results on relatives of the Hadwiger number. In particular, we prove a result of Zong [213] that the blocking number of any planar convex disc equals four. Finally, in Sect. 8 we give a systematic exposition of thin cones, introduced in [175] and rediscovered and named in [74]. We build on an idea of Füredi [74] to give an up-to-now best upper bound for the cardinality of a k-distance set in a d-dimensional normed space when k is very large compared to d (Theorem 29). (This bound has very recently been improved by Polyanskii [154].)

1.2 Terminology and Notation

For background on finite-dimensional normed linear spaces from a geometric point of view, see the survey [129] or the first five chapters of [201]. We denote a normed linear space by X, its unit ball by B_X or just B, and the unit sphere by ∂B_X. Our spaces will almost exclusively be finite dimensional. We will usually refer to these spaces as normed spaces or just spaces when there is no risk of confusion. If we want to emphasize the dimension d of a normed space, we denote the space by X^d. We will measure distances exclusively using the norm.

We write $\widehat{x}$ for the normalization $\frac{1}{\|x\|}x$ of a non-zero $x \in X$. If $A, B \subseteq X$ and $\lambda \in \mathbb{R}$, then we define, as usual, $A + B := \{a + b \colon a \in A, b \in B\}$, $\lambda A := \{\lambda a \colon a \in A\}$, $-A := (-1)A$, and $A - B := A + (-B)$. The interior, boundary, convex hull, and diameter of $A \subseteq X$ are denoted by int A, ∂A, $\operatorname{conv}(A)$, and $\operatorname{diam}(A)$, respectively. The translate of A by the vector $v \in X$ is denoted by $A + v := A + \{v\}$.

The *dual* of the normed space X is denoted by X^*. All finite-dimensional normed spaces are reflexive: $(X^*)^*$ is canonically isomorphic to X. A norm $\|\cdot\|$ is called *strictly convex* if $\|x + y\| < \|x\| + \|y\|$ whenever x and y are linearly independent, or equivalently, if ∂B_X does not contain a non-trivial line segment. A norm $\|\cdot\|$ is called *smooth* if it is C^1 away from the origin o, or equivalently, if each boundary point of the unit ball has a unique supporting hyperplane. Recall that a finite-dimensional normed space is strictly convex if and only if its dual is smooth.

For $p \in [1, \infty)$, we let $\ell_p^d = (\mathbb{R}^d, \|\cdot\|_p)$ be the d-dimensional ℓ_p space with norm

$$\|(x_1, \dots, x_d)\|_p := \left(\sum_{i=1}^{d} \|x_i\|^p\right)^{1/p}$$

and denote its unit ball by B_p^d. The space $\ell_\infty^d = (\mathbb{R}^d, \|\cdot\|)$ has norm

$$\|(x_1, \dots, x_d)\|_\infty := \max \|x_i\| .$$

We also denote the Euclidean space ℓ_2^d by $\mathbb{E}^d$, the Euclidean unit ball B_2^d by B^d, the d-cube B_∞^d by C^d, and the d-dimensional cross-polytope B_1^d by O^d. For any two normed spaces X and Y and $p \in [1, \infty]$, we define their ℓ_p-sum $X \oplus_p Y$ to be the Cartesian product $X \times Y$ with norm $\|(x, y)\|_p := \|(\|x\|, \|y\|)\|_p$.

We define $\lambda(X) = \lambda(B_X)$ to be the largest length (in the norm) of a segment contained in ∂B_X. It is easy to see that $0 \leqslant \lambda(X) \leqslant 2$, that $\lambda(X) = 0$ if and only if X is strictly convex, and if X is finite-dimensional, $\lambda(X) = 2$ if and only if X has a 2-dimensional subspace isometric to ℓ_∞^2 [34].

2 The Hadwiger Number (Translative Kissing Number) and Relatives

In the next five subsections we discuss the Hadwiger number and four of its variants: the lattice Hadwiger number, the strict Hadwiger number, the one-sided Hadwiger number and the blocking number. See also Sect. 7 for a derivation of these numbers for 2-dimensional spaces.

2.1 *The Hadwiger Number*

Let C be a convex body in a finite-dimensional vector space. A *Hadwiger family* of C is a collection of translates of C, all touching C and with pairwise disjoint interiors. The *Hadwiger number* (or *translative kissing number*) $H(C)$ of C is the maximum number of translates in a Hadwiger family of C. (The term *Hadwiger number* was introduced by L. Fejes Tóth [69].)

Denote the *central symmetral* of C by $B := \frac{1}{2}(C - C)$. By a well-known observation of Minkowski, $\{v_i + C : i \in I\}$ is a Hadwiger family if and only if $\{v_i + B : i \in I\}$ is a Hadwiger family. Also, $\{v_i + B : i \in I\}$ is a Hadwiger family if and only if $\{v_i : i \in I\}$ is a collection of unit vectors in the normed space with B as unit ball, such that $\|v_i - v_j\| \geqslant 1$ for all distinct $i, j \in I$. We define the Hadwiger number $H(X)$ of a finite-dimensional normed space X as the Hadwiger number $H(B_X)$ of the unit ball.

The Hadwiger number of X is known to be a tight upper bound for the maximum degrees of minimum-distance graphs (Sect. 4) and spanning trees (Sect. 6.1) in X, and this is why we survey what is known about this number, updating the earlier surveys of Zong [216, 217] and Böröczky Jr. [31, Sect. 9.6].

There is a recent comprehensive survey by Boyvalenkov, Dodunekov and Musin [33] on the Hadwiger number (also known as kissing number) of Euclidean balls. We only remind the reader of the following facts. Wyner [208, Sect. 5], improving on Shannon [169], determined the lower bound of $H(B^d) \geqslant (2/\sqrt{3})^{d+o(d)}$ using a greedy argument. This lower bound is essentially still the best known (see also the end of this Sect. 2.1 below), as is the upper bound $H(B^d) \leqslant 2^{0.401d+o(d)}$ by Kabatiansky and Levenshtein [100]. The following exact numbers are known: $H(B^3) = 12$ (with a long history culminating in Schütte and Van der Waerden [168]), $H(B^4) = 24$ (Musin [142]), $H(B^8) = 240$ and $H(B^{24}) = 196560$ (Levenshtein [118] and Odlyzko and Sloane [148]).

Hadwiger [90] showed the upper bound $H(C) \leqslant 3^d - 1$ for all d-dimensional convex bodies C, attained by an affine d-cube, and by a result of Groemer [82] only by affine d-cubes. In particular, the Hadwiger number of a parallelogram is 8. Grünbaum [85], answering a conjecture of Hadwiger [90], showed that $H(C) = 6$ for any planar convex body C that is not a parallelogram. The non-trivial part is showing the upper bound $H(C) \leqslant 6$. In Sect. 7 we show how this follows from the existence of an angular measure introduced by Brass [34].

Grünbaum [85] conjectured that $H(C)$ is an even number for all convex bodies, as it is in the plane, but this turned out to be false. Talata (unpublished) constructed a 3-dimensional polytope with Hadwiger number 17, and Joós [99] constructed one with Hadwiger number 15.

Robins and Salowe [162] showed that the octahedron has Hadwiger number 18 (this was also independently discovered by Larman and Zong [114] and Talata [194]). Larman and Zong [114] showed that the rhombic dodecahedron has Hadwiger number 18, and also gave results for certain elongated octahedra. Robins and Salowe [162] also obtained lower bounds for ℓ_p-balls, in particular $H(\ell_1^d) \geqslant 2^{0.0312\ldots d-o(d)}$ and $H(\ell_p^d) \geqslant (2-\varepsilon_p)^d$ for all $p \in (1,\infty)$, where $\varepsilon_p \in (0,1)$ and $\varepsilon_p \to 0$ as $p \to \infty$; the latter was rediscovered by Xu [209, Theorem 4.2], who also obtained some (weaker) constructive bounds from algebraic geometry codes. Slightly better bounds for $p \leqslant 2$ and close to 2 can be found in [176]. Larman and Zong [114] also showed $H(\ell_p^d) \geqslant (9/8)^{d+o(d)}$. It follows from the main result in Talata [195] that $H(\ell_1^d) \geqslant 1.13488^{d+o(d)}$ (see the next paragraph).

Proving a conjecture of Zong [214], Talata [191] showed that the Hadwiger number of the tetrahedron is 18. (This equals the Hadwiger number of the central symmetral of a tetrahedron, which is the affine cuboctahedron.) Talata [195] found a lower bound of $1.13488^{d+o(d)}$ for the d-dimensional simplex and more generally for *d-orthoplexes*, that is, the intersection of a $(d+1)$-dimensional cube with a hyperplane orthogonal to a diagonal). Since the difference body of a d-dimensional simplex is the hyperplane section of a $(d+1)$-dimensional cross-polytope through its centre parallel to a facet, this also gives the best-known lower bound for the ℓ_1-norm, as mentioned in the previous paragraph. Talata [195] conjectured an upper bound of $1.5^{d-o(d)}$ for the d-dimensional simplex.

The inequality

$$H(C_1 \times C_2) \geqslant (H(C_1)+1)(H(C_2)+1)-1$$

for the Cartesian product of the convex bodies C_1 and C_2 is straightforward. Zong [220] showed that equality holds if either C_1 or C_2 is at most 2-dimensional, and presented some more general conditions where equality holds. Talata [197] gave examples of convex bodies C_1 and C_2 for any dimensions larger than 2 for which this inequality is strict. In the same paper he constructed strictly convex d-dimensional bodies C such that $H(C) \geqslant \Omega(7^{d/2})$ and made the following two conjectures:

Conjecture 1 (Talata [197]) *In each pair of dimensions $d_1, d_2 \geqslant 3$ there exist d_1-dimensional convex bodies K_1, K_1' and d_2-dimensional convex bodies K_2, K_2' such that $H(K_1) = H(K_1')$ and $H(K_2) = H(K_2')$, but $H(K_1 \times K_2) \neq H(K_1' \times K_2')$.*

Conjecture 2 (Talata [197]) *There exists a constant $c > 0$ such that $H(C) \leqslant (3-c)^d$ for all strictly convex d-dimensional convex bodies.*

By an old result of Swinnerton-Dyer [189], $H(B) \geqslant d^2 + d$ for all d-dimensional B. For $d = 2, 3$ the Euclidean ball attains this bound. However, for sufficiently large d it turns out that the Hadwiger number grows exponentially in d, independent of the specific body. Bourgain (as reported in [77]) and Talata [190] showed the existence of an exponential lower bound by using Milman's Quotient-Subspace Theorem [138]. An explicit exponential lower bound of $H(B) \geqslant \Omega((2/\sqrt{3})^d)$ for any d-dimensional convex body B follows from Theorem 1 in Arias-de-Reyna, Ball, and Villa [7]. Note that this is essentially as large as the best known lower bound for the d-dimensional Euclidean ball found by Wyner [208].

2.2 Lattice Hadwiger Number

The *lattice Hadwiger number* $H_L(C)$ of a convex body C is defined to be the largest size of a Hadwiger family $\{v_i + C : i \in I\}$ of C that is contained in a lattice packing $\{v + C : v \in \Lambda\}$, where Λ is a full-dimensional lattice. By the observation of Minkowski mentioned in Sect. 2.1, $H_L(C) = H_L(B)$ where B is the central symmetral of C. We also define the lattice Hadwiger number of a finite-dimensional normed space X as $H_L(X) = H_L(B_X)$. The lattice Hadwiger number plays a role in bounding the maximum number of edges of a minimum-distance graph in X (Sect. 4.1).

Minkowski [139] already showed that $H_L(C) \leqslant 3^d - 1$ and $H_L(C) \leqslant 2(2^d - 1)$ if C is strictly convex. It is easily observed that $H(C) = H_L(C)$ for planar convex bodies C, and for the d-dimensional cube, $H_L(C^d) = H(C^d) = 3^d - 1$. The result of Swinnerton-Dyer [189] mentioned earlier, actually shows that $H_L(C) \geqslant d^2 + d$ for all d-dimensional convex bodies C. This seems to be the best-known lower bound valid for all convex bodies. Zong [217] posed the problem to show that for all d-dimensional convex bodies C, $H_L(C) \geqslant \Omega(c^d)$ for some constant $c > 1$ independent of d. The best asymptotic lower bound for the Euclidean ball is $H_L(B^d) \geqslant 2^{\Omega(\log^2 d)}$, attained by the Barnes–Wall lattice, as shown by Leech [117][1].

Zong [214] determined the lattice Hadwiger number of the tetrahedron T in 3-space: $H_L(T) = 18$, and determined a lower bound of $d^2 + d + 6\lfloor d/3 \rfloor$ for simplices. For Euclidean space these numbers are known up to dimension 9 (Watson [207]): $H_L(B^3) = 12$, $H_L(B^4) = 24$, $H_L(B^5) = 40$, $H_L(B^6) = 72$, $H_L(B^7) = 126$, $H_L(B^8) = 240$, $H_L(B^9) = 272$. In particular, $H_L(\mathbb{E}^9) = 272 < 306 \leqslant H(\mathbb{E}^9)$ is the smallest dimension where H and H_L differ for a Euclidean ball (although they are equal in dimension 24). Zong [218] showed that in each dimension $d \geqslant 3$ there exists a convex body C such that $H(C) > H_L(C)$. His example is a d-cube with two opposite corners cut off. Recall that Talata [197] constructed d-dimensional strictly convex bodies C with $H(C) \geqslant \Omega(7^{d/2})$. When compared with Minkowski's upper bound $H_L(C) \leqslant 2(2^d - 1)$ for all strictly convex bodies C, this shows that the gap between $H(C)$ and $H_L(C)$ can be very large, even for strictly convex sets (see also [193]).

[1] Very recently, Serge Vlăduţ [206] found an exponential lower bound for $H_L(B^d)$.

2.3 Strict Hadwiger Number

A *strict Hadwiger family* of C is a collection of translates of C, all touching C and all pairwise disjoint (that is, no two overlap or touch). The *strict Hadwiger number* $H'(C)$ of C is the maximum number of translates in a strict Hadwiger family of C. We also define the strict Hadwiger number of a finite-dimensional normed space X as $H'(X) = H'(B_X)$. Clearly, $H'(C) \leqslant H(C)$, and it is not difficult to see that the strict Hadwiger number of the d-dimensional cube is $H'(C^d) = 2^d$.

Doyle, Lagarias, and Randall [55] showed that $H'(C) = 5$ if C is a planar convex body that is not a parallelogram. (Robins and Salowe [162] observed that $H'(C^2) = 4$ for the parallelogram C^2). See Sect. 7 for a simple proof of this fact using angular measures.

Robins and Salowe [162] studied $H'(X)$ in connection to minimal spanning trees in a finite-dimensional normed space X; see Sect. 6.1. For the 3-dimensional Euclidean ball B^3, $H'(B^3) = 12$, as demonstrated by the many configurations of 12 pairwise non-touching balls, all touching a central ball [111]. Robins and Salowe [162] showed that for the regular octahedron O^3, $13 \leqslant H'(O^3) \leqslant 14$, and that for each $d \geqslant 3$ there exists $p \in (1, \infty)$ such that $H'(\ell_p^d) > 2^d = H'(C^d)$. Talata [190] showed that there is also an exponential lower bound for H', and the explicit exponential lower bound of $H'(C) \geqslant \Omega((2/\sqrt{3})^d)$ also follows from the results of Arias-de-Reyna, Ball, and Villa [7] mentioned at the end of Sect. 2.1.

Talata [197] studied H' for Cartesian products of convex bodies. In particular, he showed that if $C_1, \ldots, C_n$ are convex discs, with k parallelograms among them, then

$$H'(C_1 \times C_2 \times \cdots \times C_n) = 4^k(4 \cdot 6^{n-k} + 1)/5.$$

He also showed that there exist d-dimensional convex bodies K_d for which $H'(K_d) = \Omega(7^{d/2})$, from which his example of a strictly convex body with Hadwiger number $\Omega(7^{d/2})$ follows. Indeed, given any convex body with a strict Hadwiger configuration, it is easy to modify the convex body so that it becomes strictly convex and the Hadwiger configuration stays strict. Hence, Conjecture 2 would imply that $H'(C) \leqslant (3-c)^d$ for any d-dimensional convex body C.

2.4 One-Sided Hadwiger Number

The *one-sided Hadwiger number* $H_+(C)$ of a convex body C is the maximum number of translates in a Hadwiger family $\{v_i + C : i \in I\}$ such that $\{v_i : i \in I\}$ is contained in a closed half space with the origin on its boundary. We also define the one-sided Hadwiger number $H_+(X)$ of the normed space X to be the $H_+(B_X)$. Clearly, for the circular disc B^2 we have $H_+(B^2) = 4$. It is easy to show that $H_+(B) = 4$ for any convex disc B except the parallelogram C^2, where $H_+(C^2) = 5$ (see Sect. 7 for proofs). The *open one-sided Hadwiger number* $H_+^o(C)$ of C is defined similarly by replacing 'closed half space' by 'open half space' in the definition. We also define the open one-sided Hadwiger number $H_+^o(X)$ of the normed space X to be $H_+^o(B_X)$. The open one-sided Hadwiger number bounds the minimum degree of a minimum-distance graph in X (Sect. 4.2). It is not hard to show that $H_+^o(X^2) = 3$ for any normed plane X^2 with $\lambda(X^2) \leqslant 1$, and $H_+^o(X^2) = 4$ otherwise (see again Sect. 7 for proofs). G. Fejes Tóth [65] showed that for the 3-dimensional Euclidean ball B^3 we have $H_+(B^3) = 9$ (see also Sachs [164] and A. Bezdek and K. Bezdek [16] for alternative proofs). Kertész [104] showed that $H_+^o(B^3) = 8$. Musin [143] showed that for the 4-dimensional Euclidean ball, $H_+(B^4) = 18$. (K. Bezdek [18] observed that it follows from Musin's determination of $H(B^4)$ [142] that $18 \leqslant H_+(B^4) \leqslant 19$.) Bachoc and Vallentin [8] found the exact value $H_+(B^8) = 183$ and upper bounds for Euclidean spaces of dimension up to 10, improving earlier bounds by Musin [144].

K. Bezdek and Brass [21] showed that $H_+(C) \leqslant 2 \cdot 3^{d-1} - 1$ for any d-dimensional convex body C, with equality attained only by the affine d-cube C^d. They ask as an open problem for a tight upper bound of $H_+^o(C)$ valid for all d-dimensional convex bodies. Lángi and Naszódi [112] generalized some of the results in [21].

2.5 *Blocking Number*

Zong [213] introduced the blocking number of a convex body: the minimum number of non-overlapping translates of C, all touching C, and such that no other translate can touch C without overlapping some of these translates. Equivalently, the *blocking number* $B(C)$ of a convex body C is the minimum number of translates in a maximal Hadwiger family of C. The *strict blocking number* $B'(C)$ of C is the minimum size of a maximal strict Hadwiger family of C. Thus clearly, $B(C) \leqslant H(C)$ and $B'(C) \leqslant H'(C)$.

Zong [213] showed that the blocking number of any convex disc equals 4. In Sect. 7 we give a simple proof of this result, we determine the strict blocking number of all convex discs, and present some related results, all using angular measures. Dalla, Larman, Mani-Levitska, and Zong [51] determined the blocking numbers of the 3- and 4-dimensional Euclidean balls and of all cubes: $b(B^3) = 6$, $b(B^4) = 9$, $b(C^d) = 2^d$. For further results on blocking numbers, see Yu [210], Yu and Zong [211], and Zong [215–217, 219].

3 Equilateral Sets

A set of S of points in a normed space X is *equilateral* if $\|x - y\| = 1$ for any distinct $x, y \in S$. Let $e(X)$ denote the largest size of an equilateral set of points in X if it is finite. Here we emphasize results that appeared after the survey [180].

Petty [152] and Soltan [171] observed that it follows from a celebrated result of Danzer and Grünbaum [52] that $e(X) \leqslant 2^d$ for all d-dimensional X, and that equality holds iff X is isometric to ℓ_∞^d (equivalently, iff the unit ball is an affine d-cube).

The following conjecture has been made often [140, 152, 201] (see also [85]):

Conjecture 3 (Petty [152]) *For all d-dimensional X,* $e(X) \geqslant d + 1$.

It is simple to see that this conjecture holds for $d = 2$. Petty [152] established it for $d = 3$. He in fact proved that in any normed space of dimension at least 3, any equilateral set of 3 points can be extended to an equilateral set of 4 points. His proof uses the topological fact that the plane with a point removed is not simply connected. Väisälä [204] gave a more elementary proof that only uses the connectedness of the circle. (Kobos [107] also gave an alternative proof that depends on the 2-dimensional case of the Brouwer Fixed-Point Theorem.) Makeev [124] showed that the conjecture is true for $d = 4$. Brass [37] and Dekster [54] used the Brouwer Fixed-Point Theorem to show that the conjecture holds for spaces sufficiently close to $\mathbb{E}^d$. Swanepoel and Villa [187] used a variant of that argument to show that it holds for spaces sufficiently close to ℓ_∞^d. Kobos [108] showed that the conjecture holds for norms on $\mathbb{R}^d$ for which the norm is invariant under permutation of the coordinates, as well as for d-dimensional subspaces of ℓ_∞^{d+1} and spaces sufficiently close to them. There has also been work on bounding $e(X^d)$ from below in terms of d. Brass [37] and Dekster [54] combined their previously mentioned result on spaces close to Euclidean space with Dvoretzky's Theorem to show that $e(X^d)$ is bounded below by an unbounded function of the dimension. In fact, their proof, when combined with the best known dimension [165] in Dvoretzky's Theorem gives a lower bound $e(X^d) \geqslant \Omega(\sqrt{\log d}/\log\log d)$. Swanepoel and Villa [187] showed that $e(X^d) \geqslant \exp(\Omega(\sqrt{\log d}))$ by using, instead of Dvoretzky's Theorem, a theorem of Alon and Milman [3] on subspaces close to $\mathbb{E}^d$ or ℓ_∞^d, together with a version of Dvoretzky's Theorem for spaces not far from Euclidean space, due to Milman [137]. Roman Karasev (personal communication), in the hope of finding a counterexample to Conjecture 3, asked whether the above conjecture holds for $\mathbb{E}^a \oplus_1 \mathbb{E}^b$, the ℓ_1-sum of two Euclidean spaces. The special case $(a, b) = (1, d-1)$ was considered by Petty [152] (see also Sect. 3.2 below). It is not difficult to show that for Petty's space we have $e(\mathbb{R} \oplus_1 \mathbb{E}^{d-1}) \geqslant d + 1$ [180]. Joseph Ling [120] has shown that for this space, $e(\mathbb{R} \oplus_1 \mathbb{E}^{d-1}) \leqslant d + 2$ and that equality holds for all $d \leqslant 10$. Aaron Lin [119] showed that $e(\mathbb{E}^a \oplus_1 \mathbb{E}^b) \geqslant a + b + 1$ for all $a \leqslant b$ such that $a \leqslant 27$ or $b \equiv 0, 1, a \pmod{a+1}$ among other cases, with the open cases of lowest dimension being $\mathbb{E}^{28} \oplus_1 \mathbb{E}^{40}$ and $\mathbb{E}^{29} \oplus_1 \mathbb{E}^{39}$. There are other results that cast some doubt on Conjecture 3: the existence of small maximal equilateral sets (Sect. 3.2) and the existence of infinite-dimensional normed spaces that do not have infinite equilateral sets, first shown by Terenzi [199, 200]; see also Glakousakis and Mercourakis [81]. (For more on equilateral sets in infinite-dimensional space, see [72, 109, 132, 133].)

Grünbaum [86] showed that for a strictly convex space of dimension 3, $e(X) \leqslant 5$. Building on his work, Schürmann and Swanepoel [167] determined $e(X)$ for various 3-dimensional spaces, and in particular showed the existence of a smooth 3-dimensional space with $e(X) = 6$. They showed that 6 is the maximum for smooth norms in dimension 3 and characterized the 3-dimensional norms that admit equilateral sets of 6 and 7 points (see also Bisztriczky and Böröczky [26] for more general results).

We say that a set S of points in $\mathbb{R}^d$ is *strictly antipodal* if for any two distinct $x, y \in S$ there exist distinct parallel hyperplanes H_x and H_y such that $x \in H_x$, $y \in H_y$, and $A \setminus \{x, y\}$ is contained in the open slab bounded by H_x and H_y. Let $A'(d)$ denote the largest size of a strictly antipodal set in a d-dimensional space [126]. It is easy to see that $e(X^d) \leqslant A'(d)$ for all strictly convex X^d, and that there exists a strictly convex and smooth X^d such that $e(X^d) = A'(d)$. Erdős and Füredi [62] showed that $A'(d) \geqslant \Omega((2/\sqrt{3})^d)$, thus implying that there exist strictly convex d-dimensional normed spaces X^d with $e(X^d) \geqslant \Omega(2/\sqrt{3})^d$. Talata improved this by a construction (described in [31, Lemma 9.11.2]) to $A'(d) \geqslant \Omega(3^{d/3})$, and announced that $A'(d) \geqslant \Omega(5^{d/4})$ (see [31, p. 271]). Subsequently, Barvinok, Lee, and Novik [15] found another construction that shows $A'(d) \geqslant \Omega(3^{d/2})$. This is currently the best-known bound for $e(X^d)$ for strictly convex spaces[2].

Conjecture 4 (Erdős and Füredi [62]) *There exists $c > 0$ such that for all d-dimensional strictly convex spaces X^d, $e(X^d) \leqslant (2 - c)^d$.*

In fact, there is no known proof even that $e(X^d) \leqslant 2^d - 2$ for all strictly convex X^d, except in dimensions $d \leqslant 3$ (Grünbaum [86]). It might also be interesting to look at rounded cubes such as the following. For small $\varepsilon > 0$, let X^d have as unit ball the rounded d-cube $B^d_\infty + \varepsilon B^d_2$. This space is smooth, but not strictly convex. Using results from [167] it can be shown that $e(X^3) = 5$ for all sufficiently small $\varepsilon > 0$. Thus, there exist three-dimensional smooth spaces arbitrarily close to ℓ^3_∞, and with $e(\ell^3_\infty) - e(X^3) = 3$. It might be that for small ε, $e(X^d)$ is very far from $e(\ell^d_\infty) = 2^d$, possibly even linear in d.

Conjecture 5 *For some constant $C > 0$, for each $d \in \mathbb{N}$ there exists an $\varepsilon > 0$ such that $e(X^d) \leqslant Cd$, where X^d is the normed space with unit ball $B^d_\infty + \varepsilon B^d_2$.*

Note that the dual of ℓ^d_∞ is ℓ^d_1, for which Alon and Pudlák [4] has shown $e(\ell^d_1) = \mathrm{O}(d \log d)$. We propose the following conjecture:

Conjecture 6 *For any d-dimensional normed space X with dual X^*, $e(X)e(X^*) \leqslant 2^{d+\mathrm{o}(d)}$.*

It would already be interesting to show that $e(X)e(X^*) = \mathrm{o}(4^d)$.

3.1 Equilateral Sets in ℓ_∞-Sums and the Borsuk Problem

We next consider ℓ_∞-sums of normed spaces. If X and Y are normed spaces, then the unit ball of $X \oplus_\infty Y$ is the Cartesian product $B_X \times B_Y$. It is easy to see that $e(X \oplus_\infty Y) \geqslant e(X)e(Y)$. For certain X and Y it is possible to show that equality holds. The

[2] Very recently, Gerencsér and Harangi [80] proved the lower bound $A'(d) \geqslant 2^{d-1} + 1$.

Borsuk number $b(X)$ of X is defined to be the smallest k such that any subset of X of diameter 1 can be partitioned into k parts, each of diameter strictly smaller than 1. This notion was introduced by Grünbaum [84]. The Borsuk number of Euclidean space received the most attention, ever since Borsuk [30] conjectured that $b(\mathbb{E}^d)$ equals $d+1$. It is known that $b(\mathbb{E}^d) = e(\mathbb{E}^d)$ for $d = 2$ (Borsuk [30]) and $d = 3$ (Perkal [151] and Eggleston [57]), $b(\mathbb{E}^d) \geqslant (1.203\cdots + o(1))^{\sqrt{d}}$ (Kahn and Kalai [101]), $b(\mathbb{E}^d) \leqslant 2^{d-1} + 1$ (Lassak [115]), and $b(\mathbb{E}^d) \leqslant (\sqrt{3/2} + o(1))^d$ (Schramm [166] and Bourgain and Lindenstrauss [32]). Currently, the smallest dimensions for which Borsuk's conjecture is known to be false, are $b(\mathbb{E}^{65}) \geqslant 83$ (Bondarenko [29]) and $b(\mathbb{E}^{64}) \geqslant 71$ (Jenrich and Brouwer [98]). See Raigorodskii [159] and Kalai [102] for recent surveys. Clearly, $b(X) \geqslant e(X)$, although as is shown by the counterexamples to Borsuk's conjecture, these two quantities are very different already for Euclidean spaces. On the other hand, it is easy to see that $b(\ell_\infty^d) = e(\ell_\infty^d) = 2^d$. Grünbaum [84] showed that $b(X^2) = e(X^2)$ for all 2-dimensional spaces.

Zong [217] asked whether $b(X^d) \leqslant 2^d$ for all d-dimensional X^d. It is well known [163] that a d-dimensional convex body K can be covered by $O(2^d d \log d)$ translates of $-(1-\varepsilon)K$, where $\varepsilon > 0$ is arbitrarily small. It follows that $b(X^d) \leqslant O(2^d d \log d)$ for all d-dimensional X^d.

We define the following variant for finite subsets of X. Let the *finite Borsuk number* $b_f(X)$ of X be the smallest number k such that any finite subset of X of diameter 1 can be partitioned into k parts, each of diameter strictly smaller than 1. Then $b(X) \geqslant b_f(X)$, although we have no evidence either way whether these two quantities can differ for some X or not, although we note that $b_f(\ell_\infty^d) = 2^d$ and $b_f(X^2) = b(X^2) = e(X^2)$ for any two-dimensional space X^2.

Proposition 1 *For any two finite-dimensional normed spaces X and Y,*

$$e(X)e(Y) \leqslant e(X \oplus_\infty Y) \leqslant e(X)b_f(Y).$$

Proof If S is an equilateral set in X, and T an equilateral set in Y, with equal distances, then $S \times T$ is equilateral in $X \oplus_\infty Y$. This shows the first inequality. For the second inequality, let E be an equilateral set with distance 1 in $X \oplus_\infty Y$, and let $\pi_Y : X \oplus_\infty Y \to Y$ be the projection onto the second coordinate. Then $\pi_Y(E)$ has diameter at most 1 in Y, so can be partitioned into $k \leqslant b_f(Y)$ parts $E_1, \dots, E_k$, each of diameter < 1. It follows that $\pi_Y^{-1}(E_1), \dots, \pi_Y^{-1}(E_k)$ is a partition of E, and each $\pi_X(\pi_Y^{-1}(E_i))$ is equilateral. Finally, note that $\left|\pi_Y^{-1}(E_i)\right| = \left|\pi_X(\pi_Y^{-1}(E_i))\right|$ for each i. The second inequality follows. □

Corollary 2 *If X and Y are finite-dimensional normed spaces, and one of X or Y is at most 2-dimensional or Euclidean 3-space $\mathbb{E}^3$, then $e(X \oplus_\infty Y) = e(X)e(Y)$.*

Perhaps the simplest ℓ_∞-sum for which this corollary does not determine $e(X)$ is the ℓ_∞-sum of two 4-dimensional Euclidean spaces, with unit ball the Cartesian product of two 4-dimensional Euclidean balls. If $b_f(\mathbb{E}^4)$ were equal to 5, then Proposition 1 would give that $e(\mathbb{E}^4 \oplus_\infty \mathbb{E}^4) = 25$. Most likely it would be easier to determine the value of $e(\mathbb{E}^4 \oplus_\infty \mathbb{E}^4)$ than to settle Borsuk's conjecture in Euclidean 4-space.

3.2 Small Maximal Equilateral Sets

Petty [152] showed that it is not always possible to extend an equilateral set of size at least 4 to an equilateral set properly containing it. In particular, he showed that $\mathbb{R} \oplus_1 \mathbb{E}^{d-1}$ contains a maximal equilateral set of 4 points for each $d \geqslant 3$. Swanepoel and Villa [188] found many other spaces with the property of having small maximal equilateral sets. In particular, for any $p \in [1, 2)$ there exists a C_p such that ℓ_p^d and ℓ_p have maximal equilateral sets of size at most C_p.

Conjecture 7 ([188]) *Any d-dimensional normed space has a maximal equilateral set of size at most $d + 1$.*

This conjecture holds for all ℓ_p^d, $p \in [1, \infty]$, and also for all spaces sufficiently close to one of these spaces [188]. See also Kobos [107], where smooth and strictly convex spaces with maximal equilateral sets of size 4 are constructed.

3.3 Subequilateral Sets

Lawlor and Morgan [116] used the following weakening of equilateral sets. A polytope P in a normed space X is called *subequilateral* if the length of each edge of P equals the diameter of P (in the norm) [183]. We denote the maximum number of vertices in a subequilateral polytope in X^d by $e_s(X^d)$. For any equilateral set S, $\operatorname{conv}(S)$ is a subequilateral polytope, hence $e(X^d) \leqslant e_s(X^d)$. Subequilateral polytopes were used in [116] to construct certain energy-minimizing cones. These polytopes turn out to be so-called edge-antipodal polytopes, introduced by Talata [194], who conjectured that an edge-antipodal 3-polytope has a bounded number of vertices. This was proved by Csikós [48]. K. Bezdek, Bistriczky and Böröczky [20] determined the tight bound of 8, which implies that $e_s(X^3) \leqslant 8$ for any 3-dimensional normed space. Pór [155] proved the generalization of Talata's conjecture to all dimensions, by showing that for each d there exists a c_d such that any edge-antipodal d-polytope has at most c_d vertices. His proof is non-constructive, and only gives $e_s(X^d) < \infty$ for each d. In [183] it is shown that $e_s(X^d) \leqslant (1 + d/2)^d$. This in turn implies the same bound on the number of vertices of an edge-antipodal polytope.

Conjecture 8 ([183]) *A subequilateral set in a d-dimensional normed space has size at most c^d, where $c \geqslant 2$ is some absolute constant.*

The results of Bisztriczky and Böröczky [26] on edge-antipodal 3-polytopes imply that for a strictly convex X^3, $e_s(X^3) \leqslant 5$.

4 Minimum-Distance Graphs

Given any finite packing $\{C + v_i : i = 1, \ldots n\}$ of non-overlapping translates of a d dimensional convex body C, we define the *touching graph* of the packing to be the graph with a vertex for each translate, and with two translates joined by an edge if they intersect (necessarily in boundary points). By the observation of Minkowski mentioned in Sect. 2.1, if $\{C + v_i : i = 1, \ldots n\}$ is a packing of non-overlapping translates of C, then $\{B + v_i : i = 1, \ldots n\}$ is a packing of non-overlapping translates of the central symmetral $B = \frac{1}{2}(C - C)$ of C. Since B is o-symmetric, it is the unit ball of a d-dimensional normed space. We therefore make the following definition.

Given a finite set V in a normed space X with minimum distance $d = \min_{x,y \in V} \|x - y\|$, we define the *minimum-distance graph* of V to be $G_m(V) = (V, E)$ by taking all *minimum distance pairs* xy to be edges, that is, $xy \in E$ whenever $\|x - y\| = d$.

We next consider a selection of parameters of these minimum-distance graphs. As a first remark, the maximum clique number of a minimum-distance graph in X equals $e(X)$, the maximum size of an equilateral set. Note that in any 2-dimensional normed space in which the unit ball is not a parallelogram, minimum-distance graphs are always planar. In fact, no edge can intersect another edge in its relative interior [34].

4.1 *Maximum Degree and Maximum Number of Edges of Minimum-Distance Graphs*

The degree of any vertex in a minimum-distance graph is bounded above by the Hadwiger number $H(X)$ of X. This bound is sharp when taken over all minimum-distance graphs, since the minimum-distance graph of a subset of ∂B, pairwise at distance at least 1, together with the origin o, has degree exactly $H(X)$ at o.

Let $m(n, X)$ denote the maximum possible number of edges of a minimum-distance graph of n points in X. The above observation immediately gives the bound $m(n, X) \leqslant H(X)n/2$. Erdős [58] mentioned that $m(n, \mathbb{E}^2) = 3n - O(\sqrt{n})$. Harborth [94], answering a question of Reutter [161], found the exact value $m(n, \mathbb{E}^2) = \lfloor 3n - \sqrt{12n - 3} \rfloor$ for all $n \geqslant 1$. Brass [34] showed that the same upper bound holds for all norms on $\mathbb{R}^2$ except those isometric to ℓ_∞^2. A key tool in his proof is the introduction of an angular measure with various properties mimicking the Euclidean angular measure (Sect. 7). He also determined the maximum for ℓ_∞^2: $m(n, \ell_\infty^2) = \lfloor 4n - \sqrt{28n - 12} \rfloor$ for all $n \geqslant 1$.

K. Bezdek [18] considered the problem of determining $m(n, \mathbb{E}^3)$, and calls it the combinatorial Kepler problem. In [19] he showed that $6n - 7.862n^{2/3} \leqslant m(n, \mathbb{E}^3) \leqslant 6n - 0.695n^{2/3}$, and in [25], K. Bezdek and Reid improved the upper bound to $6n - 0.926n^{2/3}$. For more on Euclidean minimum-distance graphs, see the recent survey of K. Bezdek and Khan [22]. We next show how an isoperimetric argument gives a slight improvement to the bound $m(n, X) \leqslant H(X)n/2$.

Proposition 3 *For any d-dimensional normed space X^d, $m(n, X^d) \leqslant H(X)n/2 - c_d n^{1-1/d}$, where $c_d > 0$ depends only on d.*

Proof Consider a set V of n points in X^d with unit ball $B = B_X$. Let $G = (V, E)$ be the minimum-distance graph of V. Without loss of generality, the minimum distance may be taken as 1. We may identify X^d with $\mathbb{R}^d$ in such a way that the ellipsoid of maximum volume contained in B is the Euclidean ball B^d. Then a reverse isoperimetric inequality of Ball [10] states that

$$\lambda_{d-1}(\partial B) \leqslant 2d\lambda_d(B)^{1-1/d}, \tag{1}$$

where we denote k-dimensional Lebesgue measure in $\mathbb{R}^d$ by λ_k.

Let $W \subseteq V$ denote the set of all vertices of degree $< H(X)$. Let $S = \bigcup_{v \in V}(B + v)$. Then clearly, $\partial S \subseteq \bigcup_{v \in V}(\partial B + v)$. We claim that $\partial S \subseteq \bigcup_{v \in W}(\partial B + v)$. Indeed, let $x \in \partial S$, say $x \in \partial B + v_0$. Since for each neighbour v of v_0, $x \notin \operatorname{int} B + v$, we have $\|x - v\| \geqslant 1$. It follows that any two points in $\{v : vv_0 \in E\} \cup \{x\} \subset B + v_0$ are at distance at least 1. Therefore, the degree of v_0 is strictly smaller than $H(X)$, hence $v_0 \in W$.

It follows that

$$\lambda_{d-1}(\partial S) \leqslant |W| \lambda_{d-1}(\partial B). \tag{2}$$

Since the balls $\left\{\frac{1}{2}B + v : v \in V\right\}$ form a packing and are contained in S, we have

$$\lambda_d(S) \geqslant |V| \lambda_d(B)/2^d. \tag{3}$$

By the isoperimetric inequality,

$$\lambda_{d-1}(\partial S) \geqslant d\kappa_d^{1/d} \lambda_d(S)^{1-1/d}, \tag{4}$$

where $\kappa_d = \lambda_d(B^d)$ is the volume of the Euclidean unit ball. If we put (1)–(4) together, we obtain $|W| \geqslant (\kappa_d^{1/d}/2^d) |V|^{1-1/d}$, and since $|E| \leqslant \frac{1}{2}(H(X)|V| - |W|)$, the Proposition follows. □

K. Bezdek [17] derived an upper bound with an improved $n^{1-1/d}$ term which also involves the density of a densest translative packing of B_X. The main problem though, already in the Euclidean case, is the coefficient of n in this upper bound, even when the Hadwiger number is known. Indeed, if we consider a lattice packing of the unit ball, we obtain the following obvious lower bound in terms of the lattice Hadwiger number: $m(n, X) \geqslant H_L(X)n/2 - \Omega(n^{1-1/d})$. Therefore, whenever $H(X) = H_L(X)$, we have that $m(n, X) = H(X)n/2 - \Theta(n^{1-1/d})$. However, when these numbers differ, for instance in 9-dimensional Euclidean space where $H_L(\mathbb{E}^9) < H(\mathbb{E}^9)$, we do not even know the main term. When $d > 2$ and n is large, it is also not clear if point sets that maximize $m(n, X)$ have to be pieces of lattices for which H_L is attained.

4.2 Minimum Degree of Minimum-Distance Graphs

Let $\delta(X)$ denote the largest minimum degree of a minimum-distance graph in X. That is,

$$\delta(X) = \max\{\delta(G) \colon \text{G is a minimum-distance graph in X}\}.$$

We can also define $\delta(X)$ as the largest k such that all minimum-distance graphs have a vertex of degree at most k. Another description found in the literature is the following. A finite packing of translates of a convex body C is called a k^+- *neighbour packing* if each translate has at least k neighbours. Then $\delta(X)$ is the largest k such that there exists a finite k^+-neighbour packing of translates of the unit ball B_X.

By considering a vertex of the convex hull of the set of points, we see that $\delta(X) \leqslant H_+^o(X)$. Even in 2-dimensional spaces, there may be strict inequality. For example, if the unit ball is a square with two opposite corners truncated a bit, then $\delta(X^2) = 3$ by a result of Talata [196] (Theorem 21 below), but $H_+^o(X^2) = 4$ (Proposition 23 below). Also, $\delta(\mathbb{E}^3) \leqslant H_+^o(\mathbb{E}^3) = 8$ by the result of Kertész [104] mentioned in Sect. 2.4, but it is unknown whether equality holds. The best known lower bound $\delta(\mathbb{E}^3) \geqslant 6$ is due to a construction of G. Wegner of a 6-regular minimum-distance graph on 240 points in $\mathbb{E}^3$, described in [66].

Most of the results on $\delta(X)$ were obtained by Talata. In [196] he showed that $\delta(X^2) = 3$ if X^2 is not isometric to ℓ_∞^2, and $\delta(\ell_\infty^2) = 4$. In Sect. 7 we give a simple proof of this fact. He also determined $\delta(X^2 \oplus_\infty \mathbb{R}) = 10$ if X^2 is not isometric to ℓ_∞^2, and $\delta(\ell_\infty^3) = 13$. In [198] he considered $\delta(X)$ for an arbitrary finite-dimensional normed space, and showed that $\delta(X) \geqslant H_L(X)/2$, which implies the above-mentioned result of Wegner that $\delta(\mathbb{E}^3) \geqslant 6$. Talata also showed that $\delta(X) = H_L(X)/2$ if X is the ℓ_∞-sum of spaces of dimension at most 2, or equivalently, if the unit ball is the Cartesian product of segments and centrally symmetric convex discs. In [192] he showed that equality still holds if X is the ℓ_∞ sum of spaces of dimension at most 2 or ℓ_1^3. In particular, $\delta(\ell_1^3) = 9$ and $\delta(\ell_\infty^d) = (3^d - 1)/2$. As mentioned in Sect. 2.2, for high-dimensional Euclidean space the best-known lower bound[3] for the lattice Hadwiger number is not particularly strong: $H_L(\mathbb{E}^d) \geqslant 2^{\Omega(\log^2 d)}$. Alon [2] improved the corresponding bound for $\delta(\mathbb{E}^d)$ by showing that $\delta(\mathbb{E}^d) \geqslant 2^{\sqrt{d}}$ if d is a power of 4, hence $\delta(\mathbb{E}^d) \geqslant 2^{\sqrt{d}/2}$ in general. (See also the stronger conjecture of Chen [42] in Sect. 4.3 below). Talata [198] conjectured that $\delta(X) \leqslant H(X)/2$, which holds in dimension 2. In both papers [196, 198], Talata also estimated the smallest

[3] Very recently, Vlăduţ showed that $H_L(\mathbb{E}^d) \geqslant 2^{0.0219(n)-0(n)}$, which also gives an exponential lower bound on $\delta(\mathbb{E}^d)$.

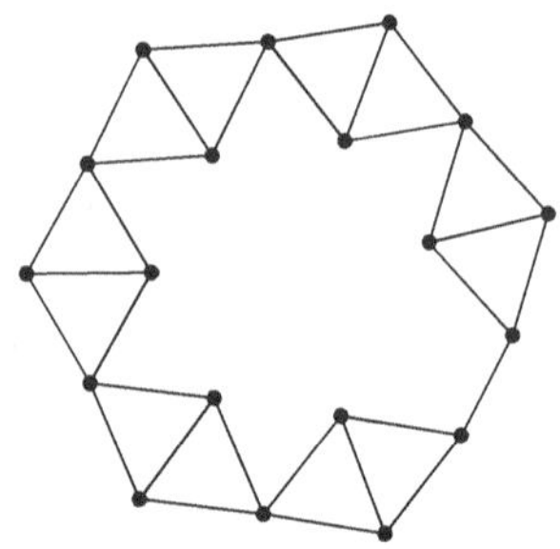

Fig. 1 Maehara's minimum-distance graph with chromatic number 4

number of points in a minimum-distance graph with minimum degree $\delta(X)$. In [198] he considered a lattice version of $\delta(X)$.

4.3 *Chromatic Number and Independence Number of Minimum-Distance Graphs*

Let $\chi_m(X)$ denote the largest chromatic number of a minimum-distance graph in X and $\alpha_m(n, X)$ the smallest independence number of a minimum-distance graph on n points in X. Then $\chi_m(X)\alpha_m(n, X) \geqslant n$. Also, $\chi_m(X) \leqslant \delta(X) + 1$, hence $\alpha_m(n, X) \geqslant n/(\delta(X) + 1)$ [24, Theorem 2]. Talata's conjecture above in Sect. 4.2 would imply the upper bound $\chi_m(X) \leqslant H(X)/2 + 1$. We have no better lower bound for the chromatic number of a general d-dimensional normed space than $\chi_m(X^d) \geqslant e(X^d) \geqslant e^{\Omega(\sqrt{\log d})}$. The Euclidean minimum-distance graph in Fig. 1 has chromatic number 4, which gives $\chi_m(\mathbb{E}^2) \geqslant 4$ (Maehara [122]). Maehara observed that the obvious generalization to higher dimensions gives $\chi_m(\mathbb{E}^d) \geqslant d + 2$. Chen [42] used strongly regular graphs to show that for any $d = q^3 - q^2 + q$, where q is a prime power, $\chi_m(\mathbb{E}^d) \geqslant q^3 + 1$. Chen conjectured that $\chi_m(\mathbb{E}^d) \geqslant c^{\sqrt{d}}$ for some constant $c > 1$.

Since any minimum-distance graph for ℓ_∞^d is a subgraph of the minimum-distance graph (in ℓ_∞^d) of the lattice $\mathbb{Z}^d$ (L. Fejes Tóth and Sauer [70]; see also Brass [34]), we obtain $\chi_m(\ell_\infty^d) \leqslant 2^d$. Since also $\chi_m(\ell_\infty^d) \geqslant e(\ell_\infty^d) = 2^d$, we obtain the exact value $\chi_m(\ell_\infty^d) = 2^d$.

By Talata's result on the minimum degree of 2-dimensional spaces mentioned above, we have $\chi_m(X^2) \leqslant \delta(X^2) + 1 = 4$ for any X^2 not isometric to ℓ_∞^2. (This also follows from the Four-Colour Theorem, since in this case the minimum-distance graph is planar.) It is easily seen that Maehara's graph in Fig. 1 can be realized in any normed plane. Since also $\chi_m(\ell_\infty^2) = 4$, we obtain $\chi_m(X^2) = 4$ for all normed planes. Consequently, $\alpha_m(n, X^2) \geqslant n/4$ for all 2-dimensional X^2. This was observed by Pollack [153] for the Euclidean plane. Csizmadia [50] improved the Euclidean lower bound to $\alpha_m(n, \mathbb{E}^2) \geqslant 9n/35$ and Swanepoel [179] to $\alpha_m(n, \mathbb{E}^2) \geqslant 8n/31$. Pach and Tóth [150] obtained the upper bound $\alpha_m(n, \mathbb{E}^2) \leqslant \lceil 5n/16 \rceil$. Swanepoel [179] also showed the lower bound $\alpha_m(n, X^2) \geqslant n/(4-\varepsilon)$, where $\varepsilon > 0$ depends on X^2, for each X^2 with $\lambda(X^2) \leqslant 1$. Most likely this assumption on X^2 is unnecessary.

Conjecture 9 *For each normed plane X^2, there exists $\varepsilon > 0$ depending only on $\lambda(X^2)$ such that the independence number of any minimum-distance graph on n points in X^2 is at least $\alpha_m(n, X^2) \geqslant n/(4-\varepsilon)$.*

K. Bezdek, Naszódi and Visy [24] introduced a quantity that they call the *k-th Petty number for packings* $P_m(k, X)$: this is the largest n such th at there exists a minimum-distance graph on n points in X with independence number $< k$. Thus, $P_m(2, X) = e(X)$, $P_m(k, X) \geqslant (k-1)e(X)$, and by Ramsey's Theorem, $P_m(k, X) < R(e(X)+1, k) \leqslant \binom{e(X)+k-1}{k-1}$ [24, Proposition 1]. Also, $P_m(k, X) \leqslant k(\delta(X)+1)-1$ [24, Corollary 2] and $P_m(k, \ell_\infty^d) = (k-1)2^d$ [24, Theorem 3].

5 Unit-Distance Graphs and Diameter Graphs

We consider unit-distance graphs and diameter graphs together, as they have similar extremal behaviour in high dimensions. Given a finite set V of points from a normed space X, we define the *unit-distance graph* on V to be the graph with vertex set V and edge set

$$E = \{ab\colon a, b \in V, \|a-b\| = 1\}.$$

We also define the *diameter graph* on V to be the graph with vertex set V and edge set

$$E = \{ab\colon a, b \in V, \|a-b\| = \operatorname{diam}(V)\}.$$

We again consider a selection of parameters of unit-distance and diameter graphs. Note that, as in the case of minimum-distance graphs, the maximum clique number of a unit-distance graph or a diameter graph in X equals $e(X)$, the maximum size of an equilateral set.

5.1 Maximum Number of Edges of Unit-Distance and Diameter Graphs

Let $U(n, X)$ denote the maximum number of edges in a unit-distance graph on n points in X, and let $D(n, X)$ denote the maximum number of edges in a diameter graph on n points in X. It is a difficult problem of Erdős [58] to show that $U(n, \mathbb{E}^2) = \mathrm{O}(n^{1+\varepsilon})$ for all $\varepsilon > 0$, with the best upper bound known $U(n, \mathbb{E}^2) = \mathrm{O}(n^{4/3})$ due to Spencer, Szemerédi and Trotter [173], and the best known lower bound $U(n, \mathbb{E}^2) = \Omega(n^{1+c/\log\log n})$ due to Erdős [58]. Erdős [61] stated that $U(n, \ell_1^2) = (n^2+n)/4$ for all $n > 4$ divisible by 4. Brass [34] determined $D(n, X^2)$ for all two-dimensional normed spaces X^2 and $U(n, X^2)$ whenever X^2 is not strictly convex:

1. $D(n, X^2) = n$ if $\lambda(X^2) = 0$,
2. $U(n, X^2) = D(n, X^2) = \lfloor n^2/4 \rfloor$ if $0 < \lambda(X^2) \leqslant 1$,
3. $U(n, X^2) = \lfloor (n^2+n)/4 \rfloor$ and $D(n, X^2) = \lfloor n^2/4 \rfloor + 1$ if $1 < \lambda(X^2) < 2$, and
4. $U(n, X^2) = \lfloor (n^2+n)/4 \rfloor$ and $D(n, X^2) = \lfloor n^2/4 \rfloor + 2$ if $\lambda(X^2) = 2$ (that is, for X^2 isometric to ℓ_∞^2 and ℓ_1^2).

Brass observed that the same proofs from geometric graph theory that give the bounds $U(n, \mathbb{E}^2) = O(n^{4/3})$ and $D(n, \mathbb{E}^2) = n$ for the Euclidean norm, still go through for all strictly convex norms. Valtr [205] constructed a strictly convex norm and examples of n points with $\Omega(n^{4/3})$ unit-distance pairs (improving earlier results of Brass [36]). This norm has a simple description: $\|(x, y)\| = |y| + \sqrt{x^2 + y^2}$. Its unit ball is bounded by two parabolic arcs with equations $y = \pm\frac{1}{2}(1 - x^2)$, $-1 \leqslant x \leqslant 1$. For this norm, the set

$$\left\{\left(\frac{i}{k}, \frac{j}{2k^2}\right) : i, j \in \mathbb{N}, -k < i \leqslant k, -k^2 < j \leqslant k^2\right\}$$

of $4k^3$ points has $\Omega(k^2)$ unit-distance pairs. The existence of such a piecewise quadratic norm suggests that improving the $O(n^{4/3})$ bound for the Euclidean norm will depend on subtler number-theoretic properties of the Euclidean norm. (Another phenomenon pointing to the difficulty is the existence of n points on the 2-sphere of radius $1/\sqrt{2}$ in $\mathbb{E}^3$ with $\Omega(n^{4/3})$ unit-distance pairs [63].)

Matoušek [131] showed the surprising result that for almost all two-dimensional X^2, $U(n, X^2) = O(n \log n \log\log n)$. Here, *almost all* means that the result holds for all norms except a meager subset of the metric space of all norms, metrized by the Hausdorff distance between their unit balls. This bound is almost best possible, as for any 2-dimensional normed space X^2, a suitable projection of the vertices and edges of a k-dimensional cube onto the plane gives a set of 2^k points with $k2^{k-1}$ unit-distance pairs, thus implying $U(n, X^2) = \Omega(n \log n)$.

In [38, Sect. 5.2, Problem 4], Brass, Moser, and Pach asks whether there is a general construction of n points with strictly more than $\Omega(n \log n)$ unit-distance pairs that can be carried out in all normed spaces of a given dimension $\geqslant 3$. It might even be that in each dimension $d \geqslant 2$, for almost all d-dimensional norms, the number of unit-distance pairs is $O_d(n \log n)$.

The determination of $U(n, \mathbb{E}^3)$ seems to be as difficult as the planar case, with the best known bounds being $O(n^{3/2})$ by Kaplan, Matoušek, Safernová, and Sharir [103] and Zahl [212], and $\Omega(n^{4/3} \log\log n)$ (Erdős [59]), although $D(n, \mathbb{E}^3) = 2n - 2$ is an old result of Grünbaum, Heppes, and Straszewicz [83, 95, 174].

For $d \geqslant 4$, Erdős [59] determined $U(n, \mathbb{E}^d)$ and $D(n, \mathbb{E}^d)$ asymptotically. By an observation of Lenz [59], in Euclidean space of dimension $d \geqslant 4$, the maximum number of unit-distance pairs in a set of n points is at least $\frac{1}{2}(1 - 1/\lfloor d/2 \rfloor)n^2 + n - \lfloor d/2 \rfloor$. By an application of the Erdős–Stone Theorem and some geometry, Erdős found asymptotically matching upper bounds. In [60] he found exact values for even $d \geqslant 4$ and all sufficiently large n divisible by $2d$, showing that for such n, $U(n, \mathbb{E}^d) = \frac{1}{2}(1 - 1/\lfloor d/2 \rfloor)n^2 + n$. Brass [35] determined $U(n, \mathbb{E}^4)$ for all $n \geqslant 1$.

Erdős and Pach [64] showed that $U(n, \mathbb{E}^d) = \frac{1}{2}(1 - 1/\lfloor d/2 \rfloor)n^2 + \Theta(n^{4/3})$ for odd $d \geqslant 5$. In [184], $U(n, \mathbb{E}^d)$ is determined exactly for all even $d \geqslant 6$ and $D(n, \mathbb{E}^d)$ for all $d \geqslant 4$, both for sufficiently large n depending on d. The Lenz construction can be adapted to give the same lower bound $U(n, \ell_p^d) \geqslant \frac{1}{2}(1 - 1/\lfloor d/2 \rfloor)n^2 + n - \lfloor d/2 \rfloor$ for all $p \in [1, \infty]$. For $p \in (1, \infty)$, this lower bound is most likely the right value asymptotically, but for $p = 1$ and $p = \infty$ the Lenz construction can be modified to give a larger lower bound. To simplify the discussion of analogues of the Lenz construction in general, we introduce the following notion. We say that a family of k sets $A_1, \dots, A_k \subset X$ is an *equilateral family* in X if for any two distinct $i, j \in \{1, \dots, k\}$ and $x \in A_i$, $y \in A_j$, $\|x - y\| = 1$. Define $a(X)$ to be the largest k such that for all $m \in \mathbb{N}$, there exists an equilateral family of k sets $A_1, \dots, A_k \subset X$, each of cardinality at least m. Note that $U(n, X) \geqslant \frac{1}{2}(1 - 1/a(X))n^2 + O(1)$.

Proposition 4 *Let $d \geqslant 2$. Then $a(\ell_2^d) = \lfloor d/2 \rfloor$, $a(\ell_1^d) \geqslant d$, $a(\ell_\infty^d) = 2^{d-1}$, and for each $p \in (1, \infty)$, $a(\ell_p^d) \geqslant \lfloor d/2 \rfloor$. For any d-dimensional normed space X^d, $a(X^d) \leqslant 2^d - 1$.*

Proof First let $1 \leqslant p < \infty$. We describe the Lenz construction [59]. Let $e_1, \dots, e_d$ be the standard basis of $\mathbb{R}^d$. Represent $\mathbb{R}^d$ as the direct sum of subspaces $V_1, \dots, V_k$, where $k = \lfloor d/2 \rfloor$, $V_i = \operatorname{span}\{e_{2i-1}, e_{2i}\}$, $i = 1, \dots, k-1$, and $V_k = \operatorname{span}\{e_{d-1}, e_d\}$ if d is even and $V_k = \operatorname{span}\{e_{d-2}, e_{d-1}, e_d\}$ if d is odd. For each $i = 1, \dots, k$, let $C_i = V_i \cap \partial(2^{-1/p} B_p^d)$, the ℓ_p-circle (or sphere if d is odd and $i = k$) in V_i around the origin and with radius $2^{-1/p}$. Let A_i consist of any m points on C_i. Then it is easy to see that $A_1, \dots, A_k$ form an equilateral family, and we obtain $a(\ell_p^d) \geqslant \lfloor d/2 \rfloor$.

The upper bound $a(\ell_2^d) \leqslant \lfloor d/2 \rfloor$ is well known [59]. Suppose $A_1, \dots, A_k \subset \ell_2^d$ form an equilateral family with three points in each set. Then a simple calculation shows that the affine hulls of the A_i are 2-dimensional, pairwise orthogonal, and have a point in common. It follows that $2k \leqslant d$.

We next consider the case $p = 1$. For $i = 1, \dots, k-1$, let A_{2i-1} consist of any m points on a segment on ∂C_i (which is the square ∂O^2), and A_{2i} any m points on the opposite segment on C_i. If d is even, do the same for $i = k$ to obtain an equilateral family of $2k = d$ sets, each of size at least m. If d is odd, then it is easy to find three edges of the octahedron $C_k = O^3$, such that the distance between any two points from different edges equals 1. (Any three pairwise disjoint edges will work.) Again we obtain an equilateral family of $2k + 1 = d$ sets, each of size m, which gives $a(\ell_1^d) \geqslant d$.

Next we consider ℓ_∞^d. As already observed by Grünbaum [87, p. 421] and Makai and Martini [123], if we choose m points on each of 2^{d-1} parallel edges of a ball of radius 1/2 in ℓ_∞^d, we obtain $a(\ell_\infty^d) \geqslant 2^{d-1}$. For the upper bound, let $A_1, \dots, A_k$ be an equilateral family with each $|A_i| > 3^d$. Since the diameter of each A_i is at most 2, we can cover each A_i with a ball of radius 1. Each such ball can be tiled with 3^d balls of radius 1/3, and by the pigeon-hole principle, there are two points of A_i inside one of these balls of radius 1/3. Therefore, we may replace each A_i by an A_i' consisting of two points at distance < 1. It follows that $\bigcup_{i=1}^k A_i'$ has diameter 1, so is contained in $[0, 1]^d$, without loss of generality. For each A_i', the d-cube $[0, 1]^d$ has a smallest face F_i that contains A_i'. (Each F_i is the join in the face lattice of $[0, 1]^d$ of the unique faces that contain the two elements of A_i' in their relative interiors.) Any two of these faces are disjoint, otherwise there would be points from different A_i that are at distance < 1. Since $|A_i'| \geqslant 2$, the dimension of each F_i is at least 1. Therefore, the vertex sets of $F_1, \dots, F_k$ partition the 2^d vertices of $[0, 1]^d$ into parts of size at least 2. It follows that $k \leqslant 2^{d-1}$.

For a general X^d, if $\{A_1, \dots, A_k\}$ is an antipodal family of $k = a(X^d)$ non-empty sets, then for any choice of $p_i \in A_i$, $\{p_i : i = 1, \dots, k\}$ is an equilateral set, hence $k \leqslant 2^d$. If $k = 2^d$, then X^d is isometric to ℓ_∞^d, and $a(X^d) \leqslant 2^{d-1}$ as shown above. Otherwise $a(X^d) < 2^d$. □

Theorem 5 *The maximum number of edges $U(n, X)$ in a unit-distance graph and the maximum number of edges $D(n, X)$ in a diameter graph on n points in a d-dimensional normed space X satisfy the following asymptotics:*

$$\lim_{n\to\infty} \frac{U(n, X)}{n^2} = \lim_{n\to\infty} \frac{D(n, X)}{n^2} = \frac{1}{2}\left(1 - \frac{1}{a(X)}\right).$$

Proof Consider an equilateral family $A_1, \dots, A_k$ in X, with each $|A_i|$ large. Since the diameter of each A_i is at most 2, and we can cover a ball of radius 2 by a finite number of balls of radius 0.49, it follows there exist subsets $A_i' \subset A_i$ such that $\operatorname{diam}(A_i') < 1$ and $|A_i'| \geqslant c_X |A_i|$. It follows that $a(X)$ is also the largest k such that there exists an equilateral family of k sets such that each set has arbitrarily large cardinality and diameter < 1. By choosing n/k points from each A_i', we obtain that $U(n, X) \geqslant D(n, X) \geqslant \frac{1}{2}(1 - 1/a(X))n^2$ for all n. It follows that

$$\liminf_{n\to\infty} \frac{U(n, X)}{n^2} \geqslant \liminf_{n\to\infty} \frac{D(n, X)}{n^2} \geqslant \frac{1}{2}\left(1 - \frac{1}{a(X)}\right).$$

Next, it follows from the Erdős–Stone Theorem that for all $\varepsilon > 0$ there exists n_0 such that $D(n, X) \leqslant U(n, X) \leqslant \frac{1}{2}(1 - a(X)^{-1} + \varepsilon)n^2$ for all $n > n_0$, that is,

$$\limsup_{n\to\infty} \frac{D(n, X)}{n^2} \leqslant \limsup_{n\to\infty} \frac{U(n, X)}{n^2} \leqslant \frac{1}{2}\left(1 - \frac{1}{a(X)}\right),$$

and the theorem follows. □

Corollary 6 *For any $d \geqslant 1$, the maximum number of edges in a unit-distance graph or a diameter graph on n points in ℓ_∞^d is asymptotically $U(n, \ell_\infty^d) = \frac{1}{2}(1 - 2^{1-d})n^2 + o(n^2)$ and $D(n, \ell_\infty^d) = \frac{1}{2}(1 - 2^{1-d})n^2 + o(n^2)$.*

Brass conjectured that ℓ_∞^d attains the maximum number of unit-distance pairs among n points in a d-dimensional normed space, for sufficiently large n. We introduce the following terminology for this maximum number. Let $U_d(n)$ denote the maximum number of edges in a unit-distance graph of n points in a d-dimensional normed space, where the maximum is taken over all norms. Let $D_d(n)$ denote the analogous quantity for the maximum number of edges in a diameter graph.

Conjecture 10 (Brass, Moser, Pach [38, Sect. 5.2, Conjecture 6]) *For each $d \in \mathbb{N}$ there exists $n_0(d)$ such that for all $n \geqslant n_0(d)$, $U_d(n) = U(n, \ell_\infty^d)$.*

We also write $U_d'(n)$ and $D_d'(n)$ for the analogues where we take the maximum only over all strictly convex d-dimensional spaces. The asymptotics as $n \to \infty$ of the four quantities $U_d(n)$, $D_d(n)$, $U_d'(n)$, $D_d'(n)$ can be described in terms of antipodal families. We say that a family $\{A_i : i = 1, \dots, k\}$ of subsets of $\mathbb{R}^d$ is an *antipodal family* if for any pair j, k of distinct indices and any $x \in A_j$, $y \in A_k$ there exist two distinct parallel hyperplanes H_x and H_y, such that $x \in H_x$, $y \in H_y$, and $\bigcup_i A_i$ is in the closed slab bounded by H_x and H_y. We denote by $a(d)$ the largest k such that for each $m \in \mathbb{N}$ there exists an antipodal family $A_1, \dots, A_k$ in $\mathbb{R}^d$ with at least m points in each A_i. We also say that the family A_i is a *strictly antipodal family* if for any pair j, k of distinct indices and any $x \in A_j$, $y \in A_k$ there exist two distinct parallel hyperplanes H_x and H_y, such that $x \in H_x$, $y \in H_y$, and $\bigcup_i A_i \setminus \{x, y\}$ is in the open slab bounded by H_x and H_y. We denote by $a'(d)$ the largest k such that for each $m \in \mathbb{N}$ there exists a strictly antipodal family $A_1, \dots, A_k$ in $\mathbb{R}^d$ with at least m points in each A_i. Note that $a'(d) \leqslant A'(d)$, the largest size of a strictly antipodal set (Sect. 3). The following two results can be proved similarly to Theorem 5, using the following two observations: An equilateral family $A_1, \dots, A_k$ with each $\operatorname{diam}(A_i) < 1$, is an antipodal family, and if the norm is strictly convex, a strictly antipodal family. Conversely, for any (strictly) antipodal family $A_1, \dots, A_k$ with $k \geqslant 2$ there exists a (strictly convex) norm that turns the antipodal family into an equilateral family with each $\operatorname{diam}(A_i) \leqslant 2$. As in Sect. 3, $\operatorname{conv}(A_i)$ can be covered by $O(3^d d \log d)$ translates of $-\frac{1}{2}\operatorname{conv}(A_i)$ [163, Eq. (6)], so by replacing the original m by $m/O(3^d d \log d)$, we may assume that each $\operatorname{diam}(A_i) \leqslant 1$.

Theorem 7 *The maximum number of edges $U_d(n)$ in a unit-distance graph and maximum number of edges $D_d(n)$ in a diameter graph on n points in a d-dimensional normed space satisfy the following asymptotics:*

$$\lim_{n\to\infty} \frac{U_d(n)}{n^2} = \lim_{n\to\infty} \frac{D_d(n)}{n^2} = \frac{1}{2}\left(1 - \frac{1}{a(d)}\right).$$

Theorem 8 (Swanepoel and Valtr [185, Theorem 8]) *The maximum number of edges $U_d'(n)$ in a unit-distance graph and maximum number of edges $D_d'(n)$ in a diameter graph on n points in a strictly convex d-dimensional normed space satisfy the following asymptotics:*

$$\lim_{n\to\infty} \frac{U_d'(n)}{n^2} = \lim_{n\to\infty} \frac{D_d'(n)}{n^2} = \frac{1}{2}\left(1 - \frac{1}{a'(d)}\right).$$

In [185] it is also shown that $W_d(n) = \frac{1}{2}(1 - 1/a'(d))n^2 + o(n^2)$, where $W_d(n)$ is the largest number of pairwise non-parallel unit distance pairs in a set of n points in some strictly convex d-dimensional normed space. Grünbaum [87, p. 421] and Makai and Martini [123] showed that $2^{d-1} \leqslant a(d) \leqslant 2^d - 1$ (see also Proposition 4). Makai and Martini [123] also showed that $a(2) = 2$. (This also follows from Theorem 5 applied to the determination of $D(n, X^2)$ for all 2-dimensional X^2 of Brass [34].)

Conjecture 11 (Grünbaum [87, p. 421], Makai and Martini [123]) *For all $d \geqslant 1$, $a(d) \leqslant 2^{d-1}$. That is, for each $d \in \mathbb{N}$ there exists m such that $\min_i |A_i| \leqslant m$ for any antipodal family $\{A_i : i = 1, \dots, 2^{d-1} + 1\}$ in $\mathbb{R}^d$.*

Csikós et al. [49] showed that $a(3) \leqslant 5$. In the light of Corollary 6 and Theorem 7, Conjecture 10 would imply Conjecture 11. Makai and Martini [123] showed that $a'(3) \geqslant 3$, and conjectured that equality holds. Barvinok, Lee, and Novik [15] showed that $a'(d) \geqslant \Omega(3^{d/2})$. The best upper bound known is the almost trivial $a'(d) \leqslant 2^d - 1$. Conjecture 4 would imply that $a'(d) \leqslant (2 - c)^d$ for some constant $c > 0$.

5.2 Chromatic Number

Denote the maximum chromatic number of all the unit-distance graphs in the normed space X by $\chi_u(X)$ and the maximum chromatic number of all the diameter graphs in X by $\chi_D(X)$.

Recall from Sect. 3.1 that the finite Borsuk number $b_f(X)$ of X is defined to be the smallest k such that any finite subset of X of diameter 1 can be partitioned into k parts of diameter smaller than 1. It is clear that $\chi_u(X) \geqslant \chi_D(X) = b_f(X) \geqslant e(X)$. In particular, as implied by the observations in Sect. 3.1, the chromatic number of any diameter graph in a d-dimensional normed space is at most $(2 + o(1))^d$. The space ℓ_∞^d is an example where 2^d is attained. Also, since $b(X^2) = b_f(X^2) = e(X^2)$ for all X^2, the maximum chromatic number of a diameter graph in X^2 is 3 if X^2 is not isometric to ℓ_∞^2.

By the De Bruijn–Erdős Theorem, $\chi_u(X)$ equals the chromatic number of the infinite unit-distance graph of the whole space X. Clearly, $\chi_u(X) \geqslant \chi_m(X)$. We are not aware of any lower bound for $\chi_u(X)$ valid for all d-dimensional norms, other than those for $\chi_m(X)$ stated in Sect. 4.3. The chromatic number of the Euclidean plane is a famously difficult problem, with the easy bounds $4 \leqslant \chi_u(\mathbb{E}^2) \leqslant 7$ still the best known estimates[4] more than 60 years after this problem was first formulated by Nelson and Hadwiger [78, 91, 170].

Chilakamarri [43] considered general two-dimensional normed spaces, and showed that the bounds $4 \leqslant \chi_u(X^2) \leqslant 7$ hold for all X^2. The lower bound follows since the so-called Moser spindle (Fig. 2) still occurs as a unit-distance graph for any norm, and the upper bound comes from an appropriate tiling of the plane by a hexagon of sides lengths $1/2$ inscribed in the circle of radius $1/2$. Chilakamarri notes that the chromatic number is exactly 4 if the unit ball is a parallelogram or a hexagon, and at

[4] Very recently, Aubrey de Grey [53] improved the lower bound to 5.

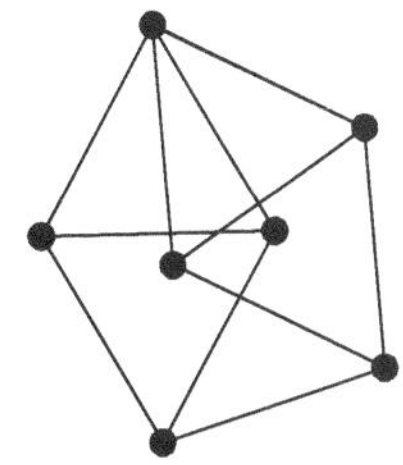

Fig. 2 The Moser spindle is a unit-distance graph for any norm in the plane

most 6 if the unit ball is an octagon. There is no known example of a normed plane for which the chromatic number is known to be more than 4, and Brass, Moser, and Pach ask as a problem to find such a plane[5] [38, Sect. 5.9, Problem 4].

For Euclidean space, Larman and Rogers [113] showed the exponential upper bound $\chi_u(\mathbb{E}^d) \leqslant (3 + o(1))^d$, which is still the best known. Frankl and Wilson [71] were the first to find a lower bound exponential in d: $\chi_u(\mathbb{E}^d) \geqslant ((1 + \sqrt{2})/2 + o(1))^d$. The currently best known lower bound of $(1.239\cdots + o(1))^d$ is due to Raigorodskii [157]. There are many specific upper and lower bounds for low-dimensional $\mathbb{E}^d$; see Raigorodskii's survey [160].

The lower bound of Frankl and Wilson uses $\{0, 1\}$-vectors, and so also gives a lower bound for all ℓ_p^d, or more generally, for any space X^d with a norm that is invariant under permuting and changing the signs of coordinates: $\chi_u(X^d) \geqslant ((1 + \sqrt{2})/2 + o(1))^d$. It is easy to see that $\chi_u(\ell_\infty^d) = e(\ell_\infty^d) = 2^d$. For ℓ_1^d, the best known lower bound is $\chi_u(\ell_1^d) \geqslant (1.365 + o(1))^d$ due to Raigorodskii [158]. For other papers on $\chi_u(\ell_p^d)$, see Broere [41] and Füredi and Kang [75].

Füredi and Kang [76] showed that $\chi_u(X^d) \leqslant 5^{d+o(d)}$ for any d-dimensional X^d. This was improved by Kupavskiy [110] to $\chi_u(X^d) \leqslant 4^{d+o(d)}$. He also showed $\chi_u(\ell_p^d) \leqslant 2^{(1+c_p+o(1))d}$ for all $p > 2$, where $0 < c_p < 1$ and $c_p \to 0$ as $p \to \infty$.

5.3 *Independence Number and Minimum Degree*

We define $\delta_u(X)$ to be the maximum over all minimum degrees of unit-distance graphs in X, if this maximum exists (otherwise we write $\delta_u(X) = \infty$). Similarly, let $\delta_D(X)$ be the maximum over all minimum degrees of diameter graphs in X, if this maximum exists (otherwise $\delta_m(X) = \infty$). In contrast to the case of minimum-distance graphs, very little is known about $\delta_u(X)$ or $\delta_D(X)$ in a normed space X. We only make the following general remarks.

If $U(n, X) = \Omega(n^2)$, or (by Theorem 5) equivalently, $D(n, X) = \Omega(n^2)$, then the Erdős–Stone Theorem implies that there is no upper bound for $\delta_u(X)$ or $\delta_D(X)$:

[5] De Grey should that the Euclidean plane is an example.

if $U(n,X) = \frac{1}{2}(1 - 1/a(X) + o(1))n^2$, equivalently, if $D(n,X) = \frac{1}{2}(1 - 1/a(X) + o(1))n^2$, then there are diameter graphs on n vertices with minimum degree $((a-1)/a + o(1))n$ (and this is sharp).

K. Bezdek, Naszódi, and Visy [24] considered the smallest independence numbers $\alpha_u(n,X)$ of a unit-distance graph on n points in X. We can also define $\alpha_D(n,X)$ to be the smallest independence number of a diameter graph on n points in X. Then $\alpha_D(n,X) \geqslant \alpha_u(n,X)$, and similarly to the case of the minimum-distance graph, $\alpha_D(n,X) \geqslant n/b_f(X) \geqslant n/(\delta_D(X)+1)$, $\chi_u(X)\alpha_u(n,X) \geqslant n$, $\chi_u(X) \leqslant \delta_u(X)+1$ and $\alpha_u(n,X) \geqslant n/(\delta_u(X)+1)$.

They [24] introduced the *k-th Petty number* $P(k,X)$ of X: the largest n such that there exists a unit-distance graph on n points in X with independence number $< k$. This is closely related to their k-th Petty number for packings (discussed in Sect. 4.3 above). Thus, $P(2,X) = P_m(2,X) = e(X)$, $P_m(k,X) \leqslant P(k,X)$, hence $(k-1)e(X) \leqslant P(k,X) < R(e(X)+1,k) \leqslant \binom{e(X)+k-1}{k-1}$ [24, Proposition 1]. They showed that $P(3,X^d) \leqslant 2 \cdot 3^d$, $P(k,X^d) \leqslant (k-1)((k-1)3^d - (k-2))$ for all $k \geqslant 4$, $P(k,X^d) \leqslant (k-1)4^d$, $P(k,X^2) \leqslant 8(k-1)$, $P(k,\ell_\infty^d) = (k-1)2^d$, and $P(k,\mathbb{E}^d) \leqslant (k-1)3^d$ for all $k \geqslant 2$; also $P(k,\ell_p^d) \leqslant (k-1)3^d$ for all $1 < p < \infty$, $k \geqslant 2$ and $d \leqslant 2^p$. They ask whether $P(k,X^d) \leqslant (k-1)2^d$ for all d-dimensional X^d and $k \geqslant 3$ (which would be sharp).

6 Other Graphs

Here we briefly mention three other graphs that are defined for finite sets of points in a finite-dimensional normed space.

6.1 Minimum Spanning Trees

For any finite subset S of a normed space X, any tree T with vertex set S and minimum total length (with the length of an edge measured in the norm) is called a *minimum spanning tree* of S. For any finite subset S of a normed space X with minimum spanning tree T (where distances between points are measured in the norm), let $\Delta(T)$ denote the maximum degree of T. Define $\Delta(S) = \max \Delta(T)$ and $\Delta'(S) = \min \Delta(T)$, where the maximum and minimum is taken over all minimum spanning trees T of S. Finally, let $\Delta(X) = \max \Delta(S)$ and $\Delta'(X) = \max \Delta'(S)$, where the maxima are taken over all finite subsets S of the normed space X. Thus, all minimum spanning trees in X have maximum degree at most $\Delta(X)$, and for each finite subset of X there exists a minimum spanning tree with maximum degree at most $\Delta'(X)$. Cieslik [44] showed that $\Delta(X) = H(X)$ for all normed spaces X. This was rediscovered by Robins and Salowe [162], who also showed that $\Delta'(\ell_p^d) = H'(\ell_p^d)$ for $1 \leqslant p < \infty$. Martini and Swanepoel [127] generalized the last result to all normed spaces: $\Delta'(X) = H'(X)$ for all finite-dimensional X. The proof needs a general position argument that is made exact by means of the Baire Category Theorem.

6.2 Steiner Minimal Trees

For any finite subset S of a finite-dimensional normed space X, any tree $T = (V, E)$ with $S \subseteq V \subset X$ and with each vertex in $V \setminus S$ of degree at least 3, is called a *Steiner tree* of S. The vertices in $V \setminus S$ are called the *Steiner points* of T. A Steiner tree of S of minimum total length is called a *Steiner minimal tree* (SMT) of S. (Since there are always at most $|S| - 2$ Steiner points in a Steiner tree, there will always exist a shortest one by compactness.) Steiner minimal trees are well studied, especially in the Euclidean plane. An overview of the extensive literature on them can be found in the monographs of Hwang, Richards and Winter [97], Cieslik [46], Prömel and Steger [156], and Brazil and Zachariasen [40]. For their history, see Boltyanski, Martini, and Soltan [28] and Brazil, Graham, Thomas, and Zachariasen [39].

Denote the maximum degree of a Steiner point in the SMT T by $\Delta_s(T)$ (and set it to 0 if there are no Steiner points). Also, denote the maximum degree of a non-Steiner points in T by $\Delta_n(T)$. Let $\Delta_s(X) = \max \Delta_s(T)$ and $\Delta_n(X) = \max \Delta_n(T)$, where both maxima are taken over all SMTs T in the normed space X. If T is an SMT of S, then T is clearly still an SMT of any S' such that $S \subseteq S' \subseteq V(T)$. It follows that $\Delta_s(X) \leqslant \Delta_n(X)$.

It is well known that $\Delta_s(\mathbb{E}^d) = \Delta_n(\mathbb{E}^d) = 3$ for all $d \geqslant 2$ [97, Sect. 6.1]. Since a Steiner minimal tree is a minimal spanning tree of its set of vertices, $\Delta_n(X) \leqslant \Delta(X) = H(X)$ (Cieslik [44]). Since any edge joining two points in an SMT can be replaced by a piecewise linear path consisting of segments parallel to the vectors pointing to the extreme points of the unit ball, we obtain the following well-known lemma, going back to Hanan [92] for $X^2 = \ell_1^2$.

Lemma 9 *If the unit ball of X^d is a polytope with v vertices, then $\Delta_s(X^d) \leqslant \Delta_n(X^d) \leqslant v$.*

This gives the upper bounds in $\Delta_s(\ell_1^d) = \Delta_n(\ell_1^d) = 2d$ and $\Delta_s(\ell_\infty^d) = \Delta_n(\ell_\infty^d) = 2^d$. For the lower bound for ℓ_1^d, note that the vertex set S of its unit ball O^d is an equilateral set with distance 2, and that $\{O^d + v : v \in S\}$ is a packing in $2O^d$. It follows that for any Steiner tree of S, the total length of edges or parts of edges in $\operatorname{int}(O^d + v)$ has to be at least 1 for each $v \in S$, hence the tree that joins each vertex in S to o is a SMT with o a Steiner point of degree $2d$. The lower bound $\Delta_s(\ell_\infty^d) \geqslant 2^d$ is shown similarly.

We denote the d-dimensional normed space on $\mathbb{R}^d$ with unit ball $\operatorname{conv}([0, 1]^d \cup [-1, 0]^d)$ by H_d. The unit ball of H^2 is an affine regular hexagon and that of H^3 an affine rhombic dodecahedron. Cieslik [44], [46, Conjecture 4.3.6] made the following conjecture:

Conjecture 12 (Cieslik [44, 46]) *The maximum degree of a vertex in an SMT in a d dimensional normed space X^d satisfies $\Delta_n(X^d) \leqslant 2^{d+1} - 2$, with equality if and only if X^d is isometric to the space H^d.*

By Lemma 9, $\Delta_n(H^d) \leqslant 2^{d+1} - 2$. Cieslik [45] proved the case $d = 2$ of Conjecture 12. In [178] the exact values of $\Delta_n(X^2)$ and $\Delta_s(X^2)$ are determined for all 2-dimensional spaces (see also Martini, Swanepoel and de Wet [128]). In particular, up to isometry, H^2 is the only 2-dimensional space that attains $\Delta_n(X^2) = 6$, with all others satisfying $\Delta_n(X^2) \leqslant 4$. In [182] it is shown that $\binom{d+1}{\lfloor (d+1)/2 \rfloor} \leqslant \Delta_s(H^d) \leqslant \Delta_n(H^d) = \binom{d+2}{\lfloor (d+2)/2 \rfloor} < 2^{d+1} - 2$ for all $d \geqslant 3$, thus partially disproving the conjecture. The spaces H^d give the largest known degrees of SMTs in Minkowski spaces of dimensions 2 to 6, with $\Delta_n(H^2) = 6$, $\Delta_n(H^3) = 10$, $\Delta_n(H^4) = 20$, $\Delta_n(H^5) = 35$, and $\Delta_n(H^6) = 70$, while $\Delta_n(\ell_\infty^d) = 2^d$ is larger for $d \geqslant 7$. It is not clear whether H^d maximises $\Delta_n(X^d)$ for $2 \leqslant d \leqslant 6$, and ℓ_∞^d maximises $\Delta_n(X^d)$ for $d \geqslant 7$.

Morgan [140, Sect. 3], [141, Chap. 10] made a related conjecture.

Conjecture 13 (Morgan [140, 141]) *The maximum degree of a Steiner point in an SMT in any d-dimensional normed space X^d satisfies $\Delta_s(X^d) \leqslant 2^d$.*

The space ℓ_∞^d shows that this conjecture would be best possible. The asymptotically best known upper bound for both conjectures is $\Delta_s(X^d) \leqslant \Delta_n(X^d) \leqslant O(2^d d^2 \log d)$ [181]. It is known that $\Delta_s(X^2) \leqslant 4$ for all X^2 [178]. There are many two-dimensional spaces attaining $\Delta_s(X^2) = 4$, some of them with a unit circle that is piecewise C^∞ [1]. They are characterised in [178].

The sharp upper bound for differentiable norms is $\Delta_s(X^d) \leqslant \Delta_n(X^d) \leqslant d + 1$ [1, 116, 177]. For the ℓ_p norm, $1 < p < \infty$, we have $3 \leqslant \Delta_s(\ell_p^d) \leqslant \Delta_n(\ell_p^d) \leqslant 7$ if $p > 2, d \geqslant 2$, and $\min\{d, \frac{p}{(p-1)\ln 2}\} \leqslant \Delta_s(\ell_p^d) \leqslant \Delta_n(\ell_p^d) \leqslant 2^{p/(p-1)}$ if $1 < p < 2$ and $d \geqslant 3$; see [177] for more detailed estimates.

Conger [47] showed that $\Delta_s(\mathbb{R}^3, \|\cdot\|_1 + \lambda \|\cdot\|_2) \geqslant 6$ for all $0 < \lambda \leqslant 1$. In [1] it is shown that $\Delta_s(\mathbb{R}^2, \|\cdot\|_1 + \lambda \|\cdot\|_2) = 4$ for all $0 < \lambda \leqslant 2 + \sqrt{2}$. The value $\lambda = 2 + \sqrt{2}$ is sharp, since it follows from the results in [178] that $\Delta_s(\mathbb{R}^2, \|\cdot\|_1 + \lambda \|\cdot\|_2) = 3$ for all $\lambda > 2 + \sqrt{2}$. In [182] it was shown that for the space $X^d = (\mathbb{R}^d, \|\cdot\|_1 + \lambda \|\cdot\|_2)$, $\Delta_s(X^d) = \Delta_n(X^d) = 2d$ if $0 < \lambda \leqslant 1$. Conger made the following conjecture [140, Sect. 3], [141, Chap. 10].

Conjecture 14 (Conger) *For any $X^d = (\mathbb{R}^d, \|\cdot\|)$ such that for some $\varepsilon > 0$, $\|\cdot\| - \varepsilon \|\cdot\|_2$ is still a norm, $\Delta_s(X^d) \leqslant 2d$.*

6.3 Sphere-of-Influence Graphs

Toussaint [202] introduced the sphere-of-influence graph of a finite set of points in Euclidean space for application to pattern analysis and image processing. See Toussaint [203] for a recent survey. This notion was later generalized to so-called closed sphere-of-influence graphs by Harary et al. [93] and to k-th closed sphere-of-influence graphs by Klein and Zachmann [105]. Some of their properties have been considered in normed spaces; see [77, 88, 134–136, 146].

Given $k \in \mathbb{N}$ and a finite set S in X, we define the *k-th closed sphere-of-influence graph* with vertex set S as follows. For each $p \in S$, let $r_k(p)$ be the smallest r such that $\{q \in S : q \neq p, \|p-q\| \leqslant r\}$ has at least k elements. Then join two points $p, q \in S$ whenever the closed balls $p + r_k(p)B_X$ and $q + r_k(q)B_X$ intersect. Although there is no upper bound on the maximum degree of a k-th closed sphere-of-influence graph, Naszódi et al. [146] showed that the minimum degree is bounded above by $k\vartheta(X)$, where $\vartheta(X)$ is the largest size of a set of points in $2B_X$ such that the distance between any two points is at least 1 and one of the points is o. A simple packing argument gives the upper bound of $\vartheta(X^d) \leqslant 5^d$, attained by ℓ_∞^d. This then also gives the upper bound of $k\vartheta(X)\,|S|\,/2$ on the number of edges.

7 Brass Angular Measure and Applications

7.1 Angular Measures

Brass [34] introduced a certain angular measure in any normed plane not isometric to ℓ_∞^2, and used it to determine the maximum number of edges in a minimum-distance graph on a set of n points in that plane (see Sect. 4.1). Here we demonstrate how some other combinatorial results on translative packings of a planar convex body can be deduced with minimal effort using this measure.

An *angular measure* on X^2 is a measure μ on the unit circle ∂B of X^2 such that $\mu(\partial B) = 2\pi$, $\mu(A) = \mu(-A)$ for all measurable $A \subseteq \partial B$, and $\mu(\{p\}) = 0$ for all $p \in \partial B$. An angular measure μ is called *proper* if $\mu(A) > 0$ for any non-trivial arc A of ∂B. We measure an angle in the obvious translation invariant way. The following is a list of easily proved properties of angular measures.

Lemma 10 *Let μ be an angular measure in any normed plane.*

1. *The sum of the measures of the interior angles of a simple closed n-gon equals $\pi(n-2)$.*
2. *Two parallel lines are cut at equal angles by a transversal. The converse is also true if the measure is proper.*
3. *Let $abcd$ be a simple quadrilateral with $\|a-b\| = \|b-c\| = \|c-d\| < \|d-a\|$. Then $\mu(\sphericalangle b) + \mu(\sphericalangle c) \geqslant \pi$, with strict inequality if the measure is proper.*

Proof Only the last statement needs proof. We first consider the case where $abcd$ is not convex. If a or d is in the convex hull of the remaining points, say $a \in \triangle bcd$, let $e = d + b - c$. Then $bedc$ is a parallelogram. Since $\|b-c\| = \|b-a\| = \|b-e\|$, we have $a \notin \operatorname{int} \triangle bce$. Then $a \in \triangle cde$, and $\|a-d\| \leqslant \max(\|c-d\|, \|d-e\|) =$

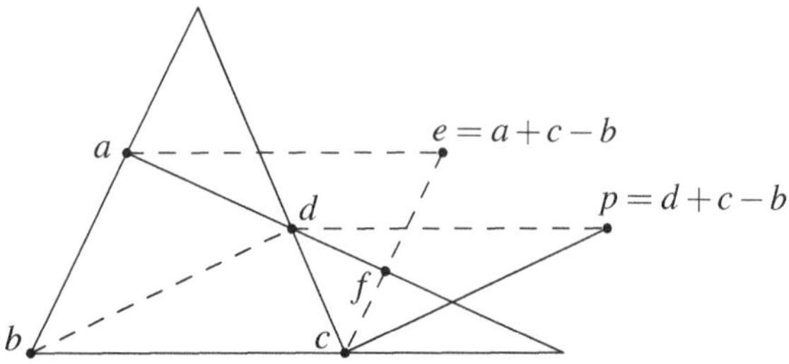

Fig. 3 Proof of Lemmas 10 and 19

$\|c-d\|$, a contradiction. Therefore, b or c is in the convex hull of the remaining points, say $b \in \triangle acd$. Then clearly $\mu(\sphericalangle b) \geqslant \pi$, and also $\mu(\sphericalangle c) > 0$ if μ is proper.

Next we consider the case where $abcd$ is convex. If $ab \parallel cd$, then $abcd$ is a parallelogram and $\|b-c\| = \|a-d\|$, a contradiction. Therefore, the lines ab and cd intersect (Fig. 3). Suppose that $ab \cap cd$ and b are on opposite sides of the line ad. Then the lines ad and bc intersect, otherwise $ad \parallel bc$ and $\|a-d\| < \|b-c\|$, a contradiction. Assume without loss of generality that $ad \cap bc$ and a are on opposite sides of cd (as in Fig. 3). Let $e = a+c-b$. Then $ecba$ is a parallelogram with d in its interior. Let $f = ad \cap ce$. Then by the triangle inequality, $\|a-d\| + \|d-c\| \leqslant \|a-f\| + \|f-c\| \leqslant \|a-e\| + \|e-c\| = \|b-c\| + \|a-b\|$, a contradiction.

Therefore, lines ab and cd intersect in a point on the same side of line ad as b. Then clearly $\mu(\sphericalangle b) + \mu(\sphericalangle c) \geqslant \pi$, with strict inequality if μ is proper. □

An angular measure is called a *Brass measure* if equilateral triangles (in the norm) are equiangular in the measure, that is, $\mu(\sphericalangle abc) = \mu(\sphericalangle bca) = \mu(\sphericalangle cab) = \pi/3$ whenever $\|a-b\| = \|b-c\| = \|c-a\| > 0$. Clearly, ℓ_∞^2 does not have a Brass measure, since in this plane we can find 8 points $a_1, a_2, \dots, a_8$ on ∂B such that $\triangle oa_ia_{i+1}$ is equilateral for each $i = 1, \dots, 8$ (with $a_9 = a_1$), and a Brass measure would give 8 angles of measure $\pi/3$ around the origin. Remarkably, any normed plane not isometric to ℓ_∞^2 has a Brass measure.

Theorem 11 (Brass [34]) *A normed plane with unit ball B admits a Brass measure iff B is not a parallelogram.*

It is not difficult to construct such a measure if the norm is strictly convex, or more generally, if $\lambda(X) \leqslant 1$ (where λ is as defined in Sect. 1.2), since then for any given point on ∂B there are exactly two points on ∂B at distance 1 in the norm from the given point. We sketch the proof of the slightly stronger Theorem 15 below.

We call a maximal segment contained in ∂B of length strictly greater than 1 a *long segment*. Thus, a normed plane X^2 has a long segment iff $\lambda(X^2) > 1$. L. Fejes Tóth [68] calls the direction of a long segment a *critical direction* of the unit ball.

Lemma 12 (Brass [34]) *Let X^2 be a normed plane with unit ball B. Then*

1. *∂B contains at most two parallel pairs of long segments, and*
2. *any long segment on ∂B has length at most 2, with equality iff B is a parallelogram.*

We define the *ends* of a long segment ab to be the two closed subsegments aa' and bb' of ab, where a' and b' are the points on ab such that $\|a - b'\| = \|b - a'\| = 1$. The following lemma is easy to prove.

Lemma 13 *For any Brass measure μ on a normed plane, any long segment has μ-measure $\pi/3$, and the ends of any long segment have μ-measure 0.*

We call a Brass measure *good* if all its non-trivial angles of measure 0 are contained in the ends of long segments. We note the following straightforward lemma.

Lemma 14 *All proper Brass measures on a normed plane X^2 are good. All good Brass measures on X^2 are proper if $\lambda(X^2) \leqslant 1$.*

Brass's proof of Theorem 11 actually gives the following strengthening.

Theorem 15 (Brass [34]) *Any normed plane X^2 for which the unit ball is not a parallelogram, admits a good Brass measure.*

Before we sketch the proof of Theorem 15, we state the following technical result.

Lemma 16 *Suppose that the unit ball B of X^2 is not a parallelogram. Let S be the union of the ends of the long segments of ∂B and the vectors parallel to long segments. Then for each $x \in \partial B \setminus S$ there exists a unique $y = f(x) \in \partial B \setminus S$ such that $\|x - y\| = 1$ and the orientation of $\sphericalangle xoy$ is positive. Furthermore, f is a bijection and satisfies $f \circ f \circ f(x) = -x$ for all $x \in \partial B \setminus S$.*

Proof For each $x \in \partial B$ there exists $y \in \partial B$ such that $\|x - y\| = 1$ and the orientation of $\sphericalangle xoy$ is positive. If $x \notin S$, then x is not parallel to a long segment, and $y =: f(x)$ is unique. It also follows from $x \notin S$ that x is not on an end of a long segment. Therefore, different $x \in \partial B \setminus S$ give different y. Thus, f is a strictly monotone function such that $f \circ f(x) = f(x) - x$, hence $-x = f(f(x)) - f(x) = f \circ f \circ f(x)$. □

Proof Sketch of Theorem 15 Choose a unit vector x not parallel to a long segment and not on an end of a long segment. (This is possible iff the unit ball B is not a parallelogram.) Consider the set S and the function f from Lemma 16. Let A be the open arc from x to $f(x)$. Choose any measure μ on $A \setminus S$ such that μ and the usual length measure on $A \setminus S$ are mutually absolutely continuous with respect to each other (thus each singleton has measure 0 and each non-trivial subarc has positive measure) and with total measure $\pi/3$. Note that f yields not only injections, but also surjections among the six parts of $\partial B \setminus S$, as can be seen by considering f^{-1}. Use the defining property of f to extend this measure to the rest of $\partial B \setminus S$. Finally, define the measure of S to be 0. □

We already mentioned the result of Petty [152] and Soltan [171] that a d-dimensional space has an equilateral set of size at most 2^d, with equality iff the unit ball is an affine d-cube. The 2-dimensional case follows easily from the existence of a Brass measure.

Lemma 17 *If the unit ball of a normed plane is not a parallelogram, then there do not exist 4 equidistant points.*

Proof Suppose that $\{a, b, c, d\}$ is an equilateral set in a normed plane with a Brass measure μ. Then no 3 of the points are collinear.

If one of the points, say d, is in the convex hull of the other 3, then on the one hand we would have $\mu(\sphericalangle adb) + \mu(\sphericalangle bdc) + \mu(cda) = 2\pi$ from the definition of an angular measure, and on the other hand $\mu(\sphericalangle adb) = \mu(\sphericalangle bdc) = \mu(cda) = \pi/3$, because μ is a Brass measure. This is a contradiction.

Otherwise, the 4 points form a convex quadrilateral $abcd$, say. Then the interior angle at each vertex equals $\pi/3$, but the sum of the 4 interior angles has to equal 2π by Lemma 10, again a contradiction. □

The following are some useful properties of Brass measures.

Lemma 18 *Let X^2 be a normed plane with a Brass measure μ. In $\triangle oab$ let $\|o - a\| = \|o - b\| = 1$.*

1 *If $\|a - b\| > 1$, then $\mu(\sphericalangle aob) \geqslant \pi/3$. If $\mu(\sphericalangle aob) = \pi/3$ and μ is a good Brass measure, then ab is contained in a long segment of ∂B_X, with a in one end and b in the other end of the long segment, both different from the inner endpoints of the ends.*
2 *If $\|a - b\| < 1$, then $\mu(\sphericalangle aob) \leqslant \pi/3$. If $\mu(\sphericalangle aob) = \pi/3$ and μ is a good Brass measure, then ab is contained in the relative interior of a long segment of ∂B_X, with a in one end and b in the other end of the long segment.*

The proof of the above lemma is straightforward, using the fact that for $a \in \partial B_X$, the function $x \mapsto \|x - a\|$ is monotone on any of the two arcs of ∂B_X from a to $-a$.

Lemma 19 *Let X^2 be a normed plane with a Brass measure μ. Let $abcd$ be a quadrilateral with*

$$\|a - b\| = \|b - c\| = \|c - d\| = \|d - a\| \leqslant \|b - d\|, \|c - a\|.$$

Then $abcd$ is convex and $\mu(\sphericalangle b) + \mu(\sphericalangle c) = \pi$.

Proof Suppose that $\|a - b\| = \|b - c\| = \|c - d\| = \|d - a\| = 1$. The quadrilateral $abcd$ must be simple, otherwise the triangle inequality would give $\|b - d\| = \|a - c\| = 1$, which would contradict Lemma 17.

Suppose that the simple quadrilateral $abcd$ is not convex. If $b \in \operatorname{int} \triangle acd$, say, then $\|b - d\| < \max(\|d - a\|, \|d - c\|)$, a contradiction. If $b \in \partial \triangle acd$, then $\|b - d\| = 1$ and ac is a long segment of length 2 on the unit circle with centre d. By Lemma 12, the unit ball is a parallelogram, which contradicts the existence of μ.

It follows that $abcd$ is convex, with all angles less than π. If $ab \parallel dc$ or $bc \parallel ad$, then the result is obvious. Assume without loss of generality that ab and cd intersect on the side of ad opposite b and c, while bc and da intersect on the side of cd opposite a and b (Fig. 3). Then, letting $e := a + c - b$, $eabc$ is a parallelogram which contains

d in its interior. Then the two unit circles with centres a and c both contain b, d and e on their boundaries. It follows that b, d, e are collinear, and be is a long segment on both circles, since $\|b-d\| \geqslant 1$. Let $p = d + c - b$. Then $\triangle cdp$ and $\triangle cep$ are equilateral, hence $\mu(\sphericalangle ecp) = \mu(\sphericalangle dcp) = \pi/3$ and $\mu(\sphericalangle dce) = 0$. It follows that $\mu(\sphericalangle b) + \mu(\sphericalangle c) = \pi - \mu(\sphericalangle dce) = \pi$. □

7.2 Applications

The most striking application of the Brass measure was the original purpose for which Brass introduced it (as mentioned in Sect. 4.1). The proof, not repeated here, follows Harborth's proof [94] for the Euclidean case.

Theorem 20 (Brass [34]) *In a normed plane for which the unit ball is not a parallelogram, the number of edges of a minimum distance graph on n points is at most* $\lfloor 3n - \sqrt{12n-3} \rfloor$.

The following theorem of Talata can be also proved using the Brass measure.

Theorem 21 (Talata [196]) *Let S be a non-empty finite set of points in a two-dimensional normed space that is not isometric to ℓ_∞^2. Then the minimum-distance graph on S has a vertex of degree at most* 3. *If $|S| \leqslant 6$ then S has a vertex of degree at most* 2.

Proof We assume that S is not collinear, otherwise the result is trivial. Then $\operatorname{conv}(S)$ is a polygon $p_1 p_2 \dots p_k$, $k \geqslant 3$. Denote the internal angle of p_i by $\sphericalangle p_i$. Let μ be any Brass measure. If p_i has degree d, then by Lemma 18, $\mu(\sphericalangle p_i) \geqslant (d-1)\pi/3$. It follows that if each p_i has degree at least 4, then $\pi(k-2) = \sum_{i=1}^{k} \mu(\sphericalangle p_i) \geqslant \pi k$, a contradiction. Therefore, some p_i has degree at most 3.

Next, suppose that each p_i has degree at least 3 and that $|S| \leqslant 6$. Then $\mu(\sphericalangle p_i) \geqslant 2\pi/3$ for all i, giving a total angle of $\pi(k-2) \geqslant 2\pi k/3$, hence $k \geqslant 6$. It follows that $S = \{p_1, p_2, \dots, p_6\}$ and each p_i has degree exactly 3. As mentioned in Sect. 4, the minimum-distance graph is planar if the space is not isometric to ℓ_∞^2. If p_i is joined to p_{i+2}, then p_{i+1} can only be joined to p_i and p_{i+2} without creating crossing edges. This contradicts that p_{i+1} has degree 3. Therefore, no p_i is joined to $p_{i\pm 2}$. Hence p_i has to be joined to $p_{i\pm 1}$ and p_{i+3}. However, the diagonals $p_i p_{i+3}$ and $p_{i+1} p_{i+4}$ of the hexagon intersect, a contradiction. □

With some more case analysis, the following can also be shown.

Theorem 22 *Let $\delta_n(X^2)$ denote the maximum value of the minimum degree $\delta(G)$ of a minimum-distance graph G of n points in the two-dimensional normed space X^2. If X^2 is not isometric to ℓ_∞^2, then*

$$\delta_n(X^2) = \begin{cases} 2, & \text{if } 3 \leqslant n \leqslant 6 \text{ or } n = 8, 9, \\ 3, & \text{if } n = 7 \text{ or } n \geqslant 10. \end{cases}$$

Also,

$$\delta_n(\ell_\infty^2) = \begin{cases} 3, & \textit{if}\ 4 \leqslant n \leqslant 11\ \textit{or}\ n = 13, 14, 15, \\ 4, & \textit{if}\ n = 12\ \textit{or}\ n \geqslant 16. \end{cases}$$

Examples demonstrating the lower bounds in Theorem 22 can be found on a triangular lattice based on an equilateral triangle when X^2 is not isometric to ℓ_∞^2, except when $n = 11$, where an example is shown in Fig. 4. Examples for ℓ_∞^2 can be found on the square lattice $\mathbb{Z}^2$.

Next we use Brass measures to give simple proofs of results on the various Hadwiger and blocking numbers of convex discs that are not parallelograms. We also show some related results that would also need elaborate proofs without using Brass measures. The assertions in the next proposition were discussed in Sect. 2.

Proposition 23 *Let C be a convex disc in the plane. The Hadwiger number $H(C)$, strict Hadwiger number $H'(C)$, and one-sided Hadwiger number $H_+(C)$ of C are given in the following table.*

	non-parallelogram		*parallelogram*	
$H(C)$	6	[85]	8	[90]
$H'(C)$	5	[55]	4	[162]
$H_+(C)$	4		5	

The open one-sided Hadwiger number of C is $H_+^o(C) = 4$ if $\lambda(\frac{1}{2}(C - C)) > 1$, and $H_+^o(C) = 3$ otherwise.

Proof By the observation of Minkowski mentioned in Sect. 2.1, two translates $v + C$ and $w + C$ overlap, touch, or are disjoint iff the same holds for the corresponding translates $v + \frac{1}{2}(C - C)$ and $w + \frac{1}{2}(C - C)$ of the central symmetral of C, so we may assume without loss of generality that C is o-symmetric and is the unit ball of the normed plane X^2. We may then reformulate each of these quantities in terms of points on the unit circle. For instance, the Hadwiger number is the largest number of points on the unit circle that are pairwise at distance at least 1.

Note that a convex disc is a parallelogram iff its central symmetral is a parallelogram. The proofs for C a parallelogram, equivalently, when X^2 is isometric to ℓ_∞^2, are straightforward, and we only give the proofs for the case when C is not a parallelogram. Let μ be a good Brass measure for X^2.

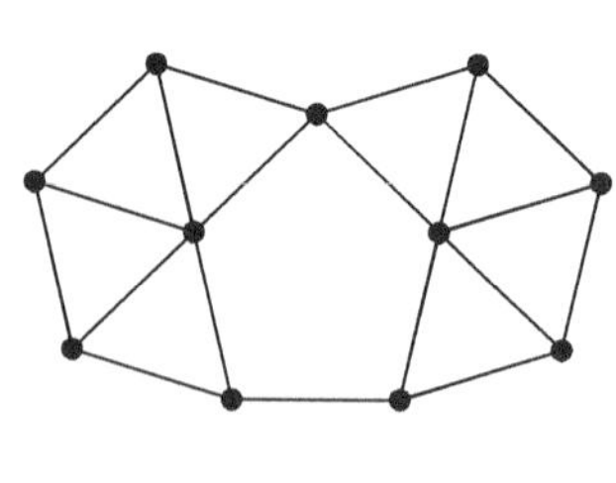

Fig. 4 A minimum-distance graph on 11 points with minimum degree 3

If there are 7 points on ∂C at mutual distances at least 1, then by Lemma 18, the sum of the angles spanned at o by consecutive points is at least $7\pi/3 > 2\pi$, a contradiction. Therefore, $H(C) \leqslant 6$. Similarly, $H_+^o(C) \leqslant H_+(C) \leqslant 4$.

The existence of 6 points on ∂C at pairwise distance $\geqslant 1$ may be established using the well-known continuity argument, or by using a good Brass measure as follows: Choose a point $x_0 \in \partial C$ not on a long segment of C, nor with ox_0 parallel to a long segment of C (there are infinitely many such points). Then choose $x_i \in \partial C$ such that $\mu(\sphericalangle x_0 o x_i) = i\pi/3$, for $i = 1, \dots, 5$. By Lemma 18, the distance between any two points is at least 1, which shows $H(C) \geqslant 6$. Similarly, the distance between any two points in $\{x_0, x_1, x_2, -x_0\}$ is at least 1, which shows that $H_+(C) \geqslant 4$ and $H_+^o(C) \geqslant 3$.

Suppose that $H_+^o(C) \geqslant 4$. Then there are 4 points on ∂C in an open half plane bounded by a line through o, at pairwise distance at least 1. It follows that the Brass measure is not proper, and by Lemma 14, ∂C contains a long segment. Conversely, if ∂C contains a long segment, then it is easy to find 4 points on ∂C in an open half plane with pairwise distances at least 1.

We show that $H'(C) = 5$ as follows. Suppose there are at least 6 points on ∂C at distance > 1. By Lemma 18, the angle spanned at o by consecutive points is $\geqslant \pi/3$, hence exactly $\pi/3$. Again by Lemma 18, the line through any two consecutive points is parallel to a long segment. Therefore, there are at least three parallel pairs of long segments on the unit circle, which contradicts Lemma 12, hence $H'(C) \leqslant 5$. To find 5 points on ∂C at distance > 1, choose any $x_1, \dots, x_5 \in \partial C$ such that $\mu(\sphericalangle x_i o x_{i+1}) = 2\pi/5 > \pi/3$ for all $i = 1, 2, 3, 4, 5$, and apply Lemma 18. □

A collection $\{v_i + C : i \in I\}$ of translates of a convex disc C that all touch C has a natural cyclic ordering determined by the cyclic ordering of the translation vectors $\{v_i : i \in I\} \subset \partial(\frac{1}{2}(C - C))$. We define a *dual Hadwiger family* of C to be a collection of translates $C + x_i$ of C, all touching C, and such that any two consecutive (in the natural ordering) translates are not disjoint (i.e. they either touch or overlap), and furthermore, o is in the convex hull of the translation vectors x_i. The last condition is to exclude trivialities. A *dual strict Hadwiger family* of C is a collection of translates of C, all touching C, and such that any two consecutive translates overlap, and furthermore, o is in the convex hull of the translation vectors. The *dual Hadwiger number* $I(C)$ of C is the minimum size of a dual Hadwiger family of C. The *dual strict Hadwiger number* $I'(C)$ of C is the minimum size of a dual strict Hadwiger family of C. As before, the dual Hadwiger number and its strict version have equivalent definitions in terms of the norm $\|\cdot\|$ with unit ball $B = \frac{1}{2}(C - C)$. The dual [strict] Hadwiger number equals the smallest number of points on ∂B containing o in their convex hull and such that consecutive points are at distance $\leqslant 1$ [< 1, respectively]. Dual Hadwiger families in the plane were considered by Grünbaum [85], where the first part of the next proposition appears without proof.

Proposition 24 *Let C be a convex disc in the plane. The dual Hadwiger number $I(C)$ and dual strict Hadwiger number $I'(C)$ are given by the following table.*

	non-parallelogram	*parallelogram*
$I(C)$	6	4
$I'(C)$	7	8

Proof The parallelogram case is easy to prove and we omit it. Without loss of generality, C is o-symmetric. Let μ be a good Brass measure on the plane X^2 with unit ball C. Suppose that there exist 5 points on ∂C with consecutive distances $\leqslant 1$. Then, by Lemma 18, the angles between consecutive vectors are all $\leqslant \pi/3$, a contradiction. Therefore, $I(C) \geqslant 6$. Equality is shown as before by inscribing a hexagon with sides of unit length to the unit circle.

Suppose there exist 6 unit vectors with consecutive distances <1. Then, by Lemma 18, all angles are $\leqslant \pi/3$, hence $= \pi/3$. As in the proof that $H'(C) \leqslant 5$, we obtain at least three parallel pairs of long segments, contradicting Lemma 12. This shows that $I'(C) \geqslant 7$. To obtain 7 unit vectors with consecutive distances <1, choose 7 unit vectors with consecutive angles all $< \pi/3$, and apply Lemma 18. □

Next, we give a simple proof of Zong's result on the blocking number of convex discs. Zong did not assume that the translates of the convex disc are non-overlapping, nor that they touch C, only that they do not overlap C. We prove this stronger result.

Lemma 25 (Zong [213]) *Let $C_1, \dots, C_m$ be translates of a convex disc C in the plane, not overlapping C, such that any translate of C that touches C overlaps with some C_i. Then $m \geqslant 4$.*

Proof We again omit the case where C is a parallelogram. As before, we may assume without loss of generality that C is o-symmetric, not a parallelogram and the unit ball of the normed plane X^2. The statement of the lemma is equivalent to the following.

Suppose that there exist points $x_1, \dots, x_m \in X^2$ such that

1. $\|x_i\| \geqslant 1$ for all $i = 1, \dots, m$,
2. and for all $x \in \partial C$ there exists an $i = 1, \dots, m$ such that $\|x - x_i\| < 1$.

Then $m \geqslant 4$.

Consider any $x_1, x_2, x_3 \in X^2$ such that $\|x_1\|, \|x_2\|, \|x_3\| \geqslant 1$. To prove the lemma, it is sufficient to find an $x \in \partial C$ such that $\|x - x_1\|, \|x - x_2\|, \|x - x_3\| \geqslant 1$.

Let μ be a good Brass measure on X^2. Let $\widehat{x_i} := \frac{1}{\|x_i\|} x_i$ $(i = 1, 2, 3)$. Then $\widehat{x_1}$, $\widehat{x_2}$, $\widehat{x_3}$ subdivide ∂C into three arcs $\widehat{x_1}\widehat{x_2}$, $\widehat{x_2}\widehat{x_3}$, $\widehat{x_3}\widehat{x_1}$.

Suppose that one of these arcs has Brass measure $> 2\pi/3$, say $\mu(\sphericalangle x_2 o x_3) > 2\pi/3$. There exists $x \in \widehat{x_2}\widehat{x_3}$ such that $\mu(\sphericalangle x_2 o x) = \mu(\sphericalangle x_3 o x) > \pi/3$. The angle $\sphericalangle x_1 o x$ contains either $\sphericalangle x_2 o x$ or $\sphericalangle x_3 o x$, hence $\mu(x_1 o x) > \pi/3$. By Lemma 18, $\|x - \widehat{x_i}\| \geqslant 1$ $(i = 1, 2, 3)$. By the triangle inequality,

$$1 \leqslant \|x - \widehat{x_i}\| = \left\| \frac{1}{\|x_i\|}(x - x_i) + (1 - \frac{1}{\|x_i\|})x \right\|$$
$$\leqslant \frac{1}{\|x_i\|} \|x - x_i\| + 1 - \frac{1}{\|x_i\|}.$$

It follows that $\|x - x_i\| \geqslant 1, i = 1, 2, 3$.

In the remaining case, $\mu(\sphericalangle x_i o x_{i+1}) = 2\pi/3$, $i = 1, 2, 3$ (modulo 3). If $\|-\widehat{x_3} - \widehat{x_1}\| \geqslant 1$ and $\|-\widehat{x_3} - \widehat{x_2}\| \geqslant 1$, then we have $\|-\widehat{x_3} - \widehat{x_i}\| \geqslant 1$ for each $i = 1, 2, 3$, and it follows from the triangle inequality as before that $\|-\widehat{x_3} - x_i\| \geqslant 1$ $(i = 1, 2, 3)$. Thus, we may assume without loss of generality that $\|-\widehat{x_3} - \widehat{x_1}\| < 1$. Since $\mu(\sphericalangle x_1 o x_3) = 2\pi/3$, $\mu(\sphericalangle x_1 o(-x_3)) = \pi/3$, and by Lemma 18, $-\widehat{x_3}$ and $\widehat{x_1}$ are in the relative interior of a long segment S.

Suppose that $\|-\widehat{x_2} - \widehat{x_1}\| < 1$. Then, similarly, $-\widehat{x_2}$ and $\widehat{x_1}$ are in the relative interior of a long segment. This long segment is necessarily S. However, since S contains the arc from $-\widehat{x_2}$ to $-\widehat{x_3}$, it follows that $\mu(S) \geqslant 2\pi/3$, which contradicts Lemma 13. Therefore, $\|-\widehat{x_2} - \widehat{x_1}\| \geqslant 1$, and similarly, $\|-\widehat{x_3} - \widehat{x_2}\| \geqslant 1$. It follows that $\|-\widehat{x_2} - \widehat{x_i}\| \geqslant 1$ for each $i = 1, 2, 3$, and we are done as before.

We conclude that $m \geqslant 4$. □

Proposition 26 (Zong [213]) *Let C be a convex disc in the plane. The blocking number $B(C)$ and the strict blocking number $B'(C)$ are given by the following table.*

	non-parallelogram	*parallelogram*
$B(C)$	4	4
$B'(C)$	3	2

Proof As before, we assume that C is o-symmetric and not a parallelogram, and μ is a good Brass measure in the normed plane X^2 with unit ball C. By Lemma 25, $B(C) \geqslant 4$. Next, we show that this is sharp. Choose any $x_1, x_2 \in \partial C$ such that $\mu(\sphericalangle x_1 o x_2) = \pi/2$. Let $x_3 = -x_1$ and $x_4 = -x_2$. Then for any $x \in \partial C$ there is an $i \in \{1, 2, 3, 4\}$ such that $\mu(\sphericalangle x o x_i) \leqslant \pi/4 < \pi/3$. Hence, $\|x - x_i\| < 1$. Thus, $\{C + x_i : i = 1, \dots, 4\}$ is a maximal Hadwiger family of C, and we conclude that $B(C) = 4$.

Given any two points $x_1, x_2 \in \partial C$ at distance 1, there exists a point x_3 outside the angular domain $\sphericalangle x_1 o x_2$ with $\mu(\sphericalangle x_1 o x_2) \leqslant \pi$ such that $\mu(\sphericalangle x_3 o x_1) > \pi/3$ and $\mu(\sphericalangle x_3 o x_2) > \pi/3$. As before, $\|x_1 - x_3\|, \|x_2 - x_3\| \geqslant 1$. Thus, $B'(C) \geqslant 3$.

To find three points on ∂C, we take the vertices x_0, x_2, x_4 of the affine regular hexagon $x_0 \cdots x_5$ from the proof of $H(C) \geqslant 6$ in the proof of Proposition 23, inscribed in the unit circle. Then the three angles $\sphericalangle x_0 o x_2$, $\sphericalangle x_2 o x_4$, $\sphericalangle x_4 o x_0$ each has Brass measure $2\pi/3$. If we take any $x \in \partial C$, it will have an angle of at most $\pi/3$, hence a distance of at most 1, to one of x_0, x_2, x_4. □

We now define the dual blocking number and its strict analogue. In the definitions of the dual blocking and strict blocking numbers we again make use of the natural ordering of the translates of C that touch C. The *dual blocking number* $A(C)$ of a convex disc C is the maximum size of a minimal dual Hadwiger family of C. The *strict dual blocking number* $A'(C)$ of C is the maximum size of a minimal strict dual Hadwiger family of C. Note that for the dual notions we do not need the non-triviality requirement that o is in the convex hull of the translation vectors, since such trivial collections of translates will not have the maximum size.

Proposition 27 *Let C be a convex disc in the plane. The dual blocking number $A(C)$ and the dual strict blocking number $A'(C)$ are given by the following table.*

	non-parallelogram	*parallelogram*
$A(C)$	11	8
$A'(C)$	12	12

Proof As before, we assume that C is o-symmetric and not a parallelogram, and μ is a good Brass measure in the normed plane X^2 with unit ball C. Suppose that we are given a set of 12 points on ∂C such that consecutive points are at distance $\leqslant 1$. If this set is minimal, then the distance between every second vector is > 1, which gives a strict Hadwiger family of 6 points, contradicting Proposition 23. Thus, $A(C) \leqslant 11$.

To find 11 points on ∂C such that consecutive points are at distance $\leqslant 1$ and non-consecutive points at distance > 1, choose any 11 points with consecutive angles $2\pi/11$ according to a Brass measure. Since $2\pi/11 < \pi/3$, the distance between consecutive points is < 1. Since $2 \cdot 2\pi/11 > \pi/3$, the distance between non-consecutive points is > 1, and we have obtained a minimal dual Hadwiger family. We conclude that $A(C) = 11$.

Next, suppose that we are given a set of 13 points on ∂C such that consecutive points are at distance < 1. For some two consecutive angles the sum of the angular measures is $\leqslant 2 \cdot 2\pi/13 < \pi/3$. It follows that we may remove the point shared among the two angles and still have all consecutive distances < 1. Therefore, the 13 points do not form a minimal strict dual Hadwiger family, hence $A'(C) \leqslant 12$.

To find 12 points on ∂C with consecutive points at distance < 1 and non-consecutive points at distance $\geqslant 1$, choose any x_1, let x_i, $i = 1, \dots, 6$, be the vertices of an inscribed affine regular hexagon with sides of length 1, let y_1 be such that $\mu(\sphericalangle x_1 o y) = \mu(\sphericalangle x_2 o y) = \pi/6$, and let y_i, $i = 1, \dots, 6$, be the vertices of an inscribed affine regular hexagon with sides of length 1. Then $\{x_i\} \cup \{y_i\}$ is a minimal strict dual Hadwiger family, and it follows that $A'(C) = 12$. □

8 Few-Distance Sets and Thin Cones

Erdős [58] asked for the minimum number $g(n)$ of distinct distances that can occur in a set of n points in the plane. We can equivalently ask for the largest number of points in a given space in which only k non-zero distances occur. We say that a subset S of a finite-dimensional normed space X is a *k- distance set* if

$$|\{\|x-y\| : x, y \in S, x \neq y\}| \leqslant k.$$

We have encountered 1-distance sets in Sect. 3 as equilateral sets. Let

$$f(k, X) = \max\{|S| : S \text{ is a } k\text{-distance subset of } X\}.$$

Thus, $f(1, X) = e(X)$.

For the Euclidean plane, Erdős [58] conjectured that $f(k, \mathbb{E}^2) = O(k^{1+\varepsilon})$ and showed that a square piece of the integer lattice gives $f(k, \mathbb{E}^2) = \Omega(k\sqrt{\log k})$. Recently, Guth and Katz [89] used a striking combination of classical algebraic geometry and topological and combinatorial methods to show that $f(k, \mathbb{E}^2) = O(k \log k)$. In higher dimensions, Erdős observed that $c_1 k^{d/2} < f(k, \mathbb{E}^d) < c_2 k^d$. It is conjectured that $f(k, \mathbb{E}^d) = O(k^{d/2+\varepsilon}))$. The current best results are due to Solymosi and Vu [172], which, when combined with the result of Guth and Katz, are $f(k, \mathbb{E}^3) = O(k^{5/3+o(1)})$ and $f(k, \mathbb{E}^d) = O(k^{(d^2+d-2)/(2d)+o(1)})$ for fixed d. Bannai, Bannai and Stanton [13] and Blokhuis [27] showed that $f(k, \mathbb{E}^d) \leqslant \binom{k+d}{k}$, which is a useful bound if d is large compared to k.

For general 2-dimensional spaces X^2 we have the bound $f(2, X^2) \leqslant 9$, with equality iff X^2 is isometric to ℓ_∞^2 [175]. Düvelmeyer [56] made a computer-assisted classification of all 2-distance sets in all 2-dimensional normed spaces. This classification is quite involved, but the following general statements can be inferred from his results.

Theorem 28 (Düvelmeyer [56]) *Let X^2 be a normed plane.*

1. *If the unit ball of X^2 is not a polygon, then $f(2, X^2) \leqslant 5$.*
2. *If the unit ball of X^2 is not a polygon and $f(2, X^2) = 5$, then any 2-distance set of 5 points is the vertex set of an affine regular pentagon, the ratio between the two distances is the golden ratio $(1+\sqrt{5})/2$, and the unit ball of X^2 has an inscribed affine regular decagon.*
3. *If $f(2, X^2) \geqslant 8$, then X^2 is isometric to ℓ_∞^2 and the 2-distance set of eight points corresponds to a subset of $\{0, 1, 2\}^2$ in ℓ_∞^2.*

For general d-dimensional normed spaces X^d, the following conjecture was made in [175].

Conjecture 15 ([175]) *For all $k \geqslant 1$ and $d \geqslant 1$, for any d-dimensional normed space X^d we have $f(k, X^d) \leqslant (k+1)^d$.*

This conjecture is known to hold for $k = 1$ and arbitrary d (by Petty and Soltan's result on equilateral sets) and for all k and all 2-dimensional spaces [175]. It is not difficult to show that $f(k, \ell_\infty^d) = (k+1)^d$, with the section $\{0, 1, \ldots, k\}^d \subset \ell_\infty^d$ of the integer lattice giving the lower bound [175].

We can partition a k-distance set in X^d into $b_f(X^d)$ many $(k-1)$-distance sets, hence $f(k, X^d) \leqslant b_f(X^d) f(k-1, X^d)$. By induction, we obtain that $f(k, X^d) \leqslant e(X^d) b_f(X^d)^{k-1} \leqslant (2 + o_k(1))^{kd}$. With a different inductive argument that involves the triangle inequality we can show that $f(k, X^d) \leqslant 2^{kd}$ [175]. This is the best general upper bound known for fixed k and large d. Next, we present a bound for fixed d and large k.

Theorem 29 *For any d-dimensional normed space X^d, $f(k, X^d) \leqslant (k+1)^{5^{d+o(d)}}$.*

The basic idea of the proof is from [175] and refined by building on an idea of Füredi [74]. In the next section we present a proof of this theorem, after introducing thin cones and their basic properties. (We note that very recently Polyanskii [154] showed that $f(k, X^d) \leqslant k^{O(d3^d)}$.)

8.1 Thin Cones

We recall that an *ordered vector space* $(V, \leqslant)$ is a vector space with a partial order compatible with the vector space structure in the following sense: If $a \leqslant b$ then $a + x \leqslant b + x$ and $\lambda a \leqslant \lambda b$ for all $a, b, x \in V$ and $\lambda \geqslant 0$. We also recall that a subset P of the vector space V is a *convex cone* if $x + y, \lambda x \in P$ whenever $x, y \in P$ and $\lambda \geqslant 0$, and that a convex cone P is called *proper* if $P \cap (-P) = \{o\}$. We then have the well-known correspondence between partial orders on V and proper convex cones in V: If $\leqslant$ is a partial order then its *positive cone* $P_{\leqslant} = \{v \in V : v \geqslant o\}$ is a proper convex cone, and conversely, if P is a proper convex cone V, we can define $a \leqslant_P b$ by $b - a \in P$, and $(V, \leqslant_P)$ will be an ordered vector space.

Note that we do not assume that the positive cones of our partial orders are closed, and so cannot deduce from $a_n \leqslant b$ and $\lim_n a_n = a$ that $a \leqslant b$. (For example, the cones defined in the proof of Theorem 32 below are not necessarily closed.)

We now connect the norm with the partial order. We say that a partial order $\leqslant$ on a normed space X is *monotone* if $\|x + y\| > \|x\|$ for all $x, y \geqslant o$ with $y \neq o$. A proper convex cone P in a normed space $(X, \|\cdot\|)$ is called a *thin cone* if $((P \cap \partial B_X) - (P \cap \partial B_X)) \cap P = \{o\}$, or equivalently, if $a - b \notin P$ for any chord ab of the unit sphere inside the cone P. Thin cones were introduced in [175] and independently in [74].

Lemma 30 *A proper convex cone P in X is thin iff $\leqslant_P$ is a monotone partial order.*

Proof Suppose that $\leqslant_P$ is monotone. Let $a, b \in P \cap \partial B_X$ such that $a - b \in P$. Let $x = b$ and $y = a - b$. If $y \neq o$, then $\|a\| = \|x + y\| > \|x\| = \|b\|$, which contradicts $\|a\| = \|b\| = 1$. Therefore, $y = a - b = o$, which shows that the cone P is thin.

Conversely, suppose that P is a thin cone. Let $x, y \in P$ with $y \neq o$ and suppose that $\|x + y\| \leqslant \|x\|$. Then $x \neq o$ and $\lambda = \frac{\|x+y\|}{\|x\|} \in [0, 1]$. Also, $x + y \neq o$, otherwise $x \in P \cap (-P) = \{o\}$, a contradiction. Let $a = \frac{1}{\|x\|} x$ and $b = \frac{1}{\|x+y\|}(x + y)$. Since

P is a convex cone, $a, b, b - a = \frac{1}{\|x+y\|}((1-\lambda)x + y) \in P$. It follows that $b - a \in ((P \cap \partial B_X) - (P \cap \partial B_X)) \cap P$, and since P is thin, $b - a = o$. However, then $(1-\lambda)x = -y \in P \cap -P$, which contradicts that P is a proper cone. Therefore, $\|x + y\| > \|x\|$. □

We call a family $\mathscr{P}$ of proper convex cones in a vector space V *separating* if

$$\bigcup_{P \in \mathscr{P}} \big(P \cup (-P)\big) = V.$$

Lemma 31 *A family $\mathscr{P}$ of proper convex cones in the vector space V is separating iff for all $x, y \in V$ there exists $P \in \mathscr{P}$ such that x and y are comparable in $\leqslant_P$.*

We omit the straightforward proof. We also say that a family $\mathscr{O}$ of partial orders on V is *separating* if $\{P_\leqslant \colon \leqslant \in \mathscr{O}\}$ is a separating family of cones. We are particularly interested in separating families of thin cones. The space ℓ_∞^d has a separating family of d thin cones, namely the cones generated by any d pairwise non-opposite facets of the unit ball B_∞^d. More generally, if the unit ball of X is a polytope with $2f$ facets, then X has a separating family of f thin cones.

Theorem 32 ([175]) *Any two-dimensional normed space has a separating family of two thin cones.*

Proof Let $a, b \in \partial B$ be chosen such that the area of the triangle $\triangle oab$ is maximized. Then B is contained in the parallelogram with vertices $\pm a \pm b$. Let P_1 be the cone generated by $\{a, b\}$, and P_2 the cone generated by $\{-a, b\}$. If ∂B does not contain a line segment parallel to oa or ob, then P_1 and P_2 are both thin cones.

If, on the other hand, ∂B contains line segments parallel to oa or ob, then we show that a and b can be chosen in such a way that no line segment on ∂B parallel to oa or ob will intersect the interiors of both P_1 and P_2. Indeed, if ∂B contains a maximal line segment cd parallel to oa, then we can replace b by either endpoint of cd without changing the area of $\triangle aob$. If ∂B furthermore contains a maximal line segment ef parallel to the new ob, then we can similarly replace a by either endpoint of this maximal segment without changing the area of the triangle. Note that changing a in this way does not create a line segment parallel to the new oa with b in its interior, since then b would be a smooth point of B, and we would also have two different lines through b that support B, namely the lines parallel to the old oa and the new oa. It follows that no line segment on ∂B parallel to the new oa or the new ob will intersect the interiors of both P_1 and P_2.

We next modify P_1 and P_2 so that they become thin. If ∂B contains a line segment parallel to oa inside P_i, then we remove the set $\{\lambda a \colon \lambda > 0\}$ from P_i. And if ∂B contains a line segment parallel to ob inside P_i, then we remove $\{\lambda b \colon \lambda > 0\}$ from P_i. The family $\{P_1, P_2\}$ will stay a separating family, since we never remove the same set from both P_1 and P_2. □

Unfortunately, there are d-dimensional spaces for which any separating family of thin cones will have size exponential in d. A simple example is the Euclidean space $\mathbb{E}^d$. It is easily seen that a proper convex cone P in $\mathbb{E}^d$ is thin iff $\langle x, y\rangle \geqslant 0$ for all $x, y \in P$. The orthants generate a separating family of 2^{d-1} thin cones for $\mathbb{E}^d$. Heppes [96] has shown that if the Euclidean unit sphere in $\mathbb{R}^3$ is partitioned into parts of angular diameter at most $\pi/2$, then at least 8 parts are needed. Therefore, any separating family of thin cones in $\mathbb{E}^3$ will contain at least 4 cones. In higher dimensions, we can make the following simple estimate. By the isodiametric inequality for the Euclidean sphere, any thin cone in $\mathbb{E}^d$ intersects the unit sphere in a set of surface measure at most that of a spherical cap of angular diameter $\pi/2$. Since such a spherical cap is easily seen to be contained in a Euclidean ball of radius $1/\sqrt{2}$, which moreover covers the convex hull of the spherical cap and the centre of the ball, it follows that any separating family of thin cones for $\mathbb{E}^d$ will contain at least $\frac{1}{2}(\sqrt{2})^d$ cones.

The following result gives a simple sufficient condition for a convex cone to be thin.

Lemma 33 *A convex cone P in X is thin if $\|x - y\| < 1$ for all $x, y \in P \cap \partial B_X$.*

Proof The hypothesis immediately implies that P is a proper cone.

Let $a, b \in P \cap \partial B_X$ such that $a - b \in P$. Suppose that $a - b \neq o$. Then

$$\widehat{a-b} := \frac{1}{\|a-b\|}(a-b) \in P \cap \partial B_X.$$

By hypothesis, $\left\|b - \widehat{a-b}\right\| < 1$. However,

$$\begin{aligned}
\left\|b - \widehat{a-b}\right\| &= \left\|b - \frac{1}{\|a-b\|}(a-b)\right\| \\
&= \left\|\left(1 + \frac{1}{\|a-b\|}\right)b - \frac{1}{\|a-b\|}a\right\| \\
&\geqslant \left\|\left(1 + \frac{1}{\|a-b\|}\right)b\right\| - \left\|\frac{1}{\|a-b\|}a\right\| \\
&= 1 + \frac{1}{\|a-b\|} - \frac{1}{\|a-b\|} = 1,
\end{aligned}$$

a contradiction. Therefore, $a - b = o$, and P is a thin cone. □

Suppose that S is a subset of the unit sphere of a d-dimensional normed space X contained in some open ball of radius $1/2$. Does it follow that for any $p, q \in \operatorname{conv}(S)$, $\|\widehat{p} - \widehat{q}\| < 1$? By the above lemma, a positive answer would imply that the convex cone generated by the intersection of an open ball of radius $1/2$ and the unit sphere is thin. However, this conclusion is false when $d \geqslant 3$ under the weaker assumption that $\|x - y\| < 1$ for all $x, y \in S$, as the following example shows.

Let X be the d-dimensional subspace $\left\{(\alpha_1, \dots, \alpha_d, \beta) \colon \alpha_i, \beta \in \mathbb{R}, \sum_{i=1}^{d} \alpha_i = 0\right\}$ of ℓ_1^{d+1}. Write $e_1, \dots, e_{d+1}$ for the standard basis of ℓ_1^{d+1}. Fix $\varepsilon \in (0, 1)$. For each $i = 1, \dots, d$, define

$$x_i = \frac{d(1-\varepsilon)}{2(d-1)} e_i - \frac{1-\varepsilon}{2(d-1)} \sum_{j=1}^{d} e_j + \varepsilon e_{d+1}.$$

Then simple calculations show that $S := \{x_1, \dots, x_d\}$ is an equilateral set of unit vectors in X where the distance between any two is

$$\left\|x_i - x_j\right\|_1 = (1-\varepsilon)d/(d-1), \quad 1 \leqslant i < j \leqslant d.$$

Also, if we let $p = \frac{1}{d-1} \sum_{i=1}^{d-1} x_i$ and $q = x_d$, then $p, q \in \operatorname{conv}(S)$ and a calculation shows that $\|p\|_1 = \frac{1+(d-2)\varepsilon}{d-1}$, $\|q\|_1 = 1$, and

$$\|\widehat{p} - \widehat{q}\|_1 = 2(1-\varepsilon) = \frac{2(d-1)}{d} \operatorname{diam}(S).$$

Thus, the diameter of $\{\widehat{p} \colon p \in \operatorname{conv}(S)\}$ is almost double that of S. This example is almost worst possible, at least for diameters up to about $1/2$, as the following theorem shows. We first estimate the distance to the origin from the convex hull of a set of unit vectors. This lemma is essentially Lemma 37 in [186]; see also the remark after the proof there.

Lemma 34 *Let X be a d-dimensional normed space and $S \subseteq \partial B_X$ with* $\operatorname{diam}(S) < 1 + 1/d$. *Then $\|p\| \geqslant 1 - (1 - 1/d)\operatorname{diam}(S)$ for all $p \in \operatorname{conv}(S)$.*

Proof By Carathéodory's Theorem, it is sufficient to prove the lemma for finite S. Thus, without loss of generality, S is finite and p is an element of $\operatorname{conv}(S)$ of minimum norm. By Carathéodory's Theorem, p is in the convex hull of $k \leqslant d+1$ points from S. Write $p = \sum_{i=1}^{k} \lambda_i x_i$ where $\sum_{i=1}^{k} \lambda_i = 1$, $\lambda_i > 0$ and $x_i \in S$. Then for any $i = 1, \dots, k$, with $D := \operatorname{diam}(S)$,

$$\begin{aligned} 1 - \|p\| = \|x_i\| - \|p\| \leqslant \|x_i - p\| &= \left\| \sum_{j=1}^{k} \lambda_j (x_i - x_j) \right\| \\ &\leqslant \sum_{j \neq i} \lambda_j \left\|x_i - x_j\right\| \leqslant (1 - \lambda_i) D. \end{aligned}$$

In particular, since $D < 1 + 1/d$, $p \neq o$. Since p minimizes the norm of all points from $C := \operatorname{conv}\{x_1, \dots, x_k\}$, it follows that $o \notin C$ and either p is in some facet of C or C lies in a hyperplane of X. Therefore, p is in the convex hull of at most d of these points, and we may suppose that $k \leqslant d$. If we sum the inequality $1 - \|p\| \leqslant$

$(1-\lambda_i)D$ over $i = 1, \dots, k$, we obtain $k(1 - \|p\|) \leqslant (k-1)D$, hence $\|p\| \geqslant 1 - (1-1/k)D \geqslant 1-(1-1/d)D$. □

Theorem 35 *Let S be a subset of the unit sphere of a d-dimensional normed space ($d \geqslant 2$) and let $0 < D \leqslant d/(2d-1)$ be given such that $\|x-y\| < D$ for all $x, y \in S$. Then $\|\widehat{p} - \widehat{q}\| < (2-\frac{1}{d})D$ for any $p, q \in \mathrm{conv}(S)$.*

Proof Without loss of generality, $\|p\| \leqslant \|q\|$. Also, since $p, q \in \mathrm{conv}(S)$, $\|p-q\| < D$ and $\|q\| \leqslant 1$. Lemma 34 and the given bound on D imply that $\|p\| \geqslant 1-(1-1/d)D \geqslant D$. The triangle inequality then gives

$$\begin{aligned}
\|\widehat{q} - \widehat{p}\| &= \left\| \frac{1}{\|q\|}(q-p) - \left(\frac{1}{\|p\|} - \frac{1}{\|q\|}\right) p \right\| \\
&\leqslant \frac{1}{\|q\|}\|q-p\| + \left(\frac{1}{\|p\|} - \frac{1}{\|q\|}\right)\|p\| \\
&< \frac{D}{\|q\|} + 1 - \frac{\|p\|}{\|q\|} \leqslant \frac{D - \|p\|}{1} + 1 \\
&\leqslant D - \left(1 - \left(1 - \frac{1}{d}\right)D\right) + 1 = \left(2 - \frac{1}{d}\right)D.
\end{aligned}$$

□

The same proof shows that for D up to $d/(d-1)$ we have $\|\widehat{p} - \widehat{q}\| \leqslant D/(1-(1-1/d)D)$, which is non-trivial for D up to about $2/3$. We do not know whether the bound of this theorem still holds for D larger than $1/2$ or if there are better counterexamples than the one described before Lemma 34. We need the theorem only for the case of $D = 1/2$, in the proof of the next corollary.

Corollary 36 *Any d-dimensional normed space has a separating family of $O(5^{d+o(d)})$ thin cones.*

Proof It is well known that the unit sphere ∂B_X can be covered by $O(5^d d \log d)$ open balls B_i ($i = 1, \dots, n$) of radius $1/4$ (see [163, Eq. (3)]). We may assume that the collection $\{B_i\}$ is minimal, hence $o \notin B_i$. Therefore, the convex cone C_i generated by $B_i \cap \partial B_X$ is proper, and $\{C_i : i = 1, \dots, n\}$ is a separating family of cones.

We next show that C_i is a thin cone for each $i = 1, \dots, n$. Let $a, b \in C_i \cap \partial B_X$. Then $a = \widehat{p}$ and $b = \widehat{q}$ for some $p, q \in \mathrm{conv}(B_i \cap \partial B_X)$. Theorem 35, applied to $S = B_i \cap \partial B_X$, $D = 1/2$ and p, q, gives that $\|a - b\| < 1$. By Lemma 33, C_i is a thin cone. □

Let P be a convex, proper cone and S a finite subset of the vector space V. Then $\leqslant_P$ restricted to S gives a finite poset with *height* $h(S, \leqslant_P)$ defined to be the largest cardinality of a chain in $(S, \leqslant_P)$.

Theorem 37 *Let X be a finite-dimensional normed space with a finite separating family $\mathscr{P}$ of thin cones. Let S be a finite subset of X. Then*

$$|S| \leqslant \prod_{P \in \mathscr{P}} h(S, \leqslant_P).$$

Proof For each $x \in S$ and $P \in \mathscr{P}$, let $h(x; S, \leqslant_P)$ denote the largest h such that there exist $x_1, \ldots, x_h \in S$ such that $x = x_1 >_P x_2 >_P \cdots >_P x_h$. Then the mapping

$$\eta \colon S \to \prod_{P \in \mathscr{P}} \{1, 2 \ldots, h(S, \leqslant P)\}$$

defined by $h(x) = (h(x, S, \leqslant_P) \colon P \in \mathscr{P})$ is injective. Indeed, for any distinct $x, y \in X$ there exists $P \in \mathscr{P}$ such that $y - x \in P \cup (-P)$. Without loss of generality, $y - x \in P \setminus \{o\}$. Let $H = h(x; S, \leqslant_P)$. There exist $x_1, \ldots, x_H \in S$ such that $x = x_1 >_P x_2 >_P \cdots >_P x_H$. However, then $y > x_1 >_P x_2 >_P \cdots >_P x_H$, hence $h(y; S, \leqslant_P) > H = h(x; S, \leqslant_P)$ and $\eta(x) \neq \eta(y)$. □

8.2 Applications

We can now prove the upper bound on the size of a k-distance set.

Proof of Theorem 29 For any k-distance set S in X^d and any thin cone P, $h(S, \leqslant_P) \leqslant k + 1$. Now apply Corollary 36 and Theorem 37 to obtain the result. □

As a second application, we obtain an upper bound on the length of a sequence of spheres $p_i + r_i \partial B_X$, $i = 1, \ldots, n$, such that $p_{i+1}, \ldots, p_n \in p_i + r_i \partial B_X$ for each $i = 1, \ldots, n-1$.

Theorem 38 *Let $p_1, p_2, \ldots, p_n \in X^d$ such that $\|p_i - p_j\| = \|p_i - p_k\|$ whenever $i < j < k$. Then $n \leqslant 2^{5^{d+o(d)}}$.*

Proof For $S = \{p_1, p_2, \ldots, p_n\}$ and any thin cone P, we have $h(S, \leqslant_P) \leqslant 2$. Then apply Corollary 36 and Theorem 37. □

Using a different technique, Naszódi, Pach and Swanepoel [147] recently obtained the much better upper bound of $O(6^d d^2 \log^2 d)$ in the above theorem, which was subsequently improved to $O(3^d d)$ by Polyanskii [154].

Acknowledgements We thank Tomasz Kobos, István Talata and a very thorough anonymous referee for providing corrections to a previous version.

References

1. M. Alfaro, M. Conger, K. Hodges, A. Levy, R. Kochar, L. Kuklinski, Z. Mahmood, K. von Haam, The structure of singularities in Φ-minimizing networks in R^2. Pac. J. Math. **149**, 201–210 (1991). MR1105695 (92d:90106)
2. N. Alon, Packings with large minimum kissing numbers. Discrete Math. **175**(1–3), 249–251 (1997). MR1475852 (98f:05040)
3. N. Alon, V.D. Milman, Embedding of l_∞^k in finite dimensional Banach spaces. Isr. J. Math. **45**(4), 265–280 (1983). MR0720303 (85f:46027)
4. N. Alon, P. Pudlák, Equilateral sets in l_p^n. Geom. Funct. Anal. **13**(3), 467–482 (2003). MR1995795 (2004h:46011)
5. J. Alonso, H. Martini, M. Spirova, Discrete geometry in Minkowski spaces, *Discrete Geometry and Optimization*, Fields Institute Communications (Springer, New York, 2013), pp. 1–15. MR3156773
6. G. Ambrus, I. Bárány, V. Grinberg, Small subset sums. Linear Algebra Appl. **499**, 66–78 (2016). MR3478885
7. J. Arias-de-Reyna, K. Ball, R. Villa, Concentration of the distance in finite-dimensional normed spaces. Mathematika **45**(2), 245–252 (1998). MR1695717 (2000b:46013)
8. C. Bachoc, F. Vallentin, Semidefinite programming, multivariate orthogonal polynomials, and codes in spherical caps. Eur. J. Comb. **30**(3), 625–637 (2009). MR2494437 (2010d:90065)
9. P. Balister, B. Bollobás, K. Gunderson, I. Leader, M. Walters, Random geometric graphs and isometries of normed spaces. Trans. Am. Math. Soc. **370**, 7361–7389 (2018). arXiv:1504.05324
10. K. Ball, Volume ratios and a reverse isoperimetric inequality. J. Lond. Math. Soc. (2) **44**(2), 351–359 (1991). MR1136445 (92j:52013)
11. H.-J. Bandelt, V. Chepoi, Embedding metric spaces in the rectilinear plane: a six-point criterion. Discrete Comput. Geom. **15**, 107–117 (1996). MR1367834 (97a:51022)
12. H.-J. Bandelt, V. Chepoi, M. Laurent, Embedding into rectilinear spaces. Discrete Comput. Geom. **19**(4), 595–604 (1998). MR1620076 (99d:51017)
13. E. Bannai, E. Bannai, D. Stanton, An upper bound for the cardinality of an s-distance subset in real Euclidean space II. Combinatorica **3**, 147–152 (1983). MR0726452 (85e:52013)
14. I. Bárány, On the power of linear dependencies, *Building Bridges*, vol. 19, Bolyai Society Mathematical Studies (Springer, Berlin, 2008), pp. 31–45. MR2484636 (2010b:05003)
15. A. Barvinok, S.J. Lee, I. Novik, Explicit constructions of centrally symmetric k-neighborly polytopes and large strictly antipodal sets. Discrete Comput. Geom. **49**(3), 429–443 (2013). MR3038522
16. A. Bezdek, K. Bezdek, A note on the ten-neighbour packings of equal balls. Beitr. Algebra Geom. **27**, 49–53 (1988). MR984401 (90a:52025)
17. K. Bezdek, On the maximum number of touching pairs in a finite packing of translates of a convex body. J. Comb. Theory, Ser. A **98**(1), 192–200 (2002). MR1897933 (2003c:52026)
18. K. Bezdek, Sphere packings revisited. Eur. J. Comb. **27**(6), 864–883 (2006). MR2226423 (2007a:52021)
19. K. Bezdek, Contact numbers for congruent sphere packings in Euclidean 3-space. Discrete Comput. Geom. **48**(2), 298–309 (2012). MR2946449
20. K. Bezdek, T. Bisztriczky, K. Böröczky, Edge-antipodal 3-polytopes, *Combinatorial and Computational Geometry*, vol. 52, Mathematical Sciences Research Institute Publications (Cambridge University Press, Cambridge, 2005), pp. 129–134. MR2178317 (2007a:52009)
21. K. Bezdek, P. Brass, On k^+-neighbour packings and one-sided Hadwiger configurations. Beitr. Algebra Geom. **44**(2), 493–498 (2003). MR2017050 (2004i:52017)
22. K. Bezdek, M.A. Khan, Contact numbers for sphere packings, this volume, 25–48 (2018). arXiv:1601.00145
23. K. Bezdek, M.A. Khan, The geometry of homothetic covering and illumination. in: *Discrete Geometry and Symmetry*, ed. by M. Conder, A. Deza , A. Weiss. GSC 2015. Springer Proceedings in Mathematics & Statistics, vol. 234 (Springer, Cham, 2018). arXiv:1602.06040

24. K. Bezdek, M. Naszódi, B. Visy, On the mth Petty numbers of normed spaces, *Discrete Geometry*, vol. 253, Monographs and Textbooks in Pure and Applied Mathematics (Dekker, New York, 2003), pp. 291–304. MR2034723 (2005a:51004)
25. K. Bezdek, S. Reid, Contact graphs of unit sphere packings revisited. J. Geom. **104**(1), 57–83 (2013). MR3047448
26. T. Bisztriczky, K. Böröczky, On antipodal 3-polytopes. Rev. Roum. Math. Pures Appl. **50**(5–6), 477–481 (2005). MR2204128 (2006k:52004)
27. A. Blokhuis, Few-distance sets, CWI Tract 7, Stichting Mathematisch Centrum, Amsterdam (1984). MR0751955 (87f:51023)
28. V. Boltyanski, H. Martini, V. Soltan, *Geometric Methods and Optimization Problems*, vol. 4, Combinatorial Optimization (Kluwer, Dordrecht, 1999). MR1677397 (2000c:90002)
29. A. Bondarenko, On Borsuk's conjecture for two-distance sets. Discrete Comput. Geom. **51**(3), 509–515 (2014). MR3201240
30. K. Borsuk, Drei Sätze über die n-dimensionale euklidische Sphäre. Fundam. Math. **20**, 177–190 (1933)
31. K. Böröczky Jr., *Finite Packing and Covering*, vol. 154, Cambridge Tracts in Mathematics (Cambridge University Press, Cambridge, 2004). MR2078625 (2005g:52045)
32. J. Bourgain, J. Lindenstrauss, On covering a set in R^N by balls of the same diameter, *Geometric Aspects of Functional Analysis (1989-1990)*, vol. 1469, Lecture Notes in Mathematics (Springer, Berlin, 1991), pp. 138–144. MR1122618 (92g:52018)
33. P. Boyvalenkov, S. Dodunekov, O. Musin, A survey on the kissing numbers. Serdica Math. J. **38**, 507–522 (2012). MR3060792
34. P. Brass, Erdős distance problems in normed spaces. Comput. Geom. **6**, 195–214 (1996). MR1392310 (97c:52036)
35. P. Brass, On the maximum number of unit distances among n points in dimension four, *Intuitive Geometry (Budapest, 1995)*, Bolyai Society Mathematical Studies (János Bolyai Mathematical Society, Budapest, 1997), pp. 277–290. MR1470764 (98j:52030)
36. P. Brass, On convex lattice polyhedra and pseudocircle arrangements, *Charlemagne and his Heritage*, vol. 2 (Aachen, 1995), 1200 Years of Civilization and Science in Europe (Brepols, Turnhout, 1998), pp. 297–302. MR1672425 (2000a:52031)
37. P. Brass, On equilateral simplices in normed spaces. Beitr. Algebra Geom. **40**, 303–307 (1999). MR1720106 (2000i:52012)
38. P. Brass, W.O.J. Moser, J. Pach, *Research Problems in Discrete Geometry* (Springer, New York, 2005). MR2163782 (2006i:52001)
39. M. Brazil, R.L. Graham, D.A. Thomas, M. Zachariasen, On the history of the Euclidean Steiner problem. Arch. Hist. Exact Sci. **68**, 327–354 (2014). MR3200931
40. M. Brazil, M. Zachariasen, *Optimal Interconnection Trees in the Plane*, vol. 29, Algorithms and Combinatorics (Springer, Cham, 2015). MR3328741
41. I. Broere, Colouring $\mathbf{R}^n$ with respect to different metrics. Geombinatorics **4**(1), 4–9 (1994). MR1279706 (95g:05044)
42. H. Chen, Ball packings with high chromatic numbers from strongly regular graphs. Discrete Math. **340**, 1645–1648 (2017). arXiv:1502.02070
43. K.B. Chilakamarri, Unit-distance graphs in Minkowski metric spaces. Geom. Dedicata **37**(3), 345–356 (1991). MR1094697 (92b:05036)
44. D. Cieslik, Knotengrade kürzester Bäume in endlichdimensionalen Banachräumen. Rostocker Math. Kolloq. **39**, 89–93 (1990). MR1090608 (92a:05039)
45. D. Cieslik, The vertex-degrees of Steiner minimal trees in Minkowski planes, in *Topics in Combinatorics and Graph Theory*, ed. by R. Bodendiek, R. Henn (Physica-Verlag, Heidelberg, 1990), pp. 201–206. MR1100038 (91m:05059)
46. D. Cieslik, *Steiner Minimal Trees*, vol. 23, Nonconvex Optimization and its Applications (Kluwer, Dordrecht, 1998). MR1617288 (99i:05062)
47. M. Conger, Energy-minimizing networks in $\mathbf{R}^n$, Honours thesis, Williams College, Williamstown MA (1989)

48. B. Csikós, Edge-antipodal convex polytopes—a proof of Talata's conjecture, *Discrete Geometry*, vol. 253, Monographs and Textbooks in Pure and Applied Mathematics (Dekker, New York, 2003), pp. 201–205. MR2034716 (2004m:52026)
49. B. Csikós, G. Kiss, K.J. Swanepoel, P. Oloff de Wet, Large antipodal families. Period. Math. Hung. **58**(2), 129–138 (2009). MR2531160 (2010m:52058)
50. G. Csizmadia, On the independence number of minimum distance graphs. Discrete Comput. Geom. **20**, 179–187 (1998). MR1637884 (99e:05044)
51. L. Dalla, D.G. Larman, P. Mani-Levitska, C. Zong, The blocking numbers of convex bodies. Discrete Comput. Geom. **24**(2–3), 267–277 (2000). MR1758049 (2001d:52011)
52. L. Danzer, B. Grünbaum, Über zwei Probleme bezüglich konvexer Körper von P. Erdős und von V. L. Klee. Math. Z. **79**, 95–99 (1962). MR0138040 (25 #1488)
53. A.D.N.J. de Grey, The chromatic number of the plane is at least 5. arXiv:1804.02385
54. B.V. Dekster, Simplexes with prescribed edge lengths in Minkowski and Banach spaces. Acta Math. Hung. **86**(4), 343–358 (2000). MR1756257 (2001b:52001)
55. P.G. Doyle, J.C. Lagarias, D. Randall, Self-packing of centrally symmetric convex bodies in $\mathbb{R}^2$. Discrete Comput. Geom. **8**, 171–189 (1992). MR1162392 (93e:52038)
56. N. Düvelmeyer, General embedding problems and two-distance sets in Minkowski planes. Beitr. Algebra Geom. **49**, 549–598 (2008). MR2468075 (2009j:52007)
57. H.G. Eggleston, Covering a three-dimensional set with sets of smaller diameter. J. Lond. Math. Soc. **30**, 11–24 (1955). MR0067473 (16,734b)
58. P. Erdős, On sets of distances of n points. Am. Math. Mon. **53**, 248–250 (1946). MR0015796 (7,471c)
59. P. Erdős, On sets of distances of n points in Euclidean space. Magyar Tud. Akad. Mat. Kutató Int. Közl. **5**, 165–169 (1960). MRMR0141007 (25 #4420)
60. P. Erdős, On some applications of graph theory to geometry. Can. J. Math. **19**, 968–971 (1967). MR0219438 (36 #2520)
61. P. Erdős, Problems and results in combinatorial geometry, *Discrete Geometry and Convexity (New York, 1982)*, vol. 440, Annals of the New York Academy of Sciences (New York Academy of Sciences, New York, 1985), pp. 1–11. MR809186 (87g:52001)
62. P. Erdős, Z. Füredi, The greatest angle among n points in the d-dimensional Euclidean space, *Combinatorial Mathematics (Marseille-Luminy, 1981)*, vol. 75, North-Holland Mathematics Studies (North-Holland, Amsterdam, 1983), pp. 275–283. MR841305 (87g:52018)
63. P. Erdős, D. Hickerson, J. Pach, A problem of Leo Moser about repeated distances on the sphere. Am. Math. Mon. **96**(7), 569–575 (1989). MR1008787 (90h:52008)
64. P. Erdős, J. Pach, Variations on the theme of repeated distances. Combinatorica **10**(3), 261–269 (1990). MR1092543 (92b:52037)
65. G. Fejes Tóth, Ten-neighbour packing of equal balls. Period. Math. Hung. **12**(2), 125–127 (1981). MR603405 (82e:52013)
66. G. Fejes Tóth, W. Kuperberg, A survey of recent results in the theory of packing and covering, in *New Trends in Discrete and Computational Geometry*, vol. 10, Algorithms and Combinatorics, ed. by J. Pach (Springer, Berlin, 1993), pp. 251–279. MR1228046 (94h:52037)
67. L. Fejes Tóth, *Lagerungen in der Ebene, auf der Kugel und im Raum, Zweite verbesserte und erweiterte Auflage*, vol. 65, Die Grundlehren der mathematischen Wissenschaften (Springer, Berlin, 1972). MR0353117 (50 #5603)
68. L. Fejes Tóth, Five-neighbour packing of convex discs. Period. Math. Hung. **4**, 221–229 (1973). MR0345006 (49 #9745)
69. L. Fejes Tóth, On Hadwiger numbers and Newton numbers of a convex body. Studia Sci. Math. Hung. **10**(1—-2), 111–115 (1975). MR0440469 (55 #13344)
70. L. Fejes Tóth, N. Sauer, Thinnest packing of cubes with a given number of neighbours. Can. Math. Bull. **20**(4), 501–507 (1977). MR0478017 (57 #17513)
71. P. Frankl, R.M. Wilson, Intersection theorems with geometric consequences. Combinatorica **1**(4), 357–368 (1981). MR0647986 (84g:05085)
72. D. Freeman, E. Odell, B. Sari, T. Schlumprecht, Equilateral sets in uniformly smooth Banach spaces. Mathematika **60**(1), 219–231 (2014). MR3164528

73. R.E. Fullerton, Integral distances in Banach spaces. Bull. Am. Math. Soc. **55**, 901–905 (1949). MR0032934 (11,369c)
74. Z. Füredi, Few-distance sets in d-dimensional normed spaces, Oberwolfach Rep. **2**(2) (2005), 947–950, Abstracts from the Discrete Geometry workshop held 10–16 April 2005, Organized by M. Henk, J. Matoušek, E. Welzl, Oberwolfach Reports **2**(2). MR2216216
75. Z. Füredi, J.-H. Kang, Distance graph on $\mathbb{Z}^n$ with l_1 norm. Theor. Comput. Sci. **319**(1–3), 357–366 (2004). MR2074960 (2005c:05079)
76. Z. Füredi, J.-H. Kang, Covering the n-space by convex bodies and its chromatic number. Discrete Math. **308**(19), 4495–4500 (2008). MR2433777 (2009c:52031)
77. Z. Füredi, P.A. Loeb, On the best constant for the Besicovitch covering theorem. Proc. Am. Math. Soc. **121**(4), 1063–1073 (1994). MR1249875 (95b:28003)
78. M. Gardner, Mathematical Games. Sci. Am. **203**(4), 172–180 (1960)
79. G.P. Gehér, A contribution to the Aleksandrov conservative distance problem in two dimensions. Linear Algebra Appl. **481**, 280–287 (2015). MR3349657
80. B. Gerencsér, V. Harangi, Acute sets of exponentially optimal site, to appear in Discrete Comput. Geom. arXiv:1709.03411
81. E. Glakousakis, S. Mercourakis, Examples of infinite dimensional Banach spaces without infinite equilateral sets. Serdica Math. J. **42**(1), 65–88 (2016). MR3523955
82. H. Groemer, Abschätzungen für die Anzahl der konvexen Körper, die einen konvexen Körper berühren. Monatsh. Math. **65**, 74–81 (1961). MR0124819 (23 #A2129)
83. B. Grünbaum, A proof of Vázsonyi's conjecture. Bull. Res. Counc. Isr. Sect. A **6**, 77–78 (1956). MR0087115 (19,304d)
84. B. Grünbaum, Borsuk's partition conjecture in Minkowski planes. Bull. Res. Counc. Isr. Sect. F **7F**, 25–30 (1957/1958). MR0103440 (21 #2209)
85. B. Grünbaum, On a conjecture of H. Hadwiger. Pac. J. Math. **11**, 215–219 (1961). MR0138044 (25 #1492)
86. B. Grünbaum, Strictly antipodal sets. Isr. J. Math. **1**, 5–10 (1963). MR0159263 (28 #2480)
87. B. Grünbaum, *Convex Polytopes*, 2nd edn., Graduate Texts in Mathematics (Springer, New York, 2003). MR1976856 (2004b:52001)
88. L. Guibas, J. Pach, M. Sharir, Sphere-of-influence graphs in higher dimensions, *Intuitive Geometry (Szeged, 1991)*, vol. 63, Studies Colloquia mathematica Societatis János Bolyai (North-Holland, Amsterdam, 1994), pp. 131–137. MR1383618 (97a:05183)
89. L. Guth, N.H. Katz, On the Erdős distinct distances problem in the plane. Ann. Math. (2) **181**(1), 155–190 (2015). MR3272924
90. H. Hadwiger, Über Treffanzahlen bei translationsgleichen Eikörpern. Arch. Math. **8**, 212–213 (1957). MR0091490 (19,977e)
91. H. Hadwiger, Ungelöste Probleme No. 40. Elem. Math. **16**, 103–104 (1961)
92. M. Hanan, On Steiner's problem with rectilinear distance. SIAM J. Appl. Math. **14**, 255–265 (1966). MR0224500 (37 #99)
93. F. Harary, M.S. Jacobson, M.J. Lipman, F.R. McMorris, Abstract sphere-of-influence graphs. Math. Comput. Modelling **17**(11), 77–83 (1993). MR1236512 (94f:05119)
94. H. Harborth, Lösung zu Problem 664A. Elem. Math. **29**, 14–15 (1974)
95. A. Heppes, Beweis einer Vermutung von A. Vázsonyi. Acta Math. Acad. Sci. Hung. **7**, 463–466 (1956). MR0087116 (19,304e)
96. A. Heppes, Decomposing the 2-sphere into domains of smallest possible diameter. Period. Math. Hung. **36**(2–3), 171–180 (1998). MR1694597 (2000f:52023)
97. F.K. Hwang, D.S. Richards, P. Winter, *The Steiner Tree Problem*, vol. 53, Annals of Discrete Mathematics (North Holland, Amsterdam, 1992). MR1192785 (94a:05051)
98. T. Jenrich, A.E. Brouwer, A 64-dimensional counterexample to Borsuk's conjecture. Electron. J. Comb. **21**(4) (2014). Paper 4.29, 3 pp. MR3292266
99. A. Joós, On a convex body with odd Hadwiger number. Acta Math. Hung. **119**(4), 307–321 (2008). MR2429292 (2009f:52044)
100. G.A. Kabatiansky, V.I. Levenshtein, Bounds for packings on the sphere and in space. Probl. Peredachi Inf. **14**(1), 3–25 (1978); English translation: Probl. Inf. Transm. **14**(1), 1–17 (1978). MR0514023 (58 #24018)

101. J. Kahn, G. Kalai, A counterexample to Borsuk's conjecture. Bull. Am. Math. Soc. (N.S.) **29**(1), 60–62 (1993). MR1193538 (94a:52007)
102. G. Kalai, Some old and new problems in combinatorial geometry I: Around Borsuk's problem, *Surveys in Combinatorics*, vol. 424, London Mathematical Society Lecture Note Series (Cambridge University Press, Cambridge, 2015), pp. 147–174. MR3497269
103. H. Kaplan, J. Matoušek, Z. Safernová, M. Sharir, Unit distances in three dimensions. Comb. Probab. Comput. **21**(4), 597–610 (2012). MR2942731
104. G. Kertész, Nine points on the hemisphere, *Intuitive geometry (Szeged, 1991)*, vol. 63, Colloquia mathematica Societatis János Bolyai (North-Holland, Amsterdam, 1994), pp. 189–196. MR1383625 (97a:52031)
105. J. Klein, G. Zachmann, Point cloud surfaces using geometric proximity graphs. Comput. Graph. **28**(6), 839–850 (2004)
106. D.J. Kleitman, On a lemma of Littlewood and Offord on the distribution of certain sums. Math. Z. **90**, 251–259 (1965). MR0184865 (32 #2336)
107. T. Kobos, An alternative proof of Petty's theorem on equilateral sets. Ann. Pol. Math. **109**(2), 165–175 (2013). MR3103122
108. T. Kobos, Equilateral dimension of certain classes of normed spaces. Numer. Funct. Anal. Optim. **35**(10), 1340–1358 (2014). MR3233155
109. P. Koszmider, Uncountable equilateral sets in Banach spaces of the form $C(K)$. Israel Journal of Mathematics **224**(1), 83–103 (April 2018). arXiv:1503.06356
110. A. Kupavskiy, On the chromatic number of $\mathbb{R}^n$ with an arbitrary norm. Discrete Math. **311**(6), 437–440 (2011). MR2799896 (2012d:52028)
111. R. Kusner, W. Kusner, J. C. Lagarias, S. Shlosman, The twelve spheres problem, this volume, 219–278 (2018), arXiv:1611.10297
112. Z. Lángi, M. Naszódi, On the Bezdek-Pach conjecture for centrally symmetric convex bodies. Can. Math. Bull. **52**(3), 407–415 (2009). MR2547807 (2010j:52068)
113. D.G. Larman, C.A. Rogers, The realization of distances within sets in Euclidean space. Mathematika **19**, 1–24 (1972). MR0319055 (47 #7601)
114. D.G. Larman, C. Zong, On the kissing numbers of some special convex bodies. Discrete Comput. Geom. **21**(2), 233–242 (1999). MR1668102 (99k:52030)
115. M. Lassak, An estimate concerning Borsuk partition problem. Bull. Acad. Pol. Sci. Sér. Sci. Math. **30**(1982)(9–10), 449–451 (1983). MR0703571 (84j:52014)
116. G. Lawlor, F. Morgan, Paired calibrations applied to soap films, immiscible fluids, and surfaces or networks minimizing other norms. Pac. J. Math. **166**(1), 55–83 (1994). MR1306034 (95i:58051)
117. J. Leech, Some sphere packings in higher space. Can. J. Math. **16**, 657–682 (1964). MR0167901 (29 #5166)
118. V.I. Levenshtein, Bounds for packings in n-dimensional Euclidean space. Dokl. Akad. Nauk SSSR **245**(6), 1299–1303 (1979). MR529659 (80d:52017)
119. A. Lin, Equilateral sets in the ℓ_1 sum of Euclidean spaces, manuscript (2016)
120. J.M. Ling, On the size of equilateral sets in spaces with the double-cone norm, manuscript (2006)
121. B. Lund, A. Magazinov, The sign-sequence constant of the plane, Acta Math. Hung. **151**, 117–123 (2017). arXiv:1510.04536
122. H. Maehara, On configurations of solid balls in 3-space: chromatic numbers and knotted cycles. Graphs Comb. **23**(1), 307–320 (2007). MR2320637 (2008c:05068)
123. E. Makai Jr., H. Martini, On the number of antipodal or strictly antipodal pairs of points in finite subsets of $\mathbf{R}^d$, *Applied geometry and discrete mathematics*, vol. 4, DIMACS Series in Discrete Mathematics and Theoretical Computer Science (American Mathematical Society, Providence, 1991), pp. 457–470. MR1116370 (92f:52020)
124. V.V. Makeev, Equilateral simplices in a four-dimensional normed space. Zap. Nauchn. Sem. S.-Peterburg. Otdel. Mat. Inst. Steklov. (POMI) Geom. i Topol. **329**(9), 88–91 (2005), 197; English translation in J. Math. Sci. (N. Y.) **140**(4), 548–550 (2007). MR2215334 (2007b:52010)

125. S.M. Malitz, J.I. Malitz, A bounded compactness theorem for L^1-embeddability of metric spaces in the plane. Discrete Comput. Geom. **8**(4), 373–385 (1992). MR1176377 (93i:51034)
126. H. Martini, V. Soltan, Antipodality properties of finite sets in Euclidean space. Discrete Math. **290**(2–3), 221–228 (2005). MR2123391 (2005i:52017)
127. H. Martini, K.J. Swanepoel, Low-degree minimal spanning trees in normed spaces. Appl. Math. Lett. **19**(2), 122–125 (2006). MR2198397 (2007f:52009)
128. H. Martini, K.J. Swanepoel, P.O. de Wet, Absorbing angles, Steiner minimal trees, and antipodality. J. Optim. Theory Appl. **143**(1), 149–157 (2009). MR2545946 (2010m:05080)
129. H. Martini, K.J. Swanepoel, G. Weiß, The geometry of Minkowski spaces—a survey. I. Expo. Math. **19**(2), 97–142 (2001). MRMR1835964 (2002h:46015a). Erratum, Expo. Math. **19**(4), 364 (2001). MR1876256 (2002h:46015b)
130. J. Matoušek, Lectures on discrete geometry, *Graduate Texts in Mathematics*, vol. 212 (Springer, New York, 2002). MR1899299 (2003f:52011)
131. J. Matoušek, The number of unit distances is almost linear for most norms. Adv. Math. **226**(3), 2618–2628 (2011). MR2739786 (2011k:52008)
132. S.K. Mercourakis, G. Vassiliadis, Equilateral sets in Banach spaces of the form $C(K)$. Stud. Math. **231**(3), 241–255 (2015). MR3471052
133. S.K. Mercourakis, G. Vassiliadis, Equilateral sets in infinite dimensional Banach spaces. Proc. Am. Math. Soc. **142**(1), 205–212 (2014). MR3119196
134. T.S. Michael, T. Quint, Sphere of influence graphs: edge density and clique size. Math. Comput. Modelling **20**(7), 19–24 (1994). MR1299482 (95i:05103)
135. T.S. Michael, T. Quint, Sphere of influence graphs in general metric spaces. Math. Comput. Modelling **29**(7), 45–53 (1999). MR1688596 (2000c:05106)
136. T.S. Michael, T. Quint, Sphere of influence graphs and the L_∞-metric. Discrete Appl. Math. **127**(3), 447–460 (2003). MR1976026 (2004g:05139)
137. V.D. Milman, A new proof of A. Dvoretzky's theorem on cross-sections of convex bodies. Funkts. Anal. Prilozh. **5**(4), 28–37 (1971); English translation in Funct. Anal. Appl. **5**, 288–295 (1971). MR0293374 (45 #2451)
138. V.D. Milman, Almost Euclidean quotient spaces of subspaces of a finite-dimensional normed space. Proc. Am. Math. Soc. **94**(3), 445–449 (1985). MR0787891 (86g:46025)
139. H. Minkowski, *Diophantische Approximationen* (Chelsea Publishing Co., New York, 1957). MR0086102 (19,124f)
140. F. Morgan, Minimal surfaces, crystals, networks, and undergraduate research. Math. Intell. **14**, 37–44 (1992)
141. F. Morgan, *Riemannian Geometry, A Beginner's Guide*, 2nd edn. (A.K. Peters, Wellesley, MA, 1998). MR1600519 (98i:53001)
142. O.R. Musin, The problem of the twenty-five spheres. Usp. Mat. Nauk **58**(4(352)), 153–154 (2003); English translation: Russ. Math. Surv. **58**(4), 794–795 (2003). MR2042912 (2005a:52016)
143. O.R. Musin, The one-sided kissing number in four dimensions. Period. Math. Hung. **53**(1–2), 209–225 (2006). MR2286472 (2007j:52019)
144. O.R. Musin, Bounds for codes by semidefinite programming, Tr. Mat. Inst. Steklova **263** (2008); Geometriya, Topologiya i Matematicheskaya Fizika. I, 143–158; reprinted in Proc. Steklov Inst. Math. **263**(1) (2008), 134–149. MR2599377 (2011c:94085)
145. M. Naszódi, Flavors of translative coverings, this volume, 335–358(2018). arXiv:1603.04481
146. M. Naszódi, J. Pach, K.J. Swanepoel, Sphere-of-influence graphs in normed spaces. in: Discrete Geometry and Symmetry, ed. by M. Conder, A. Deza , A. Weiss. GSC 2015. Springer Proceedings in Mathematics & Statistics, vol. 234 (Springer, Cham, 2018). arXiv:1603.04481
147. M. Naszódi, J. Pach, K.J. Swanepoel, Arrangements of homothets of a convex body. Mathematika **63**, 696–710(2017). arXiv:1608.04639
148. A.M. Odlyzko, N.J.A. Sloane, New bounds on the number of unit spheres that can touch a unit sphere in n dimensions. J. Comb. Theory, Ser. A **26**(2), 210–214 (1979). MR530296 (81d:52010)

149. M.I. Ostrovskii, *Metric Embeddings. Bilipschitz and Coarse Embeddings into Banach Spaces*, De Gruyter Studies in Mathematics (De Gruyter, Berlin, 2013). MR3114782
150. J. Pach, G. Tóth, On the independence number of coin graphs. Geombinatorics **6**, 30–33 (1996). MR1392795 (97d:05176)
151. J. Perkal, Sur la subdivision des ensembles en parties de diamètre inférieur. Colloq. Math. **1**(1), 45 (1947)
152. C.M. Petty, Equilateral sets in Minkowski spaces. Proc. Am. Math. Soc. **29**, 369–374 (1971). MR0275294 (43 #1051)
153. R. Pollack, Increasing the minimum distance of a set of points. J. Comb. Theory, Ser. A **40**, 450 (1985). MR0814430 (87b:52020)
154. A. Polyanskii, Pairwise intersecting homothets of a convex body. Discrete Math. **340**, 1950–1956 (2017). arXiv:1610.04400
155. A. Pór, On e-antipodal polytopes, manuscript (2003)
156. H.J. Prömel, A. Steger, *The Steiner Tree Problem. A Tour Through Graphs, Algorithms, and Complexity*, Advanced Lectures in Mathematics (Vieweg, Braunschweig, 2002). MR1891564 (2003a:05047)
157. A.M. Raigorodskii, The Borsuk problem and the chromatic numbers of some metric spaces. Usp. Mat. Nauk **56**(1(337)), 107–146 (2001); English translation in Russ. Math. Surv. **56**(1), 103–139 (2001). MR1845644 (2002m:54033)
158. A.M. Raigorodskii, On the chromatic number of a space with the metric l_q. Usp. Mat. Nauk **59**(5(359)), 161–162 (2004); English translation in Russ. Math. Surv. **59**(5), 973–975 (2004). MR2125940 (2006e:05171)
159. A.M. Raigorodskii, Around the Borsuk conjecture. Sovrem. Mat. Fundam. Napravl. **23**, 147–164 (2007); English translation in J. Math. Sci. (N. Y.) **154**(4), 604–623 (2008). MR2342528 (2008j:52035)
160. A.M. Raigorodskii, Coloring distance graphs and graphs of diameters, *Thirty Essays on Geometric Graph Theory* (Springer, New York, 2013), pp. 429–460. MR3205167
161. O. Reutter, Problem 664A. Elem. Math. **27**, 19 (1972)
162. G. Robins, J.S. Salowe, Low-degree minimum spanning trees. Discrete Comput. Geom. **14**, 151–165 (1995). MR1331924 (96f:05180)
163. C.A. Rogers, C. Zong, Covering convex bodies by translates of convex bodies. Mathematika **44**(1), 215–218 (1997). MR1464387 (98i:52026)
164. H. Sachs, No more than nine unit balls can touch a closed unit hemisphere. Stud. Sci. Math. Hung. **21**(1–2), 203–206 (1986). MR0898858 (88k:52021)
165. G. Schechtman, Two observations regarding embedding subsets of Euclidean spaces in normed spaces. Adv. Math. **200**(1), 125–135 (2006). MR2199631 (2006j:46015)
166. O. Schramm, Illuminating sets of constant width. Mathematika **35**(2), 180–189 (1988). MR0986627 (89m:52013)
167. A. Schürmann, K.J. Swanepoel, Three-dimensional antipodal and norm-equilateral sets. Pac. J. Math. **228**(2), 349–370 (2006). MR2274525 (2007m:52024)
168. K. Schütte, B.L. van der Waerden, Das problem der dreizehn Kugeln. Math. Ann. **125**, 325–334 (1953). MR0053537 (14,787e)
169. C.E. Shannon, Probability of error for optimal codes in a Gaussian channel. Bell Syst. Tech. J. **38**, 611–656 (1959). MR0103137 (21 #1920)
170. A. Soifer, *The Mathematical Coloring Book* (Springer, New York, 2009). MR2458293 (2010a:05005)
171. P.S. Soltan, Analogues of regular simplexes in normed spaces. Dokl. Akad. Nauk SSSR **222**(6), 1303–1305 (1975); English translation: Soviet Math. Dokl. **16**(3), 787–789 (1975). MR0383246 (52 #4127)
172. J. Solymosi, V.H. Vu, Near optimal bounds for the Erdős distinct distances problem in high dimensions. Combinatorica **28**(1), 113–125 (2008). MR2399013 (2009f:52042)
173. J. Spencer, E. Szemerédi, W. Trotter Jr., Unit distances in the Euclidean plane, *Graph Theory and Combinatorics (Cambridge, 1983)* (Academic Press, London, 1984), pp. 293–303. MR0777185 (86m:52015)

174. S. Straszewicz, Sur un problème géométrique de P. Erdős. Bull. Acad. Pol. Sci. Cl. III. **5**, 39–40, IV–V (1957). MR0087117 (19,304f)
175. K.J. Swanepoel, Cardinalities of k-distance sets in Minkowski spaces. Discrete Math. **197/198**, 759–767 (1999). MR1674902 (99k:52028)
176. K.J. Swanepoel, New lower bounds for the Hadwiger numbers of ℓ_p balls for $p<2$. Appl. Math. Lett. **12**(5), 57–60 (1999). MR1750139 (2001e:94024)
177. K.J. Swanepoel, Vertex degrees of Steiner Minimal Trees in ℓ_p^d and other smooth Minkowski spaces. Discrete Comput. Geom. **21**, 437–447 (1999). MR1672996 (2000g:05054)
178. K.J. Swanepoel, The local Steiner problem in normed planes. Networks **36**, 104–113 (2000). MR1793318 (2001f:05049)
179. K.J. Swanepoel, Independence numbers of planar contact graphs. Discrete Comp. Geom. **28**, 649–670 (2002). MR1949907 (2003j:52016)
180. K.J. Swanepoel, Equilateral sets in finite-dimensional normed spaces, in *Seminar of Mathematical Analysis*, vol. 71, Coleccion Abierta (Universidad de Sevilla. Secretariado de Publicaciones, 2004), pp. 195–237. MR2117069 (2005j:46009)
181. K.J. Swanepoel, Quantitative illumination of convex bodies and vertex degrees of geometric Steiner minimal trees. Mathematika **52**(1–2), 47–52 (2005). MR2261841 (2008f:52009)
182. K.J. Swanepoel, The local Steiner problem in finite-dimensional normed spaces. Discrete Comput. Geom. **37**(3), 419–442 (2007). MR2301527 (2008b:52003)
183. K.J. Swanepoel, Upper bounds for edge-antipodal and subequilateral polytopes. Period. Math. Hung. **54**(1), 99–106 (2007). MR2310370 (2008k:52020)
184. K.J. Swanepoel, Unit distances and diameters in Euclidean spaces. Discrete Comput. Geom. **41**(1), 1–27 (2009). MR2470067 (2010f:52031)
185. K.J. Swanepoel, P. Valtr, Large convexly independent subsets of Minkowski sums. Electron. J. Comb. **17**(1) (2010). Research Paper 146, 7pp. MR2745699 (2012c:52036)
186. K.J. Swanepoel, Sets of unit vectors with small subset sums. Trans. Am. Math. Soc. **368**, 7153–7188 (2016). MR3471088
187. K.J. Swanepoel, R. Villa, A lower bound for the equilateral number of normed spaces. Proc. Am. Math. Soc. **136**, 127–131 (2008). MR2350397 (2008j:46010)
188. K.J. Swanepoel, R. Villa, Maximal equilateral sets. Discrete Comput. Geom. **50**(2), 354–373 (2013). MR3090523
189. H.P.F. Swinnerton-Dyer, Extremal lattices of convex bodies. Proc. Camb. Philos. Soc. **49**, 161–162 (1953). MR0051880 (14,540f)
190. I. Talata, Exponential lower bound for the translative kissing numbers of d-dimensional convex bodies. Discrete Comput. Geom. **19**(3), 447–455 (1998). MR1615129 (98k:52046)
191. I. Talata, The translative kissing number of tetrahedra is 18. Discrete Comput. Geom. **22**(2), 231–248 (1999). MR1698544 (2000e:52021)
192. I. Talata, A legnagyobb minimális szomszédszám egy oktaéder eltoltjainak véges elhelyezéseiben [Determining the largest possible minimum number of neighbours in a finite packing of translates of an octahcdron] Tudományos Közlemények, Szent István Egyetem Műszaki Főiskolai Kar, 2006, pp. 122–125
193. I. Talata, On a lemma of Minkowski. Period. Math. Hung. **36**(2–3), 199–207 (1998). MR1694585 (2000i:52035)
194. I. Talata, On extensive subsets of convex bodies. Period. Math. Hung. **38**(3), 231–246 (1999). MR1756241 (2001b:52035)
195. I. Talata, A lower bound for the translative kissing numbers of simplices. Combinatorica **20**(2), 281–293 (2000). MR1767027 (2001d:52030)
196. I. Talata, On minimum kissing numbers of finite translative packings of a convex body. Beitr. Algebra Geom. **43**(2), 501–511 (2002). MR1957754 (2003j:52018)
197. I. Talata, On Hadwiger numbers of direct products of convex bodies, *Combinatorial and Computational Geometry*, vol. 52, Mathematical Sciences Research Institute Publications (Cambridge University Press, Cambridge, 2005), pp. 517–528. MR2178337 (2006g:52030)
198. I. Talata, Finite translative packings with large minimum kissing numbers. Stud. Univ. Žilina Math. Ser. **25**(1), 47–56 (2011). MR2963987

199. P. Terenzi, Successioni regolari negli spazi di Banach (Regular sequences in Banach spaces). Rend. Semin. Mat. Fis. Milano **57**(1987), 275–285 (1989). MR1017856 (90m:46022)
200. P. Terenzi, Equilater sets in Banach spaces. Boll. dell'Unione. Mat. Ital. A (7) **3**(1), 119–124 (1989). MR0990095 (90c:46017)
201. A.C. Thompson, *Minkowski Geometry*, Encyclopedia of Mathematics and its Applications (Cambridge University Press, Cambridge, 1996). MR1406315 (97f:52001)
202. G.T. Toussaint, A graph-theoretical primal sketch, in *Computational Morphology, A Computational Geometric Approach to the Analysis of Form*, Machine Intelligence and Pattern Recognition, ed. by G.T. Toussaint (North-Holland, Amsterdam, 1988), pp. 229–260. MR0993994 (89k:68151)
203. G.T. Toussaint, The sphere of influence graph: theory and applications. Int. J. Inf. Technol. Comput. Sci. **14**(2), 37–42 (2014)
204. J. Väisälä, Regular simplices in three-dimensional normed spaces. Beitr. Algebra Geom. **53**(2), 569–570 (2012). MR2971762
205. P. Valtr, Strictly convex norms allowing many unit distances and related touching questions, manuscript (2005)
206. S. Vlăduţ, Lattices with exponentially large kissing numbers. arXiv:1802.00886
207. G.L. Watson, The number of minimum points of a positive quadratic form. Diss. Math. **84**, 42p. (1971). MR0318061 (47 #6610)
208. A.D. Wyner, Capabilities of bounded discrepancy decoding. Bell Syst. Tech. J. **44**, 1061–1122 (1965). MR0180417 (31 #4652)
209. L. Xu, A note on the kissing numbers of superballs. Discrete Comput. Geom. **37**(3), 485–491 (2007). MR2301531 (2008b:52027)
210. L. Yu, Blocking numbers and fixing numbers of convex bodies. Discrete Math. **309**(23–24), 6544–6554 (2009). MR2558619 (2010j:52035)
211. L. Yu, C. Zong, On the blocking number and the covering number of a convex body. Adv. Geom. **9**(1), 13–29 (2009). MR2493260 (2010d:52007)
212. J. Zahl, An improved bound on the number of point-surface incidences in three dimensions. Contrib. Discrete Math. **8**(1), 100–121 (2013). MR3118901
213. C. Zong, Packing and covering, Ph.D. thesis, Technische Universität Wien (1993)
214. C. Zong, The kissing numbers of tetrahedra. Discrete Comput. Geom. **15**(3), 239–252 (1996). MR1380392 (97c:11070)
215. C. Zong, *Strange Phenomena in Convex and Discrete Geometry* (Springer, New York, 1996). MR1416567 (97m:52001)
216. C. Zong, The kissing numbers of convex bodies — a brief survey. Bull. Lond. Math. Soc. **30**, 1–10 (1998). MR1479030 (98k:52048)
217. C. Zong, The kissing number, blocking number and covering number of a convex body, *Surveys on Discrete and Computational Geometry*, vol. 453, Contemporary Mathematics (American Mathematical Society, Providence, 2008), pp. 529–548. MR2405694 (2010b:52029)
218. C.M. Zong, An example concerning the translative kissing number of a convex body. Discrete Comput. Geom. **12**, 183–188 (1994). MR1283886 (95e:52033)
219. C.M. Zong, Some remarks concerning kissing numbers, blocking numbers and covering numbers. Period. Math. Hung. **30**(3), 233–238 (1995). MR1334968 (96g:52039)
220. C.M. Zong, The translative kissing number of the Cartesian product of two convex bodies, one of which is two-dimensional. Geom. Dedicata **65**(2), 135–145 (1997). MR1451968 (98e:52022)

Correction to: New Trends in Intuitive Geometry

Gergely Ambrus, Imre Bárány, Károly J. Böröczky, Gábor Fejes Tóth and János Pach

Correction to:
G. Ambrus et al. (eds.), *New Trends in Intuitive Geometry,* Bolyai Society Mathematical Studies 27, https://doi.org/10.1007/978-3-662-57413-3

In the original version of the book, the grant number has been corrected from "174008" to "174024" in the chapter "The Topological Transversal Tverberg Theorem Plus Constraints" and Copyright Holder name has been corrected from "Springer-Verlag GmbH Germany, part of Springer Nature" to "János Bolyai Mathematical Society and Springer-Verlag GmbH Germany, part of Springer Nature" throughout the book.

The updated version of the book can be found at
https://doi.org/10.1007/978-3-662-57413-3
https://doi.org/10.1007/978-3-662-57413-3_3

G. Ambrus et al. (eds.), *New Trends in Intuitive Geometry*, Bolyai Society Mathematical Studies 27, https://doi.org/10.1007/978-3-662-57413-3_18

Kolibri 360 GmbH
Ferdinand-Jühlke-Straße 7
99095 Erfurt, Deutschland
produktsicherheit@kolibri360.de